Source Separation and Machine Learning

Source Separation and Machine Learning

Jen-Tzung Chien
National Chiao Tung University

ACADEMIC PRESS
An imprint of Elsevier

Academic Press is an imprint of Elsevier
125 London Wall, London EC2Y 5AS, United Kingdom
525 B Street, Suite 1650, San Diego, CA 92101, United States
50 Hampshire Street, 5th Floor, Cambridge, MA 02139, United States
The Boulevard, Langford Lane, Kidlington, Oxford OX5 1GB, United Kingdom

Notices

Knowledge and best practice in this field are constantly changing. As new research and experience broaden our understanding, changes
in research methods, professional practices, or medical treatment may become necessary.

Practitioners and researchers must always rely on their own experience and knowledge in evaluating and using any information,
methods, compounds, or experiments described herein. In using such information or methods they should be mindful of their own safety
and the safety of others, including parties for whom they have a professional responsibility.

To the fullest extent of the law, neither the Publisher nor the authors, contributors, or editors, assume any liability for any injury and/or
damage to persons or property as a matter of products liability, negligence or otherwise, or from any use or operation of any methods,
products, instructions, or ideas contained in the material herein.

Library of Congress Cataloging-in-Publication Data
A catalog record for this book is available from the Library of Congress

British Library Cataloguing-in-Publication Data
A catalogue record for this book is available from the British Library

ISBN: 978-0-12-817796-9

For information on all Academic Press publications
visit our website at https://www.elsevier.com/books-and-journals

Working together
to grow libraries in
developing countries

www.elsevier.com • www.bookaid.org

Publisher: Mara Conner
Acquisition Editor: Tim Pitts
Editorial Project Manager: John Leonard
Production Project Manager: Surya Narayanan Jayachandran
Designer: Mark Rogers

Typeset by VTeX

Contents

List of Figures xi
List of Tables xvii
Foreword xix
Preface xxi
Acknowledgments xxiii
Notations and Abbreviations xxv

PART 1 FUNDAMENTAL THEORIES

CHAPTER 1   **Introduction** **3**
    1.1   Source Separation 3
        1.1.1   Multichannel Source Separation 3
        1.1.2   Monaural Source Separation 6
        1.1.3   Reverberant Source Separation 8
    1.2   Applications and Challenges 12
        1.2.1   Applications for Source Separation 12
        1.2.2   Challenges in Separation and Learning 15
        1.2.3   Front-End Processing and Back-End Learning 17
    1.3   Overview of This Book 19

CHAPTER 2   **Model-Based Source Separation** **21**
    2.1   Independent Component Analysis 22
        2.1.1   Learning Procedure 23
    2.2   Nonnegative Matrix Factorization 25
        2.2.1   Learning Procedure 26
        2.2.2   Learning Objective 28
        2.2.3   Sparse Regularization 31
    2.3   Nonnegative Tensor Factorization 32
        2.3.1   Tucker Decomposition 34
        2.3.2   CP Decomposition 36
    2.4   Deep Neural Network 37
        2.4.1   Error Backpropagation Algorithm 39
        2.4.2   Practices in Deep Learning 41
    2.5   Recurrent Neural Network 45
        2.5.1   Backpropagation Through Time 45
        2.5.2   Deep Recurrent Neural Network 48
        2.5.3   Long Short-Term Memory 49
    2.6   Summary 52

CHAPTER 3   **Adaptive Learning Machine** **53**
    3.1   Learning Problems in Source Separation 53

3.2	Information-Theoretic Learning	54
	3.2.1 Divergence Measure	55
3.3	Bayesian Learning	57
	3.3.1 Model Regularization	57
	3.3.2 Bayesian Source Separation	59
3.4	Sparse Learning	60
	3.4.1 Sparse Bayesian Learning	62
3.5	Online Learning	65
	3.5.1 Online Bayesian Learning	67
3.6	Discriminative Learning and Deep Learning	68
	3.6.1 Discriminative NMF	68
	3.6.2 Discriminative DNN	73
3.7	Approximate Inference	75
	3.7.1 Maximum Likelihood and EM Algorithm	76
	3.7.2 Maximum *a Posteriori*	83
	3.7.3 Variational Inference	85
	3.7.4 Sampling Methods	91
3.8	Summary	96

PART 2 ADVANCED STUDIES

CHAPTER 4	**Independent Component Analysis**	**99**
4.1	ICA for Speech Recognition	99
	4.1.1 Construction of Independence Space	100
	4.1.2 Adaptation in Independent Space	102
	4.1.3 System Evaluation	103
4.2	Nonparametric Likelihood Ratio ICA	106
	4.2.1 ICA Objective Function	106
	4.2.2 Hypothesis Test for Independence	109
	4.2.3 Nonparametric Likelihood Ratio	110
	4.2.4 Application and Evaluation	113
4.3	Convex Divergence ICA	118
	4.3.1 Convex Divergence	118
	4.3.2 Divergence Measure for NMF	123
	4.3.3 ICA Procedure	124
	4.3.4 Simulated Experiments	127
	4.3.5 Real-World Experiments	129
4.4	Nonstationary Bayesian ICA	132
	4.4.1 Sequential and Noisy ICA	134
	4.4.2 Automatic Relevance Determination	136
	4.4.3 Sequential and Variational Learning	138
	4.4.4 System Evaluation	142
4.5	Online Gaussian Process ICA	145
	4.5.1 Temporal Structure for Separation	146

	4.5.2	Online Learning and Gaussian Process	147
	4.5.3	Sequential and Variational Learning	151
	4.5.4	Sequential Monte Carlo ICA	153
	4.5.5	System Evaluation	155
4.6	Summary		158
CHAPTER 5	**Nonnegative Matrix Factorization**		**161**
5.1	Convolutive NMF		161
	5.1.1	Nonnegative Matrix Factor Deconvolution	162
	5.1.2	Speech Dereverberation Model	165
	5.1.3	Bayesian Speech Dereverberation	166
	5.1.4	System Evaluation	171
5.2	Probabilistic Nonnegative Factorization		174
	5.2.1	Probabilistic Latent Component Analysis	174
	5.2.2	Shift-Invariant PLCA	176
	5.2.3	PLCA Versus NMF	179
	5.2.4	Interpretation and Application	181
5.3	Bayesian NMF		182
	5.3.1	Gaussian–Exponential Bayesian NMF	183
	5.3.2	Poisson–Gamma Bayesian NMF	186
	5.3.3	Poisson–Exponential Bayesian NMF	190
	5.3.4	Evaluation for Supervised Separation	196
	5.3.5	Evaluation for Unsupervised Separation	199
5.4	Group Sparse NMF		202
	5.4.1	Group Basis Representation	202
	5.4.2	Bayesian Group Sparse Learning	205
	5.4.3	Markov Chain Monte Carlo Sampling	209
	5.4.4	System Evaluation	216
5.5	Deep Layered NMF		218
	5.5.1	Deep Structural Factorization	219
	5.5.2	Layered NMF	221
	5.5.3	Discriminative Layered NMF	223
	5.5.4	System Evaluation	227
5.6	Summary		228
CHAPTER 6	**Nonnegative Tensor Factorization**		**231**
6.1	NTF for Source Separation		232
	6.1.1	Modulation Spectrograms	232
	6.1.2	Multiresolution Spectrograms	234
6.2	Convolutive NTF		237
	6.2.1	Nonnegative Tensor Factor Deconvolution	237
	6.2.2	Sparse Nonnegative Tensor Factor Deconvolution	240
6.3	Probabilistic NTF		242
	6.3.1	Probabilistic Matrix Factorization	242
	6.3.2	Probabilistic Tensor Factorization	244

        6.3.3 Probabilistic Nonnegative Tensor Factorization 246
   6.4 Bayesian Tensor Factorization 249
        6.4.1 Hierarchical Bayesian Model 249
        6.4.2 Bayesian Learning Algorithm 250
   6.5 Infinite Tensor Factorization 251
        6.5.1 Positive Semidefinite Tensor Factorization 251
        6.5.2 Gamma Process 254
        6.5.3 Variational Inference 255
        6.5.4 System Evaluation 256
   6.6 Summary 257

CHAPTER 7   Deep Neural Network **259**
   7.1 Deep Machine Learning 259
        7.1.1 Deep Spectral Masking 260
        7.1.2 Deep Ensemble Learning 265
        7.1.3 Deep Speech Segregation 268
   7.2 Deep Processing and Learning 270
        7.2.1 Deep Clustering 270
        7.2.2 Masking and Learning 274
        7.2.3 Permutation-Invariant Training 275
        7.2.4 Matrix Factorization and Neural Network 278
        7.2.5 Spectro-Temporal Neural Factorization 280
   7.3 Discriminative Deep Recurrent Neural Network 284
        7.3.1 Joint Optimization and Mask Function 284
        7.3.2 Discriminative Source Separation 287
        7.3.3 System Evaluation 290
   7.4 Long Short-Term Memory 291
        7.4.1 Physical Interpretation 291
        7.4.2 Speaker Generalization 293
        7.4.3 Bidirectional Long Short-Term Memory 295
   7.5 Variational Recurrent Neural Network 299
        7.5.1 Variational Autoencoder 299
        7.5.2 Model Construction and Inference 299
        7.5.3 System Evaluation 304
   7.6 Neural Turing Machine 307
        7.6.1 Memory Augmented Source Separation 307
        7.6.2 LSTM Controller and Addressing Mechanism 309
        7.6.3 System Evaluation 312
   7.7 End-to-End Memory Network 313
        7.7.1 Recall Neural Network 314
        7.7.2 Sequence-to-Sequence Learning 316
        7.7.3 System Evaluation 318
   7.8 Summary 319

CHAPTER 8   **Summary and Future Trends** **321**
    8.1   Machine Learning for Source Separation 321
    8.2   Potential Topics and Directions 323
Appendix A   **Basic Formulas** **325**
    A.1   Expectation 325
    A.2   Jensen's Inequality 325
    A.3   Gamma Function 325
    A.4   Trace 326
    A.5   Transpose 326
    A.6   Derivative 326
    A.7   Complete Square 327
    A.8   Tensor Algebra 327
Appendix B   **Probabilistic Distribution Functions** **329**
    B.1   Multivariate Gaussian Distribution 329
    B.2   Laplace Distribution 330
    B.3   Student's t-Distribution 331
    B.4   Gamma Distribution 331
    B.5   Wishart Distribution 333
    B.6   Poisson Distribution 334
    B.7   Exponential Distribution 334

Bibliography 337

Index 349

List of Figures

Fig. 1.1	Cocktail party problem with three speakers and three microphones.	4
Fig. 1.2	A general linear mixing system with n observations and m sources.	5
Fig. 1.3	An illustration of monaural source separation with three source signals.	7
Fig. 1.4	An illustration of singing voice separation.	8
Fig. 1.5	Speech recognition system in a reverberant environment.	9
Fig. 1.6	Characteristics of room impulse response.	9
Fig. 1.7	Categorization of various methods for speech dereverberation.	10
Fig. 1.8	Categorization of various applications for source separation.	12
Fig. 1.9	Blind source separation for electroencephalography artifact removal.	14
Fig. 1.10	Challenges in front-end processing and back-end learning for audio source separation.	18
Fig. 2.1	Evolution of different source separation methods.	21
Fig. 2.2	ICA learning procedure for finding a demixing matrix $\mathbf{W}$.	24
Fig. 2.3	Illustration for nonnegative matrix factorization $\mathbf{X} \approx \mathbf{BW}$.	25
Fig. 2.4	Nonnegative matrix factorization shown in a summation of rank-one nonnegative matrices.	26
Fig. 2.5	Supervised learning for single-channel source separation using nonnegative matrix factorization in presence of a speech signal $\mathbf{X}^s$ and a music signal $\mathbf{X}^m$.	27
Fig. 2.6	Unsupervised learning for single-channel source separation using nonnegative matrix factorization in presence of two sources.	27
Fig. 2.7	Illustration of multiway observation data.	33
Fig. 2.8	A tensor data which is composed of three ways of time, frequency and channel.	34
Fig. 2.9	Tucker decomposition for a three-way tensor.	34
Fig. 2.10	CP decomposition for a three-way tensor.	37
Fig. 2.11	Multilayer perceptron with one input layer, one hidden layer, and one output layer.	38
Fig. 2.12	Comparison of activations using ReLU function, logistic sigmoid function, and hyperbolic tangent function.	39
Fig. 2.13	Calculations in error backpropagation algorithm: (A) forward pass and (B) backward pass. The propagations are shown by arrows. In the forward pass, activations a_t and outputs z_t in individual nodes are calculated and propagated. In the backward pass, local gradients δ_t in individual nodes are calculated and propagated.	41
Fig. 2.14	Procedure in the error backpropagation algorithm: (A) compute the error function, (B) propagate the local gradient from the output layer L to (C) hidden layer $L - 1$ and (D) until input layer 1.	42
Fig. 2.15	A two-layer structure in restricted Boltzmann machine.	43
Fig. 2.16	A stack-wise training procedure for deep belief network. This model is further used as a pretrained model for deep neural network optimization.	44
Fig. 2.17	A recurrent neural network with single hidden layer.	45
Fig. 2.18	Backpropagation through time with a single hidden layer and two time steps $\tau = 2$.	46
Fig. 2.19	A deep recurrent neural network for single-channel source separation in the presence of two source signals.	49

Fig. 2.20   A procedure of single-channel source separation based on DNN or RNN where two source
             signals are present.   49
Fig. 2.21   Preservation of gradient information through a gating mechanism.   50
Fig. 2.22   A detailed view of long short-term memory. Block circle means multiplication. Recurrent state
             $\mathbf{z}_{t-1}$ and cell $\mathbf{c}_{t-1}$ in a previous time step are shown by dashed lines.   51
Fig. 3.1   Portrait of Thomas Bayes (1701–1761) who is credited for the Bayes theorem.   58
Fig. 3.2   Illustration of a scalable model in presence of increasing number of training data.   59
Fig. 3.3   Optimal weights $\widehat{\mathbf{w}}$ estimated by minimizing the ℓ_2- and ℓ_1-regularized objective functions
             which are shown on the left and right, respectively.   62
Fig. 3.4   Laplace distribution centered at zero with different control parameters λ.   63
Fig. 3.5   Comparison of two-dimensional Gaussian distribution (A) and Student's t-distribution (B) with
             zero mean.   64
Fig. 3.6   Scenario for a time-varying source separation system with (A) one male, one female and one
             music player. (B) Then, the male is moving to a new location. (C) After a while, the male
             disappears and the female is replaced by a new one.   66
Fig. 3.7   Sequential updating of parameters and hyperparameters for online Bayesian learning.   68
Fig. 3.8   Evolution of different inference algorithms for construction of latent variable models. 75
Fig. 3.9   Illustration for (A) decomposing a log-likelihood into a KL divergence and a lower bound,
             (B) updating the lower bound by setting KL divergence to zero with new distribution $q(\mathbf{Z})$, and
             (C) updating lower bound again by using new parameters $\boldsymbol{\Theta}$. Adapted from Bishop (2006).   80
Fig. 3.10   Illustration of updating lower bound of the log-likelihood function in EM iteration from τ to
             $\tau + 1$. Adapted from Bishop (2006).   81
Fig. 3.11   Bayes relation among posterior distribution, likelihood function, prior distribution and evidence
             function.   83
Fig. 3.12   Illustration of seeking an approximate distribution $q(\mathbf{Z})$ which is factorizable and maximally
             similar to the true posterior $p(\mathbf{Z}|\mathbf{X})$.   86
Fig. 3.13   Illustration of minimizing KL divergence via maximization of variational lower bound.
             KL divergence is reduced from left to right during the optimization procedure.   88
Fig. 4.1   Construction of the eigenspace and independent space for finding the adapted models from a set
             of seed models by using enrollment data.   102
Fig. 4.2   Comparison of kurtosis values of the estimated eigenvoices and independent voices.   104
Fig. 4.3   Comparison of BIC values of using PCA with eigenvoices and ICA with independent voices. 105
Fig. 4.4   Comparison of WERs (%) of using PCA with eigenvoices and ICA with independent voices
             where the numbers of basis vectors (K) and the numbers of adaptation sentences (L) are
             changed.   105
Fig. 4.5   Taxonomy of contrast functions for optimization in independent component analysis.   107
Fig. 4.6   ICA transformation and k-means clustering for speech recognition with multiple hidden
             Markov models.   115
Fig. 4.7   Generation of supervectors from different acoustic segments of aligned utterances.   116
Fig. 4.8   Taxonomy of ICA contrast functions where different realizations of mutual information are
             included.   119
Fig. 4.9   Illustration of a convex function $f(x)$.   119
Fig. 4.10   Comparison of (A) a number of divergence measures and (B) α-DIV and C-DIV for various α
             under different joint probability $P_{y_1,y_2}(A, A)$.   122

Fig. 4.11   Divergence measures of demixed signals versus the parameters of demixing matrix θ_1 and θ_2. KL-DIV, C-DIV at $\alpha = 1$ and C-DIV at $\alpha = -1$ are compared.   128

Fig. 4.12   Divergence measures versus number of learning. KL-ICA and C-ICA at $\alpha = 1$ and $\alpha = -1$ are evaluated. The gradient descent (GD) and natural gradient (NG) algorithms are compared.   129

Fig. 4.13   Comparison of SIRs of three demixed signals in the presence of an instantaneous mixing condition. Different ICA algorithms are evaluated.   130

Fig. 4.14   Comparison of SIRs of three demixed signals in the presence of instantaneous mixing condition with additive noise. Different ICA algorithms are evaluated.   131

Fig. 4.15   Integrated scenario of a time-varying source separation system from t to $t + 1$ and $t + 2$ as shown in Fig. 3.6.   133

Fig. 4.16   Graphical representation for nonstationary Bayesian ICA model.   136

Fig. 4.17   Comparison of speech waveforms for (A) source signal 1 (blue in web version or dark gray in print version), source signal 2 (red in web version or light gray in print version) and two mixed signals (black) and (B) two demixed signals 1 (blue in web version or dark gray in print version) and two demixed signals 2 (red in web version or light gray in print version) by using NB-ICA and OLGP-ICA algorithms.   143

Fig. 4.18   Comparison of variational lower bounds by using NB-ICA with (blue in web version or dark gray in print version) and without (red in web version or light gray in print version) adaptation of ARD parameter.   144

Fig. 4.19   Comparison of the estimated ARD parameters of source signal 1 (blue in web version or dark gray in print version) and source signal 2 (red in web version or light gray in print version) using NB-ICA.   144

Fig. 4.20   Evolution from NB-ICA to OLGP-ICA for nonstationary source separation.   146

Fig. 4.21   Graphical representation for online Gaussian process ICA model.   150

Fig. 4.22   Comparison of square errors of the estimated mixing coefficients by using NS-ICA (black), SMC-ICA (pink in web version or light gray in print version), NB-ICA (blue in web version or dark gray in print version) and OLGP-ICA (red in web version or mid gray in print version).   157

Fig. 4.23   Comparison of absolute errors of temporal predictabilities between true and demixed source signals where different ICA methods are evaluated.   158

Fig. 4.24   Comparison of signal-to-interference ratios of demixed signals where different ICA methods are evaluated.   159

Fig. 5.1   Graphical representation for Bayesian speech dereverberation.   167

Fig. 5.2   Graphical representation for Gaussian–Exponential Bayesian nonnegative matrix factorization.   184

Fig. 5.3   Graphical representation for Poisson–Gamma Bayesian nonnegative matrix factorization.   187

Fig. 5.4   Graphical representation for Poisson–Exponential Bayesian nonnegative matrix factorization.   190

Fig. 5.5   Implementation procedure for supervised speech and music separation.   197

Fig. 5.6   Histogram of the estimated number of bases ($\widehat{K}$) using PE-BNMF for source signals of speech, piano and violin.   198

Fig. 5.7   Comparison of the averaged SDR using NMF with a fixed number of bases (1–5 pairs of bars), PG-BNMF with the best fixed number of bases (sixth pair of bars), PG-BNMF with adaptive number of bases (seventh pair of bars), PE-BNMF with the best fixed number of bases (eighth pair of bars) and PE-BNMF with adaptive number of bases (ninth pair of bars).   198

Fig. 5.8   Implementation procedure for unsupervised singing voice separation.   199

Fig. 5.9   Comparison of GNSDR of the separated singing voices at different SMRs using PE-BNMFs with K-means clustering (denoted by BNMF1), NMF clustering (denoted by BNMF2) and shifted NMF clustering (denoted by BNMF3). Five competitive methods are included for comparison.   201

Fig. 5.10   Illustration for group basis representation.   204

Fig. 5.11   Comparison of Gaussian, Laplace and LSM distributions centered at zero.   207

Fig. 5.12   Graphical representation for Bayesian group sparse nonnegative matrix factorization.   208

Fig. 5.13   Spectrograms of "music 5" containing the drum signal (first panel), the saxophone signal (second panel), the mixed signal (third panel), the demixed drum signal (fourth panel) and the demixed saxophone signal (fifth panel).   217

Fig. 5.14   Illustration for layered nonnegative matrix factorization.   221

Fig. 5.15   Illustration for discriminative layered nonnegative matrix factorization.   224

Fig. 6.1   Categorization and evolution for different tensor factorization methods.   231

Fig. 6.2   Procedure of producing the modulation spectrograms for tensor factorization.   232

Fig. 6.3   (A) Time resolution and (B) frequency resolution of a mixed audio signal.   235

Fig. 6.4   Comparison between (A) nonnegative matrix factorization and (B) positive semidefinite tensor factorization.   253

Fig. 7.1   Categorization of different deep learning methods for source separation.   260

Fig. 7.2   A deep neural network for single-channel speech separation, where $\mathbf{x}_t$ denotes the input features of the mixed signal at time step t, $\mathbf{z}_t^{(l)}$ denotes the features in the hidden layer l, $\widehat{\mathbf{y}}_{1,t}$ and $\widehat{\mathbf{y}}_{2,t}$ denote the mask function for source one and source two, and $\widehat{\mathbf{x}}_{1,t}$ and $\widehat{\mathbf{x}}_{2,t}$ are the estimated signals for source one and source two, respectively.   261

Fig. 7.3   Illustration of stacking in deep ensemble learning for speech separation.   267

Fig. 7.4   Speech segregation procedure by using deep neural network.   268

Fig. 7.5   A deep recurrent neural network for speech dereverberation.   279

Fig. 7.6   Factorized features in spectral and temporal domains for speech dereverberation.   283

Fig. 7.7   Comparison of SDR, SIR and SAR of the separated signals by using NMF, DRNN, DDRNN-bw and DDRNN-diff.   291

Fig. 7.8   Vanishing gradients in a recurrent neural network. Degree of lightness means the level of vanishing in the gradient. This figure is a counterpart to Fig. 2.21 where vanishing gradients are mitigated by a gating mechanism.   292

Fig. 7.9   A simplified view of long short-term memory. The red (light gray in print version) line shows recurrent state z_t. The detailed view was provided in Fig. 2.22.   293

Fig. 7.10   Illustration of preserving gradients in long short-term memory. There are three gates in a memory block. ○ means gate opening while − denotes gate closing.   293

Fig. 7.11   A configuration of four stacked long short-term memory layers along three time steps for monaural speech separation. Dashed arrows indicate the modeling of the same LSTM across time steps while solid arrows mean the modeling of different LSTMs in deep structure.   294

Fig. 7.12   Illustration of bidirectional recurrent neural network for monaural source separation. Two hidden layers are configured to learn bidirectional features for finding soft mask functions for two sources. Forward and backward directions are shown in different colors.   296

Fig. 7.13   (A) Encoder and decoder in a variational autoencoder. (B) Graphical representation for a variational autoencoder.   300

Fig. 7.14   Graphical representation for (A) recurrent neural network and (B) variational recurrent neural network.   301

Fig. 7.15   Inference procedure for variational recurrent neural network.   302

Fig. 7.16   Implementation topology for variational recurrent neural network.   305

Fig. 7.17   Comparison of SDR, SIR and SAR of the separated signals by using NMF, DNN, DRNN, DDRNN and VRNN.   306

Fig. 7.18   (A) Single-channel source separation with dynamic state $\mathcal{Z}_t = \{\mathbf{z}_t, \mathbf{r}_t, \mathbf{w}_{r,t}, \mathbf{w}_{w,t}\}$ in recurrent layer $L - 1$. (B) Recurrent layers are driven by a cell $\mathbf{c}_t$ and a controller for memory $\mathbf{M}_t$ where dashed line denotes the connection between cell and memory at previous time step and bold lines denote the connections with weights.   308

Fig. 7.19   Four steps of addressing procedure which is driven by parameters $\{\mathbf{k}_t, \beta_t, g_t, \mathbf{s}_t, \gamma_t\}$.   310

Fig. 7.20   An end-to-end memory network for monaural source separation containing a bidirectional LSTM on the left as an encoder, an LSTM on the right as a decoder and an LSTM on the top as a separator.   315

List of Tables

Table 2.1	Comparison of NMF updating rules based on different learning objectives	31
Table 2.2	Comparison of updating rules of standard NMF and sparse NMF based on learning objectives of squared Euclidean distance and Kullback–Leibler divergence	32
Table 3.1	Comparison of approximate inference methods using variational Bayesian and Gibbs sampling	95
Table 4.1	Comparison of syllable error rates (SERs) (%) with and without HMM clustering and ICA learning. Different ICA algorithms are evaluated	117
Table 4.2	Comparison of signal-to-interference ratios (SIRs) (dB) of mixed signals without ICA processing and with ICA learning based on MMI and NLR contrast functions	117
Table 4.3	Comparison of different divergence measures with respect to symmetric divergence, convexity parameter, combination weight and special realization	121
Table 5.1	Comparison of multiplicative updating rules of standard NMFD and sparse NMFD based on the objective functions of squared Euclidean distance and KL divergence	164
Table 5.2	Comparison of multiplicative updating rules of standard NMF2D and sparse NMF2D based on the objective functions of squared Euclidean distance and KL divergence	164
Table 5.3	Comparison of using different methods for speech dereverberation under various test conditions (near, near microphone; far, far microphone; sim, simulated data; real, real recording) in terms of evaluation metrics of CD, LLR, FWSegSNR and SRMR (dB)	173
Table 5.4	Comparison of different Bayesian NMFs in terms of inference algorithm, closed-form solution and optimization theory	195
Table 5.5	Comparison of GNSDR (dB) of the separated singing voices (V) and music accompaniments (M) using NMF with fixed number of bases $K = 10$, 20 and 30 and PE-BNMF with adaptive $\widehat{K}$. Three clustering algorithms are evaluated	200
Table 5.6	Comparison of NMF and different BNMFs in terms of SDR and GNSDR for two separation tasks. Standard deviation is given in the parentheses	201
Table 5.7	Comparison of SIRs (in dB) of the reconstructed rhythmic signal (denoted by R) and harmonic signal (denoted by H) based on NMF, BNMF, GNMF and BGS-NMF. Six mixed music signals are investigated	218
Table 5.8	Performance of speech separation by using NMF, LNMF and DLNMF in terms of SDR, SIR and SAR (dB)	228
Table 6.1	Comparison of multiplicative updating rules of NMF2D and NTF2D based on the objective functions of squared Euclidean distance and Kullback–Leibler divergence	241
Table 7.1	Comparison of using different models for speech dereverberation in terms of SRMR and PESQ (in dB) under the condition of using simulated data and near microphone	284
Table 7.2	Comparison of STOIs using DNN, LSTM and NTM under different SNRs with seen speakers	313
Table 7.3	Comparison of STOIs using DNN, LSTM and NTM under different SNRs with unseen speakers	313
Table 7.4	Comparison of STOIs under different SNRs by using DNN, LSTM and different variants of NTM and RCNN	319

Foreword

With the use of Deep Neural Networks (DNNs) and Recursive Neural Networks (RNNs), speech recognition performance has recently improved rapidly, and speech recognition has been widely used in smart speakers and smart phones. Speech recognition performance for speech uttered in a quiet environment has become quite close to human performance. However, in situations where there is noise in the surroundings or room reverberation, speech recognition performance falls far short of human performance.

Noise and other person's voices are almost always superimposed on the voice in the house, office, conference room, etc. People can naturally extract and hear the conversation of interested persons, even while many people are chatting like in a cocktail party. This is said to be a cocktail party effect. To automatically recognize the target voice, it is necessary to develop a technology for separating the voice uttered in the actual environment from the surrounding noise and removing the influence of the room reverberation. When analyzing and processing music signals, it is also required to separate overlapped sound source signals.

Sound source separation is, therefore, a very important technology in a wide range of signal processing, particularly speech, sound, and music signal processing. Various researches have been conducted so far, but the performance of current sound source separation technology is far short of human capability. This is one of the major reasons why speech recognition performance in a general environment does not reach human performance.

Generally, speech and audio signals are recorded with one or more microphones. For this reason, the sound source separation technology can be classified into monaural sound source separation and multichannel sound source separation. This book focuses on blind source separation (BSS) which is the process of separating a set of source signals from a set of mixed signals without the aid of information or with very little information about the source signals or the mixing process. It is rare that information on the sound source signal to be separated is obtained beforehand, so it is important to be able to separate the sound source without such information. This book also addresses various challenging issues covering the single-channel source separation where the multiple source signals from a single mixed signal are learned in a supervised way, as well as the speaker and noise independent source separation where a large set of training data is available to learn a generalized model.

In response to the growing need for performance improvement in speech recognition, research on BSS has been rapidly advanced in recent years based on various machine learning and signal processing technology. This book describes state-of-the-art machine learning approaches for model-based BSS for speech recognition, speech separation, instrumental music separation, singing voice separation, music information retrieval, brain signal separation and image processing.

The model-based techniques, combining various signal processing and machine learning techniques, range from linear to nonlinear models. Major techniques include: Independent Component Analysis (ICA), Nonnegative Matrix Factorization (NMF), Nonnegative Tensor Factorization (NTF), Deep Neural Network (DNN), and Recurrent Neural Network (RNN). The rapid progress in the last few years is largely due to the use of DNNs and RNNs.

This book is unique in that it covers topics from the basic signal processing theory concerning BSS to the technology using DNNs and RNNs in recent years. At the end of this book, the direction of future research is also described. This landmark book is very useful as a student's textbook and researcher's reference book. The readers will become articulate in the state-of-the-art of model-based BSS. I would like to encourage many students and researchers to read this book to make big contributions to the future development of technology in this field.

Sadaoki Furui
President, Toyota Technological Institute at Chicago
Professor Emeritus, Tokyo Institute of Technology

Preface

In general, blind source separation (BSS) is known as a rapidly emerging and promising area which involves extensive knowledge of signal processing and machine learning. This book introduces state-of-the-art machine learning approaches for model-based blind source separation (BSS) with applications to speech recognition, speech separation, instrumental music separation, singing voice separation, music information retrieval, brain signal separation and image processing. The traditional BSS approaches based on independent component analysis were designed to resolve the mixing system by optimizing a contrast function or an independent measure. The underdetermined problem in the presence of more sources than sensors may not be carefully tackled. The contrast functions may not flexibly and honestly measure the independence for an optimization with convergence. Assuming the static mixing condition, one cannot catch the underlying dynamics in source signals and sensor networks. The uncertainty of system parameters may not be precisely characterized so that the robustness against adverse environments is not guaranteed. The temporal structures in mixing systems as well as source signals may not be properly captured. We assume that the model complexity or the dictionary size may not be fitted to the true one in source signals. With the remarkable advances in machine learning algorithms, the issues of underdetermined mixtures, optimization of contrast function, nonstationary mixing condition, multidimensional decomposition, ill-posed condition and model regularization have been resolved by introducing the solutions of nonnegative matrix factorization, information-theoretic learning, online learning, Gaussian process, sparse learning, dictionary learning, Bayesian inference, model selection, tensor decomposition, deep neural network, recurrent neural network and memory network. This book will present how these algorithms are connected and why they work for source separation, particularly in speech, audio and music applications. We start with a survey of BSS applications and model-based approaches. The fundamental theories, including statistical learning, optimization algorithm, information theory, Bayesian learning, variational inference and Monte Carlo Markov chain inference, will be addressed. A series of case studies are then introduced to deal with different issues in model-based BSS. These case studies are categorized into independent component analysis, nonnegative matrix factorization, nonnegative tensor factorization and deep neural network ranging from a linear to a nonlinear model, from single-way to multiway processing, and from a shallow feedforward model to a deep recurrent model. At last, we will point out a number of directions and outlooks for future studies.

This book is written as a textbook with fundamental theories and advanced technologies developed in the last decade. It is also shaped as a style of research monograph because some advances in source separation using machine learning or deep learning methods are extensively addressed.

The material of this book is based on a tutorial lecture on this theme at the 40th International Conference on Acoustics, Speech, and Signal Processing (ICASSP) in Brisbane, Australia, in April 2015. This tutorial was one of the most popular tutorials in ICASSP in terms of the number of attendees. The success of this tutorial lecture brought the idea of writing a textbook on this subject to promote using machine learning for signal processing. Some of the material is also based on a number of invited talks and distinguished lectures in different workshops and universities in Japan and Hong Kong. We strongly believe in the importance of machine learning and deep learning approaches to source separation, and sincerely encourage the researchers to work on machine learning approaches to source separation.

Acknowledgments

First, I want to thank my colleagues and research friends, especially the members of the Machine Learning Lab at National Chiao Tung University. Some of the studies in this book were actually conducted while discussing and working with them. I would also like to thank many people for contributing good ideas, proofreading a draft, and giving me valuable comments, which greatly improved this book, including Sadaoki Furui, Chin-Hui Lee, Shoji Makino, Tomohiro Nakatani, George A. Saon, Koichi Shinoda, Man-Wai Mak, Shinji Watanabe, Issam El Naqa, Huan-Hsin Tseng, Zheng-Hua Tan, Zhanyu Ma, Tai-Shih Chi, Shoko Araki, Marc Delcroix, John R. Hershey, Jonathan Le Roux, Hakan Erdogan and Tomoko Matsui. The research experiences were especially impressive and inspiring when working on source separation problems with my past and current students, in particular Bo-Cheng Chen, Chang-Kai Chao, Shih-Hsiung Lee, Meng-Feng Chen, Tsung-Han Lin, Hsin-Lung Hsieh, Po-Kai Yang, Chung-Chien Hsu, Guan-Xiang Wang, You-Cheng Chang, Kuan-Ting Kuo, Kai-Wei Tsou and Che-Yu Kuo. We are very grateful to the Ministry of Science and Technology of Taiwan for long-term support for our researches on machine learning and source separation. The great efforts from the editors of Academic Press at Elsevier, namely Tim Pitts, Charlie Kent, Carla B. Lima, John Leonard and Sheela Josy, are also appreciated. Finally, I would like to thank my family for supporting my whole research live.

Jen-Tzung Chien
Hsinchu, Taiwan
October 2018

Notations and Abbreviations

GENERAL NOTATIONS

This book observes the following general mathematical notations across different chapters:

$\mathbb{Z}^+ = \{1, 2, \ldots\}$	Set of positive integers
$\mathbb{R}$	Set of real numbers
$\mathbb{R}_+$	Set of positive real numbers
$\mathbb{R}^D$	Set of D dimensional real numbers
a	Scalar variable
$\mathbf{a}$	Vector variable
$\mathbf{a} = \begin{bmatrix} a_1 & \cdots & a_N \end{bmatrix}^\mathsf{T} = \begin{bmatrix} a_1 \\ \vdots \\ a_N \end{bmatrix}$	Elements of a vector, which can be described with the square brackets $[\cdots]$. $^\mathsf{T}$ denotes the transpose operation.
$\mathbf{A}$	Matrix variable
$\mathbf{A} = \begin{bmatrix} a & b \\ c & d \end{bmatrix}$	Elements of a matrix, which can be described with the square brackets $[\cdots]$.
$\mathcal{A}$	Tensor variable
$\mathbf{I}_D$	$D \times D$ identity matrix
$|\mathbf{A}|$	Determinant of a square matrix
$\mathrm{tr}[\mathbf{A}]$	Trace of a square matrix
$A = \{a_1, \ldots, a_N\} = \{a_n\}_{n=1}^N$	Elements in a set, which can be described with the curly braces $\{\cdots\}$.
$A = \{a_n\}$	Elements in a set, where the range of index n is omitted for simplicity.
$|A|$	The number of elements in a set A. For example, $|\{a_n\}_{n=1}^N| = N$.
$f(x)$ or f_x	Function of x
$p(x)$ or $q(x)$	Probabilistic distribution function of x
$\mathcal{F}[f]$	Functional of f. Note that a functional uses the square brackets $[\cdot]$ while a function uses the parentheses $(\cdot)$.
$\mathbb{E}[\,\cdot\,]$	Expectation function
$\mathbb{H}[\,\cdot\,]$	Entropy function
$\mathbb{E}_{p(x|y)}[f(x)|y] = \int f(x) p(x|y) dx$	The expectation of $f(x)$ with respect to probability distribution $p(x|y)$
$\mathbb{E}_x[f(x)|y] = \int f(x) p(x|y) dx$	Another form of the expectation of $f(x)$, where the subscript with the probability distribution and/or the conditional variable is omitted, when it is trivial.

$$\delta(a, a') = \begin{cases} 1 & a = a', \\ 0 & \text{otherwise} \end{cases}$$ Kronecker delta function for discrete variables a and a'

$\delta(x - x')$ — Dirac delta function for continuous variables x and x'

$\boldsymbol{\Theta}_{\text{ML}}, \boldsymbol{\Theta}_{\text{MAP}}, \ldots$ — The model parameters $\boldsymbol{\Theta}$ estimated by a specific criterion (e.g., maximum likelihood (ML), maximum *a posteriori* (MAP), etc.) are represented by the criterion abbreviation in the subscript.

BASIC NOTATIONS USED FOR SOURCE SEPARATION

We also list the specific notations for source separation. This book keeps the consistency by using the same notations for different models and applications. The explanations of the notations in the following list provide a general definition.

n	Number of channels or sensors
m	Number of sources
$\boldsymbol{\Theta}$	Set of model parameters
$\mathcal{M}$	Model variable including type of model, structure, hyperparameters, etc.
$\boldsymbol{\Psi}$	Set of hyperparameters
$Q(\cdot \vert \cdot)$	Auxiliary function used in EM algorithm
$\mathbf{H}$	Hessian matrix
$T \in \mathbb{Z}_+$	Number of observation frames
$t \in \{1, \ldots, T\}$	Time frame index
$\mathbf{x}_t \in \mathbb{R}^n$	n-dimensional mixed vector at time t with n channels
$\mathbf{X} = \{\mathbf{x}_t\}_{t=1}^T$	Sequence of T mixed vectors
$\mathbf{s}_t \in \mathbb{R}^m$	m-dimensional source vector at time t
$\mathbf{y}_t \in \mathbb{R}^m$	m-dimensional demixed vector at time t
$\mathbf{A} = \{a_{ij}\} \in \mathbb{R}^{n \times m}$	Mixing matrix
$\mathbf{W} = \{w_{ji}\} \in \mathbb{R}^{m \times n}$	Demixing matrix in independent component analysis (ICA)
$\mathcal{D}(\mathbf{X}, \mathbf{W})$	Contrast function for ICA using observation data $\mathbf{X}$ and demixing matrix $\mathbf{W}$. This function is written as the divergence measure to be minimized.
$\mathcal{J}(\mathbf{X}, \mathbf{W})$	Contrast function for ICA using observation data $\mathbf{X}$ and demixing matrix $\mathbf{W}$. This function is written as the probabilistic measure to be maximized.
$\mathbf{X} = \{X_{mn}\} \in \mathbb{R}_+^{M \times N}$	Nonnegative mixed observation matrix in nonnegative matrix factorization (NMF) with N frames and M frequency bins
$\mathbf{B} = \{B_{mk}\} \in \mathbb{R}_+^{M \times K}$	Nonnegative basis matrix in NMF with M frequency bins and K basis vectors
$\mathbf{W} = \{W_{kn}\} \in \mathbb{R}_+^{K \times N}$	Nonnegative weight matrix in NMF with K basis vectors and N frames
η	Learning rate
τ	Iteration or shifting index
λ	Regularization parameter
$\mathcal{X} = \{\mathcal{X}_{lmn}\} \in \mathbb{R}^{L \times M \times N}$	Three-way mixed observation tensor having dimensions L, M and N
$\mathcal{G} = \{\mathcal{G}_{ijk}\} \in \mathbb{R}^{I \times J \times K}$	Three-way core tensor having dimensions I, J and K
$\mathbf{x}_t = \{x_{td}\}$	D-dimensional mixed observation vector in a deep neural network (DNN) or a recurrent neural network (RNN). There are T vectors.

$\mathbf{r}_t = \{r_{tk}\}$	K-dimensional source vector
$\mathbf{y}_t = \{y_{tk}\}$	K-dimensional demixed vector
$\mathbf{z}_t = \{z_{tm}\}$	m-dimensional feature vector
$\{a_{tm}, a_{tk}\}$	Activations in a hidden layer and an output layer
$\{\delta_{tm}, \delta_{tk}\}$	Local gradients with respect to $\{a_{tm}, a_{tk}\}$
$\sigma(a)$	Sigmoid function using an activation a
$s(a)$	Softmax function
$\mathbf{w}^{(l)}$	Feedforward weights in the lth hidden layer
$\mathbf{w}^{(ll)}$	Recurrent weights in the lth hidden layer
$E(\mathbf{w})$	Error function of using the whole training data $\{\mathbf{X} = \{x_{td}\}, \mathbf{R} = \{r_{tk}\}\}$
$E_n(\mathbf{w})$	Error function corresponding to the nth minibatch of training data $\{\mathbf{X}_n, \mathbf{R}_n\}$

ABBREVIATIONS

BSS:	Blind Source Separation (page 3)
ICA:	Independent Component Analysis (page 22)
NMF:	Nonnegative Matrix Factorization (page 25)
NTF:	Nonnegative Tensor Factorization (page 33)
CP:	Canonical Decomposition/Parallel Factors (page 36)
PARAFAC:	Parallel Factor Analysis (page 233)
STFT:	Short-Time Fourier Transform (page 6)
GMM:	Gaussian Mixture Model (page 6)
DNN:	Deep Neural Network (page 37)
SGD:	Stochastic Gradient Descent (page 40)
MLP:	Multilayer Perceptron (page 38)
ReLU:	Rectified Linear Unit (page 39)
FNN:	Feedforward Neural Network (page 38)
DBN:	Deep Belief Network (page 43)
RBM:	Restricted Boltzmann Machine (page 43)
RNN:	Recurrent Neural Network (page 45)
DRNN:	Deep Recurrent Neural Network (page 48)
DDRNN:	Discriminative Deep Recurrent Neural Network (page 290)
BPTT:	Backpropagation Through Time (page 45)
LSTM:	Long Short-Term Memory (page 50)
BLSTM:	Bidirectional Long Short-Term Memory (page 296)
BRNN:	Bidirectional Recurrent Neural Network (page 296)
CNN:	Convolutional Neural Network (page 277)
RIR:	Room Impulse Response (page 8)
NCTF:	Nonnegative Convolutive Transfer Function (page 11)
MIR:	Music Information Retrieval (page 13)
CASA:	Computational Auditory Scene Analysis (page 14)
FIR:	Finite Impulse Response (page 11)
DOA:	Direction of Arrival (page 6)

MFCC: Mel-Frequency Cepstral Coefficient (page 27)
KL: Kullback–Leibler (page 29)
IS: Itakura–Saito (page 30)
ML: Maximum Likelihood (page 76)
RHS: Right-Hand Side (page 29)
LHS: Left-Hand Side (page 180)
ARD: Automatic Relevance Determination (page 59)
MAP: Maximum *A Posteriori* (page 83)
RLS: Regularized Least-Squares (page 62)
SBL: Sparse Bayesian Learning (page 63)
LDA: Linear Discriminant Analysis (page 69)
SNR: Signal-to-Noise Ratio (page 103)
SIR: Signal-to-Interference Ratio (page 24)
SDR: Signal-to-Distortion Ratio (page 74)
SAR: Source-to-Artifacts Ratio (page 74)
EM: Expectation Maximization (page 76)
VB: Variational Bayesian (page 85)
VB-EM: Variational Bayesian Expectation Maximization (page 85)
ELBO: Evidence Lower Bound (page 87)
MCMC: Markov Chain Monte Carlo (page 92)
HMM: Hidden Markov Model (page 100)
PCA: Principal Component Analysis (page 100)
MDL: Minimum Description Length (page 101)
BIC: Bayesian Information Criterion (page 101)
CIM: Component Importance Measure (page 102)
MLED: Maximum Likelihood Eigendecomposition (page 103)
LR: Likelihood Ratio (page 110)
NLR: Nonparametric Likelihood Ratio (page 106)
ME: Maximum Entropy (page 108)
MMI: Minimum Mutual Information (page 108)
NMI: Nonparametric Mutual Information (page 116)
WER: Word Error Rate (page 104)
SER: Syllable Error Rate (page 115)
C-DIV: Convex Divergence (page 118)
C-ICA: Convex Divergence ICA (page 118)
WNMF: Weighted Nonnegative Matrix Factorization (page 123)
NB-ICA: Nonstationary Bayesian ICA (page 132)
OLGP-ICA: Online Gaussian Process ICA (page 145)
SMC-ICA: Sequential Monte Carlo ICA (page 145)
GP: Gaussian Process (page 59)
AR: Autoregressive Process (page 147)
NMFD: Nonnegative Matrix Factor Deconvolution (page 161)
NMF2D: Nonnegative Matrix Factor 2-D Deconvolution (page 161)
NTFD: Nonnegative Tensor Factor Deconvolution (page 237)

NMF2D: Nonnegative Tensor Factor 2-D Deconvolution (page 237)
GIG: Generalized Inverse-Gaussian (page 168)
MGIG: Matrix-variate Generalized Inverse-Gaussian (page 255)
LPC: Linear Prediction Coefficient (page 172)
CD: Cepstrum Distance (page 171)
LLR: Log-Likelihood Ratio (page 172)
FWSegSNR: Frequency-Weighted Segmental SNR (page 172)
SRMR: Speech-to-Reverberation Modulation Energy Ratio (page 172)
PLCA: Probabilistic Latent Component Analysis (page 174)
PLCS: Probabilistic Latent Component Sharing (page 181)
CAS: Collaborative Audio Enhancement (page 181)
1-D: One-Dimensional (page 177)
2-D: Two-Dimensional (page 177)
BNMF: Bayesian Nonnegative Matrix Factorization (page 182)
GE-BNMF: Gaussian–Exponential BNMF (page 195)
PG-BNMF: Poisson–Gamma BNMF (page 195)
PE-BNMF: Poisson–Exponential BNMF (page 195)
SMR: Speech-to-Music Ratio (page 196)
NSDR: Normalized Signal-to-Distortion Ratio (page 199)
GNSDR: Global Normalized Signal-to-Distortion Ratio (page 199)
BGS: Bayesian Group Sparse learning (page 202)
LSM: Laplacian Scale Mixture distribution (page 206)
NMPCF: Nonnegative Matrix Partial Co-Factorization (page 202)
GNMF: Group-based Nonnegative Matrix Factorization (page 203)
DNMF: Discriminative Nonnegative Matrix Factorization (page 69)
LNMF: Layered Nonnegative Matrix Factorization (page 219)
DLNMF: Discriminative Layered Nonnegative Matrix Factorization (page 219)
FA: Factor Analysis (page 221)
PMF: Probabilistic Matrix Factorization (page 242)
PTF: Probabilistic Tensor Factorization (page 244)
PSDTF: Positive Semidefinite Tensor Factorization (page 251)
LD: Log-Determinant (page 253)
GaP: Gamma Process (page 254)
LBFGS: Limited Memory Broyden–Fletcher–Goldfarb–Shanno (page 263)
STOI: Short-Time Objective Intelligibility (page 263)
PESQ: Perceptual Evaluation of Speech Quality (page 264)
T-F: Time-Frequency (page 268)
CCF: Cross-Correlation Function (page 269)
ITD: Interaural Time Difference (page 269)
ILD: Interaural Level Difference (page 269)
GFCC: Gammatone Frequency Cepstral Coefficients (page 269)
BIR: Binaural Impulse Response (page 269)
IBM: Ideal Binary Mask (page 268)
PIT: Permutation Invariant Training (page 275)

IRM: Ideal Ratio Mask (page 274)
IAM: Ideal Amplitude Mask (page 274)
IPSM: Ideal Phase Sensitive Mask (page 274)
FC: Fully-Connected (page 283)
STF: Spectral-Temporal Factorization (page 283)
L-BFGS: Limited-memory Broyden–Fletcher–Goldfarb–Shanno (page 290)
VRNN: Variational Recurrent Neural Network (page 299)
VAE: Variational Auto-Encoder (page 299)
NTM: Neural Turing Machine (page 307)
RCNN: Recall Neural Network (page 314)

FUNDAMENTAL THEORIES

INTRODUCTION

In real world, mixed signals are received everywhere. The observations perceived by a human are degraded. It is difficult to acquire faithful information from environments in many cases. For example, we are surrounded by sounds and noises with interference from room reverberation. Multiple sources are active simultaneously. The sound effects or listening conditions for speech and audio signals are considerably deteriorated. From the perspective of computer vision, an observed image is usually blurred by noise, illuminated by lighting or mixed with the other image due to reflection. The target object becomes hard to detect and recognize. In addition, it is also important to deal with the mixed signals of medical imaging data, including magnetoencephalography (MEG) and functional magnetic resonance imaging (fMRI). The mixing interference from external sources of electromagnetic fields due to the muscle activity significantly masks the desired measurement from brain activity. Therefore, how to come up with a powerful solution to separate a mixed signal into its individual source signals is nowadays a challenging problem, which has attracted many researchers working in this direction and developing practical systems and applications.

This chapter starts with an introduction to various types of separation system in Section 1.1. We then address the separation problems and challenges in Section 1.2 where machine learning and deep learning algorithms are performed to tackle these problems. A set of practical systems and applications using source separation are illustrated. An overview of the whole book is systematically described in Section 1.3.

1.1 SOURCE SEPARATION

Blind source separation (BSS) aims to separate a set of source signals from a set of mixed signals without or with very little information about the source signals or the mixing process. BSS deals with the problem of signal reconstruction from a mixed signal or a set of mixed signals. Such a scientific domain is multidisciplinary. *Signal processing* and *machine learning* are two professional domains which have been widely explored to deal with various challenges in BSS. In general, there are three types of a mixing system or sensor network in real-world applications, namely *multichannel source separation*, *monaural source separation* and *deconvolution-based separation,* which are surveyed in what follows.

1.1.1 MULTICHANNEL SOURCE SEPARATION

A classical example of a source separation problem is the *cocktail party problem*, where a number of people are talking simultaneously in a room at a cocktail party, and a listener is trying to follow one of the discussions. As shown in Fig. 1.1, three speakers $\{s_{t1}, s_{t2}, s_{t3}\}$ are talking at the same time. Three microphones $\{x_{t1}, x_{t2}, x_{t3}\}$ are installed nearby as the sensors to acquire speech signals which

Source Separation and Machine Learning. https://doi.org/10.1016/B978-0-12-804566-4.00012-7

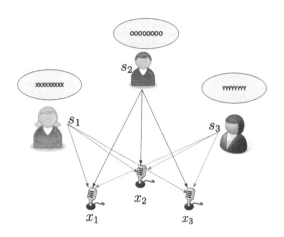

FIGURE 1.1

Cocktail party problem with three speakers and three microphones.

are mixed differently depending on the location, angle and channel characteristics of individual micro-phones. A linear mixing system is constructed as

$$
\begin{aligned}
x_{t1} &= a_{11}s_{t1} + a_{12}s_{t2} + a_{13}s_{t3}, \\
x_{t2} &= a_{21}s_{t1} + a_{22}s_{t2} + a_{23}s_{t3}, \\
x_{t3} &= a_{31}s_{t1} + a_{32}s_{t2} + a_{33}s_{t3}.
\end{aligned}
\tag{1.1}
$$

This 3×3 mixing system can be written in a vector and matrix form as $\mathbf{x}_t = \mathbf{A}\mathbf{s}_t$ where $\mathbf{x}_t = [x_{t1}\ x_{t2}\ x_{t3}]^\top$, $\mathbf{s}_t = [s_{t1}\ s_{t2}\ s_{t3}]^\top$ and $\mathbf{A} = [a_{ij}] \in \mathbb{R}^{3\times3}$. This system involves the current time t and constant mixing matrix $\mathbf{A}$ without considering the noise effect. We also call it the *instantaneous* and *noiseless* mixing system. Assuming 3×3 mixture matrix $\mathbf{A}$ is invertible, an inverse problem is then tackled to identify the source signals as $\mathbf{s}_t = \mathbf{W}\mathbf{x}_t$ where $\mathbf{W} = \mathbf{A}^{-1}$ is the demixing matrix which exactly recovers the original source signals $\mathbf{s}_t$ from the mixed observations $\mathbf{x}_t$.

More generally, the multichannel source separation is formulated as an $n \times m$ linear mixing system consisting of a set of n linear equations for n individual channels or sensors $\mathbf{x}_t \in \mathbb{R}^{n\times1}$ where m sources $\mathbf{s}_t \in \mathbb{R}^{m\times1}$ are present (see Fig. 1.2). A general linear mixing system $\mathbf{x}_t = \mathbf{A}\mathbf{s}_t$ is expressed and extended by

$$
\begin{aligned}
x_{t1} &= a_{11}s_{t1} + a_{12}s_{t2} + \cdots + a_{1m}s_{tm}, \\
x_{t2} &= a_{21}s_{t1} + a_{22}s_{t2} + \cdots + a_{2m}s_{tm}, \\
&\ \ \vdots \qquad\qquad \vdots \\
x_{tn} &= a_{n1}s_{t1} + a_{n2}s_{t2} + \cdots + a_{nm}s_{tm}
\end{aligned}
\tag{1.2}
$$

where the mixing matrix $\mathbf{A} = [a_{ij}] \in \mathbb{R}^{n\times m}$ is merged. In this problem, the mixing matrix $\mathbf{A}$ and the source signals $\mathbf{s}_t$ are unknown. Our goal is to reconstruct the source signal $\mathbf{y}_t \in \mathbb{R}^{n\times1}$ by finding a

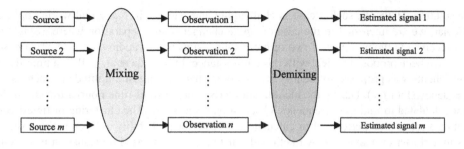

FIGURE 1.2

A general linear mixing system with n observations and m sources.

demixing matrix $\mathbf{W} \in \mathbb{R}^{m \times n}$ through $\mathbf{y}_t = \mathbf{W}\mathbf{x}_t$. We aim at estimating the demixing matrix according to an objective function $\mathcal{D}(\mathbf{X}, \mathbf{W})$ from a set of mixed signals $\mathbf{X} = \{\mathbf{x}_t\}_{t=1}^{T}$ so that the constructed signals are close to the original source signals as much as possible, i.e., $\mathbf{y}_t \approx \mathbf{s}_t$. There are three situations in multichannel source separation:

Determined System: n = m

In this case, the number of channels is the same as the number of sources. We are solving a determined system where a unique solution exists if the mixing matrix $\mathbf{A}$ is nonsingular and the invertible matrix $\mathbf{W} = \mathbf{A}^{-1}$ is tractable. The exact solution to this situation is obtained by $\mathbf{y}_t = \mathbf{W}\mathbf{x}_t$ where $\mathbf{y}_t = \mathbf{s}_t$. For the application of audio signal separation, this condition implies that the number of speakers or musical sources is the same as the number of microphones which are used to acquire audio signals. A microphone array is introduced in such a situation. This means that if more sources are present, we need to employ more microphones to estimate the individual source signals. Independent component analysis (ICA), as detailed in Section 2.1, is developed to resolve this case. A number of advanced solutions to this BSS case will be described in Chapter 4.

Overdetermined System: n > m

In this case, the number of channels n is larger than the number of sources m. In mathematics, this is an overdetermined case in a system of equations where there are more equations than unknowns. In audio source separation, each speaker or musical source is seen as an available degree of freedom while each channel or microphone is viewed as a constraint that restricts one degree of freedom. The overdetermined case appears when the system has been overconstrained. Such an overdetermined system is almost always *inconsistent* so that there is no consistent solution especially when constructed with a random mixing matrix $\mathbf{A}$. In Sawada et al. (2007), the complex-valued ICA was developed to tackle the situation, in which the number of microphones was enough for the number of sources. This method separates the frequency bin-wise mixtures. For each frequency bin, an ICA demixing matrix is estimated to optimally push the distribution of the demixed elements far from a Gaussian.

Underdetermined System: n < m

Underdetermined system in source separation occurs when the number of channels is less than the number of sources (Winter et al., 2007). This system is underconstrained. It is difficult to estimate a reliable solution. However, such a case is challenging and relevant in many real-world applications. In particular, we are interested in the case of single-channel source separation where a single mixed signal is received in the presence of two or more source signals. For audio source separation, several methods have been proposed to deal with this circumstance. In Sawada et al. (2011), a time-frequency masking scheme was proposed to identify which source had the largest amplitude in each individual time-frequency slot (f, t). During the identification procedure, a short-time Fourier transform (STFT) was first calculated to find time-frequency observation vectors $\mathbf{x}_{ft}$. The clustering of time-frequency observation vectors was performed to calculate the posterior probability $p(j|\mathbf{x}_{ft})$ that a vector $\mathbf{x}_{ft}$ belongs to a cluster or a source j. A likelihood function $p(\mathbf{x}_{ft}|j)$ based on Gaussian mixture model (GMM) was used in this calculation. A time-frequency masking function $\mathcal{M}_{ft}^{j}$ was accordingly determined to estimate the separated signals $\widehat{\mathbf{s}}_{ft}^{j} = \mathcal{M}_{ft}^{j}\mathbf{x}_{ft}$ for an individual source j.

In some cases, the number of sources m is unknown beforehand. Estimating the number of sources requires identifying the right condition and developing the right solution to overcome the corresponding mixing problem. In Araki et al. (2009a, 2009b), the authors constructed a GMM with Dirichlet prior for mixture weights to identify the direction-of-arrival (DOA) of source speech signal from individual time-frequency observations $\mathbf{x}_{ft}$ and used DOA information to learn the number of sources and develop a specialized solution for sparse source separation.

In addition, we usually assume the mixing system is time-invariant or, equivalently, the mixing matrix $\mathbf{A}$ is time independent. This assumption may not faithfully reflect the real-world source separation where sources are moving or changing, or the number of sources is also changing. In this case, the mixing matrix is time dependent, i.e., $\mathbf{A} \to \mathbf{A}(t)$. We need to find the demixing matrix, which is also time dependent as $\mathbf{W} \to \mathbf{W}(t)$. Estimating the source signals $\mathbf{s}_t$ under the *nonstationary* mixing system is relevant in practice and crucial for real-world blind source separation.

1.1.2 MONAURAL SOURCE SEPARATION

BSS is in general highly underdetermined. Many applications involve a single-channel source separation problem ($n = 1$). Among different realizations of a mixing system, it is crucial to deal with single-channel source separation because a wide range of applications involve only a single recording channel but mix or convolve with various sources or interferences. Fig. 1.3 demonstrates a scenario of monaural source separation using a single microphone with three sources. We aim to suppress the ambient noises, including bird and airplane, and identify the human voices for listening or understanding. Therefore, single-channel source separation can be generally treated as a venue to speech enhancement or noise reduction in a way that we want to enhance or purify the speech signal in the presence of surrounding noises.

There are two learning strategies in monaural source separation, *supervised learning* and *unsupervised learning*. Supervised approach conducts source separation given by the labeled training data from different sources. Namely, the separated training data are collected in advance. Using this strategy, source separation is not truly blind. A set of training data pairs with mixed signals and separated signals are provided to train a demixing system which is generalizable to decompose those unseen mixed signals. Nonnegative matrix factorization (NMF) and deep neural network (DNN) are two machine learning paradigms to deal with single-channel source separation which will be extensively described in Sections 2.2 and 2.4 with a number of advanced works organized in Chapters 5 and 7, respectively. Basically, NMF (Lee and Seung, 1999) factorizes a nonnegative data matrix $\mathbf{X} = \{\mathbf{x}_t\}_{t=1}^{T}$ into a prod-

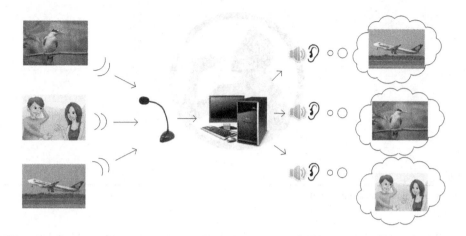

FIGURE 1.3

An illustration of monaural source separation with three source signals.

uct of a nonnegative basis (or template) matrix **B** and a nonnegative weight (or activation) matrix **W**, $\mathbf{X} \approx \mathbf{BW}$. For the application of audio source separation, NMF is implemented by using the Fourier spectrogram of audio signals for **X** which is always nonnegative. The individual separated signals can be extracted by activating the corresponding basis vectors using the corresponding activation weights. NMF is seen as a linear model. In addition, DNN is introduced as a specialized mechanism to characterize the detailed nonlinear structure in a mixed signal which is learned by solving a supervised regression problem (Wang and Wang, 2012, 2013, Grais et al., 2014). The separated signals are treated as the regression outputs during the training procedure based on minimization of regression errors. In Huang et al. (2014a), a deep recurrent neural network further captured the deep temporal information for DNN source separation.

On the other hand, unsupervised monaural source separation is conducted without the need of the separated training data. This strategy involves truly blind processing for source separation because the only data we observe is the single channel of mixed signal recorded in test time. Such a learning task is important for singing voice separation which aims at extracting the singing voice with background accompaniment (Ozerov et al., 2007, Raj et al., 2007, Durrieu et al., 2011, Li and Wang, 2007) as illustrated in Fig. 1.4. Singing voice conveys important information for different applications of music information retrieval including singer identification (Mesaros et al., 2007), music emotion annotation (Yang and Lee, 2004), melody extraction, and lyric recognition. In Rafii and Pardo (2013), the repeating structure of the spectrogram of the mixed music signal was extracted for singing voice separation. In Huang et al. (2012), a robust principal component analysis was proposed to decompose the spectrogram of mixed signal into a low-rank matrix for accompaniment signal and a sparse matrix for vocal signal. System performance was improved by imposing the harmonicity constraints (Yang, 2012). A pitch extraction algorithm was applied to extract the harmonic components of singing voice. In Virtanen (2007), Zhu et al. (2013), Yang et al. (2014b), an NMF model was developed for unsupervised separation of singing voice in a single-channel music signal. Nevertheless, the

FIGURE 1.4

An illustration of singing voice separation.

separation performance was considerably deteriorated under ill-posed condition. Model regularization was ignored in the above supervised and unsupervised methods.

1.1.3 REVERBERANT SOURCE SEPARATION

We are surrounded by sounds and noises in the presence of room reverberation. In a room environment, sound waves propagate from a speaker or a source to the microphones. These waves are repeatedly reflected at the walls inside a room. These reflections change the acoustic characteristics of the original speech. As a result, solving the instantaneous mixtures, as mentioned in Section 1.1.1, could not really reflect the real reverberant environment which structurally mixes the sources as the convolutional mixtures (Sawada et al., 2007, Yoshioka et al., 2011). In addition to source separation for instantaneous mixing system, another important source separation task is designed to handle the *convolutive mixing* system. This task is also called the *reverberant source separation* which is currently an emerging area in machine learning for signal processing. In particular, we are interested in solving the problem of *speech dereverberation* to build a robust speech recognition system as shown in Fig. 1.5. The dereverberated speech is first estimated by a preprocessing approach to dereverberation. After feature extraction, the transcription of dereverberated speech is then obtained by a back-end decoder using the acoustic, pronunciation and language models. The acoustic model is generally trained by using clean speech without degradation due to reverberation effect.

A reverberant speech signal $x(t)$ is basically expressed as a linear convolution of a clean speech signal $s(t)$ and a room impulse response (RIR) $r(t)$ in the time domain as

$$x(t) = \sum_{\tau=0}^{T_r} r(\tau)s(t - \tau) \tag{1.3}$$

where T_r is the length of RIR. Such a convolutive mixture is usually separated in the frequency domain and then reconstructed in the time domain. Typically, RIR consists of three portions: direct sound, early reflections and late reverberation as illustrated in Fig. 1.6. The sound reflections that arrive shortly after the direct sound are called early reverberation, which usually occur within 50 milliseconds (ms), and reflections that arrive after the early reverberation are called late reverberation. The property of

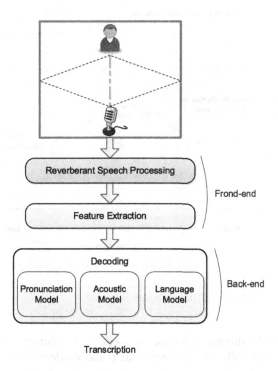

FIGURE 1.5

Speech recognition system in a reverberant environment.

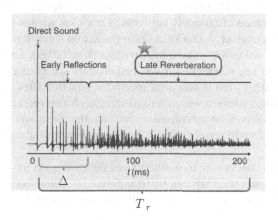

FIGURE 1.6

Characteristics of room impulse response.

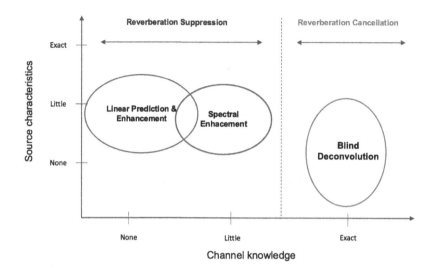

FIGURE 1.7

Categorization of various methods for speech dereverberation.

early reflections relies on the distance between the speaker and microphone. The magnitude of late reverberation decays exponentially and the decay rate is independent of the positions of the speaker and microphone. The time during which the magnitude of late reverberation reduces by 60 dB relative to the direct sound is called the reverberation time T_{60}. The length of early reflections is denoted by Δ.

Fig. 1.7 depicts different speech dereverberation methods, which are categorized into two parts according to the amount of channel knowledge. In general, the reverberation *suppression* methods use very little information about channels and sources while the reverberation *cancellation* methods are provided with configuration of channels but without prior knowledge of the mixing process and source characteristics. In Kailath et al. (2000), a linear prediction algorithm was proposed for estimating the inverse filter of an unknown system. In Kinoshita et al. (2009), a multistep linear prediction was developed to suppress late reverberation for multichannel speech dereverberation. Another approach to reverberation suppression is based on spectral subtraction (Boll, 1979). In Tachioka et al. (2013), a spectral subtraction method was proposed for speech dereverberation via an estimation of reverberation time. Reflections were subtracted to enhance the direct sound for speech recognition. The above-mentioned solutions to reverberation suppression were formulated by using different signal processing methods.

On the other hand, it is challenging to develop blind deconvolution for reverberation cancellation where machine learning based on NMF can be incorporated to learn source signal in the frequency domain. For ease of expression, we rewrite the reverberant speech signal of Eq. (1.3) under an RIR of length L as

$$x(t) = \sum_{l=0}^{L-1} r(l)s(t-l). \tag{1.4}$$

By performing STFT, this convolution is yielded by Talmon et al. (2009)

$$x_c(f,t) \approx \sum_{l=0}^{L-1} r_c(f,l)s_c(f,t-l) \tag{1.5}$$

where f denotes the frequency index and $x_c(f,t)$, $r_c(f,t)$ and $s_c(f,t)$, $1 \le f \le F$, $1 \le t \le T$, denote the corresponding complex-valued STFT coefficients. The RIR signal is also seen as a finite impulse response (FIR) of reverberation system. Basically, the phase of RIR signal $r(t)$ is sensitive to the change of reverberation condition. It is convenient to treat this phase as a random variable and marginalize it out of the model. Assuming $r(t)$ is an independent and uniformly distributed random variable, the power spectrum $|x_c(f,t)|^2$ can be accordingly approximated as the convolution of power spectra of $|r_c(f,t)|^2$ and $|s_c(f,t)|^2$ (Kameoka et al., 2009), namely

$$|x_c(f,t)|^2 \approx \sum_{l=0}^{L-1} |r_c(f,l)|^2 \, |s_c(f,t-l)|^2. \tag{1.6}$$

Different from the implementation of Eq. (1.6) in the power spectrum domain, an alternative approach was performed in Mohammadiha et al. (2015) by adopting the nonnegative convolutive transfer function (NCTF), which was constructed in the magnitude spectrum domain as

$$X_{ft} \approx \sum_{l=0}^{L-1} R_{fl}S_{f,t-l} \tag{1.7}$$

where $X_{ft} = |x_c(f,t)|$, $R_{fl} = |r_c(f,l)|$ and $S_{ft} = |s_c(f,t)|$ denote the magnitude spectra, which are all nonnegative. Here, L is seen as the length of reverberation kernel R_{fl}. In Mohammadiha et al. (2015), a speech dereverberation model, called NCTF-NMF model, was exploited by representing the clean speech signal $\mathbf{S} = \{S_{ft}\} \in \mathbb{R}_+^{F \times T}$ using NMF model

$$S_{ft} \approx [\mathbf{BW}]_{ft} = \sum_k B_{fk}W_{kt} \tag{1.8}$$

and estimating the reverberation kernel $\mathbf{R} = \{R_{fl}\}$ and weight matrix $\mathbf{W} = \{W_{kt}\} \in \mathbb{R}_+^{K \times T}$ by using the NCTF representation in Eq. (1.7). The Kullback–Leibler (KL) divergence between reverberant speech $\mathbf{X} = \{X_{ft}\} \in \mathbb{R}_+^{F \times T}$ and the reconstructed speech,

$$\{\widehat{R}_{fl}, \widehat{W}_{kt}\} = \arg \min_{\{R_{fl}, W_{kt}\}} \sum_{f,t} \mathcal{D}_{\mathrm{KL}} \left(X_{ft} \middle\| \sum_l R_{fl} \sum_k B_{fk}W_{k,t-l} \right), \tag{1.9}$$

was minimized to estimate model parameters for blind deconvolution and dereverberation. In Eq. (1.9), a set of K basis vectors $\mathbf{B} = \{B_{fk}\} \in \mathbb{R}_+^{F \times K}$ was pretrained from clean speech. The procedure of reverberant source separation was performed accordingly. In Section 5.1.2, we will describe the advanced solutions to NCTF-NMF model where Bayesian learning is merged to compensate the model uncertainty for speech dereverberation. In Section 7.2.5, an advanced algorithm based on deep recurrent neural network will be addressed for reverberant source separation as well.

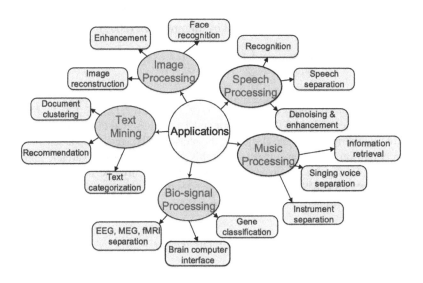

FIGURE 1.8

Categorization of various applications for source separation.

1.2 APPLICATIONS AND CHALLENGES

In what follows, we introduce a number of systems where the technologies of source separation and machine learning can be applied. We further address a variety of challenges in separation methods and learning algorithms where we need improvement to achieve desirable performance in real-world applications.

1.2.1 APPLICATIONS FOR SOURCE SEPARATION

Source separation is employed in the time domain for an instantaneous mixing system or in the frequency domain for a convolutive mixing system. More generally, source separation can be treated as a standard *unsupervised learning* problem. The demixing process in source separation is comparable with a latent component analysis in general machine learning where the target task is to find clusters from the independent components or sources we obtain by a source separation method (Bach and Jordan, 2003). Source separation is feasible for data clustering and mining. More specifically, Fig. 1.8 shows the categorization of various applications based on source separation ranging from speech processing to music processing, image processing, text mining and bio-signal processing. Speech processing applications on speech recognition, speech separation and speech enhancement are included. In music processing, source separation is developed for music information retrieval, singing voice separation and instrument signal separation. In text mining, the applications on document clustering, collaborative filtering and text categorization are included. Source separation can be also applied for image enhancement, image reconstruction, as well as face recognition. For the applications in bio-signal processing, source separation works for gene classification and EEG, MEG or fMRI separation which are useful to build brain computer interface. In what follows, we address a number of applications in detail.

Speech Separation

One of the most important applications where source separation can play a key role is in separation of speech signals. Speech separation is not only employed to decompose multichannel mixed signals into source signals for a cocktail party but also generally developed for single-channel speech enhancement or noise reduction where the target speaker and ambient noise are treated as two sources. We face the problems of multichannel and monaural source separation. As mentioned in Section 1.1.2, monaural source separation is even more crucial and challenging than multichannel source separation because monaural source separation is an underdetermined problem with substantially more applications. Speech enhancement typically involves separation of sources and noises from a single-channel recording. Enhancing speech has many applications for spoken language communication between a human and a machine, including mobile phones, hands-free devices, hearing aids, etc. Therefore, speech separation is usually combined with speech recognition or, in particular, far-talking and hands-free speech recognition to establish such a voice command controller or a human–machine interface for many practical systems, e.g., personal assistant, home automation, social robots, etc. Moreover, reverberant speech separation is important for teleconferencing systems and home automation where the solution to speech dereverberation is crucial for an effective speech communication. In other advanced spoken language systems, speech separation is even integrated into a dialog system or combined with speaker diarization, keyword spotting, language identification and many other approaches. Speech separation acts as a preprocessing component for system integration with a wide range of speech-driven natural language systems, definitely including content-based retrieval for television, radio, movie, personal video, social media, etc.

Music Separation

Music information retrieval (MIR) is known as an emerging field of research for many real-world applications where the fundamentals of signal processing and machine learning are required. The background knowledge in musicology, psychology and academic music study helps MIR researches. However, an observed music signal may be heavily mixed by different instruments, singing voices and sound effects which significantly affect the performance of MIR. Accordingly, music separation or track separation, aiming at separating the music into one track per instrument or per musical source, is seen as one of the most influential areas which considerably improves the retrieval performance for MIR applications. Instrumental music separation and singing voice separation are two key tasks in MIR systems. The goal of instrumental music separation is to separate a mixed music signal into individual source signals corresponding to different instruments, e.g., piano, flute, drum, violin, etc. To do that, knowing the characteristics of different instruments is helpful to identify different instrumental sounds. On the other hand, singing voice conveys important vocal tract information of a song. This information is practical for many music-related applications, including singer identification (Mesaros et al., 2007), singing evaluation (Yang and Lee, 2004), music emotional annotation, melody extraction, lyric recognition and lyric synchronization (Fujihara et al., 2011). But singing voice is usually mixed with background accompaniment in a music signal. Singing voice separation aims to extract the singing voice from a single-channel mixed signal without ground-truth data for pretraining.

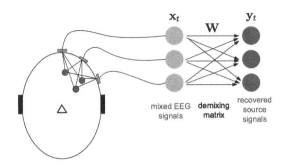

FIGURE 1.9

Blind source separation for electroencephalography artifact removal.

Audio Source Separation

Basically, speech separation and music separation are both under a broad category of source separation called the audio source separation. In addition to speech and music, the natural sound sources here are generalized to a variety of audio signals, including sounds from different animals and sounds from surrounding environments in our daily life, e.g., car passing, door slamming, people walking, running, laughing, clapping, etc. In this case, there are a lot of interesting tasks which are influential for many extended domains. An attractive purpose of audio source separation is to separate the mixed sounds for auditory scene analysis and classification. Correspondingly, the machine perception for computational auditory scene analysis (CASA) is realized to identify the environmental sounds and organize them into perceptually meaningful elements. Different from standard BSS, CASA separates the mixtures of sound sources based on the mechanism of the human auditory system.

Biomedical Source Separation

Another important field that source separation can contribute to influentially is the task of biomedical signal separation which can assist doctors in diagnosing and predicting diseases. The biomedical signals may include electroencephalography (EEG), magnetoencephalography (MEG), functional magnetic resonance imaging (fMRI) and many other methods. As we know, EEG signal is a unique neural media for brain computer communication and can be also used for disease diagnosis and therapy. But EEG artifact is serious in many recordings. How to remove the EEG artifact and acquire the clean signals is crucial in achieving high-quality communication. BSS is beneficial in dealing with EEG artifact which is formulated as a multichannel source separation system as shown in Fig. 1.9. In this case, the recovered signals $\mathbf{y}_t$ are obtained by multiplying the mixed signals by a demixing matrix $\mathbf{W}$ as $\mathbf{y}_t = \mathbf{W}\mathbf{x}_t$ at each time t. In addition, MEG and fMRI provide the functional neuroimaging of brain activity by using different physical techniques with different precisions. Again, the artifacts in the mixing measurements due to eye blinking and facial muscle movement can be removed by source separation methods. Independent component analysis, which will be detailed in Section 2.1 and Chapter 4, is known as the most popular solution to artifact removal. Different challenges in multichannel source separation should be tackled.

Image Separation

Separation of neuroimaging data using MEG and fMRI is seen as a task of image separation which removes the artifact or identifies the independent components from multiple channels. Similar to speech enhancement, source separation can be also developed for image enhancement where the target object is enhanced by canceling the background scene or removing the noise interference. Moreover, we can estimate clean image through image reconstruction and then perform image classification. In Bartlett et al. (2002), the source separation using ICA was developed for face recognition which significantly outperformed that using principal component analysis. In Cichocki and Amari (2002), a complete survey of source separation and learning algorithm for image processing was introduced.

Text Mining

Finding independent components or semantics from symbolic words is an interesting issue to build a structural text mining system. These independent components are typically extracted by a source separation algorithm which provides general solutions to different types of technical data. Therefore, source separation is feasible to extract *independent* topics which can represent text documents with minimum information redundancy. The applications can be extended to topic modeling, topic identification, document clustering, as well as recommendation system. The latent variable model in these applications consists of those independent components learned from source separation.

Although source separation is seen as a general learning framework for many applications in the presence of different technical data, this book will address those methods and solutions mainly for the tasks of speech separation and music separation under the scenarios of multichannel, single-channel and reverberant source separations. The applications to other tasks can be extended in a straightforward manner.

1.2.2 CHALLENGES IN SEPARATION AND LEARNING

There are a number of challenges in source separation and machine learning which have been attracting many researchers to work out different sophisticated solutions in the past decades. Various adaptive signal processing and machine learning algorithms have been introduced to deal with different challenges in audio source separation. These challenges are categorized and summarized in what follows.

Microphone Array Signal Processing

Multichannel source separation involves n microphones for data collection during the procedure of source separation. A microphone array is installed and operated to carry out source separation and noise suppression for speech recognition and hearing aids. In Chien et al. (2001), the microphone array signal processing (Benesty et al., 2008) was developed for speech enhancement based on delay-and-sum beamforming and then applied for speech recognition based on speaker adaptation. Basically, a microphone array is not only feasible for speech enhancement and noise cancellation but also useful for sound source localization and environmental noise monitoring. This book concerns the demixing of single- or multichannel mixing signals rather than the configuration of microphone array for other tasks. The signal decomposition can be either tackled in the time domain for instantaneous mixtures or in the frequency domain for convolutive mixtures.

Convolutive Mixtures

In audio signal applications, signals are mixed in a convolutive manner with reverberations. A mixing system contains delay, attenuation and reverberation. This makes the BSS problem difficult. We need very long FIR filters to separate the acoustic signals mixed under such conditions. Frequency-domain BSS (Sawada et al., 2007, Douglas et al., 2007) plays a key role in coping with convolutive mixtures. The overdetermined or underdetermined condition (Araki et al., 2007) is still a challenging issue. In the frequency-domain BSS, the mixed signals are transformed into the STFT domain. A multichannel demixing system is then analyzed by using ICA method (Makino et al., 2005). The demixed signals are finally transformed to the time domain to obtain individual source signals via the inverse discrete Fourier transform.

Permutation & Scaling Ambiguities

However, the multichannel source separation using ICA suffers from permutation and scaling ambiguities. It is because permuting the rows of demixing matrix or scaling the demixed source signals does not change the ICA solution. We need to align the permutation ambiguity for each frequency bin so that a separated signal in the time domain contains frequency components from the same source signal (Wang et al., 2011). In Sawada et al. (2011), the frequency bin-wise clustering and permutation alignment was proposed to cope with the complicated real-room propagation in an underdetermined mixing condition. For the issue of magnitude ambiguity when applying ICA separation, it is popular to resolve this problem by calculating the inverse matrix or the Moore–Penrose pseudo-inverse matrix of the demixing matrix after resolving the permutation ambiguity when transforming from the frequency to the time domain (Murata et al., 2001). This inverse matrix is used to smooth the magnitude of the demixed source signal for multichannel source separation.

Room Reverberation

Reverberation in acoustics is the persistence of sound after a sound is produced. A large number of reflections may be caused due to the surfaces of objects in the space which could be wall, furniture, people or even air. Reverberation is basically the occurrence of reflections that arrive in less than 50 ms. The sound of reverberation is often added to the musical instrument and to the vocals of singers. The effect of reverberation makes it difficult to retrieve original speech and music signals for speech recognition and music information retrieval, respectively. As mentioned in Section 1.1.3, reverberant source separation is an important venue to various applications, including teleconferencing system, interactive television, hands-free interface, and distant-talking speech recognition. We need to deal with the challenge of demixing or dereverberation based on the frequency-domain BSS. The NCTF reverberation representation combined with NMF separation model (Mohammadiha et al., 2015) provides a meaningful solution to speech dereverberation. In Yoshioka et al. (2012), linear filtering and spectrum enhancement methods were developed for reverberant speech processing and recognition. There are some other approaches which were surveyed in Yoshioka et al. (2011, 2012).

Unknown Number of Sources

Source separation systems usually assume that the number of sources is fixed and known. But in real-world applications this number may be missing in a target space and may be changing in time. We need to identify the number of sources in advance particularly in the case of a voice activity monitoring system. In Araki et al. (2009a, 2009b), a GMM with Dirichlet prior for mixture weight was applied to identify the DOA of the source speech signal from individual time-frequency units. This method was applied to estimate the number of sources and deal with the sparse source separation based on a Dirichlet distribution which was sparse for counting the Gaussians or, equivalently, the sources.

Unknown Model Complexity

Determining the number of sources in some sense is comparable with identifying the complexity of a demixing model which represents the mixed signals. Beyond the issue of source counting, we need to deal with the general challenge of determination of model complexity for source separation. It is because that the model complexity or model structure is always unknown in machine learning problems and the assumptions of separation model and mixing condition are basically uncertain in real-world BSS systems. This challenge is present in model-based source separation where a demixing model is assumed. We face the challenges of model selection, model uncertainty and model assumption. The model we assume may be wrong and the number of basis vectors we use in basis representation for demixed signals may be incorrect. The mixing system may be very complicated. Bayesian learning provides a probabilistic framework to deal with this challenge for BSS (Bishop, 2006, Févotte, 2007, Watanabe and Chien, 2015). Our goal is to formulate an adaptive learning algorithm to mitigate the degradation of performance due to unknown model complexity in source separation.

Heterogeneous Environments

The source separation problem is often tackled in heterogeneous environments where convolutive and additive noises are both contaminated. The mixing condition may be nonstationary, namely the sources may be moving, may be replaced, or the number of sources may be changing (Chien and Hsieh, 2013b). To deal with these issues, we need to adapt the BSS system to meet the heterogeneous environments so that the performance of the BSS system can be preserved. Online learning can capture the dynamics of the mixing process. The concept of adaptive learning can bring flexible solutions to a variety of complicated source separation systems.

1.2.3 FRONT-END PROCESSING AND BACK-END LEARNING

We therefore summarize various challenges in audio source separation and categorize them into two parts as illustrated in Fig. 1.10 (Chien et al., 2013). In front-end processing, we highlight the adaptive signal processing which analyzes the information and manipulates the signal on each source. Time-frequency modeling and spectral masking operation are implemented in a way of front-end signal processing. Source localization is identified by using a microphone array. Through signal processing, we judge how the sources are mixed. The input signals are processed to obtain the separated signals through several processing components. Therefore, front-end processing includes the frequency-domain audio source separation which could align the permutation and magnitude ambiguities (Sawada et al., 2011), separate the convolutive mixtures (Makino et al., 2005), identify the number of sources (Araki et al., 2009a), resolve the overdetermined/underdetermined problem (Araki et al., 2007, Winter et al., 2007), and compensate for the room reverberation (Yoshioka et al., 2012).

On the other hand, the back-end learning is devoted to recovering the source signals by using only the information about their mixtures observed in each microphone without additionally possessing the frequency and location information on each source. We build a statistical model for the whole system and infer the model by using the mixtures. Machine learning algorithms are developed for audio source

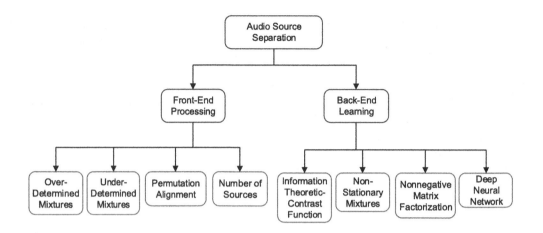

FIGURE 1.10

Challenges in front-end processing and back-end learning for audio source separation.

separation. The model-based speech separation and recognition could be jointly implemented (Rennie et al., 2010, Chien and Chen, 2006). From this perspective, we present the estimation of demixing parameters through construction and optimization of information-theoretic contrast function (Chien and Hsieh, 2012). The solutions to music source separation based on NMF (Cichocki et al., 2008, Sawada et al., 2013) and sparse learning (Chien and Hsieh, 2013a) are introduced. We also focus on the uncertainty modeling for the regularized signal separation in accordance with the Bayesian perspective (Mohammad-Djafari and Knuth, 2010). The nonstationary and temporally-correlated audio source separation (Chien and Hsieh, 2013b) is presented to overcome the heterogeneous conditions in mixing system. Basically, back-end learning is performed by applying an inference procedure or a learning algorithm. An optimization method towards meeting a specialized criterion is implemented by using a set of training signals which may be collected in online or offline manner. More detailed survey of different conditions and solutions in front-end signal processing and back-end machine learning can be found in Chien et al. (2013).

This book emphasizes machine learning approaches to different problems and conditions in source separation. We present the model-based approach which is seen as a statistical representation which can be easily incorporated with physical phenomena, measurements, uncertainties and noises in various forms of mathematical models. This approach is developed in a unified manner through different algorithms, examples, applications, and case studies. These statistical models are recognized as mainstream methods. In particular, deep learning methods are presented to discover the deep latent structure to connect the relation between mixed signal and its separated signals. Probabilistic models and deep models are both feasible to deal with different problems in source separation. Integration over two paradigms provides an insightful solution which simultaneously captures the randomness or faithfulness of representation and explores the deep abstraction and hierarchy of unknown regression for demixing. Machine learning provides a wide range of model-based approaches for blind source separation. Deep learning is an emerging branch of machine learning which involves a variety of deep model-based approaches to build high-performance systems for source separation.

1.3 **OVERVIEW OF THIS BOOK**

Source separation involves extensive knowledge of signal processing and machine learning to deal with the challenges of multichannel signal processing, convolutive mixtures, permutation ambiguity, room reverberation, unknown number of sources, unknown model complexity, heterogeneous conditions and nonstationary environment. This book emphasizes the importance of machine learning perspective and illustrates how source separation problems are tackled through adaptive learning algorithms and model-based approaches. In this back-end learning, we only use the information about mixture signals to build a source separation model which is seen as a statistical model for the whole system. The inference procedure is implemented to construct such a model from a set of observed samples. This book highlights the fundamentals and the advances in back-end learning for source separation.

In the first part of fundamental theories, we organize and survey model-based approaches according to two categories. One is the *separation models* addressed in Chapter 2 and the other is the *learning algorithms* provided in Chapter 3. In Chapter 2, we address the state-of-the-art separation models which consist of

- independent component analysis (ICA) (Section 2.1),
- nonnegative matrix factorization (NMF) (Section 2.2),
- nonnegative tensor factorization (NTF) (Section 2.3),
- deep neural network (DNN) (Section 2.4),
- recurrent neural network (RNN) (Section 2.5).

We will address how these models are evolved to deal with multichannel and single-channel source separation, and explain the weakness and the strength of these models in different mixing conditions. Multiway decomposition is presented to carry out different avenues to underdetermined source separation. A number of extensions of different models with convolutive processing, recurrent modeling, long short-term memory and probabilistic interpretation are introduced.

On the other hand, this book will present the adaptive machine learning algorithms for separation models, including

- information-theoretic learning (Section 3.2),
- Bayesian learning (Section 3.3),
- sparse learning (Section 3.4),
- online learning (Section 3.5),
- discriminative learning (Section 3.6),
- deep learning (Section 3.6).

We illustrate how the information theory is developed for optimization of ICA and NMF objective functions and why the uncertainty and the sparsity of system parameters should be faithfully characterized in ICA, NMF, NTF, DNN and RNN to assure the robustness of source separation against the adverse environments. We develop different optimization algorithms where a number of divergence measures are formulated. Bayesian learning and sparse learning are presented to pursue model regularization in model-based source separation. We embrace uncertainty and sparsity in adaptive signal processing. Structural learning for latent variable model is performed. Bayesian inference procedure is implemented. A Gaussian process is specialized to characterize temporal information in source separation. The dynamics in source signals and sensor networks are captured via the online learning and tracking. Dictionary learning and basis selection are specialized for sparse source separation. The merits of the discriminative learning for individual sources and the deep learning over the complicated mixing system are demonstrated. In addition to these machine learning methods, we also introduce the fundamentals of approximate Bayesian inference (Section 3.7) ranging from the maximum likelihood estimation based on expectation–maximization algorithm (Dempster et al., 1977) to the variational Bayesian inference (Section 3.7.3) and the Gibbs sampling algorithm (Bishop, 2006, Watanabe and Chien, 2015). These inference procedures are seen as the basics to build different probabilistic solutions to latent variable models for source separation.

In the second part of the advanced practices, this book will present a number of case studies which are recently published for model-based source separation. These studies are categorized into five modern models: ICA, NMF, NTF, DNN and RNN. A variety of learning algorithms, mentioned in Chapter 3, are employed in these models so as to carry out different solutions to overcome different issues in construction of source separation systems. Different models and algorithms are connected with further extensions. In ICA model (Chapter 4), we introduce the independent voices for speech recognition (Section 4.1), the nonparametric likelihood ratio ICA for speech separation and recognition (Section 4.2), the convex divergence ICA for speech separation (Section 4.3), the nonstationary Bayesian ICA (Section 4.4), and the online Gaussian process ICA (Section 4.5) for nonstationary speech and music separation. The criterion of independence is formulated in different ways. The noisy ICA model is also adopted. In the NMF model (Chapter 5), we survey the convolutive NMF (Section 5.1), the probabilistic nonnegative factorization (Section 5.2), the Bayesian NMF (Section 5.3), the group sparse NMF (Section 5.4), the discriminative NMF (Section 3.6.1), and the deep NMF (Section 5.5). Sparse learning, Bayesian learning, discriminative learning and deep learning are developed for NMF based source separation. This two-way matrix decomposition is further extended to the multiway tensor decomposition to allow separation of mixed signals with multiple horizons in observation space. In the NTF model (Chapter 6), we address the motivations of multiway decomposition for single-channel source separation and the solutions based on the convolutive NTF (Section 6.2), probabilistic NTF (Section 6.3), Bayesian tensor factorization (Section 6.4), and infinite NTF (Section 6.5). In the DNN model (Chapter 7), we start from the baseline DNN for single-channel BSS and then extend it to the deep recurrent neural network (DRNN) (Section 7.3), the DRNN with long short-term memory (LSTM) (Section 7.4), and the DRNN with bidirectional LSTM (Section 7.4.3). Past and future features are exploited to predict the separated source signals. The advanced studies on deep ensemble learning (Section 7.1.2), deep clustering (Section 7.2.1), joint mask and separation optimization (Section 7.2.2) and permutation invariant training (Section 7.2.3) will be also introduced. A number of modern deep sequential learning approaches will be also presented in Sections 7.5–7.7 where variational recurrent neural network, sequence-to-sequence learning, memory augmented neural network will be introduced. At last, Chapter 8 will introduce the trend of deep learning and its development for source separation. A number of the newest deep models and recurrent machines will be also introduced. Finally, various directions and outlooks will be pointed out for future studies.

MODEL-BASED SOURCE SEPARATION

2

From the historical point of view, traditional blind source separation was developed by identifying a set of independent source signals from the mixed signals which were collected from a number of sensors. The demixed signals were estimated by maximizing the measure of independence expressed in different forms. In 1994, independent component analysis (ICA) (Comon, 1994) was proposed to deal with the cocktail party problem where multiple channels or microphones were used to collect the mixed signals. A demixing matrix was estimated to decompose the mixed signals into individual sources where the resulting measure of independence was maximized. In 1999, nonnegative matrix factorization (NMF) (Lee and Seung, 1999) was proposed to carry out the so-called parts-based representation where the *parts* of a mixed signal or image were decomposed. Monaural source separation in the presence of two or more sources was implemented. Basically, NMF was performed in the supervised mode by using a number of training data. The basis parameters corresponding to target sources were obtained. In 2006, nonnegative tensor factorization (NTF) was proposed for multichannel time–frequency analysis (Mørup and Schmidt, 2006b) and also for separation of speech and music signals (Barker and Virtanen, 2013). In 2014, deep neural network (DNN) was applied to deal with single-channel source separation (Grais et al., 2014) and speech segregation (Jiang et al., 2014). Source separation was treated as a regression problem in DNN training. In 2014, deep recurrent neural network (RNN) was developed for monaural speech separation (Huang et al., 2014a), as well as singing voice separation (Huang et al., 2012). In general, ICA is seen as an unsupervised learning method. ICA source separation is truly "blind". But NMF, DNN and RNN may conduct supervised learning using training data from known speakers. We therefore don't say "blind" source separation for these methods. Fig. 2.1 illustrates the evolution of machine learning methods for source separation from ICA to NMF, NTF, DNN and RNN. Source separation models of ICA, NMF, NTF, DNN and RNN will be detailed in Sections 2.1, 2.2, 2.3, 2.4 and 2.5, respectively.

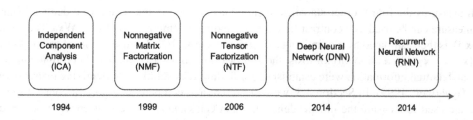

FIGURE 2.1

Evolution of different source separation methods.

Source Separation and Machine Learning. https://doi.org/10.1016/B978-0-12-804566-4.00013-9

2.1 INDEPENDENT COMPONENT ANALYSIS

Independent component analysis is known as a computational model or signal processing method which is developed to separate a set of multichannel mixed signals into a set of additive source signals or independent components. ICA (Comon, 1994) is essential for *unsupervised learning* and blind source separation (BSS). The ICA unsupervised learning procedure attempts to demix the observation vectors and identify the salient features or mixture sources. In addition to BSS, ICA is a highly effective mechanism for feature extraction and data compression. Using ICA, the demixed components can be grouped into clusters where the intracluster elements are dependent and intercluster elements are independent (Bach and Jordan, 2003, Chien and Hsieh, 2012). ICA accordingly provides a very general approach to unsupervised learning which can be applied to acoustic modeling, signal separation and many other applications which will be described in details in Chapter 4.

ICA is related to principal component analysis (PCA) and factor analysis (FA) (Hyvärinen et al., 2001), in which the former estimates a least-squares linear transformation and extracts the uncorrelated components, while the latter explores the uncorrelated common factors with residual specific factors. PCA and FA correspond to second-order methods in which the components or factors are Gaussian distributed. However, ICA aims to pursue independent components for individual sources. It is not possible to assume Gaussian distribution in ICA because the resulting components are uncorrelated and accordingly independent. As a result, there are three assumptions in ICA:

- the sources are statistically *independent*,
- each independent component has a *non-Gaussian* distribution,
- the mixing system is *determined*, i.e., $n = m$, which means that the number of sensors is the same as that of sources.

ICA finds a set of latent components which are non-Gaussian or mutually independent. Much stronger assumptions are made by using ICA when compared with the assumptions in PCA and FA where only second-order statistics are involved. Basically, non-Gaussianity is seen as a measure which is comparable with the measure of independence. The non-Gaussianity or independence can be measured based on an information-theoretic criterion using mutual information and higher-order statistics using kurtosis. The kurtosis of a zero-mean signal y is given by

$$\text{kurt}(y) = \frac{\mathbb{E}[y^4]}{\mathbb{E}^2[y^2]} - 3 \tag{2.1}$$

which measures the fourth-order statistics for non-Gaussianity or sparseness of a demixed signal y. This measure can be used as a contrast function for finding a demixed signal $\mathbf{y}_t = \mathbf{W}\mathbf{x}_t$. The demixing matrix $\mathbf{W}$ is then estimated by maximizing a contrast function $\mathcal{D}(\mathbf{X}, \mathbf{W})$ using a set of training samples $\mathbf{X} = \{\mathbf{x}_1, \ldots, \mathbf{x}_T\}$. In addition to kurtosis, the contrast functions based on likelihood function, negentropy, and mutual information were established for finding ICA solutions to demixing matrix (Comon, 1994, Hyvärinen, 1999). In Section 4.3, we will present and compare a number of contrast functions which are used to measure the independence, the non-Gaussianity or the sparseness of demixed signals. For instance, the ICA method using minimum mutual information (MMI) (Hyvärinen, 1999, Boscolo et al., 2004), as addressed in Section 4.2, is implemented by minimizing a contrast function $\mathcal{D}(\mathbf{X}, \mathbf{W})$ which measures the difference between the marginal entropy and the joint entropy of different information sources. A number of contrast functions will be addressed in Section 3.2 and optimized to conduct information-theoretic learning for ICA.

2.1.1 LEARNING PROCEDURE

In general, there is no closed-form solution to minimization of contrast function $\mathcal{D}(\mathbf{X}, \mathbf{W})$. It is popular to apply an iterative learning procedure to solve the ICA problem based on *gradient descent* or *natural gradient* algorithms (Amari, 1998). Using the gradient descent algorithm, the ICA learning rule is run for parameter updating based on

$$\mathbf{W}^{(\tau+1)} = \mathbf{W}^{(\tau)} - \eta \frac{\partial \mathcal{D}(\mathbf{X}, \mathbf{W}^{(\tau)})}{\partial \mathbf{W}^{(\tau)}} \tag{2.2}$$

where τ indicates the iteration index and η denotes the learning rate. Usually, the convergence rate is slow by using the gradient descent algorithm. A natural gradient algorithm is accordingly introduced to increase the efficiency of ICA learning rule by

$$\mathbf{W}^{(\tau+1)} = \mathbf{W}^{(\tau)} - \eta \frac{\partial \mathcal{D}(\mathbf{X}, \mathbf{W}^{(\tau)})}{\partial \mathbf{W}^{(\tau)}} (\mathbf{W}^{(\tau)})^\top \mathbf{W}^{(\tau)}. \tag{2.3}$$

These algorithms are affected when the initial demixing matrix $\mathbf{W}^{(0)}$ and learning rate η are not properly selected. This condition is serious in the case when highly nonlinear contrast functions are used in the ICA learning rule. To deal with this issue, the numerical stability was improved by step-size normalization where a sufficient condition of learning rate was satisfied (Cichocki and Amari, 2002). The scaled natural gradient algorithm (Douglas and Gupta, 2007) was developed to improve the learning process by imposing an *a posteriori* scalar gradient constraint. This algorithm was robust to different learning rates and came out with a scaled demixing matrix where the convergence rate was elevated without degrading the separation performance. In theory, there is no need to perform a whitening process in natural gradient methods since the whitening process shall result in an orthogonal matrix

$$(\mathbf{W}^{(\tau)})^\top \mathbf{W}^{(\tau)} = \mathbf{I} \tag{2.4}$$

or, equivalently, the simplification to the gradient descent algorithm in Eq. (2.2). For the standard ICA procedure using the gradient descent algorithm, a whitening process is required.

Nevertheless, Fig. 2.2 shows a standard ICA learning procedure for finding a demixing matrix $\mathbf{W}$. Starting from an initial parameter $\mathbf{W}^{(0)}$, we first conduct a data preprocessing stage which consists of centering and whitening where each original sample $\mathbf{x}_t$ is preprocessed by a mean removal operation

$$\mathbf{x}_t \leftarrow \mathbf{x}_t - \mathbb{E}[\mathbf{x}] \tag{2.5}$$

and then a whitening transformation

$$\mathbf{x}_t \leftarrow \mathbf{\Phi} \mathbf{D}^{-1/2} \mathbf{\Phi}^\top \mathbf{x}_t \tag{2.6}$$

where $\mathbf{D}$ and $\mathbf{\Phi}$ denote the eigenvalue matrix and the eigenvector matrix of $\mathbb{E}[\mathbf{x}\mathbf{x}^\top]$. Using the normalized sample vectors $\mathbf{X} = \{\mathbf{x}_t\}_{t=1}^T$, we calculate the adjustment for demixing matrix $\mathbf{W} = [w_{ij}]_{m \times m} =$

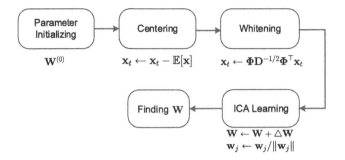

FIGURE 2.2

ICA learning procedure for finding a demixing matrix $\mathbf{W}$.

$[\mathbf{w}_1^\top, \ldots, \mathbf{w}_m^\top]^\top$ by using the differentials

$$\triangle\mathbf{W} = \{\partial\mathcal{D}(\mathbf{X}, \mathbf{W})/\partial w_{ij}\}_{(i,j)=1}^m. \tag{2.7}$$

The learning procedure is terminated when the absolute increment of contrast function $\mathcal{D}(\mathbf{X}, \mathbf{W})$ meets a predefined threshold. In each learning epoch, the normalization step

$$\mathbf{w}_j \leftarrow \mathbf{w}_j/\|\mathbf{w}_j\| \tag{2.8}$$

is performed for each row of $\mathbf{W}$ to assure orthonormal row vectors $\mathbf{w}_j$. Finally, the demixing matrix $\mathbf{W}$ is estimated to find the demixed signals from m sources via

$$\mathbf{y}_t = \{y_{tj}\}_{j=1}^m = \mathbf{W}\mathbf{x}_t \tag{2.9}$$

for individual mixed samples $\mathbf{x}_t = \{x_{ti}\}_{i=1}^m$ from m sensors or microphones. In the evaluation, the metric of signal-to-interference ratio (SIR) is widely used to access the performance of multichannel source separation system. Given the original source signals $\mathbf{S} = \{\mathbf{s}_t\}_{t=1}^T$ and demixed signals $\mathbf{Y} = \{\mathbf{y}\}_{t=1}^T$, SIR in decibels is computed by

$$\text{SIR(dB)} = 10\log_{10}\left(\frac{\sum_{t=1}^T \|\mathbf{s}_t\|^2}{\sum_{t=1}^T \|\mathbf{y}_t - \mathbf{s}_t\|^2}\right). \tag{2.10}$$

Typically, the ICA algorithm is designed for the determined system ($n = m$), but it can be further modified to find the solutions to an overdetermined system ($n > m$) and an underdetermined system ($n < m$) as described in Section 1.1.1. For the special realization of single-channel source separation, i.e., $n = 1$, it is popular to implement monaural source separation based on the nonnegative matrix factorization which is described in what follows. Different from ICA, NMF does not assume independent sources in the implementation.

FIGURE 2.3

Illustration for nonnegative matrix factorization $\mathbf{X} \approx \mathbf{BW}$.

2.2 NONNEGATIVE MATRIX FACTORIZATION

Nonnegative matrix factorization (NMF) is a group of algorithms in multivariate analysis and linear algebra where the nonnegative data matrix $\mathbf{X}$ is approximated and factorized into a product of a non-negative basis (or template) matrix $\mathbf{B}$ and a nonnegative weight (or activation) matrix $\mathbf{W}$ as follows:

$$\mathbf{X} \approx \widehat{\mathbf{X}} = \mathbf{BW}. \tag{2.11}$$

Owing to this nonnegativity property, NMF only allows additive linear interpolation which results in the so-called *parts-based representation* (Lee and Seung, 1999, Hoyer, 2004). The decomposition matrices $\mathbf{B}$ and $\mathbf{W}$ are meaningful to inspect and interpret. In Lee and Seung (1999), NMF started its application of discovering latent features in facial images and text data. NMF has been successfully developed for a wide range of learning systems such as computer vision, document clustering, chemometrics, audio signal processing and recommendation systems. Nonnegative constraint is imposed in NMF to reflect a variety of natural signals, e.g., pixel intensities, amplitude spectra, occurrence counts, which are observed in many environments. Importantly, NMF finds application in single-channel source separation. For the application of audio source separation, NMF can be implemented by using the Fourier spectrogram of audio signals, which is a nonnegative matrix $\mathbf{X} \in \mathbb{R}_+^{M \times N}$ with N frames and M frequency bins in each frame. The time-varying envelopes of audio spectrogram convey important information. Without loss of generality, the notation used in NMF $\mathbf{X} = \{X_{mn}\} \in \mathbb{R}_+^{M \times N}$ for monaural source separation corresponds to that used in NCTF $\mathbf{X} = \{X_{ft}\} \in \mathbb{R}_+^{F \times T}$ with T frames and F frequency bins for reverberant source separation as addressed in Section 1.1.3. As shown in Fig. 2.3, the decomposed basis matrix $\mathbf{B} \in \mathbb{R}_+^{M \times K}$ is formed as K basis vectors with M frequency bins in each basis element. The decomposed weight matrix $\mathbf{W} \in \mathbb{R}_+^{K \times N}$ provides the nonnegative parameters to additively combine K basis vectors to come up with N frames of individual sources. The reconstruction of mixed signals $\mathbf{X}$ due to basis vectors in $\mathbf{B} = [\mathbf{b}_1, \ldots, \mathbf{b}_K]$ and weight vectors in $\mathbf{W}$ is also seen as an approximation based on *basis representation*. Namely, the separated signals are represented by a set of basis vectors using the corresponding weight parameters in a matrix

$$\mathbf{W} = \mathbf{A}^\top = [\mathbf{a}_1, \ldots, \mathbf{a}_K]^\top \in \mathbb{R}_+^{K \times N}. \tag{2.12}$$

As shown in Fig. 2.4, NMF can be manipulated as a bilinear model based on the sum of bilinear combination of K rank-one nonnegative matrices where each matrix is calculated as the outer product of two vectors $\mathbf{b}_k$ and $\mathbf{a}_k$, namely

$$\mathbf{X} \approx \mathbf{BW} = \mathbf{BA}^\top = \sum_k \mathbf{b}_k \circ \mathbf{a}_k \tag{2.13}$$

where $\circ$ denotes the outer product.

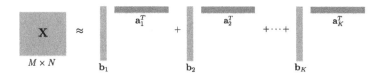

FIGURE 2.4

Nonnegative matrix factorization shown in a summation of rank-one nonnegative matrices.

In Smaragdis (2007), the 1-dimensional (1-D) convolutive NMF was proposed to extract the bases which considered the dependencies across successive columns of input spectrogram for supervised single-channel speech separation. In Schmidt and Morup (2006), 2-D NMF was developed to discover fundamental bases or notes for blind instrumental music separation in the presence of harmonic variations due to piano and trumpet with shift-invariance along the log-frequency domain. In Mørup and Schmidt (2006b), 2-D nonnegative tensor factorization was proposed for multichannel source separation. In real-world conditions, target signal is usually contaminated with a variety of interferences such as ambient noise, competing speech and background music. A microphone array was utilized to extract the spatial information for NMF-based source separation in Ozerov and Fevotte (2010). Such a multi-channel system normally works better than a single-channel system. But, in many situations, the mixed recordings are observed from a single microphone. Extracting the target signals from a mixed monaural signal is increasingly important in practical source separation systems. In what follows, we address NMF-based monaural source separation systems based on a supervised learning and an unsupervised learning.

2.2.1 LEARNING PROCEDURE

Basically, NMF is performed in a supervised fashion to find dictionaries or basis vectors $\mathbf{B}$ corresponding to different sources. Fig. 2.5 shows supervised learning of NMF for source separation in the presence of two sources. For the case of source separation in the presence of speech and music signals, the observed magnitude spectrogram $\mathbf{X}$ is viewed as an addition of speech spectrogram $\mathbf{X}^s$ and music spectrogram $\mathbf{X}^m$. We factorize the magnitude spectrogram of training data to find speech bases $\mathbf{B}^s$ and music bases $\mathbf{B}^m$ via $\mathbf{X}^s \approx \mathbf{B}^s \mathbf{W}^s$ and $\mathbf{X}^m \approx \mathbf{B}^m \mathbf{W}^m$ where the numbers of basis vector of speech, K^s, and music, K^m, meet the condition $K = K^s + K^m$. The trained speech and music bases, $\mathbf{B}^s$ and $\mathbf{B}^m$, are then fixed and applied in test phase where the mixed magnitude spectrogram of a test audio signal $\mathbf{X}$ is represented by using the trained bases, i.e.,

$$\mathbf{X} \approx [\mathbf{B}^s \ \mathbf{B}^m] \, \mathbf{W}. \tag{2.14}$$

The estimated spectrograms of speech and music are found by multiplying the basis matrix with the corresponding weight matrix $\widehat{\mathbf{W}}$ estimated from test data via $\widehat{\mathbf{X}}^s = \mathbf{B}^s \widehat{\mathbf{W}}^s$ and $\widehat{\mathbf{X}}^m = \mathbf{B}^m \widehat{\mathbf{W}}^m$. In addition, the soft mask function based on the Wiener gain is applied to improve the spectrograms for speech source $\widetilde{\mathbf{X}}^s$ and music source $\widetilde{\mathbf{X}}^m$ based on

$$\widetilde{\mathbf{X}}^s = \mathbf{X} \odot \frac{\widehat{\mathbf{X}}^s}{\widehat{\mathbf{X}}^s + \widehat{\mathbf{X}}^m}, \tag{2.15}$$

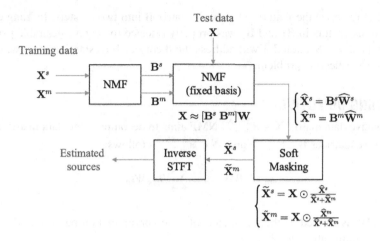

FIGURE 2.5

Supervised learning for single-channel source separation using nonnegative matrix factorization in presence of a speech signal $\mathbf{X}^s$ and a music signal $\mathbf{X}^m$.

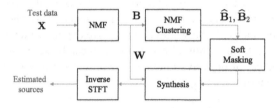

FIGURE 2.6

Unsupervised learning for single-channel source separation using nonnegative matrix factorization in presence of two sources.

$$\widetilde{\mathbf{X}}^m = \mathbf{X} \odot \frac{\widehat{\mathbf{X}}^m}{\widehat{\mathbf{X}}^s + \widehat{\mathbf{X}}^m} \tag{2.16}$$

where $\odot$ denotes the element-wise multiplication. Finally, the separated speech and music signals are obtained by the overlap-and-add method using the original phase information. An inverse STFT is performed to find the separated time signals of two sources.

On the other hand, Fig. 2.6 illustrates an implementation of unsupervised learning of NMF model for single-channel source separation in the presence of two sources. This setting will happen in many practical situations, including the application of singing voice separation where there is no training data available for finding the basis vectors for vocal source $\mathbf{B}_1$ of a singer and music source $\mathbf{B}_2$ of a background accompaniment. In the implementation, we estimate two sets of basis vectors, $\widehat{\mathbf{B}}_1$ and $\widehat{\mathbf{B}}_2$, by applying the clustering algorithm over the estimated basis vectors $\mathbf{B}$ from test data $\mathbf{X}$. In Spiertz and Gnann (2009), the k-means clustering algorithm was run based on the basis vectors in the Mel-frequency cepstral coefficient (MFCC) domain. For consistency, the clustering based on NMF could be also applied to factorize the feature set in basis matrix $\mathbf{B}$ into two clusters. In Yang et al. (2014b), the number of basis vectors in $\widehat{\mathbf{B}}_1$ and $\widehat{\mathbf{B}}_2$ was properly selected to achieve desirable performance for singing voice separation. Section 3.3 will address the theory of Bayesian learning, which provides a means to this model selection problem.

2.2.2 LEARNING OBJECTIVE

Given a nonnegative data matrix $\mathbf{X} \in \mathbb{R}_+^{M \times N}$, NMF aims to decompose the data matrix into a product of two nonnegative matrices $\mathbf{B} \in \mathbb{R}_+^{M \times K}$ and $\mathbf{W} \in \mathbb{R}_+^{K \times N}$ as follows:

$$X_{mn} \approx [\mathbf{BW}]_{mn} = \sum_k B_{mk} W_{kn} \tag{2.17}$$

where $\boldsymbol{\Theta} = \{\mathbf{B}, \mathbf{W}\}$ is formed as the parameter set. The approximation based on NMF is optimized according to the minimization problem

$$(\widehat{\mathbf{B}}, \widehat{\mathbf{W}}) = \arg \min_{\mathbf{B}, \mathbf{W} \geq 0} \mathcal{D}(\mathbf{X} \| \mathbf{BW}). \tag{2.18}$$

There are three learning objectives which are popular to measure the distance or the divergence $\mathcal{D}(\mathbf{X} \| \mathbf{BW})$ between the observed data $\mathbf{X}$ and the approximated data $\mathbf{BW}$. The closed-form solutions to these learning objectives do not exist. We would like to derive the multiplicative updating rules for parameters $\mathbf{B}$ and $\mathbf{W}$. A general form of multiplicative updating can be derived in the following way. For example, the terms in gradient of objective function $\mathcal{D}$ with respect to nonnegative matrix parameter $\boldsymbol{\Theta}$ are divided into positive terms and negative terms as

$$\frac{\partial \mathcal{D}}{\partial \boldsymbol{\Theta}} = \left[\frac{\partial \mathcal{D}}{\partial \boldsymbol{\Theta}}\right]^+ - \left[\frac{\partial \mathcal{D}}{\partial \boldsymbol{\Theta}}\right]^- \tag{2.19}$$

where $\left[\frac{\partial \mathcal{D}}{\partial \boldsymbol{\Theta}}\right]^+ > 0$ and $\left[\frac{\partial \mathcal{D}}{\partial \boldsymbol{\Theta}}\right]^- > 0$. The multiplicative update rule is yielded by

$$\boldsymbol{\Theta} \leftarrow \boldsymbol{\Theta} \odot \left[\frac{\partial \mathcal{D}}{\partial \boldsymbol{\Theta}}\right]^- \oslash \left[\frac{\partial \mathcal{D}}{\partial \boldsymbol{\Theta}}\right]^+ \tag{2.20}$$

where $\odot$ and $\oslash$ denote element-wise multiplication and division, respectively. Multiplicative updating was proved to be convergent (Lee and Seung, 2000) in the expectation-maximization (EM) algorithm (Dempster et al., 1977) which will be addressed in Section 3.7.1.

Squared Euclidean Distance
Squared Euclidean distance is a straightforward way to measure the reconstruction loss or regression loss which is expressed by

$$\mathcal{D}_{\text{EU}}(\mathbf{X} \| \mathbf{BW}) = \sum_{m,n} \left(X_{mn} - [\mathbf{BW}]_{mn}\right)^2. \tag{2.21}$$

Minimizing the loss function in Eq. (2.21) subject to constraints $\mathbf{B}, \mathbf{W} \geq 0$, we obtain the multiplicative updating rules for $\mathbf{B}$ and $\mathbf{W}$ in a form of (Lee and Seung, 2000)

$$B_{mk} \leftarrow B_{mk} \frac{[\mathbf{X}\mathbf{W}^\top]_{mk}}{[\mathbf{B}\mathbf{W}\mathbf{W}^\top]_{mk}}, \tag{2.22}$$

$$W_{kn} \leftarrow W_{kn} \frac{[\mathbf{B}^\top \mathbf{X}]_{kn}}{[\mathbf{B}^\top \mathbf{B}\mathbf{W}]_{kn}}. \tag{2.23}$$

It is obvious that the multiplicative updating rule in Eqs. (2.22)–(2.23) always finds the nonnegative parameters B_{mk} and W_{kn}. The multiplicative factor becomes unity in case of perfect reconstruction $\mathbf{X} = \mathbf{B}\mathbf{W}$ where the updating is terminated. In particular, the multiplicative updating for W_{kn} can be extended from the additive updating rule based on gradient descent algorithm

$$W_{kn} \leftarrow W_{kn} + \eta_{kn} \left([\mathbf{B}^\top \mathbf{X}]_{kn} - [\mathbf{B}^\top \mathbf{B}\mathbf{W}]_{kn} \right) \tag{2.24}$$

where the second term in right-hand side (RHS) of Eq. (2.24) is calculated by the derivative $\frac{\partial \mathcal{D}_{\mathrm{EU}}(\mathbf{X}\|\mathbf{B}\mathbf{W})}{\partial W_{kn}}$. By setting the learning rate as

$$\eta_{kn} = \frac{W_{kn}}{[\mathbf{B}^\top \mathbf{B}\mathbf{W}]_{kn}}, \tag{2.25}$$

the additive updating in Eq. (2.24) is converted to the multiplicative updating in Eq. (2.23). Such a derivation is easily extended to find the updating of basis parameter B_{mk}. The resulting EU-NMF is constructed. Alternatively, the multiplicative updating rule in Eqs. (2.22) and (2.23) can be obtained by identifying the positive and negative terms in the gradient, i.e., the second term in RHS of Eq. (2.24), and then substituting them into Eq. (2.20).

Kullback–Leibler Divergence

Importantly, the information-theoretic learning provides a meaningful objective function for machine learning where the information theory plays an important role. Information-theoretic objectives (as detailed in Section 3.2) are not only useful for ICA but also for NMF. When implementing NMF, Kullback–Leibler (KL) divergence $\mathcal{D}_{\mathrm{KL}}(\mathbf{X}\|\mathbf{B}\mathbf{W})$ (Kullback and Leibler, 1951) is used to calculate the relative entropy between $\mathbf{X}$ and $\mathbf{B}\mathbf{W}$ over individual entries (m, n), which is defined by

$$\mathcal{D}_{\mathrm{KL}}(\mathbf{X}\|\mathbf{B}\mathbf{W})$$
$$= \sum_{m,n} \left(X_{mn} \log \frac{X_{mn}}{[\mathbf{B}\mathbf{W}]_{mn}} + [\mathbf{B}\mathbf{W}]_{mn} - X_{mn} \right). \tag{2.26}$$

Minimizing this KL divergence, we can derive the multiplicative updates for parameters $\mathbf{B}$ and $\mathbf{W}$ in the resulting KL-NMF

$$B_{mk} \leftarrow B_{mk} \frac{\sum_n W_{kn}(X_{mn}/[\mathbf{B}\mathbf{W}]_{mn})}{\sum_n W_{kn}}, \tag{2.27}$$

$$W_{kn} \leftarrow W_{kn} \frac{\sum_m B_{mk}(X_{mn}/[\mathbf{BW}]_{mn})}{\sum_m B_{mk}} \qquad (2.28)$$

where the multiplicative gain is formed to assure nonnegativity in updating parameters B_{mk} and W_{kn}. Again, the updating rule of parameter W_{kn} is obtained by substituting the learning rate

$$\eta_{kn} = \frac{W_{kn}}{\sum_m B_{mk}} \qquad (2.29)$$

into the additive updating rule based on the gradient descent algorithm

$$W_{kn} \leftarrow W_{kn} + \eta_{kn} \left(\sum_m B_{mk} \frac{X_{mn}}{[\mathbf{BW}]_{mn}} - \sum_m B_{mk} \right) \qquad (2.30)$$

where the second term in RHS of Eq. (2.30) is calculated by the derivative $\frac{\partial \mathcal{D}_{\mathrm{KL}}(\mathbf{X} \| \mathbf{BW})}{\partial W_{kn}}$.

Itakura–Saito Divergence

In addition, it is popular to develop an NMF solution based on the Itakura–Saito (IS) divergence (Févotte et al., 2009) where the learning objective is constructed by (Itakura and Saito, 1968)

$$\mathcal{D}_{\mathrm{IS}}(\mathbf{X} \| \mathbf{BW})$$
$$= \sum_{m,n} \left(\frac{X_{mn}}{[\mathbf{BW}]_{mn}} - \log \frac{X_{mn}}{[\mathbf{BW}]_{mn}} - 1 \right). \qquad (2.31)$$

Minimizing this IS divergence and following the style of derivation for EU-NMF and KL-NMF, we accordingly find the solution to B_{mk} and W_{kn}:

$$B_{mk} \leftarrow B_{mk} \frac{\sum_n W_{kn}(X_{mn}/[\mathbf{BW}]_{mn}^2)}{\sum_n W_{kn}(1/[\mathbf{BW}]_{mn})}, \qquad (2.32)$$

$$W_{kn} \leftarrow W_{kn} \frac{\sum_m B_{mk}(X_{mn}/[\mathbf{BW}]_{mn}^2)}{\sum_m B_{mk}(1/[\mathbf{BW}]_{mn})}, \qquad (2.33)$$

for the resulting IS-NMF algorithm.

β Divergence

More generally, β divergence between $\mathbf{X}$ and $\mathbf{BW}$ is incorporated in finding a solution to NMF (Cichocki et al., 2006a) as

$$\mathcal{D}_\beta(\mathbf{X} \| \mathbf{BW})$$
$$= \sum_{m,n} \frac{1}{\beta(\beta-1)} \left(X_{mn}^\beta + (\beta-1)[\mathbf{BW}]_{mn}^\beta - \beta X_{mn}[\mathbf{BW}]_{mn}^{\beta-1} \right) \qquad (2.34)$$

Table 2.1 Comparison of NMF updating rules based on different learning objectives

	Standard NMF
Squared Euclidean distance ($\beta = 2$)	$\mathbf{B} \leftarrow \mathbf{B} \odot \frac{\mathbf{X}\mathbf{W}^\top}{\mathbf{B}\mathbf{W}\mathbf{W}^\top}, \mathbf{W} \leftarrow \mathbf{W} \odot \frac{\mathbf{B}^\top\mathbf{X}}{\mathbf{B}^\top\mathbf{B}\mathbf{W}}$
KL divergence ($\beta = 1$)	$\mathbf{B} \leftarrow \mathbf{B} \odot \frac{\frac{\mathbf{X}}{\mathbf{B}\mathbf{W}}\mathbf{W}^\top}{\mathbf{1}\mathbf{W}^\top}, \mathbf{W} \leftarrow \mathbf{W} \odot \frac{\mathbf{B}^T\frac{\mathbf{X}}{\mathbf{B}\mathbf{W}}}{\mathbf{B}^\top\mathbf{1}}$
IS divergence ($\beta = 0$)	$\mathbf{B} \leftarrow \mathbf{B} \odot \frac{((\mathbf{B}\mathbf{W})^{\cdot[-2]}\odot\mathbf{X})\mathbf{W}^\top}{(\mathbf{B}\mathbf{W})^{\cdot[-1]}\mathbf{W}^\top}, \mathbf{W} \leftarrow \mathbf{W} \odot \frac{\mathbf{B}^\top((\mathbf{B}\mathbf{W})^{\cdot[-2]}\odot\mathbf{X})}{\mathbf{B}^\top(\mathbf{B}\mathbf{W})^{\cdot[-1]}}$

where $\mathbf{A}^{\cdot[n]}$ denotes the matrix with entries $[\mathbf{A}]_{ij}^n$. Particularly, the squared Euclidean distance, KL divergence and IS divergence are seen as special realizations of a β divergence when $\beta = 2$, $\beta = 1$, and $\beta = 0$, respectively. In a similar way, the general solution to β-NMF can be derived as

$$\mathbf{B} \leftarrow \mathbf{B} \odot \frac{\left((\mathbf{B}\mathbf{W})^{\cdot[\beta-2]} \odot \mathbf{X}\right) \mathbf{W}^\top}{(\mathbf{B}\mathbf{W})^{\cdot[\beta-1]}\mathbf{W}^\top}, \tag{2.35}$$

$$\mathbf{W} \leftarrow \mathbf{W} \odot \frac{\mathbf{B}^\top\left((\mathbf{B}\mathbf{W})^{\cdot[\beta-2]} \odot \mathbf{X}\right)}{\mathbf{B}^\top(\mathbf{B}\mathbf{W})^{\cdot[\beta-1]}}. \tag{2.36}$$

This β-NMF is then reduced to EU-NMF, KL-NMF and IS-NMF by setting $\beta = 2$, $\beta = 1$, and $\beta = 0$, respectively. Table 2.1 shows a table of NMF updating rules based on different learning objectives or β divergence under different β. Note that the equations are consistently written in matrix form; $\mathbf{1}$ denotes a $M \times N$ matrix with all entries being one. Again, these rules are eligible to produce nonnegative parameters $\mathbf{B}$ and $\mathbf{W}$ due to the multiplicative factors.

In Gaussier and Goutte (2005), NMF with KL divergence was illustrated to yield the maximum likelihood (ML) solution to probabilistic latent semantic analysis (Hofmann, 1999). More generally, the optimization problem in NMF using different divergence measures can be converted into a probabilistic optimization based on a likelihood function $p(\mathbf{X}|\boldsymbol{\Theta})$ of observed spectra $\mathbf{X}$ using different probabilistic distributions where the parameters or latent variables $\boldsymbol{\Theta} = \{\mathbf{B}, \mathbf{W}\}$ are assumed to be fixed but unknown. The randomness in model construction due to parameter variations or ill-formed condition is not characterized. Such a maximum likelihood estimation is prone to be overtrained (Bishop, 2006, Watanabe and Chien, 2015). Section 5.3 will address how probabilistic NMF is constructed and how Bayesian learning for NMF is conducted to improve the performance of monaural separation of speech and music signals as well as monaural separation of singing voice.

2.2.3 SPARSE REGULARIZATION

In addition to Bayesian learning, sparse learning is crucial for model regularization. It is because that, in real-world, source separation problem is usually ill-conditioned. Only a few components or basis vectors in $\mathbf{B}$ are active to encode the separated signals. However, different signals may be constructed in different basis vectors. It is meaningful to impose sparsity in basis representation for NMF (Hoyer,

Table 2.2 Comparison of updating rules of standard NMF and sparse NMF based on learning objectives of squared Euclidean distance and Kullback–Leibler divergence

	NMF	Sparse NMF
Squared Euclidean distance	$\mathbf{B} \leftarrow \mathbf{B} \odot \frac{\mathbf{X}\mathbf{W}^\top}{\mathbf{B}\mathbf{W}\mathbf{W}^\top}$ $\mathbf{W} \leftarrow \mathbf{W} \odot \frac{\mathbf{B}^\top\mathbf{X}}{\mathbf{B}^\top\mathbf{B}\mathbf{W}}$	$\mathbf{B} \leftarrow \mathbf{B} \odot \frac{\mathbf{X}\mathbf{W}^\top + \mathbf{B} \odot (1(\mathbf{B}\mathbf{W}\mathbf{W}^\top \odot \mathbf{B}))}{\mathbf{B}\mathbf{W}\mathbf{W}^\top + \mathbf{B} \odot (1(\mathbf{X}\mathbf{W}^\top \odot \mathbf{B}))}$ $\mathbf{W} \leftarrow \mathbf{W} \odot \frac{\mathbf{B}^\top\mathbf{X}}{\mathbf{B}^\top\mathbf{B}\mathbf{W} + \lambda}$
KL divergence	$\mathbf{B} \leftarrow \mathbf{B} \odot \frac{\frac{\mathbf{X}}{\mathbf{B}\mathbf{W}}\mathbf{W}^\top}{1\mathbf{W}^\top}$ $\mathbf{W} \leftarrow \mathbf{W} \odot \frac{\mathbf{B}^\top\frac{\mathbf{X}}{\mathbf{B}\mathbf{W}}}{\mathbf{B}^\top 1}$	$\mathbf{B} \leftarrow \mathbf{B} \odot \frac{\frac{\mathbf{X}}{\mathbf{B}\mathbf{W}}\mathbf{W}^\top + \mathbf{B} \odot (1(1\mathbf{W}^\top \odot \mathbf{B}))}{1\mathbf{W}^\top + \mathbf{B} \odot (1(\frac{\mathbf{X}}{\mathbf{B}\mathbf{W}}\mathbf{W}^\top \odot \mathbf{B}))}$ $\mathbf{W} \leftarrow \mathbf{W} \odot \frac{\mathbf{B}^\top\frac{\mathbf{X}}{\mathbf{B}\mathbf{W}}}{\mathbf{B}^\top 1 + \lambda}$

2004). Robustness in system performance can be improved. To do so, we modify the learning objective from Eq. (2.18) to

$$(\widehat{\mathbf{B}}, \widehat{\mathbf{W}}) = \arg \min_{\mathbf{B}, \mathbf{W} \geq 0} \mathcal{D}(\mathbf{X} \| \mathbf{B}\mathbf{W}) + \lambda \cdot g(\mathbf{W}) \tag{2.37}$$

where $g(\cdot)$ is a penalty function for model regularization. This learning objective is controlled by a regularization parameter λ, which balances the tradeoff between reconstruction error and penalty function. The most common choices of penalty function are the ℓ_2 norm (also called weight decay) and ℓ_1 norm (also called Lasso) (Tibshirani, 1996). Lasso stands for "least absolute shrinkage and selection operation" which encourages sparsity in a learning machine. Section 3.4 will address the details of sparse learning which can be applied in different solutions to source separation. In addition to imposing sparsity in parameter $\mathbf{W}$, the sparse penalty can be also employed in both parameters $\mathbf{B}$ and $\mathbf{W}$ for NMF-based source separation (Cichocki et al., 2006b). Considering Lasso regularization for weight parameters $\mathbf{W}$ in EU-NMF and KL-NMF, we correspondingly derive the updating rules for sparse EU-NMF and sparse KL-NMF as compared in Table 2.2. Obviously, the multiplicative factors for updating parameters $\mathbf{W}$ are decreased due to nonnegative regularization parameter $\lambda \geq 0$.

As a result, NMF is implemented with both constraints of nonnegativity and sparsity. Many real-world data are nonnegative, and the corresponding hidden components have physical meaning only with nonnegativity, which is likely related to probability distribution for probabilistic modeling. On the other hand, sparseness is closely related to conducting the feature selection. Selecting a subset of relevant features or basis vectors for basis representation of a target signals, either mixed or separated signals, is crucial to achieve robustness in learning representation. The ultimate goal of learning representation is to find a statistically fitted model which can be physically interpreted and logically extended. It is important to seek a tradeoff between interpretability and statistical fidelity.

2.3 NONNEGATIVE TENSOR FACTORIZATION

Nonnegative matrix factorization (NMF) conducts a two-way decomposition over a matrix or a two-way tensor. Such a two-way representation via matrix factorization using NMF is invaluable but may be insufficient to reflect the nature of signal in multidimensional arrays. In many real-world environments, mixed signals are observed and collected in multiple ways. Higher-order ways such as trials, condi-

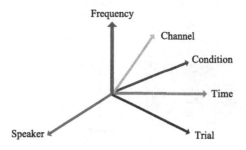

FIGURE 2.7

Illustration of multiway observation data.

tions, subjects, channels, spaces, times, frequencies are ubiquitous in different technical data. Fig. 2.7 shows an illustration of multiway or multidimensional observation data. For instance, the mixed speech signals may be collected in different trials from different speakers and recorded by using different microphones or channels in different conditions. These signals may be observed in time and frequency domains. In fact, speech signals are multiway data. Multiways provide multiple features from different horizons which are helpful for source separation. In addition to speech signals, there are many other signals containing higher-order ways or modes in data structure such as

- video: height × width × time
- color image: height × width × (red, green, blue)
- face: people × pose × illumination × angle
- stereo audio: channel × frequency × time
- electroencephalogram (EEG): channel × time × trial
- text: user × query × webpage
- social network: score × object × referee × criterion
- economics, environmental science, chemical science, biology, etc.

These structural data, generally called *tensors*, are seen as the geometric objects that describe the linear relations between vectors, scalars and other tensors (Cichocki et al., 2009, Mørup, 2011). A tensor is a multiway array or multidimensional matrix. Extending from NMF to nonnegative tensor factorization (NTF) paves an avenue to accommodate richer data structure in learning for source separation. NTF conducts the so-called multiway decomposition where both constraints of nonnegativity and sparsity are imposed to provide physical meaning for the extracted features or hidden factors and the separated signals in adverse condition. In case of audio signal processing, the learning objective is to decompose the multichannel time-frequency audio signals into multiple components with different modalities. Fig. 2.8 illustrates an example of tensor data which contains three-way information of time, frequency and channel. The channel information in audio signals may be recorded from different microphones (with different angles and positions), acquired from different views or obtained from different processing units.

To assure the performance of source separation, we would like to identify the latent components or features from a multiway observation which are common across different domains and discriminative across different conditions. Tensor factorization provides an effective solution to analyze a

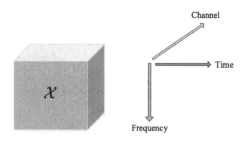

FIGURE 2.8

A tensor data which is composed of three ways of time, frequency and channel.

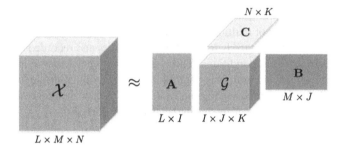

FIGURE 2.9

Tucker decomposition for a three-way tensor.

multiway structural observation. There are several solutions to tensor factorization (De Lathauwer et al., 2000a), including Tucker decomposition, canonical decomposition/parallel factors (CP) decomposition and block term decomposition (De Lathauwer, 2008). In addition to nonnegativity and sparsity, the other constraints, e.g., orthogonality or discrimination, can be also imposed to carry out NTF to meet different regularization conditions. In what follows, tensor decomposition is addressed for the case of three-way tensor. The extension to general case in presence of N-way tensor can be derived correspondingly. We introduce tensor factorization methods based on Tucker decomposition and CP decomposition. For clarity in the notation, the boldface lowercase letter denotes a vector $\mathbf{x}$, the boldface capital letter denotes a matrix $\mathbf{X}$, and the boldface Euler script letter denotes a multiway tensor $\boldsymbol{\mathcal{X}}$.

2.3.1 TUCKER DECOMPOSITION

Tucker (1966) proposed a decomposition method for three-way arrays as a multidimensional extension of factor analysis (Basilevsky, 1994). A tensor can be decomposed into a core tensor multiplied by factor matrices along the corresponding modes. Fig. 2.9 shows the Tucker decomposition for a three-way tensor $\boldsymbol{\mathcal{X}} = \{\mathcal{X}_{lmn}\} \in \mathbb{R}^{L \times M \times N}$ where L, M and N denote the dimensions in three directions. Given this three-way tensor, we perform the approximation

$$\mathcal{X} \approx \mathcal{G} \times_1 \mathbf{A} \times_2 \mathbf{B} \times_3 \mathbf{C}$$
$$= \sum_i \sum_j \sum_k \mathcal{G}_{ijk}(\mathbf{a}_i \circ \mathbf{b}_j \circ \mathbf{c}_k) \tag{2.38}$$

where $\mathcal{G} = \{\mathcal{G}_{ijk}\} \in \mathbb{R}^{I \times J \times K}$ denotes the core tensor with dimensions I, J and K, which are basically smaller than the dimensions L, M and N in the original tensor $\mathcal{X}$, respectively. In Eq. (2.38), $\mathbf{A} = \{A_{li}\} \in \mathbb{R}^{L \times I}$, $\mathbf{B} = \{B_{mj}\} \in \mathbb{R}^{M \times J}$ and $\mathbf{C} = \{C_{nk}\} \in \mathbb{R}^{N \times K}$ denote the factor matrices corresponding to three horizons where the dimensions in factor matrices are associated with the dimensions in the original tensor $\{L, M, N\}$ and core tensor $\{I, J, K\}$ in the corresponding horizons. Here, $\circ$ denotes the outer product and $\times_n$ denotes the model-n product, which is defined in Appendix A.8. This approximation is expressed as a linear combination of different outer products of column vectors $\mathbf{a}_i$, $\mathbf{b}_j$ and $\mathbf{c}_k$ of matrices $\mathbf{A}$, $\mathbf{B}$ and $\mathbf{C}$, respectively, where the entries of core tensor $\{\mathcal{G}_{ijk}\}$ are used as the interpolation weights. Alternatively, the entries of observed tensor $\{\mathcal{X}_{lmn}\}$ can be written as the following interpolation:

$$\mathcal{X}_{lmn} \approx \sum_i \sum_j \sum_k \mathcal{G}_{ijk} A_{li} B_{mj} C_{nk}, \tag{2.39}$$

by using the entries of core tensor $\{\mathcal{G}_{ijk}\}$ and three factor matrices $\{A_{li}, B_{mj}, C_{nk}\}$. The core tensor $\mathcal{G}$ is seen as the tensor weights when integrating those factors in different ways. For simplicity, this approximation can be expressed as

$$\mathcal{X} \approx [\![\mathcal{G}; \mathbf{A}, \mathbf{B}, \mathbf{C}]\!]. \tag{2.40}$$

Such a Tucker decomposition is calculated in a form of multilinear singular value decomposition (SVD) (De Lathauwer et al., 2000a). The decomposition is not unique. It is more likely to find unique solution by imposing constraints. In general, NTF is realized as a Tucker decomposition subject to nonnegativity constraint which results in a parts-based representation where each part in the integration is formed as the outer product of factor vectors in different ways.

Tucker decomposition can be solved by using the higher-order SVD (De Lathauwer et al., 2000a) or the higher-order orthogonal iteration (HOOI) (De Lathauwer et al., 2000b). HOOI is known as an efficient method based on the alternating least squares algorithm. The optimization problem is formulated as a minimization of sum-of-squares error function

$$\min_{\mathcal{G}, \mathbf{A}, \mathbf{B}, \mathbf{C}} \|\mathcal{X} - [\![\mathcal{G}; \mathbf{A}, \mathbf{B}, \mathbf{C}]\!]\|^2 \tag{2.41}$$

subject to column-wise orthogonal matrices $\mathbf{A}$, $\mathbf{B}$ and $\mathbf{C}$. Due to the property of orthogonality, the core tensor can be derived as

$$\mathcal{G} = \mathcal{X} \times_1 \mathbf{A}^\top \times_2 \mathbf{B}^\top \times_3 \mathbf{C}^\top. \tag{2.42}$$

The objective function is therefore rewritten as

$$
\begin{aligned}
&\|\mathcal{X} - [\![\mathcal{G}; \mathbf{A}, \mathbf{B}, \mathbf{C}]\!]\|^2 \\
=&\|\mathcal{X}\|^2 - \|\mathcal{X} \times_1 \mathbf{A}^\top \times_2 \mathbf{B}^\top \times_3 \mathbf{C}^\top\|^2.
\end{aligned}
\tag{2.43}
$$

Since the $\|\mathcal{X}\|^2$ is a constant value, the optimization problem can be thought as several subproblems:

$$
\max_{\mathbf{A},\mathbf{B},\mathbf{C}} \|\mathcal{X} \times_1 \mathbf{A}^\top \times_2 \mathbf{B}^\top \times_3 \mathbf{C}^\top\|^2
\tag{2.44}
$$

subject to column-wise orthogonal matrices $\mathbf{A}$, $\mathbf{B}$ and $\mathbf{C}$. In the implementation, $\mathbf{A}$, $\mathbf{B}$ and $\mathbf{C}$ are estimated alternatively according to the singular value decomposition (SVD) and then used to find $\mathcal{G}$ through Eq. (2.42) (De Lathauwer et al., 2000b). Algorithm 2.1 shows the procedure of higher-order orthogonal iteration for three-way Tucker decomposition where the maximization steps are implemented by SVD method.

Algorithm 2.1 Higher-Order Orthogonal Iteration.

Initialize the factor matrices $\mathbf{A}$, $\mathbf{B}$ and $\mathbf{C}$ with orthogonal columns.
Iterate until convergence:
    $\tilde{\mathcal{G}} \leftarrow \mathcal{X} \times_1 \mathbf{A}^\top$. Maximize over $\mathbf{A}$ with $\mathbf{A}^\top \mathbf{A} = \mathbf{I}$.
    $\tilde{\mathcal{G}} \leftarrow \mathcal{X} \times_1 \mathbf{A}^\top \times_2 \mathbf{B}^\top$. Maximize over $\mathbf{B}$ with $\mathbf{B}^\top \mathbf{B} = \mathbf{I}$.
    $\tilde{\mathcal{G}} \leftarrow \mathcal{X} \times_1 \mathbf{A}^\top \times_2 \mathbf{B}^\top \times_3 \mathbf{C}^\top$. Maximize over $\mathbf{C}$ with $\mathbf{C}^\top \mathbf{C} = \mathbf{I}$.
Use converged values $\mathbf{A}$, $\mathbf{B}$ and $\mathbf{C}$ to calculate $\mathcal{G} \leftarrow \mathcal{X} \times_1 \mathbf{A}^\top \times_2 \mathbf{B}^\top \times_3 \mathbf{C}^\top$.
Return $\mathbf{A}$, $\mathbf{B}$, $\mathbf{C}$ and $\mathcal{G}$.

2.3.2 CP DECOMPOSITION

Canonical Decomposition/Parallel Factors (CP) decomposition (Carroll and Chang, 1970) factorizes the tensor into a weighted sum of finite number of rank-one tensor. The approximation in this factorization is calculated as the weighted outer products over K column vectors $\{\mathbf{a}_k, \mathbf{b}_k, \mathbf{c}_k\}$ of factor matrices $\{\mathbf{A}, \mathbf{B}, \mathbf{C}\}$. Given an input tensor $\mathcal{X} \in \mathbb{R}^{L \times M \times N}$, CP decomposition performs the approximation

$$
\mathcal{X} \approx \sum_k \lambda_k (\mathbf{a}_k \circ \mathbf{b}_k \circ \mathbf{c}_k) \triangleq [\![\lambda; \mathbf{A}, \mathbf{B}, \mathbf{C}]\!]
\tag{2.45}
$$

where $\lambda = \{\lambda_k\}$ denotes the weights corresponding to K rank-one tensors. More specifically, the identity weights $\lambda_k = 1$ for all k are assumed or the identity tensor is assigned for core tensor, i.e., $\mathcal{G} = \mathcal{I}$. In this case, a three-way tensor $\mathcal{X} \in \mathbb{R}^{L \times M \times N}$ is decomposed as a sum of trilinear terms

$$
\mathcal{X} \approx \mathcal{I} \times_1 \mathbf{A} \times_2 \mathbf{B} \times_3 \mathbf{C} = \sum_k \mathbf{a}_k \circ \mathbf{b}_k \circ \mathbf{c}_k
$$

$$
\triangleq [\![\mathbf{A}, \mathbf{B}, \mathbf{C}]\!].
\tag{2.46}
$$

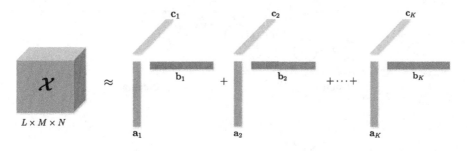

FIGURE 2.10

CP decomposition for a three-way tensor.

Each entry in an observed tensor can be approximated by a linear combination of K products of entries of factor matrices A_{lk}, B_{mk} and C_{nk} as

$$\mathcal{X}_{lmn} \approx \widehat{\mathcal{X}}_{lmn} = \sum_{k} A_{lk} B_{mk} C_{nk}. \qquad (2.47)$$

Fig. 2.10 illustrates CP approximation for a three-way tensor. Basically, CP decomposition in Eqs. (2.46)–(2.47) is a special realization of Tucker decomposition with two assumptions. First, the core tensor is assumed to be superdiagonal. Second, the number of components or columns in different ways of factor matrices is assumed to be identical, i.e., $I = J = K$. Again, CP decomposition can be realized for NTF where the observed tensor $\mathcal{X} \in \mathbb{R}_{+}^{L \times M \times N}$ and three factor matrices $\mathbf{A} = [\mathbf{a}_1, \dots, \mathbf{a}_K] \in \mathbb{R}_{+}^{L \times K}$, $\mathbf{B} = [\mathbf{b}_1, \dots, \mathbf{b}_K] \in \mathbb{R}_{+}^{M \times K}$ and $\mathbf{C} = [\mathbf{c}_1, \dots, \mathbf{c}_K] \in \mathbb{R}_{+}^{N \times K}$ are all nonnegative. The alternative least squares algorithm is applied to alternatively estimate factor matrices $\mathbf{A}$, $\mathbf{B}$ and $\mathbf{C}$ until convergence according to the HOOI algorithm shown in Algorithm 2.1.

In general, source separation based on NMF in Section 2.2 and NTF in Section 2.3 learns to estimate the latent sources through a single-layer linear factorization procedure without involving layer-wise structure. In the next section, we address a different paradigm to source separation where a nonlinear layer-wise structure is introduced to conduct deep learning to characterize the complicated relation between the mixed signal and different source signals. In Chien and Bao (2018), tensor factorization was combined with deep learning to carry out a tensor factorized neural network via the tensor factorized error backpropagation by using high-dimensional observations with multiple ways.

2.4 DEEP NEURAL NETWORK

Deep learning based on the artificial neural network is run under a deep architecture or hierarchy, consisting of multiple hidden layers, which captures the high-level abstraction behind data and characterizes the complex nonlinear relationship between inputs and targets. Such a deep neural network (DNN) has been successfully applied for a number of regression and classification systems including speech recognition (Yu et al., 2012, Hinton et al., 2012, Saon and Chien, 2012b), image classification (LeCun et al., 1998), natural language processing (Chien and Ku, 2016), music information retrieval,

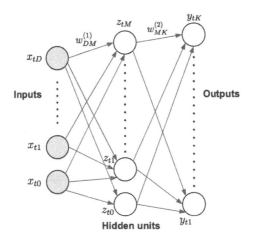

FIGURE 2.11

Multilayer perceptron with one input layer, one hidden layer, and one output layer.

to name a few. Fig. 2.11 shows an example of multilayer perceptron (MLP) (Rumelhart et al., 1986) with D-dimensional input vector $\mathbf{x}_t$, M-dimensional hidden feature vector $\mathbf{z}_t$, and K-dimensional output vector $\mathbf{y}_t$ at time t. Multiple hidden layers can be extended correspondingly. The mapping function between input $\mathbf{x} = \{x_{td}\}$ and output $\mathbf{y} = \{y_{tk}\}$ is established by

$$
\begin{aligned}
y_{tk} &= y_k(\mathbf{x}_t, \mathbf{w}) \\
&= f\left(\sum_{m=0}^{M} w_{mk}^{(2)} f\left(\sum_{d=0}^{D} w_{dm}^{(1)} x_{td} \right) \right) \\
&= f\left(\sum_{m=0}^{M} w_{mk}^{(2)} z_{tm} \right) \triangleq f(a_{tk})
\end{aligned}
\tag{2.48}
$$

where the synaptic weights of the first layer and the second layer $\mathbf{w} = \{w_{dm}^{(1)}, w_{mk}^{(2)}\}$ are introduced. In this figure, $x_{t0} = z_{t0} = 1$ and the weights $\{w_{0m}^{(1)}, w_{0k}^{(2)}\}$ denote the bias parameters in neurons. There are two calculations in this layer-wise feedforward neural network (FNN). The first one is an affine transformation, namely the layer-wise multiplication by using parameters in different layers $\{w_{dm}^{(1)}\}$ and $\{w_{mk}^{(2)}\}$. This transformation calculates the layer-wise activations from input layer to output layer in this order as

$$
a_{tm} = \sum_{d=0}^{D} w_{dm}^{(1)} x_{td} \quad \Longrightarrow \quad a_{tk} = \sum_{m=0}^{M} w_{mk}^{(2)} a_{tm}.
\tag{2.49}
$$

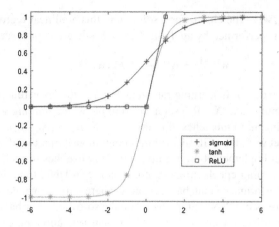

FIGURE 2.12

Comparison of activations using ReLU function, logistic sigmoid function, and hyperbolic tangent function.

The second one is the nonlinear activation function $f(\cdot)$. Fig. 2.12 shows a comparison of activation functions using rectified linear unit (ReLU)

$$f(a) = \text{ReLU}(a) = \max\{0, a\}, \tag{2.50}$$

logistic sigmoid function

$$f(a) = \sigma(a) = \frac{1}{1 + e^{-a}}, \tag{2.51}$$

and hyperbolic tangent function

$$f(a) = \tanh(a) = \frac{e^a - e^{-a}}{e^a + e^{-a}}. \tag{2.52}$$

Here, the value of logic sigmoid function is between 0 and 1 while that of hyperbolic tangent function is between -1 and 1. ReLU is the most popular activation in the implementation although there are some others shown with good performance in different tasks (Goodfellow et al., 2016).

2.4.1 ERROR BACKPROPAGATION ALGORITHM

When implementing DNN training for single-channel source separation, we collect a set of training samples $\{\mathbf{X}, \mathbf{R}\} = \{\mathbf{x}_t, \mathbf{r}_t\}_{t=1}^{T}$, including the mixed signals as an input vector $\mathbf{x}_t$ and the source vectors as an target vector $\mathbf{r}_t = \{r_{tk}\}$, and formulate a regression problem for optimization. The DNN parameters $\mathbf{w}$ are accordingly estimated by minimizing the sum-of-squares error function calculated from DNN outputs as

$$E(\mathbf{w}) = \frac{1}{2} \sum_{t=1}^{T} \|\mathbf{y}(\mathbf{x}_t, \mathbf{w}) - \mathbf{r}_t\|^2 \tag{2.53}$$

where $\mathbf{y}(\mathbf{x}_t, \mathbf{w}) = \{y_k(\mathbf{x}_t, \mathbf{w})\}$. The closed-form solution to this nonlinear regression problem does not exist. The minimization is performed by applying the stochastic gradient descent (SGD) algorithm

$$\mathbf{w}^{(\tau+1)} = \mathbf{w}^{(\tau)} - \eta \nabla E_n(\mathbf{w}^{(\tau)}) \tag{2.54}$$

where τ is the iteration index, η is learning rate, and $E_n(\cdot)$ is the error function calculated by using the nth minibatch of training data $\{\mathbf{X}_n, \mathbf{R}_n\}$ sampled from the whole training set $\{\mathbf{X}, \mathbf{R}\}$. Each learning epoch is executed by using all minibatches of training set $\{\mathbf{X}, \mathbf{R}\} = \{\mathbf{X}_n, \mathbf{R}_n\}$. Starting from an initialization with parameter set $\mathbf{w}^{(0)}$, the SGD algorithm is continuously run to reduce the error function E_n until convergence. In real implementation, we randomize minibatches in each learning epoch and run a sufficient number of learning epochs to attain convergence in DNN training. Such a SGD training basically obtains better performance than batch training where only the whole batch data is adopted. In general, there are two passes in DNN training based on the so-called error backpropagation algorithm. In the forward pass, the affine transformation and nonlinear activation are calculated in a layer-wise manner from the input layer to the output layer. In the backward pass, the derivatives of the error function with respect to individual weights are calculated from the output layer back to the input layer, namely one finds the derivatives for updating the weights in different layers in this order

$$\frac{\partial E_n(\mathbf{w}^{(\tau)})}{\partial w_{mk}^{(2)}} \implies \frac{\partial E_n(\mathbf{w}^{(\tau)})}{\partial w_{dm}^{(1)}} \tag{2.55}$$

where the minibatch samples $\{\mathbf{X}_n, \mathbf{R}_n\}$ are used. Figs. 2.13(A) and 2.13(B) illustrate the calculations of the error backpropagation algorithm in the forward and backward passes, respectively. An important trick to carry out this error backpropagation algorithm in the backward pass is to calculate the local gradient of the mth neuron in the hidden layer using the mixed signal at each time $\mathbf{x}_t$,

$$\begin{aligned} \delta_{tm} &\triangleq \frac{\partial E_t}{\partial a_{tm}} \\ &= \sum_k \frac{\partial E_t}{\partial a_{tk}} \frac{\partial a_{tk}}{\partial a_{tm}} \\ &= \sum_k \delta_{tk} \frac{\partial a_{tk}}{\partial a_{tm}}, \end{aligned} \tag{2.56}$$

which is updated from local gradients δ_{tk} from all neurons k in the output layer, namely one updates the local gradient in this order $\delta_{tk} \implies \delta_{tm}$. Having all local gradients in the output and hidden layers, the derivatives for SGD updating are calculated as

$$\frac{\partial E_n(\mathbf{w}^{(\tau)})}{\partial w_{dm}^{(1)}} = \sum_{t \in \{\mathbf{X}_n, \mathbf{R}_n\}} \delta_{tm} x_{td}, \tag{2.57}$$

$$\frac{\partial E_n(\mathbf{w}^{(\tau)})}{\partial w_{mk}^{(2)}} = \sum_{t \in \{\mathbf{X}_n, \mathbf{R}_n\}} \delta_{tk} z_{tm}, \tag{2.58}$$

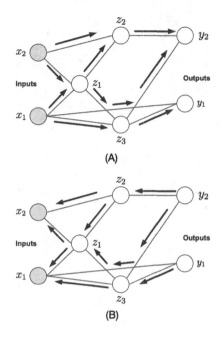

FIGURE 2.13

Calculations in error backpropagation algorithm: (A) forward pass and (B) backward pass. The propagations are shown by arrows. In the forward pass, activations a_t and outputs z_t in individual nodes are calculated and propagated. In the backward pass, local gradients δ_t in individual nodes are calculated and propagated.

which are accumulated from the error function E_t using a minibatch of time signals $t \in \{\mathbf{X}_n, \mathbf{R}_n\}$, i.e., $E_n = \sum_{t \in \{\mathbf{X}_n, \mathbf{R}_n\}} E_t$. For example, at each time t, the derivative for updating a connected weight $w_{mk}^{(2)}$ can be simply expressed as a product of the output z_{tm} of neuron m in the hidden layer and the local gradient δ_{tk} of neuron k in output layer. The same style of calculation is also applied for updating the connected weight $w_{dm}^{(1)}$ between neurons in the input layer d and hidden layer m. Fig. 2.14 shows the procedure of error backpropagation algorithm for training a general l-layer multilayer perceptron, including forward calculation of error function and backward calculations of error differentiation or local gradient from layer l to layer $l - 1$ back to layer 1.

2.4.2 PRACTICES IN DEEP LEARNING

The deep architecture in DNN is representationally efficient because the computational units in different neurons and layers follow the same functions, i.e., the affine transformation and nonlinear activation. However, it is important to address why going deep does matter when training a deep neural network (DNN) for single-channel source separation. Basically, the mapping between the spectra of mixed signal and source signals is really complicated. The performance of source separation based on linear and shallow models, e.g., ICA, NMF and NTF, is prone to be bounded. A nonlinear and deep model provides a means to deal with this weakness. In general, deep learning pursues a hierar-

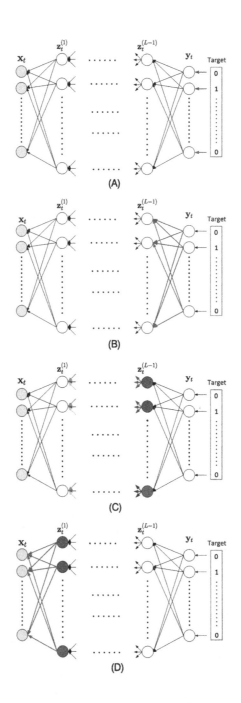

FIGURE 2.14

Procedure in the error backpropagation algorithm: (A) compute the error function, (B) propagate the local gradient from the output layer L to (C) hidden layer $L-1$ and (D) until input layer 1.

hidden

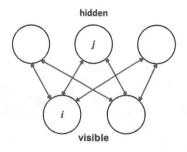

visible

FIGURE 2.15

A two-layer structure in restricted Boltzmann machine.

chical representation which allows nonlocal generalization in the time signal, as well as increases the comprehensibility in structural learning. Multiple layers of latent variables enable the combinational sharing of the strength of statistics. Hierarchical learning using DNN makes it easy to monitor what is being learned and to guide the machine to better subspaces for regression or classification. Hierarchy of representation with increasing level of abstraction is learned. Each level is seen as a kind of trainable feature transform. After training a deep model, the lower-level representation can be used across different tasks.

In addition, DNN is constructed with many fully-connected layers where the parameter space is huge. The training procedure can be very hard to assure convergence in the error backpropagation algorithm. There is no theoretical guarantee to illustrate good performance by using a deep structure. In some cases, a DNN trained with random initialization may perform even worse than a shallow model. Therefore, a critical issue in DNN training is to find a reliable initialization and a fast convergence to assure a well-trained "deep" model structure so that we can apply it to demix an unknown test signal or, equivalently, predict the corresponding source signals.

Deep belief network (DBN) (Hinton and Salakhutdinov, 2006) is a probabilistic generative model which provides a meaningful initialization or pretraining for DNN. DBN is an unsupervised learning method which learns to reconstruct the inputs. The layers of DBN act as the feature extractors. This DBN can be further combined with supervision for regression problem in the application of source separation. DBN provides a theoretical approach to pretrain each layer for DNN optimization. In the implementation, the restricted Boltzmann machine (RBM) serves as a building block to compose a DBN with multiple layers of latent variables. The building is performed in a bottom-up and stack-by-stack manner. Fig. 2.15 shows a graphical model of RBM with node i in the visible layer and node j in the hidden layer. RBM is an undirected, generative and energy-based model with the bidirectional weights between the two layers which are trained according to the contrastive divergence algorithm (Hinton et al., 2006). Each pair of layers is formed by an RBM. After training the RBM until convergence, the hidden layer is subsequently treated as the visible layer for training the next RBM to find a deeper hidden layer. In the end, a bottom-up deep learning machine is constructed according to this tandem-based and stack-wise training procedure.

Fig. 2.16 depicts a stack-wise training procedure to estimate DBN. The training samples $\{\mathbf{x}\}$ are collected to estimate the first RBM to transform each visible sample $\mathbf{x}$ into hidden feature $\mathbf{z}^{(1)}$ by using the trained parameters $\mathbf{w}^{(1)}$. The hidden features $\{\mathbf{z}^{(1)}\}$ are then treated as visible data to train the second

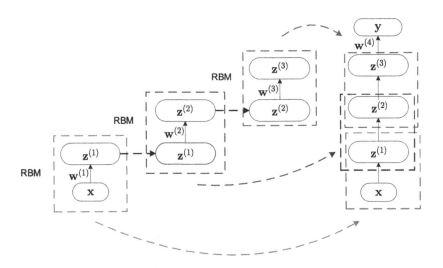

FIGURE 2.16

A stack-wise training procedure for deep belief network. This model is further used as a pretrained model for deep neural network optimization.

RBM, which projects each sample $\mathbf{z}^{(1)}$ into $\mathbf{z}^{(2)}$ in a deeper layer. In this way, we build a deep model to explore the hierarchy

$$\mathbf{x} \to \mathbf{z}^{(1)} \to \mathbf{z}^{(2)} \to \mathbf{z}^{(3)} \to \cdots \tag{2.59}$$

for DBN. The level of abstraction is naturally developed from one layer to another. Based on this greedy layer-wise training, an unsupervised model is trained to obtain a DBN. The parameters of DBN are then used as initial parameters for DNN training which are much better than those based on random initialization. Getting stuck in a local minimum can be avoided. In the final stage, a supervised top-down training is performed to refine the features in the intermediate layers by using labeled samples $\{\mathbf{r}\}$. In the source separation problem, the visible sample corresponds to the mixed spectral signal while the labeled sample means the source spectral signals. The refined features are more relevant to produce the outputs $\mathbf{y}$ in the target task. The error backpropagation algorithm is applied for this fine-tuning procedure. We accordingly implement the deep model based on the DBN-DNN method. This model is both *generative* and *discriminative* due to twofold meaning. The unlabeled data are used to find a generative model in a bottom-up and stack-wise fashion. A small amount of labeled data $\{\mathbf{x}, \mathbf{r}\}$ are then used to fine-tune DNN parameters in accordance with the error backpropagation algorithm.

Generally, the greedy layer-wise training of DNN works well from the perspectives of regularization and optimization. First, the pretraining step helps constrain the parameters in a region that is relevant to unsupervised dataset. Hence, the representations that better describe the unlabeled data are more discriminative for labeled data. The regularization issue in DNN is handled to pursue better generalization. Second, the unsupervised training initializes the lower-level parameters near the localities of better minima than random initialization can. This perspective explains how the optimization works for DNN based on DBN initialization rather than random initialization.

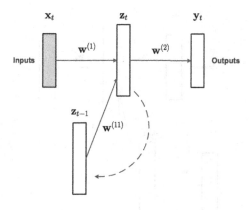

FIGURE 2.17

A recurrent neural network with single hidden layer.

2.5 RECURRENT NEURAL NETWORK

Different from feedforward neural network, the recurrent neural network (RNN) is a specialized approach to artificial neural network where the connections between neurons form a directed cycle (Williams and Zipser, 1989). RNN is developed to learn the temporal information in the time signal when FNN could not handle it. There are different architectures for using RNN, where the most popular of them was proposed in Elman (1990) and is shown in Fig. 2.17. Again, RNN is applied for source separation as a regression problem which demixes the mixed spectral signal $\mathbf{x}_t$ into the source spectral signal $\mathbf{y}_t$ at each time t. The relation between input signal $\mathbf{x}_t$ and the kth output node y_{tk} is formulated as

$$
\begin{aligned}
y_{tk} &= y_k(\mathbf{x}_t, \mathbf{w}) \\
&= f\left((\mathbf{w}^{(2)})^\top f\left((\mathbf{w}^{(1)})^\top \mathbf{x}_t + (\mathbf{w}^{(11)})^\top \mathbf{z}_{t-1}\right)\right) \\
&= f\left((\mathbf{w}^{(2)})^\top \mathbf{z}_t\right) \\
&= f(a_{tk})
\end{aligned}
\tag{2.60}
$$

where the parameter set $\mathbf{w}$ consists of input-to-hidden weights $\mathbf{w}^{(1)}$, hidden-to-hidden weights $\mathbf{w}^{(11)}$, and hidden-to-output weights $\mathbf{w}^{(2)}$. The hidden-to-hidden weights $\mathbf{w}^{(11)}$ are also known as the recurrent weights.

2.5.1 BACKPROPAGATION THROUGH TIME

In the implementation, RNN parameters are estimated by using the SGD algorithm based on backpropagation through time (BPTT) (Williams and Zipser, 1995). RNN is trained and optimized to continuously predict a sequence of source signals $\{\mathbf{y}_t\}_{t=1}^T$ from a sequence of mixed signals $\{\mathbf{x}_t\}_{t=1}^T$. Each prediction $\mathbf{y}_t$ is made by using the previous samples $\{\mathbf{x}_1, \ldots, \mathbf{x}_t\}$ which are present every τ time steps. RNN can be unfolded as seen in Fig. 2.18 where only two time steps ($\tau = 2$) are characterized.

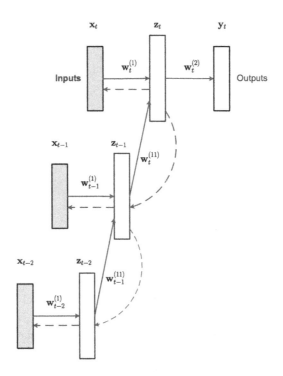

FIGURE 2.18

Backpropagation through time with a single hidden layer and two time steps $\tau = 2$.

Such an unfolded RNN is also recognized as a deep neural network with the repeated hidden layers. In this figure, solid lines show the feedforward calculation while dashed lines depict the backward calculation for local gradients, which are accumulated over τ time steps. Temporal and contextual features of sequential data are captured in RNN because the parameters are learned from the local gradients over a period of time via hidden code $\mathbf{z}_t$. In general, RNN performs better than DNN for source separation since the mixed and demixed signals are both time signals.

To carry out the updating formulas for input-to-hidden, hidden-to-hidden, and hidden-to-output weights $\mathbf{w} = \{\mathbf{w}^{(1)}, \mathbf{w}^{(11)}, \mathbf{w}^{(2)}\}$, we first calculate the sum-of-squares error function over previous τ time steps as

$$
\begin{aligned}
E_n(\mathbf{w}) &= \sum_{t=n-\tau+1}^{n} E_t(\mathbf{w}) \\
&= \frac{1}{2} \sum_{t=n-\tau+1}^{n} \|\mathbf{y}(\mathbf{x}_t, \mathbf{w}) - \mathbf{r}_t\|^2
\end{aligned}
\tag{2.61}
$$

and then perform backward calculation of local gradients from the output layer to the input layer via the error backpropagation procedure. The local gradient with respect to the activation vector in the output layer $\mathbf{a}_t^{(2)}$ at time t is calculated as

$$\delta_{tk} \triangleq \frac{\partial E_t}{\partial a_{tk}}$$
$$= (y_{tk} - r_{tk}) f'(a_{tk}). \tag{2.62}$$

This local gradient is then propagated to calculate the local gradient with respect to the activation vector in the hidden layer $\mathbf{a}_t^{(1)}$ which is divided into two cases. For the case of $\tau = 0$, there is no recurrent weight involved in updating. The local gradient is calculated by introducing the hidden-to-output weights $\mathbf{w}^{(2)} = \{w_{mk}^{(2)}\}$ in

$$\delta_{tm} \triangleq \frac{\partial E_t}{\partial a_{tm}}$$
$$= \sum_k \frac{\partial E_t}{\partial a_{tk}} \frac{\partial a_{tk}}{\partial z_{tm}} \frac{\partial z_{tm}}{\partial a_{tm}} \tag{2.63}$$
$$= \sum_k \delta_{tk} w_{mk}^{(2)} f'(a_{tm}).$$

For $\tau > 0$, the local gradient at a previous time step $t - \tau$ is calculated by using the recurrent weights or hidden-to-hidden weights $\mathbf{w}^{(11)} = \{w_{mj}^{(11)}\}$ as

$$\delta_{(t-\tau)m} = \sum_j \delta_{(t-\tau)j} w_{mj}^{(11)} f'\left(a_{(t-\tau)m}\right). \tag{2.64}$$

Having these local gradients, we accordingly calculate the derivatives for updating the input-to-hidden, hidden-to-hidden, and hidden-to-output weights in the following form:

$$\frac{\partial E_t(\tau)}{\partial w_{dm}^{(1)}} = \sum_{t'=t-\tau+1}^{t} \frac{\partial E_{t'}}{\partial a_{t'm}} \frac{\partial a_{t'm}}{\partial w_{dm}^{(1)}}$$
$$= \sum_{t'=t-\tau+1}^{t} \delta_{t'm} x_{(t'-1)d}, \tag{2.65}$$

$$\frac{\partial E_t(\tau)}{\partial w_{mj}^{(11)}} = \sum_{t'=t-\tau+1}^{t} \delta_{t'j} z_{(t'-1)m}, \tag{2.66}$$

$$\frac{\partial E_t}{\partial w_{mk}^{(2)}} = \delta_{tk} z_{tm} \tag{2.67}$$

where the gradients $\frac{\partial E_t(\tau)}{\partial w_{dm}^{(1)}}$ and $\frac{\partial E_t(\tau)}{\partial w_{mj}^{(11)}}$ for updating input-to-hidden and hidden-to-hidden weights are considerably affected by BPTT with τ steps back in time. The variables $\{x_{(t-1)d}, z_{(t-1)m}\}$ from the previous time $t - 1$ are used. The gradient for updating hidden-to-output weights $\frac{\partial E_t}{\partial w_{mk}^{(2)}}$ is not affected by recurrent weights in error backpropagation so that the formula is consistent with that of DNN in Eq. (2.58). The valuable z_{tm} from the current time t is used. When implementing the SGD algorithm, the derivatives in Eqs. (2.65), (2.66) and (2.67) are calculated by considering the error function accumulated over a minibatch

$$E_n(\mathbf{w}) = \sum_{t \in \{\mathbf{X}_n, \mathbf{R}_n\}} E_t. \tag{2.68}$$

Basically, the BPTT algorithm is run to extract the temporal information back to τ time steps and to apply it for updating the RNN parameters. The case of RNN with $\tau = 0$ is reduced to a feedforward neural network with one hidden layer. The problem of gradient vanishing (Bengio et al., 1994) happens if a too large τ is chosen. Thus, $\tau = 4$ or 5 is likely to be selected.

2.5.2 DEEP RECURRENT NEURAL NETWORK

Although RNN in Fig. 2.18 is shown as a kind of a deep unfolded model, the recurrence of hidden features is run under a shallow model where only one hidden layer is taken into account. Such an RNN architecture can be further extended to a deep recurrent neural network (DRNN) where the recurrent weights $\mathbf{w}^{(ll)}$ are applied in the lth layer with $l \in \{1, \ldots, L\}$. Fig. 2.19 illustrates an implementation of a deep recurrent neural network for single-channel source separation with one mixed signal $\mathbf{x}_t$ and two demixed source signals $\{\hat{\mathbf{x}}_{1,t}, \hat{\mathbf{x}}_{2,t}\}$. There are L layers in this DRNN where the recurrence of hidden layer $\mathbf{z}_t$ can be performed in different layers l. Such a DRNN has been successfully developed for speech separation, as well as singing voice separation (Huang et al., 2014a, 2014b, Wang et al., 2016). In the forward pass, the demixed signal $\mathbf{x}_t$ passes through multiple hidden layers $\{\mathbf{w}^{(1)}, \ldots, \mathbf{w}^{(L)}\}$ where the weights of connecting to the output layer L consist of those decomposing weights for two source signals, i.e., $\mathbf{w}^{(L)} = \{\mathbf{w}_1^{(L)}, \mathbf{w}_2^{(L)}\}$. The output signals for two sources $\mathbf{y}_{1,t}$ and $\mathbf{y}_{2,t}$ are calculated by using the activations

$$\left\{ \mathbf{a}_{1,t}^{(L)} = \{a_{1,tk}^{(L)}\}, \mathbf{a}_{2,t}^{(L)} = \{a_{2,tk}^{(L)}\} \right\} \tag{2.69}$$

in layer L. By applying the soft mask function, similar to that used in NMF, we estimate the Wiener gain function $\hat{\mathbf{y}}_{i,t} = \{\hat{y}_{i,tk}\}$ based on

$$\hat{y}_{i,tk} = \frac{|a_{i,tk}^{(L)}|}{|a_{1,tk}^{(L)}| + |a_{2,tk}^{(L)}|}. \tag{2.70}$$

The reconstructed magnitude spectrograms are then obtained by multiplying the mixed magnitude spectrogram $\mathbf{x}_t$ using the soft mask function

$$\tilde{\mathbf{x}}_{i,t} = \mathbf{x}_t \odot \hat{\mathbf{y}}_{i,t}. \tag{2.71}$$

The soft mask function applied in the demixed spectrograms using DRNN is the same as that employed in NMF and shown in Fig. 2.6 and Eq. (2.15). It is noted that the goal of the soft mask function is to

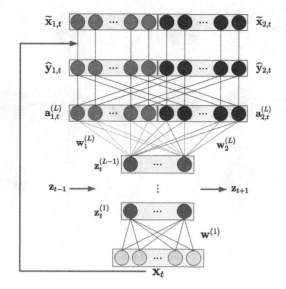

FIGURE 2.19

A deep recurrent neural network for single-channel source separation in the presence of two source signals.

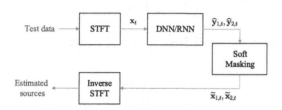

FIGURE 2.20

A procedure of single-channel source separation based on DNN or RNN where two source signals are present.

perform time-frequency masking so as to meet the constraint that the sum of the predicted spectrogram is equal to the original mixture spectrogram. Fig. 2.20 shows a procedure of single-channel source separation based on DNN, RNN or DRNN. Nevertheless, the optimization of RNN can be very hard. The extension from RNN to long short-term memory is the current trend.

2.5.3 LONG SHORT-TERM MEMORY

In general, the challenge of training a standard RNN is to deal with the problem of gradient vanishing or exploding, which happens frequently in reality. This problem is caused due to the repeated multipli-

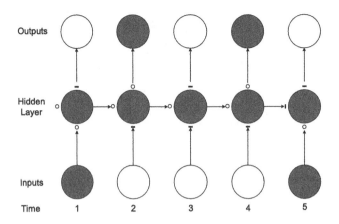

Outputs

Hidden
Layer

Inputs

Time   1   2   3   4   5

FIGURE 2.21

Preservation of gradient information through a gating mechanism.

cations by using hidden-to-hidden weights

$$
\begin{aligned}
\mathbf{z}_t &= (\mathbf{w}^{(11)})^\top \mathbf{z}_{t-1} = \left((\mathbf{w}^{(11)})^2 \right)^\top \mathbf{z}_{t-2} \\
&= \cdots = \left((\mathbf{w}^{(11)})^t \right)^\top \mathbf{z}_0
\end{aligned}
\tag{2.72}
$$

where the nonlinear activation is ignored. In Eq. (2.72), if matrix $\mathbf{w}^{(11)}$ is represented by the eigendecomposition

$$
\mathbf{w}^{(11)} = \mathbf{Q}\mathbf{\Lambda}\mathbf{Q}^\top
\tag{2.73}
$$

with eigenvalue matrix $\mathbf{\Lambda}$ and eigenvector matrix $\mathbf{Q}$, then, after running t time steps, the recurrence of hidden units may be reduced to

$$
\mathbf{z}_t = \mathbf{Q}^\top \mathbf{\Lambda}^t \mathbf{Q}\mathbf{z}_0.
\tag{2.74}
$$

The hidden units are decided by the eigenvalues to the power of t. The values of the hidden variables decay to zero if the magnitude of an eigenvalue is less than one and explode if the magnitude is greater than one. Such a decay or explosion condition not only happens in the forward pass but also in the backward pass where the local gradients $\boldsymbol{\delta}_{t-\tau} = \{\delta_{(t-\tau)m}\}$ are t times multiplied by recurrent weights $\mathbf{w}^{(11)}$. Local gradients at time t are almost impossible to propagate to the starting time. Hence long-time dependencies are hard to learn.

To tackle this problem, an improved method called the long short-term memory (LSTM) (Hochreiter and Schmidhuber, 1997) was proposed. The key idea is to preserve the gradient information in time through a gating mechanism, which is illustrated in Fig. 2.21. Along the time horizon, we have an input sequence $\{\mathbf{x}_t\}_{t=1}^T$, feature sequence $\{\mathbf{z}_t\}_{t=1}^T$ and output sequence $\{\mathbf{y}_t\}_{t=1}^T$ where $T = 5$ in this case. The input, forget and output gates are treated as the switching controllers to preserve

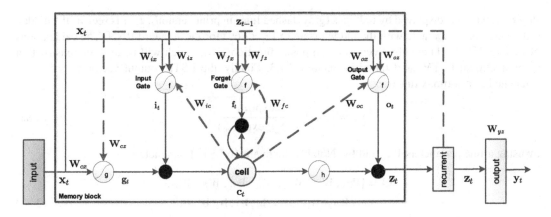

FIGURE 2.22

A detailed view of long short-term memory. Block circle means multiplication. Recurrent state $\mathbf{z}_{t-1}$ and cell $\mathbf{c}_{t-1}$ in a previous time step are shown by dashed lines.

the gradient information for input, recurrence and output, respectively. The gradients are displayed in blue (gray in print version). In this figure, the gradient at time step 1 is propagated to time steps 2 and 4 through the opened output gates (denoted by o) The closed output gates (denoted by −) prevent gradient leakage at time steps 1 and 3. In this example, the propagation of gradient at time step 1 is terminated at time step 5. Opening or closing of gates in LSTM is controlled by the training procedure.

The architecture of LSTM is formed by a memory block, which consists of a memory cell c_t and three sigmoid gates $f(a)$, including the input gate i_t, the output gate o_t, and the forget gate f_t, as depicted in Fig. 2.22. The updating of cell c_t is based on a forgetting factor f_t for previous cell c_{t-1} and a scaled input i_t with a scaling weight g_t. The activation functions in i_t, o_t and f_t use the sigmoid function (Eq. (2.51)) while that in g_t adopts the hyperbolic tangent function (Eq. (2.11)). The LSTM memory block for input x_t and output y_t is implemented in a vector form as

$$\mathbf{i}_t = \sigma(W_{ix}\mathbf{x}_t + W_{iz}\mathbf{z}_{t-1} + W_{ic}\mathbf{c}_{t-1} + \mathbf{b}_i), \tag{2.75}$$

$$\mathbf{f}_t = \sigma(W_{fx}\mathbf{x}_t + W_{fz}\mathbf{z}_{t-1} + W_{fc}\mathbf{c}_{t-1} + \mathbf{b}_f), \tag{2.76}$$

$$\mathbf{g}_t = \tanh(W_{cx}\mathbf{x}_t + W_{cz}\mathbf{z}_{t-1} + \mathbf{b}_c), \tag{2.77}$$

$$\mathbf{c}_t = \mathbf{f}_t \odot \mathbf{c}_{t-1} + \mathbf{i}_t \odot \mathbf{g}_t, \tag{2.78}$$

$$\mathbf{o}_t = \sigma(W_{ox}\mathbf{x}_t + W_{oz}\mathbf{z}_{t-1} + W_{oc}\mathbf{c}_t + \mathbf{b}_o), \tag{2.79}$$

$$\mathbf{z}_t = \mathbf{o}_t \odot \tanh(\mathbf{c}_t), \tag{2.80}$$

$$\mathbf{y}_t = s(W_{yz}\mathbf{z}_t + \mathbf{b}_y) \tag{2.81}$$

where $\odot$ denotes the element-wise product, W_{ix}, W_{fx}, W_{cx} and W_{ox} denote the weight matrices from input gate to input, forget, cell and output gates, respectively. The corresponding bias vectors are $\mathbf{b}_i$, $\mathbf{b}_f$, $\mathbf{b}_c$ and $\mathbf{b}_o$. Here, W_{ic}, W_{fc} and W_{oc} are the diagonal weight matrices for peephole connections from the cell output vector in a previous time step $\mathbf{c}_{t-1}$, which is shown by dashed lines; W_{iz}, W_{fz}, W_{cz}, W_{oz} denote the weight matrices for the output vector of the memory block in the previous time step $\mathbf{z}_{t-1}$, which is displayed by red lines (gray dashed lines in print version); $\mathbf{z}_{t-1}$ is seen as the hidden state which is comparable with the hidden state in Eq. (2.60) for the standard RNN. The calculations of Eqs. (2.75)–(2.81) are also seen as a composite function $\mathcal{Z}(\cdot)$, which will be used in construction of bidirectional LSTM as detailed in Section 7.4.3. Finally, the LSTM output vector $\mathbf{y}_t = \{y_{tk}\}$ is calculated as a softmax function

$$y_{tk} = s(a_{tk}) = \frac{\exp(a_{tk})}{\sum_m \exp(a_{tm})} \tag{2.82}$$

by using affine parameters W_{yz} and $\mathbf{b}_y$. Notably, there are 16 LSTM parameters

$$\begin{aligned}\Theta = \{&W_{ix}, W_{fx}, W_{cx}, W_{ox}, W_{ic}, W_{fc}, W_{oc}, \\ &W_{iz}, W_{fz}, W_{cz}, W_{oz}, \mathbf{b}_i, \mathbf{b}_f, \mathbf{b}_c, \mathbf{b}_o, \mathbf{b}_y\},\end{aligned} \tag{2.83}$$

which are trained by the error backpropagation algorithm. The calculation of gradients of a learning objective with respect to individual parameters in Θ can be carried out similarly to the calculation in the standard error backpropagation for DNN and RNN. Further studies and extensions of LSTM for monaural source separation will be addressed in Section 7.4.

2.6 SUMMARY

In this chapter, we have introduced linear and nonlinear source separation models, which were developed for multi- and single-channel source separation. Linear solutions based on two- and multiway decomposition paved an avenue to accommodate multimode information in the separated signals. Independent component analysis (ICA) was introduced to handle multichannel source separation. The contrast function for measuring the independence or non-Gaussianity of demixed signals was maximized to estimate the demixing matrix. Nonnegative matrix factorization (NMF) was addressed to carry out single-channel source separation based on a weight matrix and a basis matrix, which were estimated by minimizing the reconstruction error. Different divergence measures for reconstruction error were introduced and optimized. Nonnegative tensor factorization (NTF) was mentioned to allow multiway data decomposition for source separation. On the other hand, deep neural network (DNN) was addressed to deal with single-channel source separation, which was treated as a regression problem for supervised learning. The training algorithm and implementation tricks for deep model were introduced. In addition, the recurrent neural network (RNN) and deep recurrent neural network were surveyed to capture the temporal information for source separation. The extension to long short-term memory was detailed. A number of source separation models have been briefly introduced. The advanced topics of ICA, NMF, NTF, DNN and RNN will be detailed in the second part of this book which starts from Chapter 4 to Chapter 7. In what follows, we focus on a series of adaptive learning algorithms which provide fundamental theories and practices to different specialized solutions to source separation under different perspectives and learning strategies.

ADAPTIVE LEARNING MACHINE

3

This chapter addresses how machine learning algorithms are developed for source separation. First of all, we summarize different learning problems in source separation models in Section 3.1. Then, these problems are individually tackled by information-theoretic learning, Bayesian learning, sparse learning, online learning, discriminative and deep learning which will be detailed in Sections 3.2, 3.3, 3.4, 3.5, 3.6, respectively. Several algorithms are driven by Bayesian theory so as to establish the sparse Bayesian learning, sequential Bayesian learning and variational Bayesian learning. In the final part of this chapter (Section 3.7), we systematically present a number of inference algorithms for construction of latent variable models for source separation including maximum likelihood, maximum *a posteriori*, variational Bayesian and Gibbs sampling. The illustrations of how these algorithms are derived and evolved to deal with the complicated models in the presence of multiple latent variables are presented.

3.1 LEARNING PROBLEMS IN SOURCE SEPARATION

Machine learning theories and algorithms provide the foundation to carry out different source separation models subject to different constraints and topologies. Supervised learning and unsupervised learning are required to deal with the regression and clustering problems, which happen in many source separation systems. In general, the learning problems in source separation can be summarized and categorized in what follows.

Learning Objective

Source separation model based on ICA, NMF, NTM, DNN or RNN is learned by optimizing a learning objective function which drives the separation result to meet a specialized objective or contrast function. A paradigm, called the information-theoretic learning, is introduced to develop a series of learning objectives based on information theory. These learning objectives are optimized to pursue the maximally-independent sources by using ICA and the minimal reconstruction error in signal decomposition by using NMF or NTF. We would like to construct a source separation model with meaningful learning objective so that we can come out with a solution which can be clearly interpreted. Section 3.2 addresses a series of learning objectives based on the information-theoretic learning.

Regularization & Generalization

Machine learning for source separation is basically partitioned into a training phase and a test phase, which are run to learn from training data and make predictions on test data. The model we assume may be improper for the collection of training data. The model learned from training data is likely over-estimated. The prediction performance for demixing a new mixed signal is degraded. This is essentially true because the unknown test condition is always mismatched with training environment. The mismatch may come from the changing sources, moving sensors, varying domains or noise interference. Such a generalization from the training environment to the test condition is an open issue in source separation. We introduce Bayesian learning to faithfully characterize these uncertainties for model regularization in ICA, NMF and DNN. The prediction is improved due to the probabilistic modeling. Uncertainty modeling is helpful to enhance the robustness in signal demixing. Section 3.3 introduces the basics of Bayesian learning for model regularization. On the other hand, the overfitting problem in source separation using ICA, NMF and DNN is overcome by imposing sparsity for basis vectors or sensing weights. Section 3.4 illustrates why sparsity control does matter for source separation.

Nonstationary Environment

The environment in a signal separation system is usually nonstationary in real-world applications. The sources may be replaced or may be moving. The mixing condition is changing. Practically, the statistics collected for source separation should be time-varying to reflect the physical phenomenon in nonstationary environments. To deal with this issue, an incremental learning mechanism is incorporated into updating statistics for implementing an online learning procedure. A recursive Bayesian algorithm will be introduced to carry out this procedure. Section 3.5 addresses how online learning is developed to deal with different scenarios in nonstationary source separation.

Discriminative Separation

Similar to enhancing the discrimination between classes in a classification system, the discrimination in separation outputs can be also increased to improve the regression performance for source separation. The idea of discriminative training for source separation is to introduce a regularization term in the learning objective. In addition to the minimization of a sum-of-squares objective function, the discrimination between source signals is maximized. The demixed signals are optimally estimated due to this discriminative separation. Discriminative learning is focused to carry out a source separation system as described in Section 3.6.

Complicated Mixing Condition

The mapping between mixed signal and source signals can be very complicated due to two reasons. The first is caused by the underdetermined condition in single-channel source separation. This is a mixing condition with missing information, which is hard to work with. The second comes from the complicated signals in the demixed and source signals. The complicated mixing condition can be resolved by deep learning algorithms which are now the new trend in source separation algorithms. An illustration of deep learning for source separation is presented in Section 3.6. Next, the learning problems in source separation are individually handled by different machine learning strategies and algorithms.

3.2 INFORMATION-THEORETIC LEARNING

Information-theoretic learning (Principe et al., 2000) uses the descriptors from information theory, e.g., entropy and divergence, which are directly estimated from data. The supervised and unsupervised learning using these descriptors rather than using the conventional statistical descriptors based on variance and covariance is known as information-theoretic learning. Source separation based on ICA and NMF widely adopts divergence measures as learning objectives. ICA and NMF conduct the information-theoretic learning for source separation. Information-theoretic learning bridges the gap between *information theory* and *machine learning*. Adaptive learning can be achieved through information measurement from the mixed signals.

Basically, blind source separation using independent component analysis (ICA) aims to discover the latent independent sources in an unsupervised manner by minimizing the measure of dependence or Gaussianity of the demixed signals, $\mathcal{D}(\mathbf{X}, \mathbf{W})$, with respect to the demixing matrix $\mathbf{W}$ from a collection of training data $\mathbf{X} = \{\mathbf{x}_t\}$. In standard ICA, the number of microphones n is the same as the number of sources m. The demixed signals $\mathbf{y}_t = \{y_{ti}\}_{i=1}^{m}$ at time t are calculated by $\mathbf{y}_t = \mathbf{W}\mathbf{x}_t$. Building a meaningful learning objective or contrast function $\mathcal{D}(\mathbf{X}, \mathbf{W})$ is crucial for ICA. Typically, ICA involves optimization of a contrast function which is seen as an information measure originated from information theory. The optimization is fulfilled according to the gradient descent algorithm as shown in Eq. (2.2) by calculating the gradient $\frac{\partial \mathcal{D}(\mathbf{X}, \mathbf{W})}{\partial \mathbf{W}}$. For simplicity of notation, in the following illustration, we ignore time index t and evaluate the contrast function by considering only two demixed signals $\{y_1, y_2\}$ in $\mathbf{y}_t$. The extension to m sources is straightforward.

3.2.1 DIVERGENCE MEASURE

It is essential to measure the dependence between two demixed signals y_1 and y_2 as the contrast function for optimization of an ICA procedure. A popular measure is based on the mutual information. This measure is defined by using Shannon entropy (Shannon, 1948), which calculates the average amount of information contained in each message or random variable y as

$$\mathbb{H}[p(y)] = -\int p(y) \log p(y) dy. \tag{3.1}$$

Mutual information between y_1 and y_2 is measured as the relative entropy or Kullback–Leibler divergence between $p(y_1, y_2)$ and $p(y_1)p(y_2)$ given by

$$\begin{aligned}
\mathcal{D}_{\mathrm{KL}}(y_1, y_2) &= \mathbb{H}[p(y_1)] + \mathbb{H}[p(y_2)] - \mathbb{H}[p(y_1, y_2)] \\
&= \iint p(y_1, y_2) \log \frac{p(y_1, y_2)}{p(y_1)p(y_2)} dy_1 dy_2 \\
&\triangleq \mathcal{D}_{\mathrm{KL}}(p(y_1, y_2) \| p(y_1)p(y_2))
\end{aligned} \tag{3.2}$$

where $\mathcal{D}_{\mathrm{KL}}(y_1, y_2) \geq 0$ with equality if and only if y_1 and y_2 are independent of each other, i.e., $p(y_1, y_2) = p(y_1)p(y_2)$. $\mathcal{D}_{\mathrm{KL}}(y_1, y_2)$ is seen as a measure of dependence which is also denoted by $\mathcal{D}_{\mathrm{KL}}(p(y_1, y_2) \| p(y_1)p(y_2))$. We would like to minimize this contrast function to estimate the demixing matrix $\mathbf{W}$ or find the demixed signals y_1 and y_2 with the lowest dependence. In addition, we incorporate the terms $\{p(y_1, y_2), p(y_1)p(y_2)\}$ into the squared Euclidean distance and the Cauchy–Schwartz divergence to yield (Xu et al., 1998)

$$\begin{aligned}
\mathcal{D}_{\mathrm{EU}}(y_1, y_2) \\
= \iint \left[p(y_1, y_2) - p(y_1)p(y_2) \right]^2 dy_1 dy_2
\end{aligned} \tag{3.3}$$

and

$$\mathcal{D}_{CS}(y_1, y_2)$$
$$= \log \left\{ \frac{\iint p(y_1, y_2)^2 dy_1 dy_2 \cdot \iint p(y_1)^2 p(y_2)^2 dy_1 dy_2}{\left[\iint p(y_1, y_2) p(y_1) p(y_2) dy_1 dy_2 \right]^2} \right\}, \tag{3.4}$$

respectively. These quadratic divergence measures are reasonable to implement the ICA procedure. The properties $\mathcal{D}_{EU}(y_1, y_2) \geq 0$ and $\mathcal{D}_{CS}(y_1, y_2) \geq 0$ hold with the equalities if and only if y_1 and y_2 are independent. In Amari (1985), α divergence was proposed with a controllable parameter α for the *convexity* of divergence function. This divergence can be employed to measure the degree of dependence as a contrast function for the ICA learning procedure and is given by

$$\mathcal{D}_{\alpha}(y_1, y_2) = \frac{4}{1 - \alpha^2} \iint \left[\frac{1 - \alpha}{2} p(y_1, y_2) + \frac{1 + \alpha}{2} p(y_1) p(y_2) \right. $$
$$\left. - p(y_1, y_2)^{(1-\alpha)/2} \left(p(y_1) p(y_2) \right)^{(1+\alpha)/2} \right] dy_1 dy_2. \tag{3.5}$$

The convexity parameter α is adjustable for convergence speed in ICA optimization. In case $\alpha = -1$, α divergence reduces to KL divergence. Furthermore, the f divergence, addressed in Csiszar and Shields (2004), can be adopted as a measure of dependence; it is given by

$$\mathcal{D}_f(y_1, y_2)$$
$$= \iint p(y_1) p(y_2) f \left(\frac{p(y_1, y_2)}{p(y_1) p(y_2)} \right) dy_1 dy_2 \tag{3.6}$$

where $f(\cdot)$ denotes a convex function subject to $f(t) \geq 0$ for $t \geq 0$, $f(1) = 0$ and $f'(1) = 0$. The α divergence is a special case of f divergence when the function

$$f(t) = \frac{4}{1 - \alpha^2} \left[\frac{1 - \alpha}{2} + \frac{1 + \alpha}{2} t - t^{(1+\alpha)/2} \right] \tag{3.7}$$

for $t \geq 0$ is applied. Eq. (3.7) is a convex function with a real-valued control parameter α owing to $f''(t) = t^{(\alpha-3)/2} \geq 0$. In Zhang (2004), α divergence and f divergence were merged to construct a general divergence measure

$$\frac{4}{1 - \alpha^2} \left[\frac{1 - \alpha}{2} \iint f(p(y_1, y_2)) dy_1 dy_2 \right.$$
$$+ \frac{1 + \alpha}{2} \iint f(p(y_1) p(y_2)) dy_1 dy_2 \tag{3.8}$$
$$\left. - \iint f \left(\frac{1 - \alpha}{2} p(y_1, y_2) + \frac{1 + \alpha}{2} p(y_1) p(y_2) \right) dy_1 dy_2 \right].$$

This measure is nonnegative and equals to zero if and only if y_1 and y_2 are independent. Besides, the Jensen–Shannon divergence (Lin, 1991) was derived from Jensen's inequality using the Shannon entropy $\mathbb{H}[\cdot]$, which is a concave function. This divergence function is applied as a contrast function measuring the dependence between y_1 and y_2 and is defined by

$$
\begin{aligned}
\mathcal{D}_{\text{JS}}(y_1, y_2) = \mathbb{H}\big[\lambda p(y_1, y_2) + (1 - \lambda) p(y_1, y_2)\big] \\
- \lambda \mathbb{H}\big[p(y_1, y_2)\big] - (1 - \lambda)\mathbb{H}\big[p(y_1)p(y_2)\big]
\end{aligned}
\tag{3.9}
$$

where $0 \geq \lambda \geq 1$ a tradeoff weight between the joint distribution $p(y_1, y_2)$ and the product of marginals $p(y_1)p(y_2)$. Jensen–Shannon divergence is an increment of the Shannon entropy and equals to zero if and only if y_1 and y_2 are independent. This divergence measure provides both lower and upper bounds to the Bayes probability of classification error (Lin, 1991).

It is noted that the divergence measure is not only used as a contrast function for measuring the dependence between two demixed signals y_1 and y_2 for ICA procedure but also applied to construct the learning objective for optimal reconstruction of the mixed data matrix $\mathbf{X}$ using basis matrix $\mathbf{B}$ and weight matrix $\mathbf{W}$ in nonnegative matrix factorization (NMF) procedure. Both ICA and NMF are unsupervised learning machines where the targets of separated signals are not required in learning objective. Using ICA, we minimize the contrast function $\mathcal{D}_f(y_1, y_2)$ to pursue "independence" between demixed signals y_1 and y_2 while NMF minimizes the divergence measure $\mathcal{D}(\mathbf{X}\|\mathbf{BW})$ to assure the minimum "reconstruction error" as shown in Eq. (2.18). The squared Euclidean distance and KL divergence are both applied for NMF in Eqs. (2.21) and (2.26) and for ICA in Eqs. (3.3) and (3.2), respectively. In particular, α divergence has been successfully developed to build ICA (Amari, 1985) and NMF (Cichocki et al., 2008) procedures. In Section 4.3, we will introduce an example of information-theoretic learning based on a convex divergence ICA (Chien and Hsieh, 2012) and illustrate how this method was developed for separation of speech and music signals based on an instantaneous BSS task.

3.3 BAYESIAN LEARNING

Bayesian inference and statistics, pioneered by Thomas Bayes and shown in Fig. 3.1, play an essential role in dealing with the uncertainty modeling and probabilistic representation in machine learning problems, including regression, classification, supervised learning and unsupervised learning, to name a few. In this section, we start from the general description of model regularization and then present how Bayesian learning is developed to tackle regularization in source separation. The approximate Bayesian inference algorithms based on variational inference and Gibbs sampling will be addressed in Section 3.7.

3.3.1 MODEL REGULARIZATION

We are facing the challenges of big data, which are collected in adverse conditions or in heterogeneous environments. Model-based source separation methods based on ICA, NMF, DNN and RNN are learned from an online test sample or a collection of training samples. The learning strategy can be unsupervised, supervised or semisupervised. The fundamental issue of machine learning for source separation is to learn a demixing model from training data which can generalize well for separation of an unknown mixed signal. Prediction performance is an ultimate goal of a machine learning system. However, overestimation or underestimation problem is inevitable since the model we assumed

FIGURE 3.1

Portrait of Thomas Bayes (1701–1761) who is credited for the Bayes theorem.

is not always fitted to the characteristics of mixed signals and the size of training data. It is crucial to reflect the randomness of the mixed data $\mathbf{X}$, the assumed model $\mathcal{M}$ and the estimated parameters Θ in model construction. To faithfully characterize how the signals are mixed and how the relation between collected data and assumed model is connected, we need a tool to model, analyze, search, recognize and understand the real-world mixed signals under different mixing conditions. This tool should be adaptive to learn a flexible model topology according to different amount of training data, as well as construct a robust demixing system in the presence of ill-posed or mismatched conditions between the training and test environments. Also, the noise contamination in a mixing system should be taken into account to develop a noise-aware demixing process. In addition, it is challenging and cost-demanding to choose the hyperparameters Ψ or the regularization parameters of an assumed model by using additional validation data. More generally, the issue of model selection is of concern. The selection criterion does not only rely on a goodness-of-fit measure based on the likelihood function but also a regularization penalty based on model complexity. In MacKay (1992), the Occam's razor was imposed to deal with the issue of model selection based on Bayesian learning. Bayes theorem provides a complete tool or platform to resolve these model regularization issues. In summary, the capabilities of Bayesian learning are feasible to (Watanabe and Chien, 2015)

- represent the uncertainty in model construction,
- deal with overestimation or underestimation,
- reflect noise condition in observed data,
- assure robustness to ill-posed or mismatch condition,
- be automated and adaptive,
- be scalable for a large data set,
- determine the model size or select the hyperparameters.

Fig. 3.2 illustrates the concept of a scalable model with increasing amount of training data such that the complexity of model structure is accordingly increased. In general, regularization refers to a process of introducing additional information in order to represent the model uncertainty so as to solve the

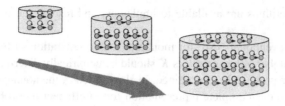

FIGURE 3.2

Illustration of a scalable model in presence of increasing number of training data.

ill-posed problem or prevent the overfitting condition. Uncertainty can be properly expressed by prior distribution $p(\Theta)$ or even a prior process for finding a drawing process for prior distributions (Teh et al., 2006).

3.3.2 BAYESIAN SOURCE SEPARATION

The real-world blind source separation encounters a number of challenging issues. First, the learning processes for source signals and mixing process are generally run in an unsupervised learning manner where there is no target information available for training. When using unsupervised learning it is sometimes hard to predict reliable performance in new test signals. The prediction is even harder if the number of sources is unknown or changed. An adaptive approach to meet the changing source number or the unknown mixing system is required. It is accordingly important and meaningful to treat the mixing process as a nonstationary environment and build a dynamic time-varying source separation system to fit such a changing environment. In addition, it is challenging to deal with the unknown mixing environment in multichannel source separation where the underdetermined and sparse sources are present. Building a balanced solution or tuning a tradeoff between overdetermined and underdetermined systems with different numbers of sources and channels is seen as a crucial and fundamental issue in source separation.

The trend of Bayesian source separation has been popular in the past decade (Févotte, 2007). In this research, the number of sources can be determined by Bayesian learning based on the scheme called automatic relevance determination (ARD) (Bishop, 2006). ARD is seen as a hyperparameter which reflects how relevant a mixed signal is to an assumed source. This ARD parameter serves as a prior for representing the relevance between the mixed and source signals, which is estimated by maximizing the marginal likelihood. The degree of relevance can be driven by forcing the sparsity of weights in a basis representation which will be discussed in Section 3.4. On the other hand, a recursive Bayes theorem application is feasible to continuously combine the likelihood function of current mixed signal given a mixing model and the prior distribution of model parameters. The resulting posterior distribution can be recursively updated to learn the changing environment for source separation. The online tracking of nonstationary condition is implemented. An example of developing the ARD scheme and recursive Bayes mechanism for source separation will be detailed in Section 4.4. Even the Gaussian process prior is useful to carry out a variant of Bayesian source separation where the temporal structure of time-varying sources is learned and explored in accordance with the methodology of a Gaussian process (GP). Section 4.5 presents a GP solution to audio source separation where the GP prior is employed to reflect the nature of time-series signals in a separation model. A number of approximate

Bayesian inference algorithms are available to implement and resolve these regularization issues in source separation.

For example, the conventional NMF for monaural source separation suffers from two regularization issues. First, the number of basis vectors K should be empirically determined. Second, the model parameters **B** and **W** are prone to be overtrained. In Hoyer (2004), the nonnegative sparse coding was proposed to learn sparse overcomplete representation for an efficient and robust NMF. Determining the regularization parameter for sparse representation plays an important role. In addition, the uncertainty modeling via probabilistic framework is helpful to improve model regularization (Bishop, 2006, Chien and Hsieh, 2013b, Saon and Chien, 2012a). The uncertainties in audio source separation may come from improper model assumption, incorrect model order, noise interference, reverberant distortion, heterogeneous data and nonstationary environment (Watanabe and Chien, 2015). In Shashanka et al. (2008), the probability latent component analysis decomposed the probability of nonnegative input data into a product of two conditional probabilities given latent components via the maximum likelihood estimation based on the expectation–maximization (EM) algorithm (Dempster et al., 1977). The nonnegative parameters could be represented by prior distributions. In Schmidt et al. (2009), Bayesian learning was introduced to conduct uncertainty decoding for robust source separation by maximizing the marginal likelihood over the randomness of model parameters. Bayesian NMF was proposed for image feature extraction based on the Gaussian likelihood and exponential prior. In Chien and Hsieh (2013a), Bayesian group sparse learning for NMF was introduced by using a Laplacian scale mixture prior for sparse coding (Chien, 2015b) based on common and individual bases for separation of rhythmic and harmonic sources, respectively. Gibbs sampling inference was implemented. In Cemgil (2009), Dikmen and Fevotte (2012), variational Bayesian (VB) inference using the Poisson likelihood for modeling error and the Gamma prior for model parameters was proposed for image reconstruction and text retrieval. Implementation cost was high due to the numerical calculation of the shape parameter. In Chien and Chang (2016), Bayesian learning was developed for speech dereverberation where exponential priors for reverberation kernel and noise signal were used to characterize the variations of a dereverberation model. A variational inference algorithm was proposed to build a Bayesian speech dereverberation system.

3.4 SPARSE LEARNING

Source separation using nonnegative matrix factorization is performed to separate a mixed signal based on a basis matrix **B** and a weight matrix **W**. NMF is seen as a dictionary learning method where the basis representation is implemented for representation learning. However, the dictionary is usually overdetermined to span a sufficiently large space for data representation. Sparse representation becomes crucial in dictionary learning for source separation in the presence of audio and music signals (Li et al., 2014, Plumbley et al., 2010). In general, basis representation of an observation $\mathbf{x} \in \mathbb{R}^D$ is expressed by

$$
\begin{aligned}
\mathbf{x} &= w_1 \mathbf{b}_1 + \cdots + w_K \mathbf{b}_K \\
&= \mathbf{B}\mathbf{w}.
\end{aligned}
\tag{3.10}
$$

In source separation, this observation $\mathbf{x}$ represents the mixed signal at a specific time moment $\mathbf{x}_t$ while the basis matrix $\mathbf{B} = [\mathbf{b}_1, \ldots, \mathbf{b}_K]$ contains the dictionaries or basis vectors corresponding to different source signals. Using basis representation, the basis vectors in a dictionary, also called atoms, are not required to be orthogonal, and they may be an overcomplete spanning set. This problem setup also allows the dimensionality K of the signals being represented, i.e., the size of basis vectors, to be higher than the dimensionality D of the mixed signals being observed. These atoms are likely to be redundant. The resulting weights $\mathbf{w} \in \mathbb{R}^K$ are prone to be sparse in ill-posed conditions. Sparse dictionary learning or sparse coding aims to find a sparse representation where the sparsity of sensing weights $\mathbf{w}$ is imposed to find a representation based on a set of overdetermined basis vectors. Sparsity constraint provides the flexibility for data representation. More specifically, the sparse representation is run by minimizing the ℓ_2-reconstruction error $\|\mathbf{x} - \mathbf{Bw}\|_2^2$ where the basis matrix $\mathbf{B}$ and weight vector $\mathbf{w}$ are both unknown. Assuming $\mathbf{B}$ is known and given, the sparse solution to optimal reconstruction is obtained by solving the minimization problem

$$\widehat{\mathbf{w}} = \arg\min_{\mathbf{w}} \frac{1}{2}\|\mathbf{x} - \mathbf{Bw}\|_2^2 + \lambda\|\mathbf{w}\|_1. \tag{3.11}$$

A lasso regularization term (Tibshirani, 1996) is imposed in the ℓ_1-regularized objective function to fulfill the sparse coding. In Eq. (3.11), λ is seen as a regularization parameter which balances the tradeoff between reconstruction error and model regularization. Fig. 3.3 illustrates the locations of optimal weight vectors $\mathbf{w}^\star$ (Bishop, 2006) estimated by minimizing the ℓ_2- and ℓ_1-regularized objective functions where the ℓ_2-norm

$$\|\mathbf{w}\|_2^2 = w_1^2 + \cdots + w_K^2 \tag{3.12}$$

and ℓ_1-norm of weight vector

$$\|\mathbf{w}\|_1 = |w_1| + \cdots + |w_K| \tag{3.13}$$

are introduced as the regularization terms, respectively. In the two-dimensional case (i.e., $K = 2$), the unit sphere in the ℓ_2-norm has the shape of a circle while the unit sphere in the ℓ_1-norm has the shape of a rectangle having vertices at the two axes with coordinates $w_1 \in \{-1, 1\}$ and $w_2 \in \{-1, 1\}$. These unit spheres for the two norms are shown by red curves (light gray in print version). The reconstruction error terms, shown in blue (dark gray in print version), are expressed as the quadratic functions of $\mathbf{w}$, which are plotted as the concentric circles with different values of variance. Compared with ℓ_2-regularization, the ℓ_1-regularized objective function is more likely to find the optimal point at vertices which simultaneously meet the curves of the reconstruction error and regularization term. As a result, the optimal solution corresponding to the ℓ_1-regularized objective function likely happens to imply either $w_1 = 0$ or $w_2 = 0$ where the sparse parameters are obtained. In a general case of basis representation with K basis vectors, the sparse coding is implemented to select a relatively small set of relevant bases to represent a target mixed signal. Those irrelevant basis vectors correspond to the estimated weights $\{\widehat{w}_k\}$ which are close to zero. Basically, sparse learning is also interpreted as a regularized least-squares optimization subject to the constraint that the ℓ_1 regularizer is sufficiently small,

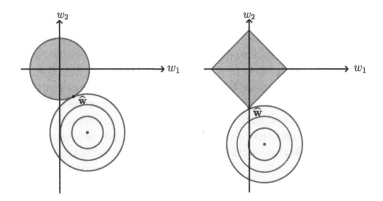

FIGURE 3.3

Optimal weights $\widehat{\mathbf{w}}$ estimated by minimizing the ℓ_2- and ℓ_1-regularized objective functions which are shown on the left and right, respectively.

e.g.,

$$\sum_{k=1}^{K} |w_k| \le \epsilon. \tag{3.14}$$

3.4.1 SPARSE BAYESIAN LEARNING

In general, the regularized least-squares (RLS) solution to data reconstruction problem is comparable with the maximum *a posteriori* (MAP) estimation where the posterior distribution, a combination of a likelihood function and a prior density, is maximized. To illustrate this comparison, we start from MAP estimation and rewrite the maximization problem as a minimization problem where the negative logarithm of the posterior distribution is minimized in a form of

$$\begin{aligned} \widehat{\mathbf{w}} &= \arg\min_{\mathbf{w}}\{-\log p(\mathbf{w}|\mathbf{x})\} \\ &= \arg\min_{\mathbf{w}}\{-\log p(\mathbf{x}|\mathbf{w}) - \log p(\mathbf{w})\}. \end{aligned} \tag{3.15}$$

Assume that the likelihood function of the reconstruction error $\mathbf{x} - \mathbf{Bw}$ is represented by a Gaussian distribution with zero mean vector and identity covariance matrix and the prior density of sensing weights is modeled by a Laplace distribution, i.e.,

$$p(\mathbf{x}|\mathbf{w}) = \mathcal{N}(\mathbf{x}|\mathbf{Bw}, \mathbf{I}) \tag{3.16}$$

and

$$p(\mathbf{w}|\lambda) = \frac{\lambda}{2}\exp(-\lambda\|\mathbf{w}\|_1). \tag{3.17}$$

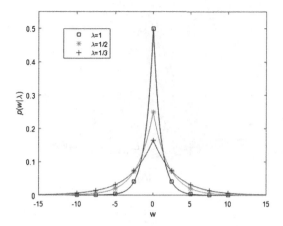

FIGURE 3.4

Laplace distribution centered at zero with different control parameters λ.

The MAP learning objective in Eq. (3.15) is equivalent to the RLS learning objective in Eq. (3.11). Fig. 3.4 shows a Laplace distribution having zero mean and different parameters λ. This Laplace distribution is a symmetric function with precision parameter 2λ. Typically, Laplace prior is seen as a sparse prior which produces a sparse solution of weight parameters $\mathbf{w}$ based on MAP estimation. This is a realization of sparse Bayesian learning (SBL) which is a combination of sparse learning and Bayesian learning. Bayesian sensing is performed to yield the error bars or distribution estimates of the true signals. In Chien (2015b), the sparse learning based on Laplace prior was proposed to construct acoustic models for speech recognition. Section 5.4 will address how the SBL based on Laplace prior was developed for single-channel separation of two music signals.

Nevertheless, there are different realizations of SBL. The earliest SBL was proposed in Tipping (2001) where a Gaussian prior with zero mean and diagonal covariance matrix, i.e.,

$$
\begin{aligned}
p(\mathbf{w}|\boldsymbol{\alpha}) &= \mathcal{N}(\mathbf{w}|\mathbf{0}, \text{diag}\{\alpha_k^{-1}\}) \\
&= \prod_{k=1}^{K} \mathcal{N}(w_k|0, \alpha_k^{-1}) \\
&\triangleq \prod_{k=1}^{K} p(w_k|\alpha_k),
\end{aligned}
\tag{3.18}
$$

was incorporated to express the sensing weights $\mathbf{w}$ in the presence of an overcomplete basis vectors. The precision parameter α_k of each weight w_k is known as an automatic relevance determination (ARD) parameter. For very large α_k, the resulting prior density $p(w_k|\alpha_k)$ approaches a Dirac delta function $\delta(w_k)$, which implies a very confident realization to the sparse solution at $w_k = 0$. This case means that the corresponding basis vector $\mathbf{b}_k$ is irrelevant to the mixed signal $\mathbf{x}$. The ARD parameter α_k naturally reflects how a mixed observation $\mathbf{x}$ is relevant to a basis vector $\mathbf{b}_k$ (Tipping, 2001). Under a overcomplete spanning set, this ARD parameter tends to have a large value. If ARD is modeled by

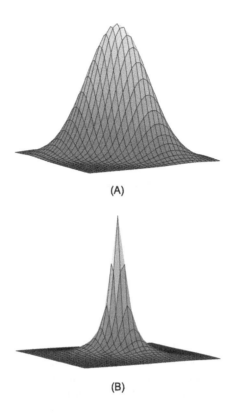

(A)

(B)

FIGURE 3.5

Comparison of two-dimensional Gaussian distribution (A) and Student's t-distribution (B) with zero mean.

a Gamma density $\mathrm{Gam}(\cdot)$ with parameters a and b, the marginal distribution of weight vector $\mathbf{w}$ turns out to be a Student's t-distribution which is verified by (Bishop, 2006, Tipping, 2001)

$$
\begin{aligned}
p(\mathbf{w}|a, b) &= \prod_{k=1}^{K} \int_{0}^{\infty} \mathcal{N}(w_k|0, \alpha_k^{-1}) \mathrm{Gam}(\alpha_k|a, b) d\alpha_k \\
&= \frac{b^a \Gamma\left(a + \frac{1}{2}\right)}{2\pi^{\frac{1}{2}} \Gamma(a)} \prod_{k=1}^{K} \left(b + \frac{w_k^2}{2}\right)^{-(a+1/2)}
\end{aligned}
\tag{3.19}
$$

where $\Gamma(\cdot)$ is the Gamma function defined by

$$
\Gamma(x) = \int_{0}^{\infty} t^{x-1} e^{-t} dt.
\tag{3.20}
$$

Although Gaussian distribution is assumed, the marginal distribution with respect to the Gamma prior is formed as a Student's t-distribution. Fig. 3.5 shows a comparison between the Gaussian and Student's

t-distributions. Student's t-distribution has a sharp peak which is easy to realize to its zero mean. This distribution is also known as a sparse distribution. In Saon and Chien (2012a), Chien (2015b), the basis matrix **B** and ARD parameters $\boldsymbol{\alpha}$ were jointly estimated for SBL by maximizing the marginal likelihood over a set of training samples **X**. SBL has been successfully developed for model-based source separation as detailed in Section 5.4.

3.5 ONLINE LEARNING

Real-world source separation involves a complicated time-varying mixing system where different scenarios of mixing conditions may happen due to the changing environments, including sources, microphones and reverberant rooms. Building a source separation system, which can truly understand the auditory scene and correctly identify the target signal, is a challenging task. Fig. 3.6 demonstrates the scenario of a nonstationary mixing system in the presence of three microphones and three sources $\{s_{t1}, s_{t2}, s_{t3}\}$. The distances of three sources relative to the centered microphone (or the second microphone) x_{t2} are denoted by $\{d_1, d_2, d_3\}$. The following scenario reflects three real-world nonstationary mixing conditions:

- moving sources or microphones,
- abruptly disappearing or appearing sources,
- source replacement,

which results in the time-varying mixing matrix $\mathbf{A} \rightarrow \mathbf{A}(t)$. In the beginning of this scenario, there are one male speaker s_{t1}, one female speaker s_{t2} and one music player s_{t3}. The mixing system of three sources relative to the second microphone is characterized by the mixing coefficients $\{a_{21}(t), a_{22}(t), a_{23}(t)\}$, which are time-varying in a nonstationary source separation system. At the next time step $t \rightarrow t+1$, the male speaker moves to a new location with an updated distance $d_1 \rightarrow d_1'$. Correspondingly, the mixing coefficient for this speaker is changed to a new value $a_{21}(t) \rightarrow a_{21}(t+1)$ while the coefficients for the other two sources remain the same, i.e., $a_{22}(t+1) = a_{22}(t)$ and $a_{23}(t+1) = a_{23}(t)$, because the distances d_2 and d_3 are unchanged. Moving to the next time step $t+1 \rightarrow t+2$, the male speaker leaves and the female speaker is replaced by a new speaker. For this case, the distance d_2 is changed to d_2' due to a new speaker. As a result, the mixing coefficient for the male speaker becomes zero, i.e., $a_{21}(t+2) = 0$, that of the female speaker is changed to a new value, i.e., $a_{22}(t+1) \rightarrow a_{22}(t+2)$, and that for the music player is unchanged, i.e., $a_{23}(t+2) = a_{23}(t+1)$. Notably, we carry out the sparse source separation for this scenario because the number of sources is decreased. The detection of missing source is based on the null mixing coefficient $a_{21}(t+2) = 0$. This is a realization of sparse source separation based on sparse learning mentioned in Section 3.4.1 where the estimated ARD parameter $\alpha_{21}(t+2)$ is sufficiently large so as to disregard the effect of the first source signal on the second microphone at time $t+2$. Basically, nonstationary source separation can be handled by continuously performing separation at different time moments. Online learning is feasible to deal with nonstationary source separation by using sequentially available observation data.

Online learning is a method of machine learning in which data becomes available sequentially and is used to update our best predictor for future data at each step, as opposed to batch learning techniques which generate the best predictor by learning the entire training data set at once. Online learning is also used in situations where it is necessary for the algorithms to dynamically adapt to new

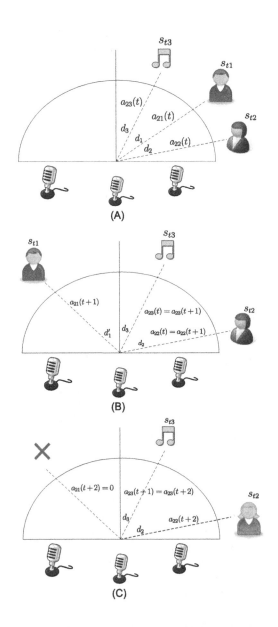

FIGURE 3.6

Scenario for a time-varying source separation system with (A) one male, one female and one music player.
(B) Then, the male is moving to a new location. (C) After a while, the male disappears and the female is replaced by
a new one.

patterns in the data. The model is updated in a scalable fashion. The more data are observed, the larger model is updated. Instead of updating parameters Θ in batch mode using cost function $E = \sum_t E_t$ from all samples $\mathbf{X} = \{\mathbf{x}_t\}_{t=1}^T$, the online or stochastic learning using gradient descent algorithm is performed according to the instantaneous cost function E_t from an individual sample $\mathbf{x}_t$ at time t or the accumulated cost function

$$E_n(\Theta) = \sum_{t \in \mathbf{X}^{(n)}} E_t(\Theta) \tag{3.21}$$

from a minibatch data $\mathbf{X}^{(n)} \subset \mathbf{X}$ as follows:

$$\Theta^{(n+1)} = \Theta^{(n)} - \eta \nabla E_n(\Theta^{(n)}). \tag{3.22}$$

3.5.1 ONLINE BAYESIAN LEARNING

In addition, Bayesian theory provides a meaningful solution to uncertainty modeling, as well as sequential learning. Using online Bayesian learning, we attempt to incrementally characterize the variations of model parameters $\Theta^{(n)}$ from the observed frames

$$\mathcal{X}^{(n)} = \{\mathbf{X}^{(1)}, \mathbf{X}^{(2)}, \dots, \mathbf{X}^{(n)}\} \tag{3.23}$$

through the stages of *prediction* and *correction*. In the prediction stage, the model parameters $\Theta^{(n)}$ using minibatch data $\mathbf{X}_n$ are predicted according to the posterior distribution given the previous minibatches $\mathcal{X}^{(n-1)}$, that is,

$$\begin{aligned} & p(\Theta^{(n)}|\mathcal{X}^{(n-1)}) \\ & = \int p(\Theta^{(n)}|\Theta^{(n-1)}) p(\Theta^{(n-1)}|\mathcal{X}^{(n-1)}) d\Theta^{(n-1)}, \end{aligned} \tag{3.24}$$

which is obtained by integrating over the uncertainty of previous parameters $\Theta^{(n-1)}$. Eq. (3.24) is known as the predictive distribution which is essential in the full Bayesian framework (Bishop, 2006). In the correction stage, when a new minibatch of mixed signals $\mathbf{X}_n$ is observed, the posterior distribution is corrected by (Spragins, 1965)

$$p(\Theta^{(n)}|\mathcal{X}^{(n)}) = \frac{p(\mathbf{X}_n|\Theta^{(n)}) p(\Theta^{(n)}|\mathcal{X}^{(n-1)})}{\int p(\mathbf{X}_n|\Theta^{(n)}) p(\Theta^{(n)}|\mathcal{X}^{(n-1)}) d\Theta^{(n)}}, \tag{3.25}$$

which is proportional to the product of a likelihood function of current minibatch $\mathbf{X}_n$ and an *a posteriori* distribution given the previous frames $\mathcal{X}^{(n-1)}$. At each minibatch n, the posterior distribution $p(\Theta^{(n)}|\mathcal{X}^{(n-1)})$ is seen as a prior distribution $p(\Theta^{(n)}|\Psi^{(n-1)})$ with hyperparameters $\Psi^{(n-1)}$ which are updated from the previous data $\mathcal{X}^{(n-1)}$. We choose a conjugate prior so that the updated posterior $p(\Theta^{(n)}|\mathcal{X}^{(n)})$ has the same distribution form as its prior density $p(\Theta^{(n)}|\Psi^{(n-1)})$. The reproducible prior/posterior distribution pair is formed for incremental learning of the posterior distribution or the prior density with hyperparameters

$$\Psi^{(0)} \to \Psi^{(1)} \to \cdots \to \Psi^{(n)}. \tag{3.26}$$

FIGURE 3.7

Sequential updating of parameters and hyperparameters for online Bayesian learning.

With the updated hyperparameters, the system parameters correspond to the modes of posterior distribution and can be realized in an online manner as

$$\Theta^{(0)} \rightarrow \Theta^{(1)} \rightarrow \cdots \rightarrow \Theta^{(n)}. \tag{3.27}$$

The newest environments are characterized for nonstationary source separation. A recursive Bayesian algorithm with three layers of variables is depicted in Fig. 3.7. Different from batch learning (Choudrey and Roberts, 2003, Hirayama et al., 2007), we present an online learning approach to detect the activities of source signals and estimate the distributions of reconstructed sources for each minibatch n. After updating the hyperparameters, the current minibatch $\mathbf{X}_n$ is abandoned and only the sufficient statistics $\mathbf{\Psi}^{(n)}$ is stored and propagated to the next learning epoch $n + 1$. Online learning is crucial for nonstationary source separation. Sections 4.4 and 4.5 will address the details how online Bayesian learning is developed for ICA-based blind source separation.

3.6 DISCRIMINATIVE LEARNING AND DEEP LEARNING

Unlike regression and classification problems, which usually have a single input and a single output, the goal of a monaural source separation system is to recover multiple source signals from a single channel of mixed signal. Treating the reconstructed signals independently may lose some crucial information during source separation. The discrimination between source signals could be learned to improve the system performance. The information from the related source signals is beneficial for source separation. This section introduces different discriminative learning methods to catch such information for monaural source separation based on nonnegative matrix factorization as well as deep neural network.

3.6.1 DISCRIMINATIVE NMF

Standard NMF is a purely generative model without taking discriminative criterion into account. The estimation of basis vectors is performed individually for various source signals. However, it is difficult to directly conduct discriminative learning because of the generative property in the learning objective based on the divergence measure which was addressed in Section 2.2.2. To pursue the discrimination in the trained basis vectors, the training criterion in NMF is modified to catch this information by characterizing the mutual relation between two source signals. NMF can be improved by imposing the regularization due to between-source discrimination. The main idea in discriminative NMF (DNMF) is that we want to estimate the basis vectors for one source, which could reconstruct this source signal very well but at the same time represent the other source signal very poorly. In other words, the individual sets of basis vectors are estimated to represent the corresponding source signals where the discrimination between source signals is enhanced. There are several related works proposed in the literature (Zafeiriou et al., 2006, Grais and Erdogan, 2013, Boulanger-Lewandowski et al., 2012, Wang and Sha, 2014, Weninger et al., 2014b) which are related to discriminative NMF. In what follows, we address three discriminative learning methods for construction of NMF models. One was developed for increasing the separability between classes for a pattern classification problem and the other two discriminative NMFs were exploited for enhancing the separability between source signals, which worked for monaural source separation with a regression problem.

DNMF-I

First of all, a discriminative NMF (Zafeiriou et al., 2006) was developed by incorporating the linear discriminant analysis (LDA) into enhancing the discrimination between classes and adopting this discriminant information in an NMF-based face recognition system. An NMF plus LDA method was proposed according to a hybrid learning objective with two terms. The first term is the reconstruction error for observation data based on NMF $\mathbf{X} \approx \mathbf{BW}$ while the second term is the Fisher discriminant function which has been successfully applied for face recognition (Belhumeur et al., 1997, Chien and Wu, 2002). The learning objective is yielded as a kind of error function written as a regularized divergence measure

$$E(\mathbf{B}, \mathbf{W}) = \mathcal{D}_{\mathrm{KL}}(\mathbf{X} \| \mathbf{BW}) + \lambda_w \mathrm{tr}[\mathbf{S}_w] - \lambda_s \mathrm{tr}[\mathbf{S}_b] \tag{3.28}$$

where λ_w and λ_b denote the regularization parameters for the within-class scatter matrix $\mathbf{S}_w$ and between-class scatter matrix $\mathbf{S}_b$ of Fisher discriminant function, respectively, defined by

$$\mathbf{S}_w = \sum_{c=1}^{C} \sum_{n=1}^{N_c} \left(\widetilde{\mathbf{x}}_n^{(c)}(\mathbf{W}) - \boldsymbol{\mu}^{(c)} \right) \left(\widetilde{\mathbf{x}}_n^{(c)}(\mathbf{W}) - \boldsymbol{\mu}^{(c)} \right)^{\top}, \tag{3.29}$$

$$\mathbf{S}_b = \sum_{c=1}^{C} N_c \left(\boldsymbol{\mu}^{(c)} - \boldsymbol{\mu} \right) \left(\boldsymbol{\mu}^{(c)} - \boldsymbol{\mu} \right)^{\top}. \tag{3.30}$$

In Eqs. (3.29)–(3.30), C denotes the number of classes, N_c denotes the number of observations of the cth class, $\widetilde{\mathbf{x}}_n^{(c)}$ denotes the transformed feature vector of the nth observation in the cth class using weight matrix $\mathbf{W}$, $\boldsymbol{\mu}^{(c)}$ denotes the mean vectors of those feature vectors $\widetilde{\mathbf{x}}_n^{(c)}$ corresponding to the cth class and $\boldsymbol{\mu}$ denotes the mean vector of the whole set of feature vectors $\widetilde{\mathbf{x}}_n^{(c)}$ in different observation n and different class c. In this implementation, the discriminant constraint is merged inside the NMF decomposition to extract the features that enforce not only the reconstruction for spatial locality (due to the first term on the RHS of Eq. (3.28)) but also the separation between classes in a discriminant manner (due to the second and third terms on the RHS of Eq. (3.28)). Notably, minimizing the learning objective in Eq. (3.28) does not only preserve the locality information by minimizing the reconstruction error but also enlarge the separation between classes by simultaneously minimizing the within-class scattering and maximizing between-class scattering. Accordingly, a decomposition of an observation set $\mathbf{X}$ to its discriminant parts is obtained and also the new updates for both the weights $\mathbf{B}$ and the bases $\mathbf{W}$ are derived. Nevertheless, such a discriminative NMF is only feasible for classification problem and could not apply for source separation.

DNMF-II

To tackle the regression problem, in Wang and Sha (2014), the learning objective in a discriminative NMF for single-channel source separation in the presence of m sources or speakers using n sentences was constructed by

$$E(\mathbf{B}_a, \mathbf{W}_{ai}) = \lambda_e E_1(\mathbf{B}_a, \mathbf{W}_{ai}) + E_2(\mathbf{B}_a, \mathbf{W}_{ai})$$
$$+ \lambda_w \alpha \sum_{a=1}^{m} \sum_{i=1}^{n} \text{tr}[\mathbf{W}_{ai} \mathbf{1}] \quad (3.31)$$

where λ_e denotes the regularization parameter for the tradeoff between $E_1(\cdot)$ and $E_2(\cdot)$ and λ_w is used to tune the regularization term (the third term of RHS) which is basically a sum over all weight parameters. The ℓ_1-norm regularization is performed to promote the sparseness in weight matrix $\mathbf{W}_{ai}$ corresponding to different source a and sentence i. In Eq. (3.31), the first term is similar to standard NMF criterion like Eq. (2.26), which is imposed to learn the basis matrix $\mathbf{B}_a$ corresponding to each source a independently as defined by

$$E_1(\mathbf{B}_a, \mathbf{W}_{ai}) = \sum_{a=1}^{m} \sum_{i=1}^{n} \mathcal{D}_{\text{KL}}(\mathbf{X}_{ai} \| \mathbf{B}_a \mathbf{W}_{ai}) \quad (3.32)$$

where $\mathbf{X}_{ai}$ represents the Mel-spectral data matrix of the ith sentence of source a and $\mathbf{B}_a$ represents the basis matrix for source a. Importantly, the second term $E_2(\cdot)$ is incorporated to conduct discriminative training. The goal using this term aims to minimize the reconstruction error between the artificial mixed signals and the reconstructed signals. This discriminative term is defined by

$$E_2(\mathbf{B}_a, \mathbf{W}_{ai})$$
$$= \sum_{a=1}^{m} \sum_{b \neq a}^{m} \sum_{i=1}^{n} \sum_{j=1}^{n} \mathcal{D}_{\text{KL}}(\mathbf{X}_{abij} \| \mathbf{B}_a \mathbf{W}_{ai} + \mathbf{B}_b \mathbf{W}_{bj}) \quad (3.33)$$

where $\mathbf{X}_{abij}$ represents the Mel-spectra of the artificial mixed signal using the ith sentence of source a and the jth sentence of source b. In Eq. (3.31), α is also a scaling factor to balance the contribution from each term. In Wang and Sha (2014), $\alpha = nm$ was chosen to reflect the number of times any $\mathbf{B}_a \mathbf{W}_{ai}$ occurring in $E(\cdot)$. The multiplicative updating rules for speaker-dependent parameters based on this discriminative NMF are yielded by (Wang and Sha, 2014)

$$\mathbf{B}_a \leftarrow \mathbf{B}_a \odot \frac{\sum_{i=1}^{n} \sum_{b=1}^{m} \sum_{j=1}^{n} (\mathbf{V}_{abij} \mathbf{W}_{ai}^{\top} + \mathbf{1}(\mathbf{1}\mathbf{W}_{ai}^{\top} \odot \mathbf{B}_a) \odot \mathbf{B}_a)}{\sum_{i=1}^{n} \sum_{b=1}^{m} \sum_{j=1}^{n} (\mathbf{1}\mathbf{W}_{ai}^{\top} + \mathbf{1}(\mathbf{V}_{abij} \mathbf{W}_{ai}^{\top} \odot \mathbf{B}_a) \odot \mathbf{B}_a)}, \quad (3.34)$$

$$\mathbf{W}_{ai} \leftarrow \mathbf{W}_{ai} \odot \frac{\mathbf{B}_a^{\top} \sum_{b=1}^{m} \sum_{j=1}^{n} \mathbf{V}_{abij}}{nm(\mathbf{B}_a^{\top} \mathbf{1} + \lambda_w)} \quad (3.35)$$

where

$$V_{abij} = \frac{X_{abij}}{(BW)_{abij}}. \tag{3.36}$$

Basically, the solution to minimization of Eq. (3.31) physically pursues the optimal reconstruction of individual utterances from individual speakers and at the same time the optimal reconstruction of mixed utterances from different sentences and different speakers. All combinations of mixing conditions between different sources using different sentences are taken into account so as to enforce the degree of discrimination among different speakers. Basically, the speaker-specific basis vectors are jointly learned. The discriminative learned basis vectors are able to reconstruct signals when they are clean as well as when they are mixed. Algorithm 3.1 demonstrates the learning procedure of this second type of discriminative NMF.

Algorithm 3.1 Discriminative Nonnegative Matrix Factorization-II.

Initialize with bases and weights $\{B_a^{(0)}, W_{ai}^{(0)}\}$
For each iteration τ
    For each source a
        Update $B_a^{(\tau)}$ of DNMF using Eq. (3.34)
        For each sentence i
            Update $W_{ai}^{(\tau)}$ of DNMF using Eq. (3.35)
        $i \leftarrow i + 1$
    $a \leftarrow a + 1$
    Check convergence
    $\tau \leftarrow \tau + 1$
Return $\{B_a, W_{ai}\}$

In system evaluation, this discriminative NMF was evaluated for speech separation in the presence of 34 speakers, each speaking 1000 short sentences (Wang and Sha, 2014). Basis vectors in **B** and weight vectors in **V** were randomly initialized. The maximum number of updating iterations was set to 100. The evaluation was performed in terms of signal-to-interference ratio (SIR) on Mel-spectra and SIR on the reconstructed speech waveforms which were inverted from Mel-spectra. SIR was defined in Eq. (2.10). Compared with the sparse NMF (Hoyer, 2004) as addressed in Section 2.2.3, this DNMF improved SIRs by a large margin in different settings.

DNMF-III

On the other hand, another type of discriminative learning for NMF was proposed in Weninger et al. (2014b). This method is implemented in two stages. The first is to perform standard NMF training which was mentioned in Section 2.2 based on KL divergence. The second is to tune the base matrices B_a for different sources a in order to make them more discriminative. The objective function is constructed by

$$E(B_a) = \sum_{a=1}^{m} \lambda_a \mathcal{D}_{KL}\left(X_a \middle\| X \odot \frac{B_a \widehat{W}_a}{B\widehat{W}}\right) \tag{3.37}$$

where $\{\widehat{\mathbf{B}}, \widehat{\mathbf{W}}\}$ and $\{\widehat{\mathbf{B}}_a, \widehat{\mathbf{W}}_a\}$ are obtained by standard NMF using the mixed signal $\mathbf{X}$ and the source signal $\mathbf{X}_a$, respectively. Here, the estimated weight matrices $\{\widehat{\mathbf{W}}, \widehat{\mathbf{W}}_a\}$ for mixed signal $\mathbf{X}$ and source signal $\mathbf{X}_a$ are provided. This work was developed for speech enhancement (Weninger et al., 2014b). In general, the learning objective based on Eq. (3.37) promotes a solution which minimizes the regression error by applying Wiener filtering for reconstruction over individual source signals $\{\mathbf{X}_a\}_{a=1}^m$. Such a Wiener filtering with Wiener weight $\frac{\mathbf{B}_a\widehat{\mathbf{W}}_a}{\mathbf{B}\widehat{\mathbf{W}}}$ is constructed and related to different source signals $\{\mathbf{X}_a\}_{a=1}^m$. The Wiener filtering is used to reconstruct each source signal $\mathbf{S}_a$ while ensuring that the source estimates sum to the mixture signal $\mathbf{X}$, that is,

$$\widehat{\mathbf{Y}}_a = \mathbf{X} \odot \frac{\mathbf{B}_a\widehat{\mathbf{W}}_a}{\sum_{a=1}^m \mathbf{B}_a\widehat{\mathbf{W}}_a} \tag{3.38}$$

such that $\widehat{\mathbf{Y}}_a \approx \mathbf{S}_a$. This is a solution to discriminative NMF because the separation for different sources is *jointly* optimized and the individuality or the *discrimination* between different sources is preserved for source separation. Discriminative NMF bases are trained such that a desired source is optimally recovered in test time. In Weninger et al. (2014b), although both the speech ($a = 1$) and noise signals ($a = 2$) are estimated, we are only concerned about the reconstructed speech signal. By setting $\lambda_a = 1$, $\lambda_{\backslash a} = 0$ and defining

$$\mathbf{Y} = \sum_a \mathbf{B}_a\mathbf{W}_a, \tag{3.39}$$

$$\mathbf{Y}_a = \mathbf{B}_a\mathbf{W}_a \tag{3.40}$$

for $a \in \{1, \ldots, m\}$ and

$$\mathbf{Y}_{\backslash a} = \mathbf{Y} - \mathbf{Y}_a, \tag{3.41}$$

the updating rules for basis matrices were derived by (Weninger et al., 2014b)

$$\mathbf{B}_a \leftarrow \mathbf{B}_a \odot \frac{\frac{\mathbf{X}_a \odot \mathbf{Y}_{\backslash a}}{\mathbf{Y} \odot \mathbf{Y}_{\backslash a}}\mathbf{W}_a^\top}{\frac{\mathbf{X} \odot \mathbf{Y}_{\backslash a}}{\mathbf{Y}^2}\mathbf{W}_a^\top}, \tag{3.42}$$

$$\mathbf{B}_{\backslash a} \leftarrow \mathbf{B}_{\backslash a} \odot \frac{\frac{\mathbf{X} \odot \mathbf{Y}_a}{\mathbf{Y}^2}\mathbf{W}_{\backslash a}^\top}{\frac{\mathbf{X}_a}{\mathbf{Y}}\mathbf{W}_{\backslash a}^\top} \tag{3.43}$$

where

$$\mathbf{B}_{\backslash a} \triangleq [\mathbf{B}_1 \cdots \mathbf{B}_{a-1}\mathbf{B}_{a+1} \cdots \mathbf{B}_m] \tag{3.44}$$

denotes the bases of all sources except a, and $\mathbf{W}_{\backslash a}$ is defined accordingly. Algorithm 3.2 illustrates the learning procedure of this third type of discriminative NMF (DNMF-III).

In system evaluation, the task of single-channel source separation was evaluated by using the speech corpus of the 2nd CHiME Speech Separation and Recognition Challenge (Weninger et al., 2014b). This task is to separate the pure speech signals from the noisy and reverberated mixture signals where

---

Algorithm 3.2 Discriminative Nonnegative Matrix Factorization-III.

Initialize with bases and weights $\{\mathbf{B}_a^{(0)}, \mathbf{W}_a^{(0)}\}$

For each iteration τ

    For each source a

        Update $\mathbf{B}_a^{(\tau)}$ of NMF using Eq. (2.27)

        Update $\mathbf{W}_a^{(\tau)}$ of NMF using Eq. (2.28)

        $a \leftarrow a + 1$

    Check convergence

    $\tau \leftarrow \tau + 1$

For each iteration τ

    Update $\{\mathbf{W}_a^{(\tau)}\}_{a=1}^m$ of NMF using Eq. (2.28)

    Update $\mathbf{B}_a^{(\tau)}$ of DNMF using Eq. (3.42)

    Update $\mathbf{B}_{\backslash a}^{(\tau)}$ of DNMF using Eq. (3.43)

    Check convergence

    $\tau \leftarrow \tau + 1$

Return $\{\mathbf{B}_a, \mathbf{W}_a\}$

---

the sources of speech and noise signals are present. The training set consists of 7138 Wall Street Journal (WSJ) utterances at six signal-to-noise ratios (SNRs) from -6 to 9 dB, in steps of 3 dB. The development and test sets consist of 410 and 330 utterances at each of these SNRs for a total of 2460 and 1980 utterances, respectively. The experimental results indicate that DNMF-III consistently performs better than sparse NMF (Hoyer, 2004) in terms of signal-to-distortion ratio (SDR) (Vincent et al., 2006).

3.6.2 DISCRIMINATIVE DNN

DNN is basically a discriminative model, so it can be applied to conduct discriminative learning straightforwardly. For monaural source separation in the presence of two sources, the regression error criterion is formed by

$$E(\mathbf{w}) = \frac{1}{2} \sum_{t=1}^{T} \left\{ \|\widehat{\mathbf{x}}_{1,t}(\mathbf{w}) - \mathbf{x}_{1,t}\|^2 + \|\widehat{\mathbf{x}}_{2,t}(\mathbf{w}) - \mathbf{x}_{2,t}\|^2 \right\} \tag{3.45}$$

where $\{\widehat{\mathbf{x}}_{1,t}(\mathbf{w}), \widehat{\mathbf{x}}_{2,t}(\mathbf{w})\}$ are the magnitude spectrograms of the reconstructed signals of two sources at time t using DNN parameters $\mathbf{w}$, and $\{\mathbf{x}_{1,t}, \mathbf{x}_{2,t}\}$ are the magnitude spectrograms of the corresponding original signals. Fig. 2.19 shows an example of deep recurrent neural network (DRNN) for single-channel source separation with two sources. Eq. (3.45) is a standard regression error function from two sources based on the DNN model. However, this objective function only considers the regression errors from the reconstructed and original signals of the same source. If two sources have similar spectra, the source separation based on Eq. (3.45) could not perform well in terms of signal-to-interference ratio (SIR). Discriminative learning between two sources is ignored.

To compensate for this issue, a discriminative DRNN was proposed by minimizing a new discriminative objective function (Huang et al., 2014a, Huang et al., 2014b) given by

$$E(\mathbf{w}) = -(1 - \lambda) \log p_{12}(\mathbf{x}, \mathbf{w}) - \lambda \mathcal{D}(p_{12} \| p_{21}) \tag{3.46}$$

where λ is the regularization parameter for tuning the tradeoff between the two terms. In Eq. (3.46), p_{12} denotes the likelihood of the training data of two target source signals $\mathbf{x} = \{\mathbf{x}_{1,t}, \mathbf{x}_{2,t}\}$ under the assumption that the neural network computes the *within-source* regression error of each feature vector based on a Gaussian distribution of reconstruction error with zero mean and unit variance (like in Eq. (3.16))

$$\log p_{12}(\mathbf{x}, \mathbf{w})$$
$$\propto -\frac{1}{2} \sum_{t=1}^{T} \left\{ \|\widehat{\mathbf{x}}_{1,t}(\mathbf{w}) - \mathbf{x}_{1,t}\|^2 + \|\widehat{\mathbf{x}}_{2,t}(\mathbf{w}) - \mathbf{x}_{2,t}\|^2 \right\}, \tag{3.47}$$

which is in accordance with Eq. (3.45). Importantly, the second term $\mathcal{D}(p_{12} \| p_{21})$ is seen as a discriminative term which is calculated as a point estimate of KL divergence between likelihood models p_{12} and p_{21}. Different from p_{12}, p_{21} is computed as two *between-source* regression errors from two sources. As a result, we obtain

$$\mathcal{D}(p_{12} \| p_{21}) = \frac{1}{2} \sum_{t=1}^{T} \left\{ \|\widehat{\mathbf{x}}_{1,t}(\mathbf{w}) - \mathbf{x}_{2,t}\|^2 + \|\widehat{\mathbf{x}}_{2,t}(\mathbf{w}) - \mathbf{x}_{1,t}\|^2 \right.$$
$$\left. - \|\widehat{\mathbf{x}}_{1,t}(\mathbf{w}) - \mathbf{x}_{1,t}\|^2 - \|\widehat{\mathbf{x}}_{2,t}(\mathbf{w}) - \mathbf{x}_{2,t}\|^2 \right\}. \tag{3.48}$$

Combining Eqs. (3.46)–(3.48), the discriminative objective function is obtained as

$$E(\mathbf{w}) = \frac{1}{2} \sum_{t=1}^{T} \left\{ \|\widehat{\mathbf{x}}_{1,t}(\mathbf{w}) - \mathbf{x}_{1,t}\|^2 + \|\widehat{\mathbf{x}}_{2,t}(\mathbf{w}) - \mathbf{x}_{2,t}\|^2 \right.$$
$$\left. - \lambda \|\widehat{\mathbf{x}}_{1,t}(\mathbf{w}) - \mathbf{x}_{2,t}\|^2 - \lambda \|\widehat{\mathbf{x}}_{2,t}(\mathbf{w}) - \mathbf{x}_{1,t}\|^2 \right\}. \tag{3.49}$$

The first two terms are the same as in Eq. (3.45), which are used to calculate the within-source reconstruction error. The third and fourth terms can be regarded as a calculation of the between-source reconstruction error. These two terms were added to improve the separation results in terms of SIR, and maintain similar or better performance in terms of signal-to-distortion ratio (SDR) and source-to-artifacts ratio (SAR) (Vincent et al., 2006, Huang et al., 2014a). It is obvious that the discriminative objective function of DRNN in Eq. (3.49) fulfills a discriminative source separation requirement, which not only minimizes the within-source error function but also maximizes the between-source error function. Notably, such a discriminative learning for DRNN is comparable with that performed for NMF, which was applied for the classification problem addressed in the first discriminative NMF method (DNMF-I) in Section 3.6.1. Nevertheless, an advanced study on discriminative learning for DRNN will be further performed in Section 7.3.2.

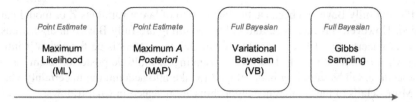

FIGURE 3.8

Evolution of different inference algorithms for construction of latent variable models.

3.7 APPROXIMATE INFERENCE

Machine learning for source separation involves an inference procedure to find parameters and hyperparameters in different models, including ICA, NMF, NTF, DNN and RNN. All these models can be represented by probabilistic graphical models and are recognized as the so-called latent variable models where a set of latent variables $\mathbf{Z}$ are introduced to express the generation of mixed signals $\mathbf{X}$ using the likelihood function $p(\mathbf{X}|\Theta)$. Parameters Θ and hyperparameters Ψ are themselves latent variables. However, some other latent variables may exist in model construction. For example, the Gaussian mixture model (GMM) and hidden Markov model (HMM) are both latent variable models where the likelihood functions using GMM and HMM parameters are expressed conditioning on mixture and state labels, which are seen as discrete latent variables. In model-based source separation, there are a number of latent variables in different models which should be considered during the model inference procedure. For example, using ICA, the demixing matrix $\mathbf{W}$ and source signals $\mathbf{S}$ are latent variables. The noise signal $\mathbf{E}$ and ARD parameter α in a noisy ICA model (Section 4.4) are also latent variable. Using NMF, both the basis matrix $\mathbf{B}$ and weight matrix $\mathbf{W}$ are latent variables. The factorized components $\mathbf{Z} = \{Z_{mkn}\}$ in NMF are also latent variables. Using DNN and RNN, the weight parameters $\mathbf{w}$ are latent variables. Model inference is carried out by maximizing the marginal likelihood where the latent variables are integrated by summing with respect to discrete latent variables $\mathbf{Z}$ or taking integral with respect to continuous latent variables $\mathbf{Z}$. The posterior distribution of latent variables should be analytical and factorizable during the inference procedure. The fundamentals of approximate inference are introduced to acquire analytical solutions for a latent variable model. In the following subsections, we will first address the maximum likelihood method and then focus on the Bayesian inference methods, including maximum *a posteriori*, variational inference and Gibbs sampling. Without loss of generality, the mixed signals $\mathbf{X}$, latent variables $\mathbf{Z}$, model parameters Θ and hyperparameters or variational parameters Ψ are expressed to illustrate general solutions to model-based source separation. The final separated signals are obtained by using the inferred latent variables or model parameters. The realizations of specialized solutions by using different source separation models will be detailed in Chapters 4, 5, 6 and 7.

In this section, we will address a number of approximate inference algorithms ranging from maximum likelihood (ML) estimation, maximum *a posteriori* (MAP) estimation, variational Bayesian (VB) inference and Gibbs sampling as shown in Fig. 3.8. Basically, ML and MAP estimation finds the point estimates Θ_{ML} and Θ_{MAP} by maximizing the likelihood function and posterior distribution, respectively. These point estimates are treated as deterministic values in the test phase for prediction of separated signals. MAP estimation is a Bayesian method due to the incorporation of a prior density

$p(\Theta)$. MAP is not fully Bayesian because the integration of latent variables $\mathbf{Z}$ or model parameters Θ is disregarded. However, VB inference and Gibbs sampling are fully Bayesian algorithms where the integration over uncertainty of latent variables or model parameters is performed. VB inference is an extension of ML or MAP, which deals with the coupling of multiple posterior latent variables in the inference procedure. Gibbs sampling is developed to take into account the uncertainty via a sampling procedure. More details of these inference algorithms are revealed in what follows.

3.7.1 MAXIMUM LIKELIHOOD AND EM ALGORITHM

This subsection addresses the general principle of maximum likelihood (ML) estimation and the detailed procedure of how Jensen's inequality is used to derive the expectation–maximization (EM) algorithm (Dempster et al., 1977). This algorithm provides an efficient training approach to tackle the incomplete data problem in ML estimation and derive a general solution of model parameter estimation problem using ML. The specialized ML solution to NMF-based source separation is addressed.

Maximum Likelihood Estimation

We assume that the matrix of a mixed signal spectra $\mathbf{X}$ is observed and is generated by some distribution $p(\mathbf{X}|\Theta)$ given by a set of model parameters Θ. The optimal ML parameters Θ_{ML} of a source separation model are estimated by maximizing the accumulated likelihood function from T frames of mixed signal spectra $\mathbf{X} = \{\mathbf{x}_t\}_{t=1}^{T}$, that is, by solving

$$\Theta_{\text{ML}} = \arg\max_{\Theta} p(\mathbf{X}|\Theta). \tag{3.50}$$

However, such ML estimation suffers from an incomplete data problem because the latent variables $\mathbf{Z}$ are missing in the optimization problem of Eq. (3.50). An incomplete data problem is difficult to solve. To circumvent this difficulty, the incomplete data $\mathbf{X}$ are combined with the latent variables $\mathbf{Z}$ to form the complete data $\{\mathbf{X}, \mathbf{Z}\}$ for model inference. The incomplete data problem in Eq. (3.50) is tackled by solving

$$\Theta_{\text{ML}} = \arg\max_{\Theta} \sum_{\mathbf{Z}} p(\mathbf{X}, \mathbf{Z}|\Theta) \tag{3.51}$$

where the discrete latent variables $\mathbf{Z}$ are assumed. For example, the NMF model with nonnegative parameters $\Theta = \{\mathbf{B}, \mathbf{W}\}$ can be expressed as a latent variable model in a form of

$$X_{mn} \approx [\mathbf{BW}]_{mn}$$
$$= \sum_{k} B_{mk} W_{kn} \triangleq \sum_{k} Z_{mkn}. \tag{3.52}$$

In Eq. (3.52), the temporal-frequency component X_{mn} of a mixed signal $\mathbf{X}$ is decomposed into K latent components $\{Z_{mkn}\}_{k=1}^{K}$ where $Z_{mkn} \in \mathbb{R}_+$ are not observed. The NMF learning problem is constructed by

$$\{\mathbf{B}_{\text{ML}}, \mathbf{W}_{\text{ML}}\}$$
$$= \arg\max_{\{\mathbf{B}, \mathbf{W}\} > 0} \sum_{\mathbf{Z}} p(\mathbf{X}|\mathbf{Z}) p(\mathbf{Z}|\mathbf{B}, \mathbf{W}). \tag{3.53}$$

Actually, the model parameters **B** and **W** are also latent variables, which contain continuous values. If we need to estimate the prior parameters of **B** and **W**, the marginal distribution in Eq. (3.53) is further integrated in a broad sense by taking an integral with respect to **B** and **W**. Section 5.3 will address this Bayesian NMF for monaural source separation.

EM Algorithm

The global optimum solution to Eq. (3.51) does not exist due to the missing data **Z**. In Dempster et al. (1977), the expectation–maximization (EM) algorithm was proposed to resolve the ML estimation problem by iteratively and alternatively performing the expectation step (E-step) and the maximization step (M-step), which result in a local optimum solution yielding model parameters Θ_{ML}. In the E-step, we calculate an auxiliary function $Q(\Theta|\Theta^{(\tau)})$, which is an expectation of the logarithm of the likelihood function using new separation parameters Θ given by the current parameters $\Theta^{(\tau)}$ at iteration τ, that is,

$$
\begin{aligned}
Q(\Theta|\Theta^{(\tau)}) &= \mathbb{E}_{\mathbf{Z}}\left[\log p(\mathbf{X}, \mathbf{Z}|\Theta)|\mathbf{X}, \Theta^{(\tau)}\right] \\
&= \sum_{\mathbf{Z}} p(\mathbf{Z}|\mathbf{X}, \Theta^{(\tau)}) \log p(\mathbf{X}, \mathbf{Z}|\Theta).
\end{aligned}
\tag{3.54}
$$

In the M-step, we maximize this auxiliary function instead of Eq. (3.51) with respect to source separation parameters Θ and estimate the new parameters at the new iteration $\tau + 1$ by

$$
\Theta^{(\tau+1)} = \arg\max_{\Theta} Q(\Theta|\Theta^{(\tau)}).
\tag{3.55}
$$

The updated parameters $\Theta^{(\tau+1)}$ are then treated as the current parameters for the next iteration of EM steps $\tau + 1 \rightarrow \tau + 2$. This iterative estimation only produces a local optimum solution, but not a global optimum solution. However, a careful setting of the initial model parameters would help the solution to reach appropriate parameter values and, moreover, the algorithm theoretically guarantees that the likelihood value is always increased as the number of iterations increases (Dempster et al., 1977). This property is very useful in the implementation, since we can easily debug the training source codes based on the EM algorithm by checking the likelihood values.

Now, we briefly prove how this indirect optimization for the auxiliary function $Q(\Theta|\Theta^{(\tau)})$ increases the likelihood function $p(\mathbf{X}|\Theta)$. Let us define the complete data as

$$
\mathbf{Y} \triangleq \{\mathbf{X}, \mathbf{Z}\}
\tag{3.56}
$$

and the log-likelihood function as

$$
L(\Theta) \triangleq \log p(\mathbf{X}|\Theta).
\tag{3.57}
$$

Since the logarithm function is monotonically increasing, evaluating the increase of $p(\mathbf{X}|\Theta)$ is the same as evaluating that of $L(\Theta)$. First, we find the equation

$$
p(\mathbf{Y}|\mathbf{X}, \Theta) = \frac{p(\mathbf{Y}, \mathbf{X}|\Theta)}{p(\mathbf{X}|\Theta)} = \frac{p(\mathbf{Y}|\Theta)}{p(\mathbf{X}|\Theta)}.
\tag{3.58}
$$

Then, we take the logarithm and take the expectation of both sides of Eq. (3.58) with respect to $p(\mathbf{Y}|\mathbf{X}, \mathbf{\Theta}^{(\tau)})$ to obtain

$$
\begin{aligned}
L(\mathbf{\Theta}) &= \log p(\mathbf{Y}|\mathbf{\Theta}) - \log p(\mathbf{Y}|\mathbf{X}, \mathbf{\Theta}) \\
&= \underbrace{\mathbb{E}_{\mathbf{Y}}\left[\log p(\mathbf{Y}|\mathbf{\Theta})|\mathbf{X}, \mathbf{\Theta}^{(\tau)}\right]}_{=Q(\mathbf{\Theta}|\mathbf{\Theta}^{(\tau)})} - \underbrace{\mathbb{E}_{\mathbf{Y}}\left[\log p(\mathbf{Y}|\mathbf{X}, \mathbf{\Theta})|\mathbf{X}, \mathbf{\Theta}^{(\tau)}\right]}_{\triangleq H(\mathbf{\Theta}|\mathbf{\Theta}^{(\tau)})}
\end{aligned}
\tag{3.59}
$$

where

$$
H(\mathbf{\Theta}|\mathbf{\Theta}^{(\tau)}) \triangleq \mathbb{E}_{\mathbf{Y}}\left[\log p(\mathbf{Y}|\mathbf{X}, \mathbf{\Theta})|\mathbf{X}, \mathbf{\Theta}^{(\tau)}\right].
\tag{3.60}
$$

Here, the second RHS in Eq. (3.59) is derived because

$$
\int p(\mathbf{Y}|\mathbf{X}, \mathbf{\Theta}^{(\tau)})L(\mathbf{\Theta})d\mathbf{Y} = L(\mathbf{\Theta}).
\tag{3.61}
$$

By applying Jensen's inequality for the convex function

$$
f(x) = -\log x,
\tag{3.62}
$$

we derive

$$
\begin{aligned}
H(\mathbf{\Theta}^{(\tau)}|\mathbf{\Theta}^{(\tau)}) - H(\mathbf{\Theta}|\mathbf{\Theta}^{(\tau)}) &= \mathbb{E}_{\mathbf{Y}}\left[\log \frac{p(\mathbf{Y}|\mathbf{X}, \mathbf{\Theta}^{(\tau)})}{p(\mathbf{Y}|\mathbf{X}, \mathbf{\Theta})}\middle|\mathbf{X}, \mathbf{\Theta}^{(\tau)}\right] \\
&= \int p(\mathbf{Y}|\mathbf{X}, \mathbf{\Theta}^{(\tau)})\left(-\log \frac{p(\mathbf{Y}|\mathbf{X}, \mathbf{\Theta})}{p(\mathbf{Y}|\mathbf{X}, \mathbf{\Theta}^{(\tau)})}\right)d\mathbf{Y} \\
&= \mathcal{D}_{\mathrm{KL}}\left(p(\mathbf{Y}|\mathbf{X}, \mathbf{\Theta}^{(\tau)})\middle\|p(\mathbf{Y}|\mathbf{X}, \mathbf{\Theta})\right) \\
&\geq -\log\left(\int p(\mathbf{Y}|\mathbf{X}, \mathbf{\Theta}^{(\tau)})\frac{p(\mathbf{Y}|\mathbf{X}, \mathbf{\Theta})}{p(\mathbf{Y}|\mathbf{X}, \mathbf{\Theta}^{(\tau)})}d\mathbf{Y}\right) = 0.
\end{aligned}
\tag{3.63}
$$

Substituting Eq. (3.63) into Eq. (3.59) yields

$$
\begin{aligned}
L(\mathbf{\Theta}) &= Q(\mathbf{\Theta}|\mathbf{\Theta}^{(\tau)}) - H(\mathbf{\Theta}|\mathbf{\Theta}^{(\tau)}) \\
&\geq Q(\mathbf{\Theta}^{(\tau)}|\mathbf{\Theta}^{(\tau)}) - H(\mathbf{\Theta}^{(\tau)}|\mathbf{\Theta}^{(\tau)}) = L(\mathbf{\Theta}^{(\tau)}).
\end{aligned}
\tag{3.64}
$$

This result is obtained because of the following property:

$$
\begin{aligned}
&Q(\mathbf{\Theta}|\mathbf{\Theta}^{(\tau)}) \geq Q(\mathbf{\Theta}^{(\tau)}|\mathbf{\Theta}^{(\tau)}) \\
&\Rightarrow L(\mathbf{\Theta}) \geq L(\mathbf{\Theta}^{(\tau)}) \\
&\Rightarrow p(\mathbf{X}|\mathbf{\Theta}) \geq p(\mathbf{X}|\mathbf{\Theta}^{(\tau)}).
\end{aligned}
\tag{3.65}
$$

Namely, an increase of the auxiliary function $Q(\Theta|\Theta^{(\tau)})$ implies an increase of the log-likelihood function $L(\Theta)$ and hence an increase of the likelihood function $p(\mathbf{X}|\Theta)$. Maximizing the auxiliary function is bound to increase the likelihood function using updated parameters

$$\Theta^{(\tau)} \rightarrow \Theta^{(\tau+1)}. \tag{3.66}$$

Note that since it is not a direct optimization of the original likelihood function, the optimization of the auxiliary function leads to a local optimum solution to the ML parameter estimation problem.

Alternative View of EM

We introduce a distribution $q(\mathbf{Z})$ defined over the latent variables $\mathbf{Z}$. For any choice of $q(\mathbf{Z})$, it is straightforward to find the decomposition

$$\log p(\mathbf{X}|\Theta) = \mathcal{D}_{\mathrm{KL}}(q\|p) + \mathcal{L}(q, \Theta) \tag{3.67}$$

where

$$\begin{aligned} \mathcal{D}_{\mathrm{KL}}(q\|p) &= -\sum_{\mathbf{Z}} q(\mathbf{Z}) \log \left\{ \frac{p(\mathbf{Z}|\mathbf{X}, \Theta)}{q(\mathbf{Z})} \right\} \\ &= -\mathbb{E}_q[\log p(\mathbf{Z}|\mathbf{X}, \Theta)] - \mathbb{H}_q[\mathbf{Z}], \end{aligned} \tag{3.68}$$

$$\begin{aligned} \mathcal{L}(q, \Theta) &= \sum_{\mathbf{Z}} q(\mathbf{Z}) \log \left\{ \frac{p(\mathbf{X}, \mathbf{Z}|\Theta)}{q(\mathbf{Z})} \right\} \\ &= \mathbb{E}_q[\log p(\mathbf{X}, \mathbf{Z}|\Theta)] + \mathbb{H}_q[\mathbf{Z}]. \end{aligned} \tag{3.69}$$

Here,

$$\mathbb{H}_q[\mathbf{Z}] \triangleq -\sum_{\mathbf{Z}} q(\mathbf{Z}) \log q(\mathbf{Z}) \tag{3.70}$$

is the entropy of $q(\mathbf{Z})$. Maximizing the log-likelihood $\log p(\mathbf{X}|\Theta)$ is equivalent to maximizing the sum of a KL divergence between $q(\mathbf{Z})$ and the posterior distribution $p(\mathbf{Z}|\mathbf{X}, \Theta)$ and a lower bound $\mathcal{L}(q, \Theta)$ of the log-likelihood function $\log p(\mathbf{X}|\Theta)$. The term $\mathcal{L}(q, \Theta)$ is called a lower bound because KL divergence is always nonnegative, i.e.,

$$\mathcal{D}_{\mathrm{KL}}(q\|p) \geq 0 \tag{3.71}$$

as illustrated in Eq. (3.63). Fig. 3.9(A) illustrates a decomposition of the log-likelihood function into a KL divergence and a lower bound. The maximum likelihood estimation of model parameters Θ is indirectly solved through seeking an approximate distribution $q(\mathbf{Z})$. This indirect estimation procedure is divided into two steps where the distribution of latent variables $q(\mathbf{Z})$ and the model parameters Θ are alternatively and iteratively updated. Given the current estimate $\Theta^{(\tau)}$, the first step is to estimate

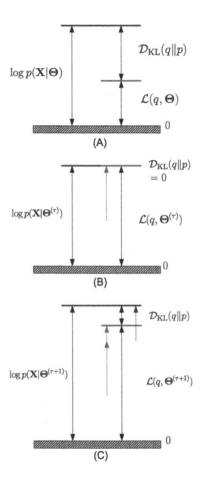

FIGURE 3.9

Illustration for (A) decomposing a log-likelihood into a KL divergence and a lower bound, (B) updating the lower bound by setting KL divergence to zero with new distribution $q(\mathbf{Z})$, and (C) updating lower bound again by using new parameters $\boldsymbol{\Theta}$. Adapted from Bishop (2006).

the distribution $q(\mathbf{Z})$ by setting $\mathcal{D}_{\text{KL}}(q \| p) = 0$, which results in

$$q(\mathbf{Z}) = p(\mathbf{Z}|\mathbf{X}, \boldsymbol{\Theta}^{(\tau)}). \tag{3.72}$$

This step is to specify the distribution of latent variables as the posterior distribution using the current parameters $\boldsymbol{\Theta}^{(\tau)}$ at iteration τ. The resulting lower bound is increased as shown in Fig. 3.9(B). Next, the second step is to fix $q(\mathbf{Z})$ and update the model parameters as

$$\boldsymbol{\Theta}^{(\tau+1)} \leftarrow \boldsymbol{\Theta}^{(\tau)} \tag{3.73}$$

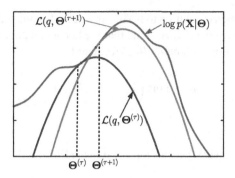

FIGURE 3.10

Illustration of updating lower bound of the log-likelihood function in EM iteration from τ to $\tau + 1$. Adapted from Bishop (2006).

through maximizing the lower bound $\mathcal{L}(q, \boldsymbol{\Theta})$ with respect to $\boldsymbol{\Theta}$. Importantly, this lower bound using $q(\mathbf{Z})$ in Eq. (3.72) is equivalent to the auxiliary function in the EM algorithm

$$
\begin{aligned}
\mathcal{L}(q, \boldsymbol{\Theta}) &= \mathbb{E}_q[\log p(\mathbf{X}|\mathbf{Z}, \boldsymbol{\Theta}) + \log p(\mathbf{Z}|\boldsymbol{\Theta})] + \text{const} \\
&= \mathbb{E}_{\mathbf{Z}}[\log p(\mathbf{X}|\mathbf{Z}, \boldsymbol{\Theta})|\mathbf{X}, \boldsymbol{\Theta}^{(\tau)}] + \text{const} \\
&= Q(\boldsymbol{\Theta}|\boldsymbol{\Theta}^{(\tau)}) + \text{const}
\end{aligned}
\tag{3.74}
$$

where the terms independent of $\boldsymbol{\Theta}$ are absorbed in the constant term. Using the new parameters $\boldsymbol{\Theta}^{(\tau+1)}$, the distribution of latent variables is further updated for next EM iteration of updating parameters, i.e.,

$$
\boldsymbol{\Theta}^{(\tau+2)} \leftarrow \boldsymbol{\Theta}^{(\tau+1)}.
\tag{3.75}
$$

In the continuous updating

$$
\boldsymbol{\Theta}^{(0)} \rightarrow \boldsymbol{\Theta}^{(1)} \rightarrow \boldsymbol{\Theta}^{(2)} \rightarrow \cdots,
\tag{3.76}
$$

the lower bound $\mathcal{L}(q, \boldsymbol{\Theta})$ is incrementally increased and the log-likelihood function $\log p(\mathbf{X}|\boldsymbol{\Theta})$ is respectively improved. Without loss of generality, we also call the first step of finding $q(\mathbf{Z})$ in Eq. (3.72) as the E-step. The second step of maximizing the lower bound $\mathcal{L}(q, \boldsymbol{\Theta})$ corresponds to the M-step in the EM algorithm. Basically, the original log-likelihood function $\log p(\mathbf{X}|\boldsymbol{\Theta})$ is nonconvex or nonconcave. Assuming the likelihood function comes from a Gaussian distribution, the lower bound or the auxiliary function in Eqs. (3.54) and (3.74) is concave. Fig. 3.10 illustrates how EM iterations pursue the maximum of log-likelihood for parameters $\boldsymbol{\Theta}_{\text{ML}}$ via continuous updates of the lower bound or, equivalently, the model parameters. The convergence of the EM procedure is assured owing to the nondecreasing auxiliary function.

ML Solution to NMF

Monaural source separation using NMF can be implemented in a nonparametric way minimizing the reconstruction error based on different divergence measures between $\mathbf{X}$ and $\mathbf{BW}$ as shown in Section 2.2.2. However, the parametric solution to NMF can be also formulated as the maximum likelihood estimation as addressed in Section 3.7.1. Assume that the latent component Z_{mkn} in Eq. (3.52) is represented by a Poisson distribution with mean $B_{mk} W_{kn}$, that is,

$$Z_{mkn} \sim \text{Pois}\left(Z_{mkn} \middle| B_{mk} W_{kn} \right) \tag{3.77}$$

where

$$\text{Pois}(x|\theta) = \exp(x \log \theta - \theta - \log \Gamma(x + 1)) \tag{3.78}$$

with a Gamma function $\Gamma(x + 1) = x!$. The log-likelihood function of the data matrix $\mathbf{X}$ given parameters Θ is then

$$
\begin{aligned}
\log p(\mathbf{X}|\mathbf{B}, \mathbf{W}) &= \log \prod_{m,n} \text{Pois}\left(X_{mn} \middle| \sum_k B_{mk} W_{kn} \right) \\
&= \sum_{m,n} \left(X_{mn} \log[\mathbf{BW}]_{mn} - [\mathbf{BW}]_{mn} \log \Gamma(X_{mn} + 1) \right).
\end{aligned} \tag{3.79}
$$

It is interesting to see that maximizing the log-likelihood function in Eq. (3.79) with respect to NMF parameters $\mathbf{B}$ and $\mathbf{W}$ is equivalent to minimizing the KL divergence between $\mathbf{X}$ and $\mathbf{BW}$ in Eq. (2.26). Such ML estimation with incomplete data problem $\mathbf{Z} = \{Z_{mkn}\}$ could be resolved by applying EM steps. In the E-step, an auxiliary function or expectation function of the log-likelihood of data $\mathbf{X}$ and latent variable $\mathbf{Z}$ given new parameters

$$\Theta = \{\mathbf{B}, \mathbf{W}\} \tag{3.80}$$

is calculated over $\mathbf{Z}$ with the current parameters

$$\Theta^{(\tau)} = \{\mathbf{B}^{(\tau)}, \mathbf{W}^{(\tau)}\} \tag{3.81}$$

at iteration τ

$$
\begin{aligned}
&\mathcal{Q}(\mathbf{B}, \mathbf{W}|\mathbf{B}^{(\tau)}, \mathbf{W}^{(\tau)}) \\
&\quad = \mathbb{E}_{\mathbf{Z}}[\log p(\mathbf{X}, \mathbf{Z}|\mathbf{B}, \mathbf{W})|\mathbf{X}, \mathbf{B}^{(\tau)}, \mathbf{W}^{(\tau)}].
\end{aligned} \tag{3.82}
$$

In the M-step, we maximize the auxiliary function, i.e., find

$$
\begin{aligned}
&(\mathbf{B}^{(\tau+1)}, \mathbf{W}^{(\tau+1)}) \\
&\quad = \arg \max_{\mathbf{B}, \mathbf{W} > 0} \mathcal{Q}(\mathbf{B}, \mathbf{W}|\mathbf{B}^{(\tau)}, \mathbf{W}^{(\tau)}),
\end{aligned} \tag{3.83}
$$

to obtain the updated solution at iteration $\tau + 1$, which is equivalent to that of standard NMF (Lee and Seung, 2000, Cemgil, 2009, Shashanka et al., 2008) as given in Eqs. (2.27)–(2.28) where Kullback–Leibler (KL) divergence between the mixed and reconstructed signals is minimized.

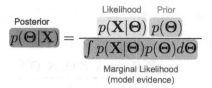

FIGURE 3.11

Bayes relation among posterior distribution, likelihood function, prior distribution and evidence function.

3.7.2 MAXIMUM *a Posteriori*

The optimization with respect to the posterior distribution $p(\Theta|\mathbf{X})$ is reasonable from a Bayesian perspective. This section first presents a general discussion of MAP approximation to obtain model parameters Θ. For simplicity, we first review the posterior distribution of model parameters given observations $\mathbf{X}$ without latent variables $\mathbf{Z}$. The MAP estimation focuses on the following parameter estimation:

$$\Theta_{\text{MAP}} = \arg\max_{\Theta} p(\Theta|\mathbf{X})$$
$$= \arg\max_{\Theta} \sum_{\mathbf{Z}} p(\mathbf{X}, \mathbf{Z}|\Theta) p(\Theta). \tag{3.84}$$

This corresponds to estimating the mode Θ_{MAP} of the posterior distribution given training data $\mathbf{X}$. The discrete latent variables $\mathbf{Z}$ are considered in a latent variable model. Using Bayes theorem, the posterior distribution, likelihood function, prior distribution and evidence function are related in Fig. 3.11. Since the evidence function

$$p(\mathbf{X}) = \int p(\mathbf{X}|\Theta) p(\Theta) d\Theta \tag{3.85}$$

does not depend on Θ, we can avoid computing this integral in MAP estimation. Thus, the estimation is based on the maximum likelihood function with the additional contribution from the prior distribution. The prior distribution behaves as a regularizer for model parameters in ML estimation. This is a Bayesian advantage over ML. Furthermore, if we use an exponential family distribution for a likelihood function and a conjugate distribution for a prior distribution, the MAP estimate is represented as the mode of the corresponding conjugate posterior distribution. This is an advantage of using conjugate distributions. In other words, it is not simple to obtain the mode of the posterior distribution, if we don't use the conjugate distribution, since we cannot obtain the posterior distribution analytically.

EM Algorithm

Similar to ML estimation, we develop EM algorithm for MAP estimation in the presence of missing data $\mathbf{Z}$ and illustrate how EM steps solve the missing data problem and lead to a local optimum Θ_{MAP}. There are two steps (E-step and M-step) in the MAP-EM algorithm. In each EM iteration, the current parameters $\Theta^{(\tau)}$ at iteration τ are updated to find new parameters Θ. First, the E-step is to calculate the posterior auxiliary function as

$$
\begin{aligned}
R(\Theta|\Theta^{(\tau)}) &= \mathbb{E}_{\mathbf{Z}}[\log p(\mathbf{X}, \mathbf{Z}, \Theta)|\mathbf{X}, \Theta^{(\tau)}] \\
&= \underbrace{\mathbb{E}_{\mathbf{Z}}[\log p(\mathbf{X}, \mathbf{Z}|\Theta)|\mathbf{X}, \Theta^{(\tau)}]}_{Q(\Theta|\Theta^{(\tau)})} \\
&\quad + \log p(\Theta)
\end{aligned}
\tag{3.86}
$$

where $\log p(\Theta)$ does not depend on $\mathbf{Z}$, and can be separated from the expectation function. Compared to the ML auxiliary function $Q(\Theta|\Theta^{(\tau)})$, we have an additional term $\log p(\Theta)$, which comes from a prior distribution of model parameters. Next, the M-step is performed to update the model parameters as

$$
\Theta^{(\tau)} \rightarrow \Theta^{(\tau+1)}
\tag{3.87}
$$

by maximizing the posterior auxiliary function

$$
\Theta^{(\tau+1)} = \arg\max_{\Theta} R(\Theta|\Theta^{(\tau)}).
\tag{3.88}
$$

The updated parameters $\Theta^{\tau+1}$ are then treated as the current parameters for the next iteration of EM steps.

To evaluate how the optimization of auxiliary function $R(\Theta|\Theta^{(\tau)})$ leads to a local optimum of $p(\mathbf{X}|\Theta)p(\Theta)$ or $p(\Theta|\mathbf{X})$, we first define a logarithm of the joint distribution

$$
L_{\text{MAP}}(\Theta) \triangleq \log p(\mathbf{X}|\Theta)p(\Theta).
\tag{3.89}
$$

Using the property

$$
p(\mathbf{X}|\Theta) = \frac{p(\mathbf{X}, \mathbf{Z}|\Theta)}{p(\mathbf{Z}|\mathbf{X}, \Theta)},
\tag{3.90}
$$

we obtain a new expression

$$
L_{\text{MAP}}(\Theta) = \log p(\mathbf{X}, \mathbf{Z}|\Theta) - \log p(\mathbf{Z}|\mathbf{X}, \Theta) + \log p(\Theta).
\tag{3.91}
$$

Similar to Eq. (3.59), an expectation operation is performed with respect to $p(\mathbf{Z}|\mathbf{X}, \Theta^{(\tau)})$ on both sides of Eq. (3.91) to yield

$$
\begin{aligned}
L_{\text{MAP}}(\Theta) &= \mathbb{E}_{\mathbf{Z}}[\log p(\mathbf{X}, \mathbf{Z}|\Theta)|\mathbf{X}, \Theta^{(\tau)}] - \mathbb{E}_{\mathbf{Z}}[\log p(\mathbf{Z}|\mathbf{X}, \Theta)|\mathbf{X}, \Theta^{(\tau)}] \\
&\quad + \log p(\Theta) \\
&= \underbrace{\mathbb{E}_{\mathbf{Z}}[\log p(\mathbf{X}, \mathbf{Z}|\Theta)|\mathbf{X}, \Theta^{(\tau)}] + \log p(\Theta)}_{=R(\Theta|\Theta^{(\tau)})} \\
&\quad - \underbrace{\mathbb{E}_{\mathbf{Z}}[\log p(\mathbf{Z}|\mathbf{X}, \Theta)|\mathbf{X}, \Theta^{(\tau)}]}_{\triangleq H(\Theta|\Theta^{(\tau)})}
\end{aligned}
\tag{3.92}
$$

where $L_{\text{MAP}}(\boldsymbol{\Theta})$ and $\log p(\boldsymbol{\Theta})$ are not changed since both terms do not depend on $\mathbf{Z}$. Thus, the posterior auxiliary function is expressed by

$$R(\boldsymbol{\Theta}|\boldsymbol{\Theta}^{(\tau)}) = L_{\text{MAP}}(\boldsymbol{\Theta}) + H(\boldsymbol{\Theta}|\boldsymbol{\Theta}^{(\tau)}). \tag{3.93}$$

Since $H(\boldsymbol{\Theta}|\boldsymbol{\Theta}^{(\tau)})$ has been shown to be a decreasing function, i.e.,

$$H(\boldsymbol{\Theta}|\boldsymbol{\Theta}^{(\tau)}) \leq H(\boldsymbol{\Theta}^{(\tau)}|\boldsymbol{\Theta}^{(\tau)}) \tag{3.94}$$

in Eq. (3.63), we prove that an increase of the posterior auxiliary function $R(\boldsymbol{\Theta}|\boldsymbol{\Theta}^{(\tau)})$ will result in an increase of the logarithm of the posterior distribution $L(\boldsymbol{\Theta})$ or an increase of the posterior distribution $p(\boldsymbol{\Theta}|\mathbf{X})$

$$R(\boldsymbol{\Theta}|\boldsymbol{\Theta}^{(\tau)}) \geq R(\boldsymbol{\Theta}^{(\tau)}|\boldsymbol{\Theta}^{(\tau)})$$
$$\Rightarrow L_{\text{MAP}}(\boldsymbol{\Theta}) \geq L_{\text{MAP}}(\boldsymbol{\Theta}^{(\tau)}) \tag{3.95}$$
$$\Rightarrow p(\boldsymbol{\Theta}|\mathbf{X}) \geq p(\boldsymbol{\Theta}^{(\tau)}|\mathbf{X}).$$

The MAP-EM algorithm is accordingly constructed and checked for convergence in a sense of local optimum for MAP parameters.

3.7.3 VARIATIONAL INFERENCE

Variational Bayesian (VB) inference has been developed in machine learning community around 1990s (Jordan et al., 1999, Attias, 1999) and is now becoming a standard technique to perform Bayesian inference for latent variable models based on the EM-like algorithm. This section starts with a general latent variable model having observation data $\mathbf{X} = \{\mathbf{x}_t\}_{t=1}^{T}$ and the set of all variables, including latent variables $\mathbf{Z}$, parameters $\boldsymbol{\Theta}$ and hyperparameters $\boldsymbol{\Psi}$. These variables are used to specify an assumed model topology $\mathcal{T}$ for data representation. The fundamental issue in Bayesian inference is to calculate the joint posterior distribution of any variables introduced in the problem

$$p(\mathbf{Z}|\mathbf{X}) \qquad \text{where } \mathbf{Z} = \{Z_j\}_{j=1}^{J}. \tag{3.96}$$

Once $p(\mathbf{Z}|\mathbf{X})$ is obtained, we can calculate the expectation functions Eq. (3.54) and Eq. (3.86) in the E-step of the EM algorithm and estimate various information in the ML and MAP estimation procedures, respectively. However, in real-world source separation, there are a number of latent variables which are coupled in the calculation of posterior distribution. For example, the components of the basis matrix $\{B_{mk}\}$ and weight matrix $\{W_{kn}\}$ are latent variables, which are correlated in the posterior distribution. An analytical calculation for the exact posterior distribution $p(\mathbf{Z}|\mathbf{X})$ does not exist. An approximate inference is introduced to deal with this problem based on the variational approach. The idea of variational inference is to use an approximate posterior distribution $q(\mathbf{Z})$, which can be factorized to act as a proxy to the true distribution $p(\mathbf{Z}|\mathbf{X})$. The concept of approximate inference using variational distribution $q(\mathbf{Z})$ is illustrated in Fig. 3.12. In what follows, we address how the variational inference is derived to find variational distribution $q(\mathbf{Z})$ which is close to the true posterior $p(\mathbf{Z}|\mathbf{X})$. The variational Bayesian expectation–maximization VB-EM algorithm is formulated to deal with ML or MAP estimation in the presence of multiple latent variables coupled in the posterior distribution.

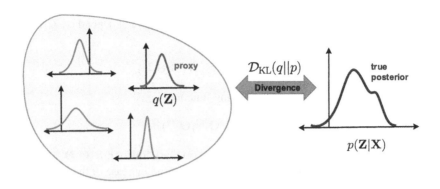

FIGURE 3.12

Illustration of seeking an approximate distribution $q(\mathbf{Z})$ which is factorizable and maximally similar to the true posterior $p(\mathbf{Z}|\mathbf{X})$.

Without loss of generality, the following formulation is shown with real-valued latent variables for simplicity. Integration of a probabilistic function with respect to latent variables is performed by using $\int d\mathbf{Z}$ instead of $\sum_{\mathbf{Z}}$.

Joint Posterior Distribution

As a measure of evaluating the distance between two distributions, Kullback–Leibler divergence (Kullback and Leibler, 1951) was introduced in the ML-EM algorithm as described in Section 3.7.1. Here, KL divergence between $q(\mathbf{Z})$ and $p(\mathbf{Z}|\mathbf{X})$ is defined by

$$
\begin{aligned}
\mathcal{D}_{\mathrm{KL}}(q(\mathbf{Z}) \| p(\mathbf{Z}|\mathbf{X})) &\triangleq \int q(\mathbf{Z}) \log \frac{q(\mathbf{Z})}{p(\mathbf{Z}|\mathbf{X})} d\mathbf{Z} \\
&= -\mathbb{E}_q[\log p(\mathbf{Z}|\mathbf{X})] - \mathbb{H}_q[\mathbf{Z}] \\
&= \int q(\mathbf{Z}) \log \frac{q(\mathbf{Z})}{\frac{p(\mathbf{X},\mathbf{Z})}{p(\mathbf{X})}} d\mathbf{Z} \\
&= \log p(\mathbf{X}) - \underbrace{\int q(\mathbf{Z}) \log \frac{p(\mathbf{X},\mathbf{Z})}{q(\mathbf{Z})} d\mathbf{Z}}_{\triangleq \mathcal{L}(q(\mathbf{Z}))}
\end{aligned}
\tag{3.97}
$$

where

$$
\begin{aligned}
\mathcal{L}(q(\mathbf{Z})) &\triangleq \int q(\mathbf{Z}) \log \frac{p(\mathbf{X},\mathbf{Z})}{q(\mathbf{Z})} d\mathbf{Z} \\
&= \mathbb{E}_q[\log p(\mathbf{X},\mathbf{Z})] + \mathbb{H}_q[\mathbf{Z}].
\end{aligned}
\tag{3.98}
$$

This functional is called the variational lower bound of $\log p(\mathbf{X})$ or the evidence lower bound (ELBO), denoted by $\mathcal{L}(q(\mathbf{Z}))$, because KL divergence is always nonnegative and therefore

$$\log p(\mathbf{X}) \geq \mathcal{L}(q(\mathbf{Z})). \tag{3.99}$$

This relation is consistent to what we had derived in Eqs. (3.67), (3.68) and (3.69). We can obtain the optimal variational distribution $q(\mathbf{Z})$ closest to the joint posterior distribution $p(\mathbf{Z}|\mathbf{X})$ by maximizing the variational lower bound $\mathcal{L}(q(Z|\mathbf{X}))$, which corresponds to minimizing the KL divergence since the log-evidence $\log p(\mathbf{X})$ does not depend on $\mathbf{Z}$. That is,

$$\widehat{q}(\mathbf{Z}) = \arg \max_{q(\mathbf{Z})} \mathcal{L}(q(\mathbf{Z})). \tag{3.100}$$

To implement an analytical inference algorithm, the solution to Eq. (3.100) should be derived subject to the constraint of a factorized posterior distribution. Therefore, the resulting inference procedure is known as the VB inference where an alternative posterior distribution $q(\mathbf{Z})$ is estimated as a surrogate of the true posterior $p(\mathbf{Z}|\mathbf{X})$ for finding ML and MAP model parameters, Θ_{ML} and Θ_{MAP}, respectively. This VB inference is also called the factorized variational inference because of the approximate inference for the factorization constraint.

Factorized Posterior Distribution

Assuming that the variational posterior distribution is conditionally independent over J latent variables, we have

$$q(\mathbf{Z}) = \prod_{j=1}^{J} q(Z_j). \tag{3.101}$$

For source separation using NMF, Z_i is any latent variable in $\{Z_{mkn}, B_{mk}, W_{kn}\}$. Note that we don't assume the factorization form for the true posterior $p(\mathbf{Z}|\mathbf{X})$. Instead, the true posterior of a target latent variable $p(Z_i|\mathbf{X})$ can be represented as the marginalized distribution of $p(\mathbf{Z}|\mathbf{X})$ over all Z_j except for Z_i as

$$p(Z_i|\mathbf{X}) = \int \cdots \int p(\mathbf{Z}|\mathbf{X}) \prod_{j \neq i}^{J} dZ_j \triangleq \int p(\mathbf{Z}|\mathbf{X}) dZ_{\backslash i} \tag{3.102}$$

where $Z_{\backslash i}$ denotes the complementary set of Z_i. By using Eq. (3.102), the KL divergence between $q(Z_i)$ and $p(Z_i|\mathbf{X})$ is represented as follows:

$$
\begin{aligned}
\mathcal{D}_{\mathrm{KL}}(q(Z_i)\|p(Z_i|\mathbf{X})) &= \int q(Z_i) \log \frac{q(Z_i)}{\int p(\mathbf{Z}|\mathbf{X}) dZ_{\backslash i}} dZ_i \\
&= \int q(Z_i) \log \frac{q(Z_i)}{\int \frac{p(\mathbf{X},\mathbf{Z})}{p(\mathbf{X})} dZ_{\backslash i}} dZ_i \\
&= \log p(\mathbf{X}) - \int q(Z_i) \log \frac{\int p(\mathbf{X},\mathbf{Z}) dZ_{\backslash i}}{q(Z_i)} dZ_i.
\end{aligned}
\tag{3.103}
$$

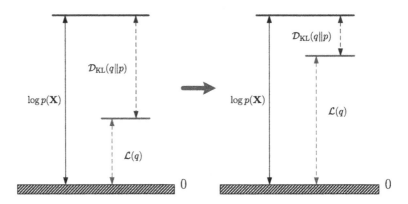

FIGURE 3.13

Illustration of minimizing KL divergence via maximization of variational lower bound. KL divergence is reduced from left to right during the optimization procedure.

Again, similar to Eq. (3.63), we can apply Jensen's inequality for the concave function $f(x) = \log x$ to get

$$
\begin{aligned}
\int q(Z_i) \log & \left(\frac{\int p(\mathbf{X}, \mathbf{Z}) dZ_{\backslash i}}{q(Z_i)} \right) dZ_i \\
&= \int q(Z_i) \log \left(\int q(Z_{\backslash i}) \frac{p(\mathbf{X}, \mathbf{Z})}{q(\mathbf{Z})} dZ_{\backslash i} \right) dZ_i \\
&\geq \int q(Z_i) \int q(Z_{\backslash i}) \log \left(\frac{p(\mathbf{X}, \mathbf{Z})}{q(\mathbf{Z})} \right) dZ_{\backslash i} dZ_i \\
&= \int q(\mathbf{Z}) \log \left(\frac{p(\mathbf{X}, \mathbf{Z})}{q(\mathbf{Z})} \right) d\mathbf{Z} = \mathcal{L}(q(\mathbf{Z}))
\end{aligned}
\tag{3.104}
$$

or, equivalently, to obtain the inequality

$$
\mathcal{D}_{\mathrm{KL}}(q(Z_i) \| p(Z_i|\mathbf{X})) \leq \log p(\mathbf{X}) - \mathcal{L}(q(\mathbf{Z})).
\tag{3.105}
$$

As a result, the variational posterior distribution of the individual latent variable $q(Z_i)$ is estimated by minimizing the KL divergence $\mathcal{D}_{\mathrm{KL}}(q(Z_i) \| p(Z_i|\mathbf{X}))$, which is the same as maximizing the variational lower bound

$$
\widehat{q}(Z_i) = \arg\max_{q(Z_i)} \mathcal{L}(q(\mathbf{Z})).
\tag{3.106}
$$

After finding J variational posterior distributions $\{\widehat{q}(Z_i)\}_{i=1}^{J}$, the joint posterior distribution $\widehat{q}(\mathbf{Z})$ is obtained by Eq. (3.101). Fig. 3.13 shows the estimation of variational distribution $q(\mathbf{X})$ by maximizing the variational lower bound $\mathcal{L}(q(\mathbf{Z}))$, which is equivalent to minimizing the KL divergence $\mathcal{D}_{\mathrm{KL}}(q(\mathbf{Z}) \| p(\mathbf{Z}|\mathbf{X}))$.

The optimization problem is expressed by

$$\max_{q(Z_i)} \; \mathbb{E}_q[\log p(\mathbf{X}, \mathbf{Z})] + \mathbb{H}_q[\mathbf{Z}]$$

$$\text{subject to} \quad \int_{\mathbf{Z}} q(d\mathbf{Z}) = 1. \tag{3.107}$$

To solve this problem, we manipulate the ELBO of Eq. (3.98) in a form of

$$
\begin{aligned}
\mathcal{L}(q(\mathbf{Z})) &= \int \prod_j q_j \left\{ \log p(\mathbf{X}, \mathbf{Z}) - \sum_j \log q_j \right\} d\mathbf{Z} \\
&= \int q_i \underbrace{\int \log p(\mathbf{X}, \mathbf{Z}) \prod_{j \neq i} q_j d Z_j \, d Z_i}_{\triangleq \log \widetilde{p}(\mathbf{X}, Z_i)} \\
&\quad - \int q_i \log q_i \, d Z_i + \text{const} \\
&= -\mathcal{D}_{\mathrm{KL}}(q(Z_i) \| \widetilde{p}(\mathbf{X}, Z_i)) + \text{const}
\end{aligned} \tag{3.108}
$$

where $q(Z_i)$ is denoted by q_i and the term

$$\log \widetilde{p}(\mathbf{X}, Z_i) = \mathbb{E}_{\mathbf{Z}_{\backslash i}}[\log p(\mathbf{X}, \mathbf{Z})] + \text{const} \tag{3.109}$$

is specified. The optimal VB posterior is finally derived as

$$\widehat{q}(Z_i) \propto \exp\left(\mathbb{E}_{\mathbf{Z}_{\backslash i}}\left[\log p(\mathbf{X}, \mathbf{Z})\right]\right). \tag{3.110}$$

It is worth noting that this form of solution provides the basis for applications of variational methods. By considering the normalization constant, it is derived as

$$\widehat{q}(Z_i) = \frac{\exp\left(\mathbb{E}_{\mathbf{Z}_{\backslash i}}\left[\log p(\mathbf{X}, \mathbf{Z})\right]\right)}{\int \exp\left(\mathbb{E}_{\mathbf{Z}_{\backslash i}}\left[\log p(\mathbf{X}, \mathbf{Z})\right]\right) d Z_i}. \tag{3.111}$$

The general form of a variational posterior distribution $\widehat{q}(Z_i)$ is derived. Eq. (3.110) tells us that if we want to infer an approximate posterior for a target latent variable Z_i, we first need to prepare the joint distribution of observations $\mathbf{X}$ and all latent variables $\mathbf{Z}$. This inference says that the log of the optimal solution to approximate posterior of a target variable Z_i is obtained by considering the log of the joint distribution over all hidden and visible variables and then taking the expectation with respect to all the other variables $\mathbf{Z}_{\backslash i}$. Note that the target posterior distribution $\widehat{q}(Z_i)$ and the other posterior distributions

$$q(\mathbf{Z}_{\backslash i}) = \prod_{j \neq i}^{J} q(Z_j) \tag{3.112}$$

depend on each other due to the expectation in Eq. (3.110). Therefore, this optimization can be performed iteratively from the initial posterior distributions for all $q(Z_i)$.

VB-EM Algorithm

From an alternative view of the EM algorithm in Section 3.7.1, the estimation of the factorized posterior distributions $\{\widehat{q}(Z_j)\}_{j=1}^{J}$ in the VB algorithm is completed to fulfill an expectation step similar to Eq. (3.72) in the standard EM algorithm. Different from the EM algorithm, the VB-EM algorithm is implemented with an additional factorization procedure due to the coupling of multiple latent variables in the posterior distribution $p(\mathbf{Z}|\mathbf{X}, \mathbf{\Theta}^{(r)})$. The factorized posterior distribution

$$q(Z_i) = \widetilde{p}(\mathbf{X}, Z_i) \tag{3.113}$$

does not result in $\mathcal{L}(q) = 0$ in Eq. (3.108) while the joint posterior distribution

$$q(\mathbf{Z}) = p(\mathbf{Z}|\mathbf{X}, \mathbf{\Theta}^{(r)}) \tag{3.114}$$

is set to force $\mathcal{L}(q, \mathbf{\Theta}) = 0$ in Eq. (3.72). Such a difference is also observed by comparing Figs. 3.13 and 3.9(A)–(B). The key reason behind this difference is caused by the factorization in the posterior distribution, which is inevitable in the implementation in the presence of multiple latent variables. More generally, we address the VB-EM algorithm for ML or MAP estimation where the model parameters $\mathbf{\Theta}$ in latent variable model are provided in ELBO $\widetilde{\mathcal{L}}(q(\mathbf{Z}), \mathbf{\Theta})$ similar to Eq. (3.69) in the EM algorithm. But, the lower bound of VB-EM $\widetilde{\mathcal{L}}(q, \mathbf{\Theta})$ is different from that of EM $\mathcal{L}(q, \mathbf{\Theta})$. For clarity, the lower bound of VB-EM inference is specifically denoted by $\widetilde{\mathcal{L}}(\cdot)$ which is different from $\mathcal{L}(\cdot)$ in basic EM. There are two steps in the VB-EM algorithm. In the VB-E step, the variational posterior distribution is estimated by maximizing the variational lower bound by

$$q^{(\tau+1)}(Z_i) = \arg\max_{q(Z_i)} \widetilde{\mathcal{L}}(q(\mathbf{Z}), \mathbf{\Theta}^{(\tau)}) \tag{3.115}$$

for each of the individual latent variables $\{Z_i\}_{i=1}^{J}$. In the VB-M step, we use the updated variational distribution

$$q^{(\tau+1)}(\mathbf{Z}) = \prod_{j=1}^{J} q^{(\tau+1)}(Z_j) \tag{3.116}$$

to estimate ML or MAP parameters by maximizing the resulting lower bound

$$\mathbf{\Theta}^{(\tau+1)} = \arg\max_{\mathbf{\Theta}} \widetilde{\mathcal{L}}(q^{(\tau+1)}(\mathbf{Z}), \mathbf{\Theta}). \tag{3.117}$$

During the VB-EM steps, the updated parameters

$$\mathbf{\Theta}^{(0)} \to \mathbf{\Theta}^{(1)} \to \mathbf{\Theta}^{(2)} \cdots \tag{3.118}$$

improve the lower bound $\widetilde{\mathcal{L}}(q(\mathbf{Z}), \mathbf{\Theta})$ or, equivalently, the likelihood function $p(\mathbf{X}|\mathbf{\Theta})$ or posterior distribution $p(\mathbf{\Theta}|\mathbf{X})$ for ML or MAP estimation, respectively. The learning procedure based on the VB-EM algorithm is also called the variational Bayesian learning.

3.7.4 SAMPLING METHODS

For most latent variable models of interest, exact inference is intractable, and so we have to resort to some form of approximation. Markov chain Monte Carlo (Neal, 1993, Gilks et al., 1996) is another realization of full Bayesian treatment for practical solutions. MCMC is known as a stochastic approximation which acts differently from the deterministic approximation based on variational Bayesian (VB) as addressed in Section 3.7.3. Variational inference using VB approximates the posterior distribution through factorization of the distribution over multiple latent variables and scales well to large applications. MCMC uses numerical sampling computation rather than solving integrals and expectations analytically. Since MCMC can use any distributions in principle, it is capable of providing highly flexible models for various applications, including many latent variable models for source separation in complicated environments. The fundamental problem in MCMC involves finding the expectation of some function $f(\mathbf{Z})$ with respect to a probability distribution $p(\mathbf{Z})$ where the components of $\mathbf{Z}$ might comprise discrete or continuous latent variables which are some factors or parameters to be inferred under a probabilistic source separation model. In case of continuous variables, we would like to evaluate the expectation

$$\mathbb{E}_{\mathbf{Z}}[f(\mathbf{Z})] = \int f(\mathbf{Z})p(\mathbf{Z})d\mathbf{Z} \tag{3.119}$$

where the integral is replaced by a sum in case of discrete variables. We suppose that such expectations are too complicated to be evaluated analytically. The general idea behind sampling methods is to obtain a set of samples $\{\mathbf{Z}^{(l)}, l = 1, \ldots, L\}$ drawn independently from the distribution $p(\mathbf{Z})$. We may approximate the integral by a sample mean of function $f(\boldsymbol{\theta})$ over these samples $\mathbf{Z}^{(l)}$

$$\widehat{f} = \frac{1}{L}\sum_{l=1}^{L} f(\mathbf{Z}^{(l)}). \tag{3.120}$$

Since the samples $\mathbf{Z}^{(l)}$ are drawn from the distribution $p(\mathbf{Z})$, the estimator $\widehat{f}$ has the correct mean, i.e., $\widehat{f} = \mathbb{E}_{\mathbf{Z}}[f(\mathbf{Z})]$. In general, 10 to 20 independent samples may suffice to estimate an expectation. However, the samples $\{\mathbf{Z}^{(l)}\}_{l=1}^{L}$ may not be drawn independently. The effective sample size might be much smaller than the apparent sample size L. This implies that relatively large sample size will be required to achieve sufficient accuracy.

The technique of *importance sampling* provides a framework for approximating the expectation in Eq. (3.119) directly but does not provide the mechanism for drawing samples from the distribution $p(\mathbf{Z})$. Suppose we wish to sample from a distribution $p(\mathbf{Z})$ that is not a simple and standard distribution. Sampling directly from $p(\mathbf{Z})$ is difficult. Importance sampling is based on the use of a *proposal distribution* $q(\mathbf{Z})$ from which it is easy to draw samples. The expectation in Eq. (3.119) is expressed in the form of a finite sum over samples $\{\mathbf{Z}^{(l)}\}$ drawn from $q(\mathbf{Z})$ as

$$\mathbb{E}_{\mathbf{Z}}[f(\mathbf{Z})] = \int f(\mathbf{Z})\frac{p(\mathbf{Z})}{q(\mathbf{Z})}q(\mathbf{Z})d\mathbf{Z}$$
$$\simeq \frac{1}{L}\sum_{l=1}^{L} \frac{p(\mathbf{Z}^{(l)})}{q(\mathbf{Z}^{(l)})} f(\mathbf{Z}^{(l)}). \tag{3.121}$$

The quantity $r_l = p(\mathbf{Z}^{(l)})/q(\mathbf{Z}^{(l)})$ is known as the importance weight which is used to correct the bias introduced by sampling from the wrong distribution.

Markov Chain Monte Carlo

A major weakness in evaluation of expectation function based on the importance sampling strategy is the severe limitation in spaces of high dimensionality. We accordingly turn into a very general and powerful framework called Markov chain Monte Carlo (MCMC), which allows sampling from a large class of distributions and which scales well with the dimensionality of the sample space. Before discussing MCMC methods in more detail, it is useful to study some general properties of Markov chains and investigate under what conditions a Markov chain can converge to the desired distribution. A first-order Markov chain is defined for a series of latent variables $\{\mathbf{Z}^{(1)}, \ldots, \mathbf{Z}^{(\tau)}\}$ in different states, iterations or learning epochs τ such that the following *conditional independence* holds for all states $\tau \in \{1, \ldots, T-1\}$:

$$p(\mathbf{Z}^{(\tau+1)}|\mathbf{Z}^{(\tau)}, \ldots, \mathbf{Z}^{(\tau)}) = p(\mathbf{Z}^{(\tau+1)}|\mathbf{Z}^{(\tau)}). \tag{3.122}$$

This Markov chain starts from the probability distribution for an initial sample $p(\mathbf{Z}^{(0)})$ and operates with the transition probability

$$T(\mathbf{Z}^{(\tau)}, \mathbf{Z}^{(\tau+1)}) \triangleq p(\mathbf{Z}^{(\tau+1)}|\mathbf{Z}^{(\tau)}). \tag{3.123}$$

A Markov chain is *homogeneous* if the transition probabilities $T(\mathbf{Z}^{(\tau)}, \mathbf{Z}^{(\tau+1)})$ are unchanged for all epochs τ. The marginal probability for variables $\mathbf{Z}^{(\tau+1)}$ at epoch $\tau+1$ is expressed in terms of the marginal probabilities over the previous epochs $\{\mathbf{Z}^{(1)}, \ldots, \mathbf{Z}^{(\tau)}\}$ in the chain

$$p(\mathbf{Z}^{(\tau+1)}) = \sum_{\mathbf{Z}^{(\tau)}} p(\mathbf{Z}^{(\tau+1)}|\mathbf{Z}^{(\tau)})p(\mathbf{Z}^{(\tau)}). \tag{3.124}$$

A distribution is said to be *invariant* or stationary with respect to a Markov chain if each step in the chain keeps the distribution invariant. For a homogeneous Markov chain with transition probability $T(\mathbf{Z}', \mathbf{Z})$, the distribution $p(\mathbf{Z})$ is invariant as

$$p(\mathbf{Z}) \rightarrow p^\star(\mathbf{Z}) \tag{3.125}$$

if the following property is held

$$p^\star(\mathbf{Z}) = \sum_{\mathbf{Z}'} T(\mathbf{Z}', \mathbf{Z})p^\star(\mathbf{Z}'). \tag{3.126}$$

Our goal is to use Markov chains to sample from a given distribution. We can achieve this goal if we set up a Markov chain such that the desired distribution is invariant. It is required that for $\tau \rightarrow \infty$, the distribution $p(\mathbf{Z}^{(\tau)})$ converges to the required invariant distribution $p^\star(\mathbf{Z})$ which is obtained irrespective of the choice of initial distribution $p(\mathbf{Z}^{(0)})$. Such an invariant distribution is also called the *equilibrium* distribution. A sufficient condition for an invariant distribution $p^\star(\mathbf{Z})$ is to choose the transition probabilities to satisfy the *detailed balance*, i.e.,

$$p^\star(\mathbf{Z})T(\mathbf{Z}, \mathbf{Z}') = p^\star(\mathbf{Z}')T(\mathbf{Z}', \mathbf{Z}) \tag{3.127}$$

for a particular distribution $p^\star(\mathbf{Z})$.

Metropolis–Hastings Algorithm

As mentioned in importance sampling, we keep sampling from a proposal distribution and maintain a record of the current state $\mathbf{Z}^{(\tau)}$. The proposal distribution $q(\mathbf{Z}|\mathbf{Z}^{(\tau)})$ depends on this current state $\mathbf{Z}^{(\tau)}$. The sequence of samples $\{\mathbf{Z}^{(1)}, \mathbf{Z}^{(2)}, \ldots, \mathbf{Z}^{(\tau)}, \ldots\}$ forms a Markov chain. The proposal distribution is chosen to be sufficiently simple to draw samples directly. At each sampling cycle, we generate a candidate sample $\mathbf{Z}^{\star}$ from the proposal distribution and then accept the sample according to an appropriate criterion. In a basic Metropolis algorithm (Metropolis et al., 1953), the proposal distribution is assumed to be symmetric as

$$q(\mathbf{Z}_a|\mathbf{Z}_b) = q(\mathbf{Z}_b|\mathbf{Z}_a) \tag{3.128}$$

for all values of $\mathbf{Z}_a$ and $\mathbf{Z}_b$. A candidate sample is accepted with the probability

$$A(\mathbf{Z}^{\star}, \mathbf{Z}^{(\tau)}) = \min\left(1, \frac{\widetilde{p}(\mathbf{Z}^{\star})}{\widetilde{p}(\mathbf{Z}^{(\tau)})}\right) \tag{3.129}$$

where an unnormalized distribution $\widetilde{p}(\mathbf{Z})$ in the target distribution

$$p(\mathbf{Z}) = \widetilde{p}(\mathbf{Z})/Z_p \tag{3.130}$$

is evaluated. In the implementation, we choose a random number u from the uniform distribution on the unit interval $(0, 1)$ and accept the sample if

$$A(\mathbf{Z}^{\star}, \mathbf{Z}^{(\tau)}) > u. \tag{3.131}$$

If the learning from $\mathbf{Z}^{(\tau)}$ to $\mathbf{Z}^{\star}$ causes an increase in the value of $p(\mathbf{Z})$, this candidate point is accepted. Once the candidate sample is accepted, then

$$\mathbf{Z}^{(\tau+1)} = \mathbf{Z}^{\star} \tag{3.132}$$

otherwise the candidate sample $\mathbf{Z}^{\star}$ is discarded, $\mathbf{Z}^{(\tau+1)}$ remains unchanged as $\mathbf{Z}^{(\tau)}$. The next candidate sample is drawn from the distribution $q(\mathbf{Z}|\mathbf{Z}^{(\tau+1)})$. This leads to multiple copies of samples in the final list of samples. As long as $q(\mathbf{Z}_a|\mathbf{Z}_b)$ is positive for any values of $\mathbf{Z}_a$ and $\mathbf{Z}_b$, the distribution of $\mathbf{Z}^{(\tau)}$ tends to $p(\mathbf{Z})$ as $\tau \to \infty$.

The Metropolis algorithm is further generalized to the Metropolis–Hastings algorithm (Hastings, 1970) which is widely adopted in MCMC inference. This generalization is developed by relaxing the assumption in the Metropolis algorithm that the proposal distribution be a symmetric function of its arguments. Using this algorithm, at epoch τ with current state $\mathbf{Z}^{(\tau)}$, we draw a sample $\mathbf{Z}^{\star}$ from the proposal distribution $q_i(\mathbf{Z}|\mathbf{Z}^{(\tau)})$ and then accept it with the probability

$$A_i(\mathbf{Z}^{\star}, \mathbf{Z}^{(\tau)}) = \min\left(1, \frac{\widetilde{p}(\mathbf{Z}^{\star})q_i(\mathbf{Z}^{(\tau)}|\mathbf{Z}^{\star})}{\widetilde{p}(\mathbf{Z}^{(\tau)})q_i(\mathbf{Z}^{\star}|\mathbf{Z}^{(\tau)})}\right) \tag{3.133}$$

where i denotes the members of the set of possible transitions. For the case of symmetric proposal distribution, the Metropolis–Hastings criterion in Eq. (3.133) reduces to the Metropolis criterion in Eq. (3.129). We can show that $p(\mathbf{Z})$ is an invariant distribution of the Markov chain generated by the

Metropolis–Hastings algorithm by investigating the property of the detailed balance in Eq. (3.127) as
follows:

$$p(\mathbf{Z})q_i(\mathbf{Z}|\mathbf{Z}')A_i(\mathbf{Z}',\mathbf{Z}) = \min(p(\mathbf{Z})q_i(\mathbf{Z}|\mathbf{Z}'), p(\mathbf{Z}')q_i(\mathbf{Z}'|\mathbf{Z}))$$
$$= \min(p(\mathbf{Z}')q_i(\mathbf{Z}'|\mathbf{Z}), p(\mathbf{Z})q_i(\mathbf{Z}|\mathbf{Z}')) \qquad (3.134)$$
$$= p(\mathbf{Z}')q_i(\mathbf{Z}'|\mathbf{Z})A_i(\mathbf{Z},\mathbf{Z}').$$

Gibbs Sampling

Gibbs sampling (Geman and Geman, 1984, Liu, 2008) is a simple and widely applicable realization of
an MCMC algorithm. This method is a special case of the Metropolis–Hastings algorithm (Hastings,
1970). Consider the distribution of J latent variables $p(\mathbf{Z}) = p(Z_1,\ldots,Z_J)$ and suppose that we have
an initial state $p(\mathbf{Z}^{(0)})$ for the Markov chain. Each step of the Gibbs sampling procedure replaces the
value of one of the variables by a value drawn from the distribution of that variable conditioned on
the values of the remaining variables. We replace the ith component Z_i by a value drawn from the
distribution $p(Z_i|\mathbf{Z}_{\backslash i})$ where $\mathbf{Z}_{\backslash i}$ denotes $\{Z_1,\ldots,Z_J\}$ but with Z_i excluded. The sampling procedure
is repeated by cycling through the variables in a particular order or in a random order from some
distribution. This procedure samples the required distribution $p(\mathbf{Z})$ which should be invariant at each
epoch of Gibbs sampling or in the whole Markov chain. It is because the marginal distribution $p(\mathbf{Z}_{\backslash i})$ is
invariant and the conditional distribution $p(Z_i|\mathbf{Z}_{\backslash i})$ is corrected at each sampling step. Gibbs sampling
of J variables for T steps is operated in the following procedure:

- Initialize $\{Z_i^{(1)} : i = 1,\ldots,J\}$,
- For $\tau = 1,\ldots,T$:
  - Sample $Z_1^{(\tau+1)} \sim p(Z_1|Z_2^{(\tau)}, Z_3^{(\tau)},\ldots,Z_J^{(\tau)})$,
  - Sample $Z_2^{(\tau+1)} \sim p(Z_2|Z_1^{(\tau+1)}, Z_3^{(\tau)},\ldots,Z_J^{(\tau)})$,
  - $\vdots$
  - Sample $Z_i^{(\tau+1)} \sim p(Z_i|Z_1^{(\tau+1)},\ldots,Z_{i-1}^{(\tau+1)}, Z_{i+1}^{(\tau)},\ldots,Z_J^{(\tau)})$,
  - $\vdots$
  - Sample $Z_J^{(\tau+1)} \sim p(Z_J|Z_1^{(\tau+1)}, Z_2^{(\tau+1)},\ldots,Z_{J-1}^{(\tau+1)})$.

When applying the Metropolis–Hastings algorithm for sampling a variable Z_i given that the remaining
variables $\mathbf{Z}_{\backslash i}$ are fixed, the transition probability from $\mathbf{Z}$ to $\mathbf{Z}^\star$ in Gibbs sampling is determined by

$$q_i(\mathbf{Z}^\star|\mathbf{Z}) = p(Z_i^\star|\mathbf{Z}_{\backslash i}). \qquad (3.135)$$

Since the remaining variables are unchanged by a sampling step,

$$(\mathbf{Z}^\star)_{\backslash i} = \mathbf{Z}_{\backslash i}. \qquad (3.136)$$

By using the property

$$p(\mathbf{Z}) = p(Z_i|\mathbf{Z}_{\backslash i})p(\mathbf{Z}_{\backslash i}), \qquad (3.137)$$

Table 3.1 Comparison of approximate inference methods using variational Bayesian and Gibbs sampling

Variational Bayes	Gibbs sampling
• deterministic approximation	• stochastic approximation
• find an analytical proxy $q(\mathbf{Z})$ that is maximally similar to $p(\mathbf{Z}|\mathbf{X})$	• design an algorithm that draws samples $\mathbf{Z}^{(1)}, \ldots, \mathbf{Z}^{(\tau)}$ from $p(\mathbf{Z}|\mathbf{X})$
• inspect distribution statistics	• inspect sample statistics
• never generates exact results	• asymptotically exact
• fast	• computationally expensive
• often hard to derive	• tricky engineering concerns
• convergence guarantees	• no convergence guarantees
• need a specific parametric form	• no need for a parametric form

the acceptance probability in the Metropolis–Hastings algorithm is obtained by

$$
\begin{aligned}
A_i(\mathbf{Z}^\star, \mathbf{Z}) &= \frac{p(\mathbf{Z}^\star) q_i(\mathbf{Z}|\mathbf{Z}^\star)}{p(\mathbf{Z}) q_i(\mathbf{Z}^\star|\mathbf{Z})} \\
&= \frac{p(Z_i^\star|(\mathbf{Z}^\star)_{\backslash i}) p((\mathbf{Z}^\star)_{\backslash i}) p(Z_i|(\mathbf{Z}^\star)_{\backslash i})}{p(Z_i|\mathbf{Z}_{\backslash i}) p(\mathbf{Z}_{\backslash i}) p(Z_i^\star|\mathbf{Z}_{\backslash i})} = 1.
\end{aligned}
\tag{3.138}
$$

This result is derived since $(\mathbf{Z}^\star)_{\backslash i} = \mathbf{Z}_{\backslash i}$ is applied. Eq. (3.138) indicates that the Gibbs sampling steps in the Metropolis–Hastings algorithm are always accepted.

When constructing latent variable models for source separation, it is popular to carry out approximate inference methods based on variational Bayesian (VB) and Gibbs sampling. Table 3.1 illustrates the complimentary differences between variational Bayesian (VB) and Gibbs sampling from different perspectives. Basically, the alternative and iterative estimation of variational distribution $q(\mathbf{Z})$ and model parameters $\mathbf{\Theta}$ is seen as a kind of *deterministic* approximation because either variable is estimated by fixing the other variable. But Gibbs sampling conducts a *stochastic* approximation since the latent variables $\mathbf{Z}$ or the model parameters $\mathbf{\Theta}$ are alternatively and iteratively estimated according to a sampling procedure. A target variable $Z_i^{(\tau+1)}$ at a new iteration $\tau + 1$ is drawn using the posterior distribution $p(Z_i|Z_1^{(\tau+1)}, \ldots, Z_{i-1}^{(\tau+1)}, Z_{i+1}^{(\tau)}, \ldots, Z_J^{(\tau)})$ of this variable Z_i given the remaining latent variables with the updated values including the preceding variables $\{Z_1^{(\tau+1)}, \ldots, Z_{i-1}^{(\tau+1)}\}$ at the new iteration $\tau + 1$ and the subsequent variables $\{Z_{i+1}^{(\tau)}, \ldots, Z_J^{(\tau)}\}$ at the previous iteration τ. VB inference is implemented through finding a proxy $q(\mathbf{Z})$ which is closest to the posterior distribution $p(\mathbf{Z}|\mathbf{X})$ while Gibbs sampling aims to draw latent variables $\mathbf{Z}$ by using posterior distribution $p(\mathbf{Z}|\mathbf{X})$. VB emphasizes finding distribution statistics while Gibbs sampling inspects the statistics from sampling. VB tends to seek a local optimum of model parameters but the solution using Gibbs sampling is asymptotically exact. VB inference is prone to converge to a local optimum within a few iterations. Gibbs sampling may find the global optimum but require infinite sampling procedure $\tau \to \infty$. This results in fast implementation using VB but slow convergence using Gibbs sampling. The difficulty in VB is the mathematically-intensive derivation of decomposed distributions $q(\mathbf{Z})$ and model parameters $\mathbf{\Theta}$.

However, only tricky engineering work is of concern in Gibbs sampling. Using VB, the specific parametric form of complicated model is assumed to assure analytical form of model parameters while Gibbs sampling does not need to follow specific parametric distributions.

3.8 SUMMARY

This chapter has introduced a series of fundamental machine learning theories which are useful to tackle different issues and address different topics in real-world source separation systems. Information-theoretic learning was developed to construct meaningful learning objectives based on information theory. The divergence measure was calculated to evaluate the independence and the reconstruction error for source separation based on ICA and NMF, respectively. Bayesian learning was emphasized to address the robustness issue in source separation to mitigate the regularization problems due to over-trained model, noise interference and mixing dynamics. Sparse learning was mentioned for an alternative way to model regularization where sparsity in basis representation was controlled to identify the scenario of a mixing system. Sparse Bayesian learning was introduced to carry out sparse learning in a probabilistic way. Furthermore, online learning was presented to deal with source separation in a nonstationary environment where the sources were missing or moving. The online learning strategy was handled by using Bayes theorem where the posterior distribution was continuously updated by combining the likelihood of new data and the prior of history data. In addition, discriminative learning was addressed to improve the separation performance by incorporating the discrimination information between source signals. Deep learning was also investigated. The examples of discriminative NMF and DNN were addressed. Importantly, we introduced the fundamentals of approximate inference algorithms from maximum likelihood to maximum *a posteriori*, variational inference and sampling method. The theories of EM algorithm, variational Bayesian EM algorithm, Markov chain Monte Carlo and Gibbs sampling were illustrated and detailed.

Basically, we have presented the first part of this book. The general discussion about the basics of source separation systems (Chapter 1) and the fundamentals of source separation theories has been addressed. Different source separation models (Chapter 2) and different machine learning algorithms (Chapter 3) have been introduced for source separation under different mixing conditions and applied environments. In what follows, we are moving to the second part of this book which addresses the advanced topics and solutions to individual source separation models, including independent component analysis, nonnegative matrix and tensor factorization, deep neural network and recurrent neural network. A number of source separation applications including speech recognition, speech separation, music signal separation, singing voice separation will be included.

ADVANCED STUDIES

2

INDEPENDENT COMPONENT ANALYSIS

4

The early blind source separation systems were constructed in the presence of multiple sensors based on independent component analysis (ICA) (Comon, 1994). The key to develop a successful ICA algorithm is the construction of a meaningful contrast function which faithfully measures the independence or the non-Gaussianity of demixed signals under a target mixing condition (Hyvärinen et al., 2001). This chapter presents a variety of ICA theories and applications. First of all, in Section 4.1, an ICA method is proposed for the construction of basis vectors where the information redundancy for representation of speaker-specific acoustic model is minimized. The minimization is driven according to the measure of non-Gaussianity. In addition, a nonparametric ICA is presented for multichannel source separation in Section 4.2. This method investigates the independence of demixed signals based on a hypothesis test. A log-likelihood function is derived for evaluating independence by avoiding Gaussianity assumption using a nonparametric method. This algorithm is developed for blind source separation of speech and music, as well as for unsupervised learning of acoustic models for speech recognition. Moreover, in Section 4.3, a flexible ICA is derived to implement the convex divergence ICA with a tunable convexity parameter. The independence measure is calculated over the demixed signals. Such a measure is optimized by minimizing the divergence between the joint distribution and the product of marginal distributions in the presence of demixed signals. To overcome the complicated scenarios in nonstationary source separation, we further present an advanced ICA method in Section 4.4 which conducts a nonstationary Bayesian learning for ICA model parameters and hyperparameters based on a sequential Bayesian algorithm. In Section 4.5, the nonstationary source separation is continuously performed by updating ICA parameters using individual minibatches and simultaneously performing temporal modeling of the mixing system and source signals using a Gaussian process. Sequential and variational Bayesian learning is implemented. The first advanced topic is focused on how the ICA method is developed for speech recognition.

4.1 ICA FOR SPEECH RECOGNITION

This section addresses an application of ICA for unsupervised learning of basis vectors when constructing the acoustic space of speakers or noise conditions for speaker or noise adaptation, respectively. Speaker or noise adaptation provides an effective way to improve the robustness for speech recognition. Finding the basis vectors is comparable to seeking the source signals where the measure of independence is maximized. Using this method, the speech observations are decomposed to find independent components based on ICA. These components represent the specific speaker, gender, accent, noise or environment, and act as the basis functions to span the vector space of human voices in different conditions. In this section, we will present how the independent voices are trained by applying the

Source Separation and Machine Learning. https://doi.org/10.1016/B978-0-12-804566-4.00016-4

ICA method and will explain how these basis vectors are adopted to construct a speaker adaptive model for speech recognition. Further, we explain the reason why information redundancy is reduced by imposing independence in the estimated basis vectors. Before addressing the ICA method, the basics of speaker or noise adaptation for speech recognition are introduced.

Traditionally, a hidden Markov model (HMM) (Rabiner and Juang, 1986) is a popular paradigm to build an acoustic model for automatic speech recognition. The HMM parameters estimated from training data may not catch the variations in test conditions. For example, the pronunciation variations from different speakers are inevitable in speaker-independent spontaneous speech recognition system. To compensate these variations and tackle the mismatch between training and test conditions, it is important to express the variety of speaking styles in different speakers and adapt or transfer the existing HMM parameters to a new test environment. The adapted acoustic models can improve the generalization for speech recognition. However, an essential concern in model adaptation is to carry out a fast adaptation solution by using sparse data. In Kuhn et al. (2000), rapid adaptation was performed in the eigenspace by finding a linear interpolation of basis vectors or eigenvoices where the interpolation weights were learned from adaptation data and the eigenvoices were trained by applying principal component analysis (PCA) (Jolliffe, 1986) based on a set of reference HMMs. Since the eigenvoices are orthogonal with the most expressive capability over speaker information, it is feasible to learn the interpolation coefficients from very limited enrollment data and find the adapted model parameters in the eigenspace. A related work was proposed to build the kernel eigenspace for fulfilling a nonlinear PCA for speaker adaptation (Mak et al., 2005). Nonetheless, there still exists redundant information in the estimated eigenvectors for learning representation. In Xu and Golay (2006), this information redundancy was reduced by decreasing the number of components in a factor model where the model uncertainty was simplified as well. Different from eigenvoices built by PCA, the ICA algorithm is introduced to estimate the independent voices (Hsieh et al., 2009), which are applied for efficient coding of an adapted acoustic model. Since information redundancy is considerably reduced in independent voices, we can effectively calculate a coordinate vector in the independent voice space, and estimate the adaptive HMMs for speech recognition. Before constructing the independent space, we investigate the superiority of ICA to PCA in data representation, in particular in speaker-adaptive acoustic modeling and illustrate how ICA is developed to reduce the information redundancy in implementation of model adaptation.

4.1.1 CONSTRUCTION OF INDEPENDENCE SPACE

Both PCA and ICA are feasible to build a vector space for data expression. PCA is trained to find latent components which are orthogonal while ICA is learned to identify the independent components which could serve as a minimum set of basis vectors for spanning a vector space. It is crucial to investigate why PCA does not perform as well as ICA in reduction of information redundancy for learning representation. In general, PCA aims to extract K zero-mean principal components $\{\mathbf{b}_1, \mathbf{b}_2, \ldots, \mathbf{b}_K\}$ which are uncorrelated to meet the condition for the first moments in the joint distribution $p(\mathbf{b}_1, \ldots, \mathbf{b}_K)$ as well as in the product of marginal distributions $\{p(\mathbf{b}_k)\}_{k=1}^K$, namely

$$\mathbb{E}[\mathbf{b}_1 \mathbf{b}_2 \cdots \mathbf{b}_K] = \mathbb{E}[\mathbf{b}_1]\mathbb{E}[\mathbf{b}_2] \cdots \mathbb{E}[\mathbf{b}_K] \tag{4.1}$$

is satisfied. On the other hand, ICA is developed to pursue K zero-mean independent components $\{s_1, s_2, \ldots, s_K\}$ which are in accordance with

$$\mathbb{E}[s_1^r s_2^r \cdots s_K^r] = \mathbb{E}[s_1^r]\mathbb{E}[s_2^r] \cdots \mathbb{E}[s_K^r] \tag{4.2}$$

for an integer r or for different moments in the joint and marginal distributions. Independent components in ICA are uncorrelated in higher-order moments. Principal components are *not* independent, but independent components are sufficiently uncorrelated. ICA paves an avenue to explore the salient features based on higher-order statistics while PCA extracts the uncorrelated components using second-order statistics. PCA runs a linear decorrelation process, which only calculates the second-order statistics. ICA is specialized to identify the independent factors or sources from a set of mixed signals. An ICA algorithm is developed to estimate the independent sources for blind source separation where non-Gaussian sources with higher-order statistics are separated by using a demixing matrix $\mathbf{W}$. This matrix is estimated by maximizing the learning objective, which measures the independence or non-Gaussianity in demixed signals. ICA is a fundamental machine learning approach to unsupervised learning where the sparse coding is fulfilled to apply for feature extraction and data compression. A sparse component is distributed with heavy tails and a sharp peak. Sparse distribution in latent sources is helpful to find the clearly clustered sources. The sparsity of distribution in independent components is larger than that in principal components. The significance of sparseness of the reconstructed signals proportionally reflects the amount of conveyed information in the construction of basis vectors for a vector space (Lee and Jang, 2001).

Owing to the strength of information preservation, the ICA method is feasible to build a set of independent voices which are uncorrelated in higher-order statistics. Using these representative voices, one can calculate the interpolated acoustic models for speech recognition. The interpolated model using independent voices is advantageous to that using eigenvoices in terms of representation capability. To evaluate the representation capability of PCA and ICA, a traditional measurement of using either Bayesian information criterion (BIC) (Schwarz, 1978) or minimum description length (MDL) (Rissanen, 1978) can be applied. The components and the size of components in PCA or ICA are evaluated. Both BIC and MDL have the same style of measurement for model selection although they came from different perspectives. A qualified or regularized model can either achieve a higher BIC value or a sparser description length for sparse representation or construction of an unknown space. The robustness in learning representation can be guaranteed. Considering either PCA or ICA model with K components in model parameters Θ estimated from training data $\mathbf{X} = \{\mathbf{x}_t\}_{t=1}^T$, the BIC value is calculated by

$$\text{BIC}(\Theta, K) = \log p(\mathbf{X}|\Theta, K) - \frac{1}{2}\xi K \log T \tag{4.3}$$

with a tuning parameter ξ. On the other hand, in basis representation using PCA or ICA, the sparseness of representation sufficiently reflects the efficiency of the information decoded and conveyed through coordinate coefficients by using different basis functions in a vector space. The *kurtosis*, defined as the fourth-order statistics in Eq. (2.1), can be used to evaluate how significantly the PCA and ICA transformations reduce the information redundancy in learning representation. Basically, the ICA method is used to construct an independent space where the information redundancy is minimized.

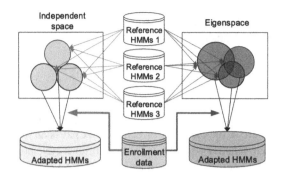

FIGURE 4.1

Construction of the eigenspace and independent space for finding the adapted models from a set of seed models by using enrollment data.

4.1.2 ADAPTATION IN INDEPENDENT SPACE

We present an ICA solution to train the independent voices and construct an independent space. The learning representation over the variety of acoustic conditions can be assured. Due to this compact vector space, the adaptive HMMs are estimated or encoded by means of a sparse coding scheme. The acoustic models are represented in the independent space, which spans different environmental conditions, including speakers and noises. The redundancy in the estimated independent voices is effectively reduced. The interpolated acoustic models can be obtained by using sparse enrollment data. Fig. 4.1 illustrates examples of model adaptation based on the eigenspace and independent space. In a learning process, we collect a set of reference HMM parameters, which were trained from stereo recordings with various noisy environments. In the implementation, an N-dimensional supervector $\mathbf{x}_m$ is composed of Gaussian mean vectors under different words, HMM states and mixture components. A data matrix $\mathbf{X}$ is formed by concatenating those supervectors from a set of M reference HMMs. Next, the redundancy information in the $M \times N$ data matrix $\mathbf{X} = \{\mathbf{x}_m\}_{m=1}^{M}$ is reduced by applying a standard ICA algorithm using an $M \times M$ demixing matrix $\mathbf{W}$ or a mixing matrix

$$\mathbf{A} = \{a_{mn}\} = \mathbf{W}^{-1} \tag{4.4}$$

is obtained. The $M \times N$ demixed matrix is calculated by

$$\mathbf{Y} = \mathbf{WX}. \tag{4.5}$$

Using eigenvoices for speaker adaptation (Kuhn et al., 2000), the eigenvectors with K largest eigenvalues are calculated to construct an eigenspace. Similar to the construction of the eigenspace, the independent space is built by ranking the columns of $\mathbf{A}$ and selecting the independent voices in terms of the component importance measure (CIM) (Xu and Golay, 2006)

$$\mathrm{CIM}(n) = \frac{1}{M} \sum_{m=1}^{M} |a_{mn}|. \tag{4.6}$$

In the selection process, we pick the independent voices $\mathbf{S} = \{\mathbf{s}_1, \ldots, \mathbf{s}_K\}$ from $\mathbf{Y}$ corresponding to K largest CIM values in the mixing vectors $\mathbf{a}_m$. We construct a minimal spanning set of basis vectors to find a set of adapted HMMs for robust speech recognition. Fig. 4.1 illustrates the concept that the independent components are salient and sufficient to reflect the acoustic variations and conditions in comparison with the eigenvoices. The information redundancy is reduced so that a small amount of adaptation data is enough to perform speaker adaptation by estimating the coordinate vector of the adapted HMM mean vectors in the independent space. By referring to the maximum likelihood eigendecomposition (MLED) in the eigenvoice method (Kuhn et al., 2000), ICA is able to carry out the so-called *maximum likelihood independent decomposition* for finding the supervector of target HMM parameters

$$\hat{\mathbf{s}} = w_1 \mathbf{s}_1 + \cdots + w_K \mathbf{s}_K \qquad (4.7)$$

where the interpolation or coordinate coefficients $\mathbf{w} = [w_1 \cdots w_K]^\top$ are calculated by maximum likelihood using adaptation data $\widetilde{\mathbf{X}}$ and independent voices $\mathbf{S}$ via

$$\mathbf{w}_{\mathrm{ML}} = \arg \max_{\mathbf{w}} p(\widetilde{\mathbf{X}} | \mathbf{w}, \mathbf{S}). \qquad (4.8)$$

Notably, the adaptation data $\widetilde{\mathbf{X}}$ for finding decomposition weights $\mathbf{w}_{\mathrm{ML}}$ are different from the training data of HMM parameters $\mathbf{X}$ used to estimate the independent voices $\mathbf{S}$.

4.1.3 SYSTEM EVALUATION

The performance of ICA for speech recognition using eigenvoices and independent voices was evaluated by using Aurora2 speech database which was a speech recognition task of connected digits in noisy environments (Hsieh et al., 2009). The number of latent components K was varied for comparison. There are three evaluation measurements, including word error rate, kurtosis and BIC, which evaluate the classification performance, information redundancy reduction and the quality of selected components, respectively. Speech recognition was assessed under multiconditional training, which contained four noise materials, including subway, babble, car and exhibition hall, and five signal-to-noise ratios (SNR) levels of 5, 10, 15, 20 dB and clean data. There were 35 supervectors, i.e., $M = 35$, consisting of HMM mean vectors from the aligned Gaussian mixture components. These supervectors came from 34 gender-dependent sets of HMMs and one set of multiconditional HMMs. In speech recognition using connected digits, each digit was modeled by 16 states where each state had three mixture components. The silence segment and short pause were modeled by three states and one state, respectively. There were 6 mixture components in each state. Each frame was characterized by 13 MFCCs and their first- and second-order derivatives. Speech recognition performance was evaluated by the test set A in Aurora2 dataset. Baseline result was obtained by using the multiconditional training method where no adaptation was performed. The Fast ICA algorithm (Hyvärinen, 1999) was implemented to estimate the demixing matrix $\mathbf{W}$.

Basically, the sparseness of the transformed signals using PCA or ICA is highly related to the conveyed information due to interpolation of basis components. The sparseness can be measured by kurtosis. This measure reflects how much information is preserved in a basis representation by applying the PCA or ICA method. Such an evaluation is conducted by calculating the kurtosis over 34 individual supervectors consisting of transformed mean vectors of acoustic HMMs in different noise conditions.

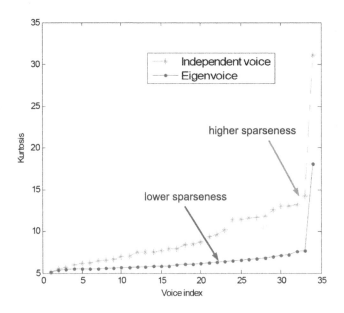

FIGURE 4.2

Comparison of kurtosis values of the estimated eigenvoices and independent voices.

Fig. 4.2 compares the PCA and ICA methods by showing the kurtosis values corresponding to 34 sorted eigenvoices and independent voices, respectively. We can see that independent voices obtain higher kurtosis values than eigenvoices for different basis components. The reduction of information redundancy using ICA is much better than that using PCA. Correspondingly, ICA is able to explore higher-order statistics by using independent voices. The increasing kurtosis sufficiently implies the peaky and sparse shape in the resulting distribution.

Fig. 4.3 compares BIC values when applying the adapted acoustic models based on eigenvoices and independent voices for the case of SNR being 10 dB. BIC was measured according to Eq. (4.3) where $\xi = 0.008$ was fixed and the log-likelihood function of adaptation utterances was calculated. We find that BIC was increased until K increased to a range between 10 and 15. ICA obtains higher BIC values than PCA under different size of hidden components K. ICA is selected as a better model than PCA in terms of model regularization.

The idea of independent voices was evaluated by the experiments on noisy speech recognition by changing the number of adaptation utterances L from 5 to 10 and 15, and the number of basis vectors K from 10 to 15. On average, there were 176 speech frames ($T = 176$) in an utterance. Fig. 4.4 displays a comparison of word error rates (WERs) (%) without and with noise adaptation by using eigenvoices based on PCA and independent voices based on ICA. The averaged WERs over different noise types and noise SNRs were calculated. It is obvious that the averaged WERs are substantially reduced by using adaptation methods based on PCA as well as ICA. WERs of using ICA are consistently lower than those of using PCA. This finding is obtained under different numbers of adaptation utterances L and basis vectors K. The more adaptation utterances enrolled, the lower the WERs achieved.

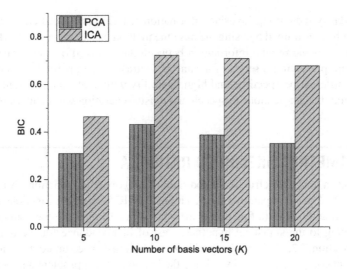

FIGURE 4.3

Comparison of BIC values of using PCA with eigenvoices and ICA with independent voices.

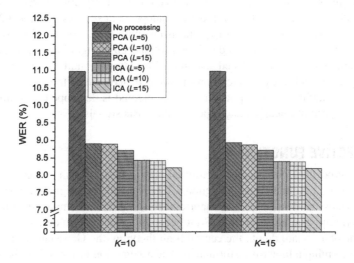

FIGURE 4.4

Comparison of WERs (%) of using PCA with eigenvoices and ICA with independent voices where the numbers of basis vectors (K) and the numbers of adaptation sentences (L) are changed.

We have presented an application of ICA solution to speech recognition which constructs an independent space of acoustic models. The information redundancy was sufficiently reduced to carry out the most informative basis vectors for noise adaptation based on maximum likelihood decomposition using sparse adaptation data. The superiority of ICA over PCA was demonstrated in terms of information redundancy reduction, Bayesian information criterion and word error rate. However, the performance might be constrained by using the baseline method based on Fast ICA (Hyvärinen, 1999), which simply minimizes the mutual information between the demixed or transformed signals. In the next sections, we are presenting a series of advanced contrast functions for ICA optimization with a wide variety from different perspectives and high flexibility with controllable parameters. In addition to speech recognition, the applications to speech and music separation will be investigated.

4.2 NONPARAMETRIC LIKELIHOOD RATIO ICA

This section presents a nonparametric likelihood ratio (NLR) objective function for independent component analysis (ICA). For a comparative study, a number of ICA objective functions are first surveyed in Section 4.2.1. A likelihood ratio function is then developed to measure the confidence towards independence according to the statistical hypothesis test of independence for the transformed signals as mentioned in Section 4.2.2. An ICA algorithm is developed to estimate a demixing matrix and build an independent component space by solving the hypothesis test problem and optimizing the log-likelihood ratio of demixed signals. The traditional way to test the independence based on Gaussian assumption is *not* suitable to realize ICA solution. We will address how ICA solution is formulated by avoiding Gaussian assumption and using the nonparametric distribution based on a kernel density function to deal with the hypothesis test problem for independent sources. Section 4.2.3 will describe an ICA method based on a nonparametric hypothesis test with a meaningful objective function. Section 4.2.4 addresses how this approach is developed for speech recognition and separation. The ICA unsupervised learning method is introduced and applied for learning multiple clusters of acoustic hidden Markov models. The multiple pronunciation variations of acoustic units are compensated for robust speech recognition. In addition, the application of ICA method is developed for multichannel source separation in the presence of linearly-mixed speech and music signals.

4.2.1 ICA OBJECTIVE FUNCTION

The ICA learning procedure is basically performed with a preprocessing calculation based on the whitening transformation as mentioned in Section 2.1.1. The whitened signals are uncorrelated. If the mixed signals are Gaussian distributed, the uncorrelated signals are automatically independent, which is useless to perform any source separation. So it becomes impossible to derive the demixing matrix $\mathbf{W}$ for ICA. On the other hand, due to the central limit theorem, the Gaussianity of random variables is increased by performing a linear transformation. The assumption of Gaussian distribution provides no additional information for finding $\mathbf{W}$. Therefore, it is required to maximize the non-Gaussianity to implement an ICA solution (Hyvärinen et al., 2001) where the degree of independence is maximized as well. A simple way to measure the non-Gaussianity of a signal is to calculate its fourth-order cumulant which is also called the kurtosis and defined in Eq. (2.1). An alternative way to evaluate non-Gaussianity is based on the cumulative density function as proposed in Blanco and Zazo (2003). The generalized Gaussian distribution (Jang et al., 2002) was adopted in the evaluation. In addition, the characteristic function was applied to evaluate the degree of independence of source variables in Eriksson and Koivunen (2003), Murata (2001). The sparser shaped the distribution, the higher the measured non-Gaussianity. In general, the measures of independence, non-Gaussianity and sparseness are

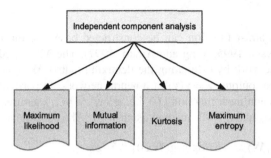

FIGURE 4.5

Taxonomy of contrast functions for optimization in independent component analysis.

comparable. These measures are sufficient and representative to be optimized to estimate the demixing matrix $\mathbf{W}$ for blind source separation. Fig. 4.5 shows a number of objective functions which are popular for optimization in the ICA procedure. In addition to the measure of non-Gaussianity using kurtosis, we introduce ICA learning objectives or contrast functions based on the maximum likelihood, maximum entropy and minimum mutual information as addressed in what follows.

Maximum Likelihood

The maximum likelihood (ML) method is straightforward and considers the likelihood function as a contrast function for ICA implementation (Cardoso, 1999, Hyvärinen and Oja, 2000, Vlassis and Motomura, 2001). Applying ML to calculate the optimal demixing matrix $\mathbf{W}_{ML}$ is implemented by maximizing log-likelihood of mixed signals $\mathbf{X} = \{\mathbf{x}_t\}_{t=1}^{T}$. Because the unknown independent source signal $\mathbf{s}_t$ at time t is related to the mixed signal $\mathbf{x}_t$ via $\mathbf{x}_t = \mathbf{A}\mathbf{s}_t$, we obtain the independence property for the transformed data $\mathbf{y}_t$ as

$$\begin{aligned}
\mathbf{y}_t &= \mathbf{W}\mathbf{x}_t \\
&= \mathbf{W}\mathbf{W}^{-1}\mathbf{s}_t = \mathbf{s}_t.
\end{aligned} \tag{4.9}$$

The ICA contrast function based on ML is formed by summing up the log-likelihoods from T individual training frames $\mathbf{X}$ as

$$\begin{aligned}
\mathcal{J}_{ML}(\mathbf{X}, \mathbf{W}) &= \sum_{t=1}^{T} \log p(\mathbf{x}_t) = \sum_{t=1}^{T} \log \left(|\det \mathbf{W}| p(\mathbf{s}_t) \right) \\
&= T \log |\det \mathbf{W}| + \sum_{t=1}^{T} \sum_{i=1}^{m} \log p(\mathbf{w}_i \mathbf{x}_t)
\end{aligned} \tag{4.10}$$

where $\mathbf{w}_i$ is the ith row of the demixing matrix $\mathbf{W}$. Here, we consider the determined mixing system in multichannel source separation where the number of channels is the same as the number of sources, $n = m$. In this case, the demixing matrix $\mathbf{W}$ is square and nonsingular.

Maximum Entropy

Alternatively, the ICA contrast function can be constructed based on the maximum entropy (ME) method (Bell and Sejnowski, 1995, Yang and Amari, 1997). The ME based ICA method meets the perspective of a neural network by estimating the demixing matrix $\mathbf{W}_{ME}$, which maximizes the entropy $\mathbb{H}(\cdot)$ of the nonlinear outputs $\mathbf{z} = \mathbf{g}(\mathbf{y})$ of a neural network. In particular, using the cumulative distribution function for a nonlinear function $g(\cdot)$, i.e., $g_i'(y_i) = p(y_i)$, assures the equivalence between the ME method using $\mathbf{W}_{ME}$ and the ML method using $\mathbf{W}_{ML}$. The equivalence is shown by

$$\begin{aligned}
\mathcal{J}_{ME}(\mathbf{X}, \mathbf{W}) &= -\mathbb{E}[\log p(\mathbf{g}(\mathbf{W}\mathbf{x}))] \triangleq \mathbb{H}(\mathbf{z}) \\
&= \mathbb{E}\left[\sum_{i=1}^{m} \log p(y_i)\right] + \log|\det \mathbf{W}| - \mathbb{E}[\log p(\mathbf{x})] \\
&\propto T \log|\det \mathbf{W}| + \sum_{t=1}^{T}\sum_{i=1}^{m} \log p(\mathbf{w}_i \mathbf{x}_t) \\
&= \mathcal{J}_{ML}(\mathbf{X}, \mathbf{W}).
\end{aligned} \tag{4.11}$$

Eq. (4.11) is derived by considering two Jacobian factors caused by the variable transformations $\mathbf{z} = \mathbf{g}(\mathbf{y})$ and $\mathbf{y}_t = \mathbf{W}\mathbf{x}_t$. From the perspectives of neural network and information theory, this solution is in accordance with the *infomax* principle for the reason that the entropy is maximized to convey the largest information in the outputs of a neural network system.

Minimum Mutual Information

Nevertheless, it is more intuitive to carry out the information maximization by following the maximum mutual information (MMI) method (Boscolo et al., 2004, Hyvärinen and Oja, 2000, Yang and Amari, 1997). The MMI-ICA algorithm is developed to estimate the demixing matrix $\mathbf{W}_{MMI}$ by maximizing mutual information among different transformed or demixed signals $\mathbf{y} = [y_1 \cdots y_m]^\top$

$$\begin{aligned}
I(y_1, \ldots, y_m) &= \sum_{i=1}^{m} \mathbb{H}(y_i) - \mathbb{H}(\mathbf{y}) \\
&= \int p(\mathbf{y}) \log \frac{p(\mathbf{y})}{\widetilde{p}(\mathbf{y})} \\
&\triangleq \mathcal{D}_{KL}(p(\mathbf{y}) \| \widetilde{p}(\mathbf{y})).
\end{aligned} \tag{4.12}$$

This mutual information follows the concept of differential entropy and is known as the Kullback–Leibler divergence between the joint likelihood function $p(\mathbf{y})$ and the product of individual likelihood functions

$$\widetilde{p}(\mathbf{y}) = \prod_{i=1}^{m} p(y_i). \tag{4.13}$$

Interestingly, such a mutual information is reduced to the negentropy in case of uncorrelated variables $\{y_i\}_{i=1}^{m}$ (Hyvärinen and Oja, 2000). Equivalently, the MMI contrast function can be expressed as the negative of mutual information, which is then maximized,

$$\mathcal{J}_{\text{MMI}}(\mathbf{X}, \mathbf{W}) = -I(y_1, \ldots, y_m)$$

$$= \mathbb{H}(\mathbf{x}) + \log |\det \mathbf{W}| + \sum_{i=1}^{m} \mathbb{E}[\log p(\mathbf{w}_i \mathbf{x})] \qquad (4.14)$$

$$= \mathcal{J}_{\text{ML}}(\mathbf{X}, \mathbf{W}).$$

Here, $\mathbb{H}(\mathbf{x})$ is seen as a constant when calculating $\mathbf{W}$. The MMI contrast function is shown to be the same as the ML contrast function. The MMI-ICA method is implemented by calculating the derivative of Eq. (4.14) with respect to $\mathbf{w}$, which is given by

$$\nabla_{\mathbf{w}} \mathcal{J}_{\text{MMI}}(\mathbf{X}, \mathbf{W}) = (\mathbf{W}^{-1})^{\top} - \mathbb{E}[\boldsymbol{\phi}(\mathbf{y})\mathbf{x}^{\top}] \qquad (4.15)$$

where $\boldsymbol{\phi}(\mathbf{y}) = [\phi(y_1) \cdots \phi(y_m)]^{\top}$ and

$$\phi(y_i) = -p'(y_i)/p(y_i). \qquad (4.16)$$

A computationally efficient approach to MMI-ICA learning algorithm was proposed by Yang and Amari (1997) and can be written as

$$\mathbf{W}^{(\tau+1)} = \mathbf{W}^{(\tau)} - \eta \left(\mathbf{I} - \mathbb{E}[\boldsymbol{\phi}(\mathbf{y})\mathbf{y}^{\top}] \right) \mathbf{W}, \qquad (4.17)$$

where a positive-definite operator $\mathbf{W}^{\top}\mathbf{W}$ was used in the stochastic gradient descent learning procedure. Next, the statistical hypothesis testing is performed to construct a new contrast function for the ICA procedure.

4.2.2 HYPOTHESIS TEST FOR INDEPENDENCE

Traditionally, in multivariate statistical analysis (Anderson, 1984), a hypothesis test problem is formulated to investigate the independence of an m-dimensional vector $\mathbf{y} = [y_1 \cdots y_m]^{\top}$ which is assumed to be Gaussian distributed. The hypothesis test problem is basically setup to evaluate the null hypothesis $\mathbf{H}_0$ against an alternative hypothesis H_1 which are defined as

$H_0 : y_1, y_2, \ldots, y_m$ are mutually independent
$H_1 : y_1, y_2, \ldots, y_m$ are *not* mutually independent

Assuming that $\mathbf{y}$ is Gaussian distributed with an unknown mean vector $\boldsymbol{\mu} = [\mu_1 \cdots \mu_m]^{\top}$ and covariance matrix Σ, the null hypothesis is equivalent to seeing the covariance between two components y_i and y_j, that is,

$$H_0 : \sigma_{ij}^2 = \mathbb{E}[(y_i - \mu_i])(y_j - \mu_j)] = 0, \qquad \text{for all } i \neq j. \qquad (4.18)$$

When any two distinct components $\{(y_i, y_j)\}_{i \neq j}$ are uncorrelated, this implies that $\{y_i\}_{i=1}^{m}$ are mutually independent. Such a property is preserved only for Gaussian distributions. For the case of the null hypothesis H_0, the covariance matrix Σ becomes a diagonal matrix with diagonal entries $\{\sigma_{ii}^2\}_{i=1}^m$, which is expressed by

$$\Sigma_D = \text{diag}\{\sigma_{ii}^2\}. \tag{4.19}$$

Given the training samples $\mathbf{Y} = \{\mathbf{y}_t\}_{t=1}^T$, the solution to optimal hypothesis test problem is derived by formulating the *ratio of likelihoods* between the null, $p(\mathbf{Y}|H_0)$, and alternative, $p(\mathbf{y}|H_1)$, hypotheses as given by

$$\lambda_{\text{LR}} = \frac{p(\mathbf{Y}|H_0)}{p(\mathbf{Y}|H_1)}$$
$$= \frac{\max\limits_{\boldsymbol{\mu}, \Sigma_D} p(\mathbf{Y}|\boldsymbol{\mu}, \Sigma_D)}{\max\limits_{\boldsymbol{\mu}, \Sigma} p(\mathbf{Y}|\boldsymbol{\mu}, \Sigma)}. \tag{4.20}$$

In the calculation of this likelihood ratio (LR), the maximal likelihoods in the numerator and denominator are searched over all possible parameter values in $(\boldsymbol{\mu}, \Sigma_D)$ and $(\boldsymbol{\mu}, \Sigma)$ under the null and alternative hypotheses, H_0 and H_1, respectively. The random test statistics are finally determined by λ_{LR}. By setting a significance level α, we derive a decision threshold λ_α according to the distribution of the test statistics λ_{LR}. We therefore accept the null hypothesis H_0 if λ_{LR} is evaluated to be higher than the decision threshold $\lambda_{\text{LR}} \geq \lambda_\alpha$. In general, the likelihood ratio λ_{LR} is a meaningful confidence measure towards accepting the null hypothesis or verifying sufficient independence in individual components $\{y_i\}_{i=1}^m$. This likelihood ratio sufficiently reflects the confidence towards independent components.

For the application of source separation, we present an ICA algorithm by evaluating or testing the independence among the transformed or demixed signals $\mathbf{y}_t = \mathbf{W}\mathbf{x}_t$ through the contrast function using a likelihood ratio objective function. A meaningful contrast function for estimating the demixing matrix is constructed in accordance with the likelihood ratio test, which is originated by the evidence from the fundamentals of statistics (Anderson, 1984). We would like to estimate the demixing matrix $\mathbf{W}_{\text{LR}}$ through optimizing the likelihood ratio criterion or, equivalently, by maximizing the confidence measure for independence among the demixed components $\{y_i\}_{i=1}^m$. The estimated matrix is capable of optimally separating the mixed signal $\mathbf{x}_t$ into the demixed signal $\mathbf{y}_t$, which is closest to the true independent signal $\mathbf{s}_t$. The optimum in the sense of likelihood ratio is obtained. We accordingly construct the contrast function

$$\mathcal{J}_{\text{LR}}(\mathbf{X}, \mathbf{W}) \triangleq \lambda_{\text{LR}}. \tag{4.21}$$

However, assuming Gaussian likelihoods for the null and alternative hypotheses is prohibited in implementation of this likelihood ratio based ICA for source separation. To deal with this issue, the parametric likelihood ratio objective function using Gaussian distributions should be modified to carry out the estimation of the demixing matrix λ_{LR}.

4.2.3 NONPARAMETRIC LIKELIHOOD RATIO

Data preprocessing using a whitening transformation in the ICA procedure will produce uncorrelated signals. If the demixed signals are assumed to be Gaussian, the uncorrelated signals will be independent so that there is no possibility to improve independence in ICA learning. Assuming Gaussianity in the transformed signals will cause unexpected performance for source separation. In Blanco and Zazo (2003), Jang et al. (2002), Eriksson et al. (2000), the generalized Gaussian distribution and the generalized Lambda distribution were introduced to construct an ICA contrast function which avoided the Gaussian assumption. More interestingly, a nonparametric learning objective was setup for source separation (Boscolo et al., 2004) where a mutual information criterion was constructed by using a nonparametric distribution based on a kernel density estimator. A similar work was also proposed for the ICA algorithm where the quantized distribution function was used (Meinicke and Ritter, 2001). Furthermore, the maximum likelihood contrast function was developed for the ICA procedure where a Gaussian mixture model was used as a smoothed kernel density function in the calculation of the likelihood function (Vlassis and Motomura, 2001).

We are presenting a nonparametric likelihood ratio (NLR) based contrast function as a measure of independence or non-Gaussianity in the ICA procedure for source separation (Chien and Chen, 2006). Gaussian assumption is prevented in fulfillment of NLR-ICA algorithm. The distribution of the transformed signals $\{y_{ti}\}_{i=1}^{m}$ from the observed samples $\{x_{ti}\}_{i=1}^{m}$ under $\mathbf{y}_t = \mathbf{W}\mathbf{x}_t$ is investigated. The nonparametric density of a component y_i is expressed by a Parzen window density function

$$p(y_i) = \frac{1}{Th} \sum_{t=1}^{T} \varphi\left(\frac{y_i - y_{ti}}{h}\right), \qquad i = 1, \ldots, m. \tag{4.22}$$

In Eq. (4.22), φ denotes the Gaussian kernel with a kernel bandwidth h,

$$\varphi(u) = \frac{1}{\sqrt{2\pi}} e^{-\frac{u^2}{2}}. \tag{4.23}$$

The centroid of the Gaussian kernel y_{ti} is given by the ith component of the sample $\mathbf{y}_t$, i.e.,

$$y_{ti} = \mathbf{w}_i \mathbf{x}_t = \sum_{j=1}^{m} w_{ij} x_{tj}. \tag{4.24}$$

The matrix $\mathbf{W}$ is expressed as $\mathbf{W} = [w_{ij}]_{m \times m} = [\mathbf{w}_1^\top \cdots \mathbf{w}_m^\top]^\top$. Definitely, the nonparametric distribution is a non-Gaussian distribution.

The NLR learning objective is derived by incorporating the nonparametric distribution into the maximum likelihood ratio criterion. By following the independence among separated signals $\{y_i\}_{i=1}^{m}$ under the null hypothesis H_0, we come out with a joint distribution of $\mathbf{y}$, which meets the product of individual distributions of $\{y_i\}_{i=1}^{m}$, that is,

$$p(\mathbf{y}|H_0) = \tilde{p}(\mathbf{y}) = \prod_{i=1}^{m} p(y_i). \tag{4.25}$$

Given this property, the NLR objective function is calculated and accumulated from a set of transformed samples $\mathbf{Y} = \{\mathbf{y}_t\}_{t=1}^{T}$ in a form of

$$
\begin{aligned}
\lambda_{\text{NLR}} &= \frac{p(\mathbf{Y}|H_0)}{p(\mathbf{Y}|H_1)} \\[2mm]
&= \frac{\displaystyle\prod_{t=1}^{T}\prod_{i=1}^{m} p(y_{ti})}{\displaystyle\prod_{t=1}^{T} p(\mathbf{y}_t)} \\[2mm]
&= \frac{\displaystyle\prod_{t=1}^{T}\prod_{i=1}^{m}\left[\frac{1}{Th}\sum_{k=1}^{T}\varphi\left(\frac{y_{ti}-y_{ki}}{h}\right)\right]}{\displaystyle\prod_{t=1}^{T}\left[\frac{1}{Th^m}\sum_{k=1}^{T}\psi\left(\frac{\mathbf{y}_t-\mathbf{y}_k}{h}\right)\right]}
\end{aligned}
\tag{4.26}
$$

where ψ denotes the Gaussian kernel at $\mathbf{v} \in \mathbb{R}^m$ which is expressed by

$$
\psi(\mathbf{v}) = \frac{1}{\sqrt{(2\pi)^m}}\exp\left\{-\frac{1}{2}\mathbf{v}^\top \mathbf{v}\right\}.
\tag{4.27}
$$

Notably, this function λ_{NLR} is calculated without specifying any parametric distributions for the component y_i and vector $\mathbf{y}$. We therefore construct the NLR contrast function λ_{NLR} which is a difference of log-likelihoods of the null hypothesis $L_0(\mathbf{W})$ against an alternative hypothesis $L_1(\mathbf{W})$ where the nonparametric likelihood functions are applied as follows:

$$
\begin{aligned}
\mathcal{J}_{\text{NLR}}(\mathbf{X}, \mathbf{W}) &= \log \lambda_{\text{NLR}} \\
&= L_0(\mathbf{W}) - L_1(\mathbf{W}).
\end{aligned}
\tag{4.28}
$$

This contrast function $\mathcal{J}_{\text{NLR}}(\mathbf{X}, \mathbf{W})$ is optimized to estimate the demixing matrix $\mathbf{W}_{\text{NLR}}$. In the implementation, the gradient descent algorithm is run by calculating the gradients $\nabla_{\mathbf{W}} L_0(\mathbf{W})$ and $\nabla_{\mathbf{W}} L_1(\mathbf{W})$ of the null and alternative hypotheses as follows:

$$
\begin{aligned}
\frac{\partial L_0(\mathbf{W})}{\partial w_{ij}} &= \sum_{t=1}^{T}\left\{\frac{\frac{1}{Th}\sum_{k=1}^{T}\frac{\partial}{\partial w_{ij}}\varphi\left(\frac{1}{h}\sum_{j=1}^{m} w_{ij}(x_{tj}-x_{kj})\right)}{\frac{1}{Th}\sum_{k=1}^{T}\varphi\left(\frac{\mathbf{w}_i(\mathbf{x}_t-\mathbf{x}_k)}{h}\right)}\right\} \\[3mm]
&= -\sum_{t=1}^{T}\left\{\frac{\sum_{k=1}^{T}\mathbf{w}_i(\mathbf{x}_t-\mathbf{x}_k)(x_{tj}-x_{kj})\varphi\left(\frac{\mathbf{w}_i(\mathbf{x}_t-\mathbf{x}_k)}{h}\right)}{h^2\sum_{k=1}^{T}\varphi\left(\frac{\mathbf{w}_i(\mathbf{x}_t-\mathbf{x}_k)}{h}\right)}\right\}
\end{aligned}
\tag{4.29}
$$

and

$$\nabla_{\mathbf{W}} L_1(\mathbf{W}) = -\sum_{t=1}^{T} \left\{ \frac{\sum_{k=1}^{T} \mathbf{W}(\mathbf{x}_t - \mathbf{x}_k)(\mathbf{x}_t - \mathbf{x}_k)^\top \psi\left(\frac{\mathbf{W}(\mathbf{x}_t - \mathbf{x}_k)}{h}\right)}{h^2 \sum_{k=1}^{T} \psi\left(\frac{\mathbf{W}(\mathbf{x}_t - \mathbf{x}_k)}{h}\right)} \right\}, \qquad (4.30)$$

respectively. The adjustment for iterative learning of $\mathbf{W}_{\text{NLR}}$ is obtained by

$$\mathbf{W}^{(\tau+1)} = \mathbf{W}^{(\tau)} - \eta \left(\nabla_{\mathbf{W}} L_0(\mathbf{W}) - \nabla_{\mathbf{W}} L_1(\mathbf{W}) \right). \qquad (4.31)$$

We equivalently optimize the confidence towards demixing $\mathbf{x}_t$ into $\mathbf{y}_t$ using the derived parameter $\mathbf{W}_{\text{NLR}}$. The NLR-ICA algorithm is developed and implemented for multichannel source separation. The NLR contrast function is originated from a hypothesis test which is different from information maximization in ML, ME and MMI contrast functions.

In the literature, the test of independence based on the characteristic function was performed to identify the independent sources in the presence of crosstalk interference (Murata, 2001). An MMI contrast function was proposed to carry out an ICA procedure where the nonparametric likelihood function was applied in calculation of mutual information (Boscolo et al., 2004). This section presents the NLR contrast function to estimate a demixing matrix and fulfill an ICA solution by following hypothesis test theory. Although the MMI and NLR contrast functions are derived from different perspectives, it is attractive to explore their relation. To do so, the negative of mutual information in Eq. (4.12) can be rewritten by

$$\mathbb{E}[\log(\widetilde{p}(\mathbf{y})/p(\mathbf{y}))], \qquad (4.32)$$

which is seen as an expectation of the log-likelihood ratio for the null hypothesis against alternative hypothesis. The NLR-ICA algorithm is implemented by running the ICA learning procedure as shown in Fig. 2.2 where the stopping criterion for learning process is setup by $\lambda_{\text{NLR}} \geq \lambda_\alpha$. There are two properties in this procedure. First, the value of λ_{NLR} is bounded by 1. This bound is reached when $p(\mathbf{y}) = \widetilde{p}(\mathbf{y}) = \prod_{i=1}^{m} y_i$. Second, we preserve the robustness in the ICA transformation by performing a normalization $\mathbf{w}_i \leftarrow \mathbf{w}_i / \|\mathbf{w}_i\|$.

4.2.4 APPLICATION AND EVALUATION

We present a couple of tasks and experiments to evaluate how the ICA algorithm is developed for speech applications and how the nonparametric likelihood ratio performs when compared with other ICA algorithms. The first task is an application of ICA for speech recognition. Different from the independent voices for noise adaptive speech recognition in Section 4.1, NLR-ICA adopts an advanced contrast function in the ICA learning procedure. In particular, the ICA method is applied for clustering of acoustic models. The pronunciation variations are characterized by these clusters and accordingly compensated for robust speech recognition. In addition, in the second task, NLR-ICA is evaluated for blind source separation of speech and music signals where multichannel source separation is performed.

Model Clustering

The first task is presented to investigate the effectiveness of ICA unsupervised learning for multiple acoustic models based on hidden Markov models (HMMs) which is evaluated by speech recognition performance. The ICA method is applied to find the independent components or estimate the representative clusters. The motivations of constructing multiple sets of HMMs for robust speech recognition are addressed below. First of all, the traditional speaker-independent acoustic models trained from a large pool of utterances and speakers are prone to be too complicated and heterogeneous. Inhomogeneous pronunciation sources are mixed during training of acoustic models. Speech recognition performance will degrade. Second, the individuality and inhomogeneity in speech data are seen as *a priori* information which is valuable to conduct sophisticated modeling. Multiple clusters of acoustic models or speech HMMs are accordingly constructed to compensate pronunciation variations for improved speech recognition (Singh et al., 2002). In general, model clusters contain the statistics for pronunciation variations in speech signals due to gender, accent, emotion, age, to name a few. Multiple clusters can be generated by applying the k-means algorithm using the aligned speech utterances corresponding to an acoustic unit. The nearest model cluster $\Theta_{\widehat{k}}$ from a total of K clusters is selected by

$$\widehat{k} = \arg\max_k p(\mathbf{X}|\Theta_k), \tag{4.33}$$

and used to recognize test speech in a new environment. It is more likely that the matching between test speech and HMM clusters brings better performance in speaker-independent speech recognition.

The speech data $\mathbf{X}$ are usually degraded by noise interference and mixed with various distortion sources, e.g., alignment errors, speaker variabilities and accent changes. Source separation using ICA can be introduced to identify the independent sources $\mathbf{S}$ or estimate the demixed components $\mathbf{Y}$ from the mixed data $\mathbf{X}$. The model clusters or HMM clusters are then estimated via the k-means algorithm to characterize unknown variations. The ICA algorithm is feasible to analyze an acoustic model as well as compensate varying sources. We would like to construct an independent space by using different columns of the demixing matrix $\mathbf{W}$. The observed vectors $\mathbf{x}_t = \{x_{ti}\}_{i=1}^m$ are transformed to that space to simulate the independent vectors $\mathbf{s}_t = \{s_{ti}\}_{i=1}^m$. Fig. 4.6 illustrates how a multiple-HMM speech recognition system is constructed by applying the ICA unsupervised learning algorithm. The ICA transformation and k-means clusters are run so as to estimate the model clusters

$$\Theta = \{\Theta_1, \ldots, \Theta_K\} \tag{4.34}$$

for each acoustic unit. Multiple sets of HMMs are trained for generalization to unknown test speakers and experimental conditions. Correspondingly, we build an extended lexicon due to multiple clusters of acoustic models. Augmenting the lexicon is helpful to deal with unknown pronunciation mismatches in speech recognition (Fukada et al., 1999, Singh et al., 2002).

An NLR-ICA algorithm is developed for the HMM clustering procedure. In the beginning, the forced alignment is performed for all training data via dynamic programming based on the Viterbi decoder (Rabiner and Juang, 1986). The feature vectors aligned to the same acoustic unit or segment are put together. For the case of using an HMM unit, we calculate sample means $\bar{\mathbf{x}}_{s_n}$ corresponding to all HMM states $\{s_n\}_{n=1}^N$ under this HMM unit. We then generate supervectors for each segment, which consist of sample means across different HMM states as

$$\mathbf{x} = [\bar{\mathbf{x}}_{s_1}^\top \bar{\mathbf{x}}_{s_2}^\top \cdots \bar{\mathbf{x}}_{s_N}^\top]^\top. \tag{4.35}$$

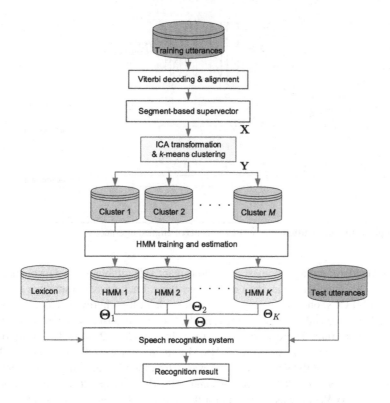

FIGURE 4.6

ICA transformation and k-means clustering for speech recognition with multiple hidden Markov models.

Fig. 4.7 shows how supervectors $\mathbf{X} = \{\mathbf{x}_t\}_{t=1}^{T}$ belonging to the same target acoustic unit are generated from different segments and utterances. After constructing a supermatrix $\mathbf{X}$, we run an ICA algorithm (or specifically, the NLR-ICA algorithm) to estimate the demixing matrix $\mathbf{W}_{\text{NLR}}$ and transform $\mathbf{X}$ into the independent space by

$$\mathbf{Y} = \mathbf{W}_{\text{NLR}}\mathbf{X}. \tag{4.36}$$

It is intuitive that the data samples in $\mathbf{Y}$ are geometrically clustered due to the ICA transformation. The significance of clustering in $\mathbf{Y}$ will be better than that in $\mathbf{X}$. The clustering membership of different acoustic segments can be obtained. We finally estimate K sets of continuous-density HMMs (Rabiner and Juang, 1986) for different acoustic units by using the feature vectors in the corresponding segments belonging to the same clusters. The mixture coefficients, mixture mean vectors and mixture covariance matrices are estimated for different states and HMM clustered.

ICA algorithms are applied for HMM clustering and evaluated for speech recognition by using a Mandarin speech corpus containing training sentences from 80 speakers and test utterances from the other 20 speakers. Gender information is balanced in training and test conditions. Detailed description of HMM construction is given in Chien and Chen (2006). The syllable error rates (SERs) (%) are

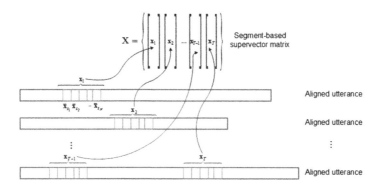

FIGURE 4.7

Generation of supervectors from different acoustic segments of aligned utterances.

reported. Each Mandarin syllable consists of a consonant subsyllable and a vowel subsyllable. Each acoustic unit was characterized by an individual subsyllable. Each speech frame was represented by 12 MFCCs, 12 delta MFCCs, one log-energy and one delta-delta-log-energy. The consonants and vowels were represented by three and five states, respectively. Model clustering was performed on the basis of an acoustic or subword unit. There were two or four clusters ($K = 2$ or 4) depending on the number of segments aligned to a specific subword or HMM unit. K was varied and controlled by a decision threshold. The dimension of the supervector, m, was determined by the dimension of features and the number of states, namely, $m = 26 \times 3$ for consonant HMMs and $m = 26 \times 5$ for vowel HMMs. The $m \times m$ demixing matrix $\mathbf{W}_{\mathrm{NLR}}$ was estimated for the ICA transformation of $\mathbf{x}_t \in \mathbb{R}^m$.

Model clustering and ICA learning are evaluated by the experiments on speech recognition. The traditional method without model clustering was treated as a baseline system. The effect of model clustering using the samples in the original space $\mathbf{X}$ and independent space $\mathbf{Y}$ was examined. The ICA procedure of Fig. 4.6 was carried out for model clustering where the MMI contrast function in Eq. (4.14) and NLR contrast function in Eq. (4.28) were optimized. MMI-ICA was implemented to find the demixing matrix for the transformation from $\mathbf{X}$ to $\mathbf{Y}$ where the mutual information among demixed components $\{y_i\}_{i=1}^m$ was minimized. NLR-ICA was implemented to achieve the highest confidence measure for finding independent components. The ICA algorithm based on nonparametric mutual information (NMI) (Boscolo et al., 2004) (also denoted as NMI-ICA) was implemented for a comparative study.

Table 4.1 reports a comparison of SERs with and without model clustering and ICA learning. The ICA algorithms using NMI, NMI and NLR are compared. The baseline result of 37.4% is reduced to 36.5% when model clustering is performed in the original space $\mathbf{X}$. The SERs are substantially reduced by model clustering or HMM clustering in the independent space $\mathbf{Y}$ using different ICA methods. The lowest SER 31.4% was achieved by unsupervised learning of multiple models using NLR-ICA. This result is lower than 33.6% and 32.5% by using MMI-ICA and NMI-ICA methods, respectively. In general, the ICA transformation is helpful for model clustering. Nonparametric distribution when realizing the ICA method performs better than the parametric approach in terms of SER. The SER of using the likelihood ratio objective is lower than that of using the mutual information objective in model clustering.

Table 4.1 Comparison of syllable error rates (SERs) (%) with and without HMM clustering and ICA learning. Different ICA algorithms are evaluated

Model clustering	ICA learning	SER (%)
No	No	37.4
Yes	No	36.5
Yes	MMI	33.6
Yes	NMI	32.5
Yes	NLR	31.4

Table 4.2 Comparison of signal-to-interference ratios (SIRs) (dB) of mixed signals without ICA processing and with ICA learning based on MMI and NLR contrast functions

ICA learning	No processing	MMI	NLR
SIR	−1.76	5.80	17.93

Signal Separation

Next, the task of multichannel source separation was completed and evaluated for blind source separation (BSS) using different ICA methods where two-channel speech and music signals were collected. The benchmark data from ICA'99 BSS Test Sets (Schobben et al., 1999) were included. The speech utterance sampled from a male (http://sound.media.mit.edu/ica-bench/sources/mike.wav) and the music signal sampled from the Beethoven-Symphony 5 (http://sound.media.mit.edu/ica-bench/sources/beet.wav) were collected for evaluation. Two source signals were mixed by a specific 2×2 mixing matrix **A** to generate the mixed signals where the delayed and convolved sources were not considered. The experimental setup is detailed in Chien and Chen (2006). The signal-to-interference ratio (SIR) (dB), defined in Eq. (2.10), was measured to examine the performance of blind source separation in the presence of speech and music signals. Table 4.2 shows the SIRs of mixed signal without ICA processing and with ICA learning based on the MMI and NLR objective functions. The SIR value is significantly improved from -1.76 dB without ICA processing to 5.8 and 17.93 dB with ICA methods based on the MMI and NLR contrast functions, respectively.

A number of information-theoretic learning algorithms based on ML, ME and MMI contrast functions have been surveyed for construction of ICA approaches to multichannel source separation. In particular, a statistical solution to a hypothesis test has been introduced to calculate the measure or the confidence of independence over the transformed signals based on a nonparametric distribution function. The nonparametric likelihood ratio is accordingly presented and applied for speech recognition and separation. The ICA procedure is shown effective for clustering of acoustic models or, equivalently, finding the acoustic clusters or sources to improve speech recognition performance. In the next section, information-theoretic learning is further explored to construct an ICA model for multichannel source separation. An advanced ICA algorithm is created employing a flexible and general contrast function based on convex divergence.

4.3 CONVEX DIVERGENCE ICA

Independent component analysis (ICA) is a crucial mathematical tool for unsupervised learning and blind source separation. ICA is capable of identifying the salient features or estimating the mutually-independent sources. A key success to ICA learning depends on a meaningful contrast function which measures the independence of the demixed or the transformed signals. This section presents an information-theoretic contrast function where the independence is measured from the perspective of information theory. A convex divergence measure is developed by adopting the inequality property in a divergence measure where a general convex function and Jensen's inequality are applied. The contrast function is therefore controlled by a convexity factor which enriches a variety of learning curves in finding an optimum of the ICA process as addressed in Section 4.3.1. Different divergence measures are evaluated. This convexity-based contrast function is also feasible to implement a new type of nonnegative matrix factorization. In Section 4.3.3, the nonparametric solution to convex divergence ICA (also called C-ICA) is addressed. Non-Gaussian source signals are expressed by the Parzen window density function in an implementation of nonparametric C-ICA. Blind source separation of speech and audio signals is investigated under the convolutive mixing condition with additive noise. Section 4.3.2 presents the implementation of C-ICA for finding a solution to the problem of nonnegative matrix factorization. Sections 4.3.4 and 4.3.5 address the experiments on different datasets with various mixing conditions in terms of convergence rate and SIR. The instantaneous, noisy and convolutive mixing conditions in the presence of speech and audio signals are evaluated.

4.3.1 CONVEX DIVERGENCE

Information-theoretic learning plays a crucial role for ICA-based source separation. In general, information-theoretic contrast functions are meaningful to evaluate the independence or non-Gaussianity of demixed signals. Following up the discussion on the ICA objective function in Section 4.2.1, we continue the survey of information-theoretic contrast functions based on mutual information. Mutual information is basically calculated as the difference between the marginal and joint entropies over various information sources as defined in Eqs. (3.2) and (4.12). Considering the case of two demixed signals y_1 and y_2, the standard mutual information is defined and yielded as the Kullback–Leibler (KL) divergence between joint distribution and product of marginal distributions of y_1 and y_2, that is,

$$
\begin{aligned}
I(y_1, y_2) &= \mathbb{H}[p(y_1)] + \mathbb{H}[p(y_2)] - \mathbb{H}[p(y_1, y_2)] \\
&= \int \int p(y_1, y_2) \log \frac{p(y_1, y_2)}{p(y_1)p(y_2)} dy_1 dy_2 \\
&\triangleq \mathcal{D}_{\mathrm{KL}}(y_1, y_2).
\end{aligned}
\tag{4.37}
$$

Here $\mathcal{D}_{\mathrm{KL}} \geq 0$ with equality if and only if y_1 and y_2 are independent with $p(y_1, y_2) = p(y_1)p(y_2)$. As addressed in Section 3.2.1, different divergence measures were presented. In addition to KL divergence (KL-DIV), the Euclidean divergence (E-DIV), Cauchy–Schwartz divergence (CS-DIV), α divergence (α-DIV), f divergence (f-DIV) and Jensen–Shannon divergence (JS-DIV) have been described in Eqs. (3.3), (3.4), (3.5), (3.6) and (3.9), respectively. The categorization of ICA contrast functions with different realizations of mutual information is illustrated in Fig. 4.8. This section presents a new diver-

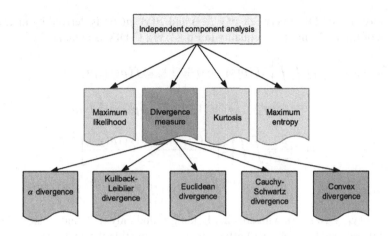

FIGURE 4.8

Taxonomy of ICA contrast functions where different realizations of mutual information are included.

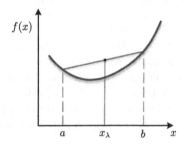

FIGURE 4.9

Illustration of a convex function $f(x)$.

gence measure, called the convex divergence, and addresses how this convex divergence is merged in an ICA procedure for multichannel source separation.

Fig. 4.9 illustrates a convex function $f(x)$ with an interpolated input x_λ between a and b which satisfies the inequality

$$f(\lambda a + (1 - \lambda)b) \leq \lambda f(a) + (1 - \lambda) f(b). \tag{4.38}$$

Importantly, the divergence measure based on a convex function can be generalized by incorporating the joint distribution $p(y_1, y_2)$ and the product of marginals $p(y_1)p(y_2)$ into a convex function $f(\cdot)$ with a convex combination weight $0 \leq \lambda \leq 1$. Jensen's inequality is yielded by

$$\begin{aligned} f(\lambda p(y_1, y_2) + (1 - \lambda) p(y_1)p(y_2)) \\ \leq \lambda f(p(y_1, y_2)) + (1 - \lambda) f(p(y_1)p(y_2)). \end{aligned} \tag{4.39}$$

The convex divergence (C-DIV) between $p(y_1, y_2)$ and $p(y_1)p(y_2)$ is derived by merging a general convex function of Eq. (3.7) into the inequality in Eq. (4.39). C-DIV is obtained by

$$
\mathcal{D}_C(y_1, y_2, \alpha) = \int \int \Big\{ \lambda f(p(y_1, y_2)) + (1 - \lambda) f(p(y_1)p(y_2))
$$

$$
- f(\lambda p(y_1, y_2) + (1 - \lambda)p(y_1)p(y_2)) \Big\} dy_1 dy_2
$$

$$
= \frac{2}{1 - \alpha^2} \int \int \Big\{ 2 \Big[\frac{p(y_1, y_2) + p(y_1)p(y_2)}{2} \Big]^{(1+\alpha)/2}
$$

$$
- \Big[p(y_1, y_2)^{(1+\alpha)/2} + (p(y_1)p(y_2))^{(1+\alpha)/2} \Big] \Big\} dy_1 dy_2,
$$

(4.40)

which satisfies $\mathcal{D}_C(y_1, y_2, \alpha) \geq 0$, with equality holding if and only if y_1 and y_2 are independent. It is meaningful that the second RHS of Eq. (4.40) is obtained by treating the contributions of $p(y_1, y_2)$ and $p(y_1)p(y_2)$ equally, i.e., setting the condition of $\lambda = 1/2$. Under this condition, the terms of the joint distribution and product of marginals can be permuted. However, the permutation property is *not* valid for the other divergence measures including C-DIV with other settings of λ. C-DIV is controlled by a tunable convexity parameter α which is also included in α-DIV. But C-DIV in Eq. (4.40) is derived from the inequality given in Eq. (4.39) while α-DIV in Eq. (3.5) is obtained by using the relative entropy provided in Eqs. (4.37) and (3.6). C-DIV balances the tradeoff between joint distribution and the product of marginals through the tunable parameter $0 \leq \lambda \leq 1$. This tunable parameter is seen as a convex combination coefficient to calculate the weighted sum for a convex hull. Nevertheless, the affine combination is performed if a real number $\lambda \in \mathbb{R}$ is considered (Boyd and Vandenberghe, 2004). The problem of using an affine combination is that the affine hull may contain the functions which are out of the domain in Zhang's divergence (Zhang, 2004). The convex hull does not suffer from this problem.

We address different realizations of a general convex divergence. Different convex functions are considered. Eq. (3.9) shows the JS-DIV, which is also denoted as the convex-Shannon divergence where the convex function $-\mathbb{H}[\cdot]$ is adopted. The convex-logarithm divergence is implemented by using the convex function $-\log(\cdot)$ while the convex-exponential divergence is realized by the applying convex function $\exp(\cdot)$. Importantly, C-DIV in Eq. (4.40) is a general contrast function which is adjustable by a control parameter α for convexity in a convex function. Special realizations of C-ICA using $\alpha = 1$ and $\alpha = -1$ can be implemented. Owing to zero values in the numerator and denominator for special cases of $\alpha = 1$ and $\alpha = -1$, L'Hopital's rule is employed to take derivatives with respect to α and calculate the result by passing to the limit $\lim_{\alpha \to \pm 1}$. C-DIV in the case of $\alpha = 1$ is realized by

$$
\mathcal{D}_C(y_1, y_2, 1) = - \int \int \Big\{ \Big[\frac{p(y_1, y_2) + p(y_1)p(y_2)}{2} \Big] \log \Big[\frac{p(y_1, y_2) + p(y_1)p(y_2)}{2} \Big]
$$

$$
- \frac{1}{2} p(y_1, y_2) \log p(y_1, p_2) - \frac{1}{2} p(y_1)p(y_2) \log(p(y_1)p(y_2)) \Big\} dy_1 dy_2
$$

(4.41)

$$
\triangleq \mathcal{D}_{JS}(y_1, y_2).
$$

This special case corresponds to implementing the convex-Shannon divergence or Jensen–Shannon divergence (JS-DIV) as given in Eq. (3.9). On the other hand, the convex divergence (C-DIV) with

Table 4.3 Comparison of different divergence measures with respect to symmetric divergence, convexity parameter, combination weight and special realization

	Symmetric divergence	Convexity parameter	Combination weight	Realization from other DIV
KL-DIV	no	no	no	α-DIV ($\alpha = -1$)
α-DIV	no	α	no	f-DIV
f-DIV	no	α	no	no
JS-DIV	$\lambda = \frac{1}{2}$	no	λ	C-DIV ($\lambda = \frac{1}{2}, \alpha = 1$)
Convex-log DIV	$\lambda = \frac{1}{2}$	no	λ	C-DIV ($\lambda = \frac{1}{2}, \alpha = -1$)
C-DIV	$\lambda = \frac{1}{2}$	α	λ	no

$\alpha = -1$ can be expressed as

$$\mathcal{D}_C(y_1, y_2, -1) = \int \int \left\{ \log \left[\frac{p(y_1, y_2) + p(y_1)p(y_2)}{2} \right] \right. $$
$$\left. - \frac{1}{2} \log p(y_1, y_2) - \frac{1}{2} \log(p(y_1)p(y_2)) \right\} dy_1 dy_2. \tag{4.42}$$

This expression is seen as a convex-logarithm divergence which incorporates the convex function $-\log(\cdot)$ into the divergence measure based on Jensen's inequality. Table 4.3 compares the properties of different divergence measures. The properties of symmetric divergence, convexity parameter, combination weight and special realization are investigated. It is intuitive that C-DIV can be extended to evaluate the dependence $\mathcal{D}_C(y_1, \ldots, y_m, \alpha)$ among m channel observations in $\mathbf{y} = [y_1 \cdots y_m]^\top$. More generally, C-DIV is used to measure the divergence between any two distribution functions. The measurement of dependence using C-ICA can be employed in the ICA procedure.

It is interesting to investigate different divergence measures ranging from KL-DIV to E-DIV, CS-DIV, α-DIV, JS-DIV and C-DIV. The cases of $\alpha = -1$, $\alpha = 0$ and $\alpha = 1$ are included in α-DIV and C-DIV. In this evaluation, we consider two binomial variables $\{y_1, y_2\}$ with two events $\{A, B\}$ (Lin, 1991, Principe et al., 2000). The joint probabilities of $P_{y_1, y_2}(A, A)$, $P_{y_1, y_2}(A, B)$, $P_{y_1, y_2}(B, A)$ and $P_{y_1, y_2}(B, B)$, and the marginal probabilities of $P_{y_1}(A)$, $P_{y_1}(B)$, $P_{y_2}(A)$ and $P_{y_2}(B)$ are characterized. Different divergence measures are calculated by fixing the marginal probabilities $\{P_{y_1}(A) = 0.6, P_{y_1}(B) = 0.4\}$. The joint probabilities $P_{y_1, y_2}(A, A)$ and $P_{y_1, y_2}(B, A)$ are evaluated in the ranges of $(0, 0.6)$ and $(0, 0.4)$, respectively. Fig. 4.10(A) illustrates the relations among a number of divergence measures by varying joint probability $P_{y_1, y_2}(A, A)$ where the condition $\{P_{y_2}(A) = 0.5, P_{y_2}(B) = 0.5\}$ is fixed. Different divergence measures reach the same minimum at $P_{y_1, y_2}(A, A) = 0.3$ where y_1 and y_2 are independent. Fig. 4.10(B) compares α-DIV and C-DIV for various parameters α. This tuning parameter adjusts the convexity or changing rate of the curves. Basically, the changing rate (or slope) of the curves in α-DIV due to different α is less sensitive when compared with that in C-DIV. In this comparison, C-DIV with $\alpha = -1$ has the steepest curve while C-DIV with $\alpha = 1$ (or JS-DIV) shows the flattest curve. The changing rate of the curves in C-DIV is increased by reducing the values of α. However, the slope is steeper with a larger α in α-DIV. Among different divergence measures, the slope of curve in C-DIV with $\alpha = -1$ is the steepest so that the minimum divergence can be achieved efficiently. Basically, the evaluation of the probability model mimics that of the contrast function in the

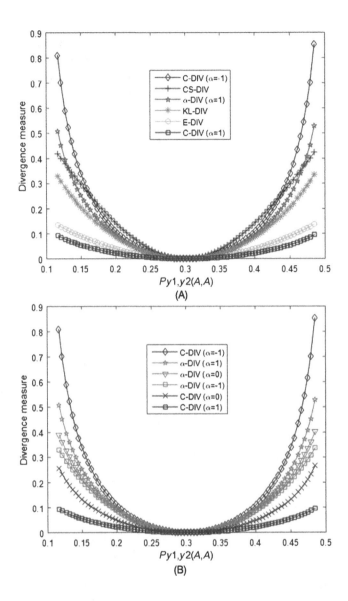

FIGURE 4.10

Comparison of (A) a number of divergence measures and (B) α-DIV and C-DIV for various α under different joint probability $P_{y_1,y_2}(A, A)$.

ICA procedure when finding the demixing matrix. The contrast function based on C-DIV with $\alpha = -1$ is favorable for ICA learning. In general, C-DIV and other divergence measures are not only useful for estimating the demixing matrix $\mathbf{W}_{\text{C-DIV}}$ in an ICA procedure but also feasible for finding the basis and weight matrices for the NMF procedure, which is described in what follows.

4.3.2 DIVERGENCE MEASURE FOR NMF

This section addresses how C-DIV is developed for nonnegative matrix factorization (NMF). Basically, NMF aims to factorize a nonnegative matrix $\mathbf{X} \in \mathbb{R}_+^{M \times N}$ as a multiplication of nonnegative matrix $\mathbf{B} \in \mathbb{R}_+^{M \times K}$ and nonnegative matrix $\mathbf{W} \in \mathbb{R}_+^{K \times N}$ where $K \leq \min(M, N)$. It is popular to develop NMF for single-channel source separation in the nonnegative spectrum domain. The learning objectives in the standard NMF are measured as the error functions based on Euclidean distance and KL divergence (Lee and Seung, 1999, 2000):

$$
\begin{aligned}
\mathcal{D}_{\text{EU}}(\mathbf{X} \| \mathbf{BW}) &= \| \mathbf{X} - \mathbf{BW} \|^2 \\
&= \sum_{m=1}^{M} \sum_{n=1}^{N} ([\mathbf{X}]_{mn} - [\mathbf{BW}]_{mn})^2,
\end{aligned}
\tag{4.43}
$$

$$
\begin{aligned}
&\mathcal{D}_{\text{KL}}(\mathbf{X} \| \mathbf{BW}) \\
&= \sum_{m=1}^{M} \sum_{n=1}^{N} \left[[\mathbf{X}]_{mn} \log \frac{[\mathbf{X}]_{mn}}{[\mathbf{BW}]_{mn}} - [\mathbf{X}]_{mn} + [\mathbf{BW}]_{mn} \right]
\end{aligned}
\tag{4.44}
$$

where $[\mathbf{BW}]_{mn}$ reveals the element (m, n) in $\mathbf{BW}$. Similar to ICA learning based on minimization of dependence or divergence measure between the demixed signals, NMF learning is performed to minimize the reconstruction loss of $\mathbf{X}$ due to factorization $\mathbf{BW}$ where the loss function is measured by a divergence measure. In Cichocki et al. (2008), α divergence was used as a loss function for NMF based on

$$
\begin{aligned}
\mathcal{D}_{\alpha}(\mathbf{X} \| \mathbf{BW}) = \frac{4}{1 - \alpha^2} \sum_{m=1}^{M} \sum_{n=1}^{N} &\left[\frac{1-\alpha}{2}[\mathbf{X}]_{mn} \right. \\
&\left. + \frac{1+\alpha}{2}[\mathbf{BW}]_{mn} - [\mathbf{X}]_{mn}^{(1-\alpha)/2}[\mathbf{BW}]_{mn}^{(1+\alpha)/2} \right].
\end{aligned}
\tag{4.45}
$$

However, the NMF learning procedure is prone to encounter an incomplete data matrix with the problem of missing elements. To tackle this problem, the weighted NMF (WNMF) (Kim and Choi, 2009) was developed by merging the nonnegative weights $\{\lambda_{mn}\}$ in a loss function as given by

$$
\begin{aligned}
&\mathcal{D}_{\text{EU}}(\mathbf{X} \| \mathbf{BW}, \lambda) \\
&= \sum_{m=1}^{M} \sum_{n=1}^{N} \lambda_{mn} ([\mathbf{X}]_{mn} - [\mathbf{BW}]_{mn})^2
\end{aligned}
\tag{4.46}
$$

with $\lambda_{mn} = 1$ if $[\mathbf{X}]_{mn}$ is observed and $\lambda_{mn} = 0$ if $[\mathbf{X}]_{mn}$ is unobserved. Attractively, WNMF can be implemented by substituting $\mathbf{X}$ and $\mathbf{BW}$ into C-DIV in Eq. (4.40) to form

$$
\begin{aligned}
\mathcal{D}_C(\mathbf{X}\|\mathbf{BW}, \alpha) = \sum_{m=1}^{M}\sum_{n=1}^{N} & \left\{ \frac{4\lambda_{mn}}{1-\alpha^2} \left[\frac{1-\alpha}{2} + \frac{1+\alpha}{2}[\mathbf{X}]_{mn} - [\mathbf{X}]_{mn}^{(1+\alpha)/2} \right] \right. \\
& + \frac{4(1-\lambda_{mn})}{1-\alpha^2} \left[\frac{1-\alpha}{2} + \frac{1+\alpha}{2}[\mathbf{BW}]_{mn} - [\mathbf{BW}]_{mn}^{(1+\alpha)/2} \right] \\
& - \frac{4}{1-\alpha^2} \left[\frac{1-\alpha}{2} + \frac{1+\alpha}{2}(\lambda_{mn}[\mathbf{X}]_{mn} + (1-\lambda_{mn})[\mathbf{BW}]_{mn}) \right. \\
& \left. \left. - (\lambda_{mn}[\mathbf{X}]_{mn} + (1-\lambda_{mn})[\mathbf{BW}]_{mn})^{(1+\alpha)/2} \right] \right\}.
\end{aligned}
\tag{4.47}
$$

The nonnegative weights in WNMF can be seen as the convex coefficients $\{\lambda_{mn}\}$, namely $\lambda_{mn} = \frac{1}{2}$ when $[\mathbf{X}]_{mn}$ is observed and $\lambda_{mn} = 0$ when $[\mathbf{X}]_{mn}$ is unobserved. The NMF based on C-DIV is realized if all weights in Eq. (4.47) are the same as $\lambda_{mn} = \frac{1}{2}$. Due to the nonnegativity constraint, NMF is implemented when audio signals are represented in either the magnitude or power spectrum domain. The details of NMF for single-channel source separation will be described in Chapter 5. This chapter presents ICA learning and multichannel source separation by only considering the real-valued speech and audio signals in the time domain. The convex divergence ICA procedure is addressed in the next subsection.

4.3.3 ICA PROCEDURE

The ICA procedure based on convex divergence ICA (C-ICA) is implemented for multichannel source separation when the number of channels, n, is the same as the number of sources, m. Following the standard ICA learning procedure as addressed in Section 2.1.1, C-ICA is established by finding an $m \times m$ demixing matrix $\mathbf{W}_{C\text{-DIV}}$, which achieves the maximum of independence or the minimum of C-DIV in Eq. (4.40) for the demixed signals $\mathbf{y} = \{y_i\}_{i=1}^{m}$. The implementation needs to calculate the differential $\partial \mathcal{D}_C(\mathbf{X}, \mathbf{W}, \alpha)/\partial w_{ij}$. Considering the demixed signals $\mathbf{y}_t = \mathbf{W}\mathbf{x}_t$ at time t, where $y_{ti} = \mathbf{w}_i\mathbf{x}_t$, the contrast function based on C-DIV is calculated with a control or convexity parameter α as

$$
\begin{aligned}
\mathcal{D}_C(\mathbf{X}, \mathbf{W}, \alpha) = \frac{2}{1-\alpha^2}\sum_{t=1}^{T} & \left\{ 2\left[\frac{1}{2}\left(p(\mathbf{W}\mathbf{x}_t) + \prod_{i=1}^{m} p(\mathbf{w}_i\mathbf{x}_t) \right) \right]^{(1+\alpha)/2} \right. \\
& \left. - \left[p(\mathbf{W}\mathbf{x}_t)^{(1+\alpha)/2} + \left(\prod_{i=1}^{m} p(\mathbf{w}_i\mathbf{x}_t) \right)^{(1+\alpha)/2} \right] \right\}.
\end{aligned}
\tag{4.48}
$$

The Lebesgue measure is employed to approximate the integral in Eq. (4.40) over the distribution of $\mathbf{y}_t = \{y_{ti}\}_{i=1}^{m}$. We therefore build the C-ICA learning procedure by deriving the derivative

$\partial \mathcal{D}_C(\mathbf{X}, \mathbf{W}, \alpha)/\partial w_{ij}$ as

$$
\frac{1}{1-\alpha} \sum_{t=1}^{T} \left\{ \underbrace{\left[\left[\frac{1}{2} \left(\frac{p(\mathbf{x}_t)}{|\det \mathbf{W}|} + \prod_{i=1}^{m} p(\mathbf{w}_i \mathbf{x}_t) \right) \right]^{(\alpha-1)/2} - \left[\frac{p(\mathbf{x}_t)}{|\det \mathbf{W}|} \right]^{(\alpha-1)/2} \right]}_{a_{t1}} \right.
$$

$$
\times \underbrace{\left[-\frac{p(\mathbf{x}_t)}{|\det \mathbf{W}|^2} W_{ij} \operatorname{sign}(\det \mathbf{W}) \right]}_{a_{t2}} + \underbrace{\left[\frac{dp(\mathbf{w}_i \mathbf{x}_t)}{d(\mathbf{w}_i \mathbf{x}_t)} x_{tj} \prod_{l \neq i}^{m} p(\mathbf{w}_l \mathbf{x}_t) \right]}_{a_{t3}}
$$

$$
\left. \times \underbrace{\left[\left[\frac{1}{2} \left(\frac{p(\mathbf{x}_t)}{|\det \mathbf{W}|} + \prod_{i=1}^{m} p(\mathbf{w}_i \mathbf{x}_t) \right) \right]^{(\alpha-1)/2} - \left[\prod_{i=1}^{m} p(\mathbf{w}_i \mathbf{x}_t) \right]^{(\alpha-1)/2} \right]}_{a_{t4}} \right\} \tag{4.49}
$$

where $\operatorname{sign}(\cdot)$ is a sign function, W_{ij} is the cofactor of element (i, j) of matrix $\mathbf{W}$, and x_{tj} is the jth element of $\mathbf{x}_t$. In Eq. (4.49), the joint distribution $p(\mathbf{y}_t) = p(\mathbf{x}_t)/|\det \mathbf{W}|$ is used, and four terms a_{t1}, a_{t2}, a_{t3} and a_{t4} are derived. The term a_{t1} denotes the difference between the joint distribution $p(\mathbf{y}_t)$ and the product of marginals $\prod_{i=1}^{m} p(y_{ti})$ while a_{t4} reveals the average over $p(\mathbf{y}_t)$ and $\prod_{i=1}^{m} p(y_{ti})$. An exponent of $(\alpha - 1)/2$ is incorporated in each of the four brackets. The term a_{t2} denotes the derivative $dp(\mathbf{y}_t)/dw_{ij}$ while a_{t3} reveals the derivative $d\left(\prod_{i=1}^{m} p(y_{ti})\right)/dw_{ij}$. It is reasonable that the C-ICA algorithm is driven by weighting the derivative terms $dp(\mathbf{y}_t)/dw_{ij}$ and $d\left(\prod_{i=1}^{m} p(y_{ti})\right)/dw_{ij}$ using the corresponding weights a_{t1} and a_{t4}. In the starting period of adaptive learning, a_{t1} and a_{t4} provide strong weights for rapid learning. The learning speed is gradually and stably reduced when the weights a_{t1} and a_{t4} are decreased and the parameter updating approaches independence. The ICA procedure is terminated when the parameter $\mathbf{W}_{\text{C-ICA}}$ projects $\mathbf{X}$ onto an independent space with $a_{t1} = 0$ and $a_{t4} = 0$.

The estimation accuracy of a demixing matrix in the ICA procedure is inevitably deteriorated with the wrongly assumed distribution for demixed sources. Such a circumstance happens when Gaussian distribution is assumed. A meaningful idea is to use the Parzen window density function and estimate the nonparametric distribution for flexible modeling of data distribution using the kernel basis function with a width parameter h. The nonparametric ICA algorithms using NMI-ICA (Boscolo et al., 2004) and NLR-ICA (Chien and Chen, 2006) have been described and evaluated in Sections 4.2.3 and 4.2.4. This section demonstrates how convex divergence is minimized to carry out a nonparametric C-ICA for multichannel source separation. The Parzen window density function in Eq. (4.22) is used to express the univariate distribution $p(y_{ti})$ for the demixed signals $\mathbf{y}_t = \{y_{ti}\}_{i=1}^{m}$. The multivariate distribution $p(\mathbf{y}_t)$ is represented by

$$
p(\mathbf{y}_t) = \frac{1}{Th^m} \sum_{k=1}^{T} \psi\left(\frac{\mathbf{y}_t - \mathbf{y}_k}{h} \right). \tag{4.50}
$$

Owing to the smooth and soft function using a Gaussian kernel, the convergence of the learning algorithm using this nonparametric ICA (Boscolo et al., 2004) is basically faster than that using the parametric ICA (Cichocki and Amari, 2002, Douglas and Gupta, 2007) based on parametric distributions. By incorporating Eqs. (4.22) and (4.50) with $\mathbf{y}_t = \mathbf{W}\mathbf{x}_t$ and $y_{ti} = \mathbf{w}_i \mathbf{x}_t$ and substituting them into Eq. (4.48), the contrast function based on nonparametric C-DIV is obtained as

$$
\mathcal{D}_{\mathrm{C}}(\mathbf{X}, \mathbf{W}, \alpha) = \frac{2}{1-\alpha^2} \sum_{t=1}^{T} \left\{ 2 \left[\frac{p(\mathbf{x}_t)}{2|\det \mathbf{W}|} \right. \right.
$$
$$
+ \prod_{i=1}^{m} \frac{1}{2Th} \sum_{k=1}^{T} \varphi \left(\frac{\mathbf{w}_i(\mathbf{x}_t - \mathbf{x}_k)}{h} \right) \Bigg]^{(1+\alpha)/2} \tag{4.51}
$$
$$
- \left[\left[\frac{p(\mathbf{x}_t)}{|\det \mathbf{W}|} \right]^{(1+\alpha)/2} + \left[\prod_{i=1}^{m} \frac{1}{2Th} \sum_{k=1}^{T} \varphi \left(\frac{\mathbf{w}_i(\mathbf{x}_t - \mathbf{x}_k)}{h} \right) \right]^{(1+\alpha)/2} \right] \Bigg\}.
$$

The learning algorithm of nonparametric C-ICA is employed by substituting Eqs. (4.22) and (4.50) into Eq. (4.49) to derive the derivative $\partial \mathcal{D}_{\mathrm{C}}(\mathbf{X}, \mathbf{W}, \alpha)/\partial w_{ij}$ as

$$
\frac{1}{1-\alpha} \sum_{t=1}^{T} \left\{ \left[\frac{1}{2} \left(\frac{p(\mathbf{x}_t)}{|\det \mathbf{W}|} + \prod_{i=1}^{m} p(\mathbf{w}_i \mathbf{x}_t) \right) \right]^{(\alpha-1)/2} \left[-\frac{p(\mathbf{x}_t)}{|\det \mathbf{W}|^2} W_{ij} \operatorname{sign}(\det \mathbf{W}) \right. \right.
$$
$$
\left. - \frac{1}{Th} \sum_{k=1}^{T} \varphi \left(\frac{\mathbf{w}_i(\mathbf{x}_t - \mathbf{x}_k)}{h} \right) \left(\frac{\mathbf{w}_i(\mathbf{x}_t - \mathbf{x}_k)}{h} \right) \left(\frac{x_{tj} - x_{kj}}{h} \right) \prod_{l \neq i}^{m} p(\mathbf{w}_l \mathbf{x}_t) \right]
$$
$$
+ \left[\frac{p(\mathbf{x}_t)}{|\det \mathbf{W}|} \right]^{(\alpha-1)/2} \frac{p(\mathbf{x}_t)}{|\det \mathbf{W}|^2} W_{ij} \operatorname{sign}(\det \mathbf{W}) + \left[\prod_{i=1}^{m} p(\mathbf{w}_i \mathbf{x}_t) \right]^{(\alpha-1)/2} \tag{4.52}
$$
$$
\times \frac{1}{Th} \sum_{k=1}^{T} \varphi \left(\frac{\mathbf{w}_i(\mathbf{x}_t - \mathbf{x}_k)}{h} \right) \left(\frac{\mathbf{w}_i(\mathbf{x}_t - \mathbf{x}_k)}{h} \right) \left(\frac{x_{tj} - x_{kj}}{h} \right) \prod_{l \neq i}^{m} p(\mathbf{w}_l \mathbf{x}_t) \Bigg\}.
$$

Eq. (4.52) provides a general approach to nonparametric C-ICA with a tunable parameter α. The nonparametric solutions to convex-Shannon ICA using $\alpha = 1$, as well as convex-logarithm ICA using $\alpha = -1$, are derived accordingly. The lower the convexity parameter α, the larger the derivative $\partial \mathcal{D}_{\mathrm{C}}(\mathbf{X}, \mathbf{W}, \alpha)/\partial w_{ij}$ (Chien and Hsieh, 2012). Correspondingly, a lower parameter α in C-ICA results in a higher changing rate in the learning objective. C-ICA is sensitive to parameter updating by changing α. This property is coincident with what we have shown in Figs. 4.10(A)–(B). In the following, an advanced ICA procedure based on C-ICA is evaluated for two sets of experiments. The convergence speed of the ICA procedure is investigated.

4.3.4 SIMULATED EXPERIMENTS

This set of experiments is designed to evaluate the sensitivity of divergence measures with respect to the demixing matrix $\mathbf{W}$. Different from the simulated experiment in Fig. 4.10, the mixed signals here are continuous variables. The simulated experiment (Boscolo et al., 2004, Chen, 2005) is run for the case of two sources ($m = 2$) with a demixing matrix

$$\mathbf{W} = \begin{bmatrix} \cos\theta_1 & \sin\theta_1 \\ \cos\theta_2 & \sin\theta_2 \end{bmatrix}. \tag{4.53}$$

This parametric matrix corresponds to a counterclockwise rotation with angles θ_1 and θ_2 in a polar coordinate space. Two rows $\mathbf{w}_1$ and $\mathbf{w}_2$ of the matrix $\mathbf{W}$ are normalized in magnitude and used as an orthogonal basis to span this space where the corresponding angles satisfy $\theta_2 = \theta_1 \pm \pi/2$. The demixing matrix is changed by using different angle parameters θ_1 and θ_2. Source separation is not varied by the radius. In this evaluation, the simulation data are prepared by mixing two zero-mean continuous sources. One is the uniform distribution

$$p(s_1) = \begin{cases} \frac{1}{2\tau_1}, & s_1 \in [-\tau_1, \tau_1], \\ 0, & \text{otherwise}, \end{cases} \tag{4.54}$$

which is a kind of a sub-Gaussian distribution. The other is the Laplace distribution

$$p(s_2) = \frac{1}{2\tau_2} \exp\left(-\frac{|s_2|}{\tau_2} \right), \tag{4.55}$$

also known as a type of super-Gaussian distribution. There were 1000 training samples drawn from these distributions with $\tau_1 = 3$ and $\tau_2 = 1$. The kurtoses of the two distributions were measures from the training samples, i.e., -1.23 for the sub-Gaussian signal s_1 and 2.99 for the super-Gaussian signal s_2. Here, the mixing matrix was assumed to be the identity matrix so that $x_1 = s_1$ and $x_2 = s_2$. Fig. 4.11 illustrates the relation between the divergence measures of demixed signals and the sensitivity of demixing matrix due to various parameters of the polar system θ_1 and θ_2. The range of angles $[0, \pi]$, producing sufficient diversity of demixing matrices, was evaluated. The cases of KL-DIV, C-DIV with $\alpha = 1$ and C-DIV with $\alpha = -1$ were investigated. It is obvious that the lowest divergence measures at zero consistently happen at the same setting $\{\theta_1 = 0, \theta_2 = \pi/2\}$. In this comparison, we obtain global minimum points by applying KL-DIV and C-DIV without being trapped in local minima. The changing rate of C-DIV at $\alpha = 1$ is low or C-DIV is insensitive when θ_2 is between 0.5 and 2.5. Nevertheless, KL-DIV and C-DIV at $\alpha = -1$ produce steep learning curves with high convexity in the same range of θ_2. Among these three divergence measures, the changing rate using convex divergence at $\alpha = -1$ with various θ_1 and θ_2 has the largest value, which equivalently implies the largest adjustment or achieves the fastest estimation in a learning process for ICA or other learning machines.

By using the same simulated data, we would like to evaluate and compare the convergence rates of learning curves by using different ICA algorithms with various divergence measures. The Kullback–Leibler divergence and convex divergence using nonparametric distributions were optimized to implement the nonparametric solutions to KL-ICA (Boscolo et al., 2004) and C-ICA (Chien and Hsieh, 2012) for blind source separation, respectively. In this evaluation, the kernel bandwidth of a nonparametric distribution was selected according to the number of training samples as $h = 1.06 \cdot T^{-1/5}$. The

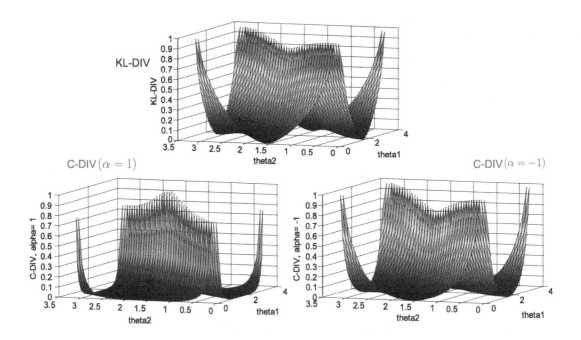

FIGURE 4.11

Divergence measures of demixed signals versus the parameters of demixing matrix θ_1 and θ_2. KL-DIV, C-DIV at $\alpha = 1$ and C-DIV at $\alpha = -1$ are compared.

divergence measures of the demixed signals were investigated. KL-ICA and two C-ICAs at $\alpha = 1$ and $\alpha = -1$ were compared. Fig. 4.12 compares the learning curves by using different ICA algorithms with different divergence measures and learning algorithms. The gradient descent (GD) learning in Eq. (2.2) and the natural gradient (NG) learning in Eq. (2.3) were included for a comparative study. The results of KL-ICA at $\eta = 0.17$, C-ICA at $\alpha = 1$, $\eta = 0.1$, and C-ICA at $\alpha = -1$, $\eta = 0.008$ were compared. Random initialization using nonsingular matrix was considered in ICA learning. The stopping criterion for the convergence check was specified when the absolute increase of a divergence measure was below 0.02. In this comparison, the flattest curve is obtained by using C-ICA at $\alpha = 1$ while the steepest curve is achieved by using C-ICA at $\alpha = -1$. In terms of the learning curve, KL-ICA is steeper than C-ICA with $\alpha = 1$ but flatter than C-ICA with $\alpha = -1$. In general, GD and NG algorithms perform similarly in their learning curves. In terms of convergence speed, the NG learning algorithm is more efficient than the GD learning algorithm. Learning curves in this comparison are evaluated to be in accordance with the results of examining the changing rates of learning objectives by varying the probability functions (Fig. 4.10) and the angle parameters of demixing matrices (Fig. 4.11). The convergence rate sufficiently reflects the sensitivity due to varying model parameters. The convergence behavior, investigated by a simple problem based on a mixture signal of super-Gaussian and sub-Gaussian distributions, can be extended to a general case with another mixture of non-Gaussian distributions having a different number of sources.

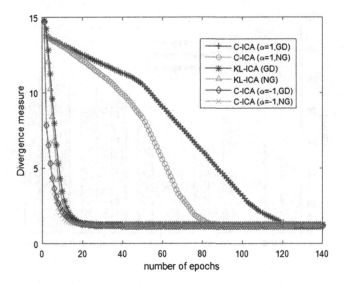

FIGURE 4.12

Divergence measures versus number of learning. KL-ICA and C-ICA at $\alpha = 1$ and $\alpha = -1$ are evaluated. The gradient descent (GD) and natural gradient (NG) algorithms are compared.

4.3.5 REAL-WORLD EXPERIMENTS

Real-world experiments are conducted for multichannel source separation where different mixing conditions are investigated with different source signals. First of all, three observation signals mixed with three source signals, consisting of two speech signals from two different males and one music source, were collected. ICA algorithms were developed for blind source separation where a 3×3 mixing matrix

$$\mathbf{A} = \begin{bmatrix} 0.8 & 0.2 & 0.3 \\ 0.3 & -0.8 & 0.2 \\ -0.3 & 0.7 & 0.3 \end{bmatrix} \tag{4.56}$$

was adopted in an instantaneous mixing system with source signals recorded at 8 kHz sampling rate. These experimental signals were collected from ICA'99 BSS dataset accessible at http://sound.media. mit.edu/ica-bench/ (Schobben et al., 1999). The nonparametric C-ICA algorithms at $\alpha = 1$ and $\alpha = -1$ were carried out. In this evaluation, the JADE algorithm (Cardoso, 1999) (using cumulant as the contrast function), Fast ICA algorithm (Hyvärinen, 1999) (using negentropy as the contrast function) and NMI-ICA algorithm (Boscolo et al., 2004) (using Shannon mutual information as the contrast function) were implemented for comparison. NMI-ICA is also known as a nonparametric variant of KL-ICA. The stopping threshold in the convergence condition was specified as 0.05. The signal-to-interference ratios (SIRs), as defined in Eq. (2.10), were calculated by using the demixed signals obtained by various ICA algorithms. Fig. 4.13 shows a comparison of SIRs of three demixed signals in an instantaneous BSS task. These SIRs are calculated for two speech signals from two speakers and one music signal. The

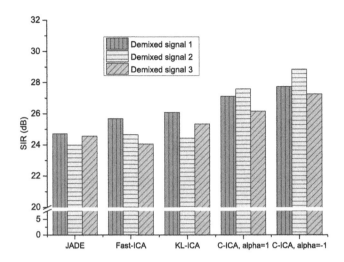

FIGURE 4.13

Comparison of SIRs of three demixed signals in the presence of an instantaneous mixing condition. Different ICA algorithms are evaluated.

SIRs of demixed signals using Fast-ICA are comparable with those using KL-ICA. Fast-ICA and KL-ICA obtains higher SIRs than JADE. Nevertheless, two C-ICA methods obtain much higher SIRs than other methods, especially for demixed signals 1 and 2, which belong to speech utterances from the two males. C-ICA at $\alpha = -1$ achieves higher SIRs than C-ICA at $\alpha = 1$. When evaluating convergence, C-ICA at $\alpha = 1$ converged in 31 learning epochs while KL-ICA converged in 39 learning epochs and C-ICA with $\alpha = 1$ converged in 124 learning epochs.

In addition to instantaneous mixing condition, there are two other tasks evaluated for different ICAs. The first task is introduced to evaluate the separation of mixed signals which are contaminated with additive noise based on (Cichocki et al., 1998)

$$\mathbf{x} = \mathbf{As} + \boldsymbol{\varepsilon} \tag{4.57}$$

where $\boldsymbol{\varepsilon}$ is a noise vector. This task involves a noisy ICA model which will be further described in Section 4.4.1. In general, the same source signals are used but with a different mixing matrix

$$\mathbf{A} = \begin{bmatrix} 0.6 & 0.5 & -0.4 \\ 0.8 & -0.6 & 0.3 \\ -0.1 & 0.5 & 0.3 \end{bmatrix}. \tag{4.58}$$

In this set of experiments, the noise signal was represented by a zero-mean Gaussian distribution with a covariance matrix $\sigma^2 \mathbf{I}$. The noise signal was independent of the three source signals. Three mixed signals were observed to evaluate the performance of different methods under noisy ICA model. Fig. 4.14 illustrates a comparison of SIRs of demixed speech and music signals in the presence of instantaneous mixing condition with additive noise. In this comparison, SNR was fixed at 20 dB. Obviously, the SIRs

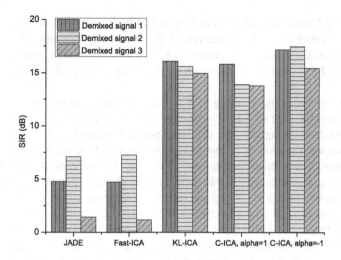

FIGURE 4.14

Comparison of SIRs of three demixed signals in the presence of instantaneous mixing condition with additive noise. Different ICA algorithms are evaluated.

of demixed signals in this instantaneous and noisy condition are decreased. However, the SIRs using C-ICAs at $\alpha = 1$ and $\alpha = -1$ are higher than those using other ICA methods; even C-ICA was not designed for noisy ICA environment. JADE and Fast-ICA performed poorly in SIRs. Both methods are degraded under noisy ICA condition in the presence of additive Gaussian noise. This is because these two methods employ different contrast functions of non-Gaussianity, which is mismatched with the assumption in a noisy ICA model. Such a circumstance is avoided by using KL-ICA and C-ICA which are developed according to different contrast functions of mutual information.

Secondly, the C-ICA algorithm is evaluated for blind source separation when convolutive mixture signals (Araki et al., 2003) are observed. In this evaluation, a convolutive mixing condition with two source signals was simulated. The speech signal from the first male speaker was mixed with the music signal based on a 2×2 transfer matrix $\mathbf{A}(z)$ in the z domain. The head-related transfer function, expressed by a 64th-order finite impulse response filter, was used in this transfer matrix, which synthesized a binaural source signal coming from a point in geometric space. Basically, the convolutive mixing system in the time domain corresponds to the instantaneous mixing system in the frequency (or z) domain. We have the property $\widetilde{\mathbf{x}}(z) = \mathbf{A}(z)\widetilde{\mathbf{s}}(z)$ where $\widetilde{\mathbf{s}}(z)$ and $\widetilde{\mathbf{x}}(z)$ denote the source and mixed signals in the z domain, respectively. To deal with the convolutive mixing condition, different ICA methods were performed to decompose the instantaneous mixed signals in the frequency domain. Notably, the ICA algorithms were implemented to estimate the complex-valued matrix $\mathbf{W}(z)$ for source separation performed individually in each frequency bin. For this convolutive mixing system, the demixed time signals were finally obtained by transforming the demixed frequency signals based on 1024-point inverse STFT. Assuming the frequency signals were characterized a super-Gaussian distribution as mentioned in Sawada et al. (2003), a parametric C-ICA procedure was realized by implementing Eq. (4.49). The distributions of complex-valued demixed signals in the frequency domain were modeled. Experimental results show that the SIRs of demixed speech and music signals are elevated from 22.6 and 22.2 dB using KL-ICA to 24.2 and 22.9 dB using C-ICA with $\alpha = 1$, and 24.5 and 23.4 dB using C-ICA with $\alpha = -1$, respectively. SIRs of Fast-ICA are only 13.08 and 12.69 dB for demixed speech and music signals, respectively. C-ICA achieves desirable performance in multichannel source separation for the conditions of instantaneous, noisy and convolutive mixtures of speech and music signals.

In summary, the advanced model using C-ICA is established in an information-theoretic learning algorithm where a general contrast function based on convex divergence (C-DIV) measures the independence among the demixed signals. The demixing matrix is derived by minimizing C-DIV between the joint distribution and the product of marginal distributions where the nonparametric distributions using Parzen window density functions are used. The convex divergence ICA (C-ICA) procedure is presented with a controllable convexity parameter α. Experiments demonstrate the merit of C-ICA with $\alpha = -1$ in terms of convergence speed and signal-to-interference ratio when compared with other ICA methods and other convexity parameters of C-ICA. In what follows, we focus on Bayesian learning, sparse learning and online learning for a noisy ICA model where multichannel source separation is tackled in nonstationary mixing environments. A sparsity parameter is introduced and estimated to detect the changing number of sources in an auditory scene model. Model regularization is taken into account via a probabilistic framework where the uncertainty of mixed signals and system parameters is sufficiently represented to discover statistics of latent source signals for source separation. There are two related Bayesian solutions to noisy ICA model. The first one is the nonstationary Bayesian ICA which carries out an online learning algorithm based on a recursive Bayesian formula. The second one is the online Gaussian process ICA where the nonstationary and temporally-correlated mixing system is captured by an online Gaussian process.

4.4 NONSTATIONARY BAYESIAN ICA

Traditional independent component analysis was developed for blind source separation by assuming a stationary mixing condition where the mixing matrix $\mathbf{A}$ is deterministic and the statistics of source signals $\mathbf{S} = \{\mathbf{s}_t\}_{t=1}^{T}$ are unchanged in time. However, the mixing system and source signals are nonstationary in many real-world applications, e.g., the source signals may abruptly appear or disappear, the sources may be replaced by new ones or be even moving in time. The number of sources can also be changing. Fig. 4.15 shows an integrated scenario of a dynamic time-varying source separation system in Fig. 3.6, which involves (A) three microphones and three sources with one male, one female and one music player, (B) moving a male to a new location, and (C) exclusion of a male and replacement of a new female, as addressed in Section 3.5. The condition of moving sources is comparable with that of moving microphones. Typically, such an integrated scenario is complicated in general and is difficult to solve by using traditional stationary ICA methods. To deal with such a time-varying mixing condition in source separation, a meaningful approach is to adaptively identify the status of the mixing system and capture the newest distributions of source signals at each time using the up-to-date minibatch data.

In the literature, a number of methods have been developed to tackle time-varying source separation in the presence of dynamic changes of source signals and mixing conditions. Two scenarios are considered separately. The first scenario addresses moving sources or sensors. For this condition, an adaptive learning algorithm is required to capture the changes of the mixing matrix time by time. In Everson and Roberts (2000), the compensation of these changes was performed via a

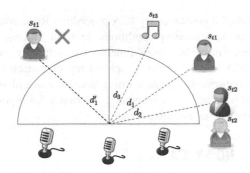

FIGURE 4.15

Integrated scenario of a time-varying source separation system from t to $t + 1$ and $t + 2$ as shown in Fig. 3.6.

Markov process which captured the mixture signals in the presence of source signals with temporal correlation. In addition, a three-dimensional tracker was introduced to capture the dynamics of sources (Naqvi et al., 2009). A beamforming method was used for source separation with moving sources where source distributions were fixed and the number of sources was unchanged. On the other hand, the second scenario reflects the circumstance that the sources or speakers may be changed, leaving or coming at various time steps. It is necessary to represent time-varying characteristics in source distributions. A useful trick to identify the dynamics of signal activity of sources was implemented according to the automatic relevance determination (MacKay, 1995, Tipping, 2001, Hsieh and Chien, 2010). In Hirayama et al. (2007), the switching ICA adopted an indicator to detect the changing status when source signals suddenly appear or disappear. The dynamics of sources were represented by a fixed generative model using HMM. The activity of source signals was modeled accordingly (Choudrey and Roberts, 2003). Moreover, an ICA model was constructed via an online variational Bayesian (VB) learning (Honkela and Valpola, 2003) by assuming that the unknown source signal was a time-varying parameter and the unknown mixing matrix was a time independent parameter. Although the time-varying source distributions were sequentially captured, the assumption of a fixed number of sources makes it impossible to detect abrupt absence or presence of source signals. In Koldovský et al. (2009), a piecewise stationary solution to ICA procedure was explored to learn time-varying non-Gaussian distribution for blind separation of audio source signals.

This section presents an advanced model, called the nonstationary Bayesian ICA (NB-ICA), to fulfill nonstationary source separation where a time dependent mixing system with $\mathbf{A} \rightarrow \mathbf{A}^{(n)}$ is used to characterize the mixing process at different time frame n using a minibatch. System parameters $\boldsymbol{\Theta}^{(n)}$ are updated by using individual minibatch samples or mixed signals $\mathbf{X}^{(n)} = \{\mathbf{x}_t^{(n)}\}_{t=1}^{L}$ from the whole dataset $\mathbf{X} = \{\mathbf{x}_t\}_{t=1}^{T}$. Using the NB-ICA procedure, the evolved statistics are characterized through online Bayesian learning from sequential signals

$$\mathcal{X}^{(n)} = \{\mathbf{X}^{(1)}, \mathbf{X}^{(2)}, \dots, \mathbf{X}^{(n)}\}. \tag{4.59}$$

Stochastic learning is performed to sequentially update an automatic relevance determination (ARD) parameter from one minibatch to another. The nonstationary source signals are considerably reflected. Based on the variational Bayesian method, the true posterior for detecting the dynamics of source signals is approximated to fulfill a nonstationary ICA procedure. Recursive Bayesian learning is performed for online tracking of nonstationary conditions. In what follows, Section 4.4.1 addresses a noisy ICA model for calculation of the likelihood function for each individual observation $\mathbf{x}_t$ at time t. Sequential learning of NB-ICA is constructed. Graphical representation is illustrated. Section 4.4.3 presents a sequential and variational Bayesian learning and inference algorithm for NB-ICA. Variational parameters and model parameters are estimated. Section 4.4.4 reports a set of experiments to evaluate the effect of ARD parameter and mixing matrix.

4.4.1 SEQUENTIAL AND NOISY ICA

In real-world mixing environments, source signals are changed time by time in their distributions. In Choudrey and Roberts (2003), Ichir and Mohammad-Djafari (2006), Snoussi and Mohammad-Djafari (2004a), the time-varying distributions of source signals were characterized by a nonstationary model based on an HMM, which represented the temporal dynamics of source distributions by following a noisy ICA model

$$\mathbf{x}_t = \mathbf{A}\mathbf{s}_t + \boldsymbol{\varepsilon}_t \tag{4.60}$$

where $\boldsymbol{\varepsilon}_t$ denotes a noise signal. The jth source signal s_{tj} of $\mathbf{s}_t = \{s_{tj}\}_{j=1}^m$ was modeled by a Gaussian mixture model with K mixture components in

$$p(\mathbf{s}_t | \boldsymbol{\Theta}) = \prod_{j=1}^m \left[\sum_{k=1}^K \pi_{jk} \mathcal{N}\left(s_{tj} \middle| \mu_{jk}, \gamma_{jk}^{-1} \right) \right] \tag{4.61}$$

where the state-dependent Gaussian mixture parameters

$$\boldsymbol{\Theta} = \{\pi_{jk}, \mu_{jk}, \gamma_{jk}\} \tag{4.62}$$

contain K sets of Gaussian parameters, including mixture weights $\{\pi_{jk}\}$, means $\{\mu_{jk}\}$ and precisions $\{\gamma_{jk}\}$. In Eq. (4.61), m mutually independent source signals were assumed. Nonstationary source signals were characterized by explicitly representing their temporal dynamics based on HMM states using Bayesian learning.

The switching ICA in Hirayama et al. (2007) addressed a solution to relax the constraint of fixed temporal condition in source signals and deal with the scenario of dynamic presence and absence of source signals s_{tj}. A switching indicator variable was incorporated to reveal the activity of source signals where $z_{tj} = 0$ implies inactive and $z_{tj} = 1$ denotes active. An active source signal was therefore represented by

$$\begin{aligned} &p(s_{tj} | z_{tj} = 1) \\ &= \pi_j \mathcal{N}(s_{tj} | 0, \gamma_{ja}^{-1}) + (1 - \pi_j)\mathcal{N}(s_{tj} | 0, \gamma_{jb}^{-1}) \end{aligned} \tag{4.63}$$

where π_j denotes the mixture weight in source j and $\{\gamma_{ja}, \gamma_{jb}\}$ denote the precisions of Gaussian distributions. An inactive source signal was setup by changing state to $z_{tj} = 0$. The switching variable z_{tj} was introduced to form a Markov process by estimating the state parameters $p(z_{1j})$ and $p(z_{tj} | z_{t-1,j})$ for initial state and state transition probabilities, respectively. This switching ICA was specialized HMM due to a switching variable. Computational cost was substantially increased by the number of states, as well as the number of sources m.

Traditional ICA procedures, constructed by maximizing kurtosis or minimizing mutual information of demixed signals, did not take into account the nonstationary mixing conditions. The noise signal and the uncertainties of ICA parameters were not considered either. It is crucial to incorporate these considerations in a practical source separation system. The traditional ICA methods assuming the fixed mixing matrix, as well as the independently and identically distributed source signals, do *not* reflect the nonstationary operating environments. A possible treatment to these assumptions for nonstationary ICA was based on a second-order Markov process (Miskin, 2000). In this section, nonstationary source separation is seen as a problem with nonstationary source signals s_t and mixing coefficients $\mathbf{A}$. As addressed in Section 3.5.1, online Bayesian learning is feasible to develop a sequential learning approach to deal with nonstationary source separation. A recursive Bayesian framework is implemented by sequentially accumulating and combining sufficient statistics from history frames to on-site current frame. Such an updating of posterior distribution is continued and propagated at each learning epoch to implement an adaptive blind source separation. This assures the adaptation of dynamic source signals and mixing matrix to fit a nonstationary environment at the frame level. Therefore, the training signals are chopped into minibatches where each minibatch n contains the mixed signals $\mathbf{X}^{(n)} = \{\mathbf{x}_t^{(n)}\}_{t=1}^L$ with L time samples, which are mixed by m unknown source signals $\mathbf{S}^{(n)} = \{\mathbf{s}_t^{(n)}\}$ using an instantaneous mixing matrix $\mathbf{A}^{(n)}$ with additive noise signals $\mathbf{E}^{(n)} = \{\boldsymbol{\varepsilon}_t^{(n)}\}$ expressed as

$$\mathbf{x}_t^{(n)} = \mathbf{A}^{(n)} \mathbf{s}_t^{(n)} + \boldsymbol{\varepsilon}_t^{(n)}. \tag{4.64}$$

The sufficient statistics are assumed to be unchanged within a minibatch but varied in different minibatches. Equivalently, the variations of the mixing matrix $\mathbf{A}^{(n)}$ and source signals $\mathbf{S}^{(n)}$ are sequentially characterized and compensated at each minibatch n by using

$$\mathcal{X}^{(n)} = \{\mathbf{X}^{(1)}, \mathbf{X}^{(2)}, \ldots, \mathbf{X}^{(n)}\}.$$

Without loss of generality, the determined system is considered by using $m \times m$ mixing matrix $\mathbf{A}^{(n)}$ where the numbers of sources and channels are both m. Fig. 4.16 displays the graphical model for implementation of nonstationary source separation using NB-ICA. A sequential and noisy ICA procedure is carried out as detailed in what follows.

Considering the noisy ICA model in Eq. (4.64) and referring to the GMM distribution in Eq. (4.61), we express the source signals $\mathbf{s}_t^{(n)}$ by using a GMM (or mixture of Gaussian, MoG) with K Gaussian components

$$p\left(\mathbf{s}_t^{(n)} \middle| \mathbf{\Pi}^{(n)} = \{\pi_{jk}^{(n)}\}, \mathbf{M}^{(n)} = \{\mu_{jk}^{(n)}\}, \mathbf{R}^{(n)} = \{\gamma_{jk}^{(n)}\}\right)$$
$$= \prod_{j=1}^m \left[\sum_{k=1}^K \pi_{jk}^{(n)} \mathcal{N}\left(s_{tj}^{(n)} \middle| \mu_{jk}^{(n)}, (\gamma_{jk}^{(n)})^{-1}\right) \right]. \tag{4.65}$$

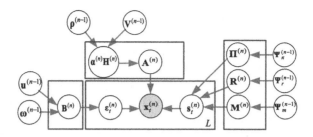

FIGURE 4.16

Graphical representation for nonstationary Bayesian ICA model.

A Gaussian noise vector $\boldsymbol{\varepsilon}_t^{(n)}$ with a zero mean vector and diagonal precision matrix $\mathbf{B}^{(n)} = \mathrm{diag}\{\beta_i^{(n)}\}$ is assumed and given by

$$\mathcal{N}\left(\boldsymbol{\varepsilon}_t^{(n)} \Big| 0, (\mathbf{B}^{(n)})^{-1}\right). \tag{4.66}$$

Or, equivalently, the likelihood function of an observation frame or minibatch $\mathbf{x}_t^{(n)}$ is calculated by

$$p(\mathbf{x}_t^{(n)}|\mathbf{A}^{(n)}, \mathbf{s}_t^{(n)}, \boldsymbol{\varepsilon}_t^{(n)}, \mathbf{B}^{(n)}) = \mathcal{N}(\mathbf{x}_t^{(n)}|\mathbf{A}^{(n)}\mathbf{s}_t^{(n)}, (\mathbf{B}^{(n)})^{-1}). \tag{4.67}$$

It is crucial to specify the prior distributions for online Bayesian learning. The scheme of automatic relevance determination can be implemented.

4.4.2 AUTOMATIC RELEVANCE DETERMINATION

In the implementation, we express the prior distribution of the $m \times m$ mixing matrix $\mathbf{A}^{(n)} = \{a_{ij}^{(n)}\}$ using

$$\begin{aligned}
p(\mathbf{A}^{(n)}|\boldsymbol{\alpha}^{(n)}) &= \prod_{i=1}^{m}\left[\prod_{j=1}^{m}\mathcal{N}\left(a_{ij}^{(n)}\Big|0, (\alpha_j^{(n)})^{-1}\right)\right] \\
&= \prod_{j=1}^{m}\mathcal{N}\left(\mathbf{a}_j^{(n)}\Big|0, (\alpha_j^{(n)})^{-1}\mathbf{I}_m\right)
\end{aligned} \tag{4.68}$$

where $\boldsymbol{\alpha}^{(n)} = \{\alpha_j^{(n)}\}$ and $\mathbf{I}_m$ is the m-dimensional identity matrix. In this expression, a zero-mean isotropic Gaussian with precision $\alpha_j^{(n)}$ is used to represent the distribution for each column $\mathbf{a}_j^{(n)}$ of the mixture matrix $\mathbf{A}^{(n)}$. Attractively, if the precision $\alpha_j^{(n)}$ in Eq. (4.68) is distributed by Gamma prior, as addressed in Section 3.4.1, a Student's t distribution can be obtained as the marginal distribution of the mixing coefficient $a_{ij}^{(n)}$. Such a distribution is peaky with zero mean and is popular for sparse Bayesian learning (Tipping, 2001). The hyperparameter $\alpha_j^{(n)}$ is known as an ARD parameter (MacKay,

1995). This parameter is used as an indicator variable to reflect the status of activity of source signal $s_{tj}^{(n)}$ using noisy ICA model. The mixing matrix $\mathbf{A}^{(n)}$ under this assumption will be sparse, i.e., the elements at the jth column will approach zero

$$\mathbf{a}_j^{(n)} = \{a_{ij}^{(n)}\} \to 0 \tag{4.69}$$

where the estimated ARD parameter $\alpha_j^{(n)}$ is prone to be large. In other words, the jth source signal is likely absent at frame n using minibatch signals $\mathbf{X}^{(n)}$. The redundant sources are disregarded automatically. Sparse Bayesian learning is implemented. In particular, ARD parameters and mixing coefficients are related as

$$\alpha_j^{(n)} = \begin{cases} \infty, & \mathbf{a}_j^{(n)} = \{a_{ij}^{(n)}\} \to 0, \\ < \infty, & \mathbf{a}_j^{(n)} = \{a_{ij}^{(n)}\} \neq 0. \end{cases} \tag{4.70}$$

The number of active sources m can be determined accordingly. The function of this ARD parameter $\alpha_j^{(n)}$ is comparable with that of the switching variable z_{jt} in switching ICA (Hirayama et al., 2007).

In general, a mixture coefficient $a_{ij}^{(n)}$ is sufficient to measure the mixing correlation or relation between source j and sensor i. This coefficient is continuously changed under nonstationary mixing conditions. Using a shared ARD parameter $\alpha_j^{(n)}$ for m different sensors is *not* distinguishable to reflect the relevance between source j and each individual sensor i. To deal with this issue, we introduce an adaptation matrix $\mathbf{H}_j^{(n)}$ to compensate such a mismatch condition by transforming the precision matrix of prior distribution of $\mathbf{a}_j^{(n)}$ and modifying the prior density of $\mathbf{A}^{(n)}$ in Eq. (4.68) with a new form

$$p(\mathbf{A}^{(n)}|\boldsymbol{\alpha}^{(n)}, \mathbf{H}^{(n)}) = \prod_{j=1}^{m} \mathcal{N}\left(\mathbf{a}_j^{(n)}\middle|0, (\alpha_j^{(n)}\mathbf{H}_j^{(n)})^{-1}\right) \tag{4.71}$$

where $\mathbf{H}^{(n)} = \{\mathbf{H}_j^{(n)}\}$. Compensation of nonstationary mixing condition is therefore performed. The parameter set for nonstationary source separation is formed by

$$\boldsymbol{\Theta}^{(n)} = \{\mathbf{A}^{(n)}, \boldsymbol{\alpha}^{(n)}\mathbf{H}^{(n)}, \mathbf{E}^{(n)}, \mathbf{B}^{(n)}, \mathbf{S}^{(n)}, \boldsymbol{\Pi}^{(n)}, \mathbf{M}^{(n)}, \mathbf{R}^{(n)}\}. \tag{4.72}$$

An online Bayesian learning is developed to carry out a recursive Bayesian solution to nonstationary source separation. The Dirichlet distribution with hyperparameter $\boldsymbol{\Psi}_\pi^{(n-1)}$, Gaussian distribution with hyperparameter $\boldsymbol{\Psi}_m^{(n-1)}$ and Gamma distribution with hyperparameter $\boldsymbol{\Psi}_r^{(n-1)}$ are introduced as *conjugate priors* (Bishop, 2006, Watanabe and Chien, 2015) for individual parameters $\boldsymbol{\Pi}^{(n)}$, $\mathbf{M}^{(n)}$ and $\mathbf{R}^{(n)}$ in online Bayesian learning, respectively. The precision parameter $\beta_i^{(n)}$ is generated by a Gamma prior with hyperparameters $\{u_{\beta_i}^{(n-1)}, \omega_{\beta_i}^{(n-1)}\}$

$$p(\beta^{(n)}|u_{\beta_i}^{(n-1)}, \omega_{\beta_i}^{(n-1)}) = \text{Gam}(\beta^{(n)}|u_{\beta_i}^{(n-1)}, \omega_{\beta_i}^{(n-1)}). \tag{4.73}$$

The new hyperparameters $\Psi^{(n)}$ are calculated by using current minibatch $\mathbf{X}^{(n)}$ and previous hyperparameters $\Psi^{(n-1)}$. The detailed solution to

$$\Psi^{(n)} = \{\mathbf{u}^{(n)} = \{u_{\beta_i}^{(n)}\}, \boldsymbol{\omega}^{(n)} = \{\omega_{\beta_i}^{(n)}\}, \Psi_\pi^{(n)}, \Psi_m^{(n)}, \Psi_r^{(n)}\} \tag{4.74}$$

was derived in Choudrey et al. (2000). Nevertheless, there are two advantages in nonstationary source separation by using NB-ICA. The first is the online learning or sequential learning with each minibatch n while the second is the adaptation or compensation of precision parameter $\mathbf{a}_j^{(n)}$ based on

$$\alpha_j^{(n)} \mathbf{I}_m \to \alpha_j^{(n)} \mathbf{H}_j^{(n)}. \tag{4.75}$$

Importantly, Wishart distribution is adopted to represent the uncertainty of ARD parameter using

$$p(\alpha_j^{(n)} \mathbf{H}_j^{(n)} | \rho_j^{(n-1)}, \mathbf{V}_j^{(n-1)}) = \mathcal{W}(\alpha_j^{(n)} \mathbf{H}_j^{(n)} | \rho_j^{(n-1)}, \mathbf{V}_j^{(n-1)})$$

$$\propto |\alpha_j^{(n)} \mathbf{H}_j^{(n)}|^{(\rho_j^{(n-1)} - m - 1)/2} \exp\left[-\frac{1}{2}\mathrm{Tr}\left[(\mathbf{V}_j^{(n-1)})^{-1} \alpha_j^{(n)} \mathbf{H}_j^{(n)}\right]\right]. \tag{4.76}$$

The hyperparameters consisting of

$$\Psi^{(n-1)} = \{\boldsymbol{\rho}^{(n-1)} = \{\rho_j^{(n-1)}\}, \mathbf{V}^{(n-1)} = \{\mathbf{V}_j^{(n-1)}\}\} \tag{4.77}$$

using previous data $\mathcal{X}^{(n-1)}$ are set. Adopting this conjugate prior for ARD matrix $\alpha_j^{(n)} \mathbf{H}_j^{(n)}$, the marginal distribution of ICA parameter $\mathbf{a}_j^{(n)}$ is derived as a multivariate sparse distribution based on Student's t distribution. The relevance or correlation between source j and sensor i is revealed by the derived ith diagonal element of the precision matrix $\alpha_j^{(n)} \mathbf{H}_j^{(n)}$. According to variational Bayesian learning, a solution to the hyperparameter selection problem

$$\Psi^{(n)} = \{\boldsymbol{\rho}^{(n)}, \mathbf{V}^{(n)}\} \tag{4.78}$$

can be formulated as addressed in the next section.

4.4.3 SEQUENTIAL AND VARIATIONAL LEARNING

Online Bayesian learning for a latent variable model is accomplished through variational Bayesian (VB) inference as addressed in Section 3.7.3. Latent variables consist of

$$\{\mathbf{A}^{(n)}, \boldsymbol{\alpha}^{(n)} \mathbf{H}^{(n)}, \mathbf{E}^{(n)}, \mathbf{B}^{(n)}, \mathbf{S}^{(n)}, \boldsymbol{\Pi}^{(n)}, \mathbf{M}^{(n)}, \mathbf{R}^{(n)}\} \tag{4.79}$$

in the NB-ICA model. Sequential and variational Bayesian learning is performed by maximizing the likelihood function $p(\mathbf{X}^{(n)} | \Psi^{(n-1)})$, which is marginalized over latent variables or model parameters $\Theta^{(n)}$. As summarized in Algorithm 4.1, at each segment or minibatch with index n, the variational Bayesian expectation-maximization (VB-EM) algorithm is run by adopting training samples $\mathbf{X}^{(n)}$ at each minibatch. Using this VB-EM algorithm, the VB-E step is first performed to estimate the variational distributions of individual parameters or, equivalently, update the hyperparameters or variational

---

Algorithm 4.1 Online Variational Bayesian Learning.

Initialize with $\boldsymbol{\Theta}^{(0)}$ and $\boldsymbol{\Psi}^{(0)}$
For each frame or minibatch data $\mathbf{X}^{(n)} = \{\mathbf{x}_t^{(n)}\}$
  For each VB-EM iteration
    VB-E step: estimate variational distributions
      For each time sample $\mathbf{x}_t^{(n)}$, $t = 1, 2, \ldots, L$
        Accumulate sufficient statistics
      Find $q(\boldsymbol{\Theta}^{(n)}|\boldsymbol{\Psi}^{(n)})$ and update $\boldsymbol{\Psi}^{(n)}$
    VB-M step: update model parameters
      Estimate distribution mode and update $\boldsymbol{\Theta}^{(n)}$
    Check if $\boldsymbol{\Theta}^{(n)}$ converged
  $n \leftarrow n + 1$
Return $\boldsymbol{\Theta}^{(n)}$ and $\boldsymbol{\Psi}^{(n)}$

---

parameters $\boldsymbol{\Psi}^{(n)}$ at each iteration. Next, in the VB-M step, a new set of model parameters $\boldsymbol{\Theta}^{(n)}$ is constructed and estimated. Sequential and variational learning algorithm is addressed in details as follows.

In the implementation, sequential learning and variational inference are carried out for blind source separation based on the NB-ICA procedure where model parameters consist of

$$\boldsymbol{\Theta}^{(n)} = \{\mathbf{A}^{(n)}, \boldsymbol{\alpha}^{(n)}\mathbf{H}^{(n)}, \mathbf{E}^{(n)}, \mathbf{B}^{(n)}, \mathbf{S}^{(n)}, \boldsymbol{\Pi}^{(n)}, \mathbf{M}^{(n)}, \mathbf{R}^{(n)}\} \tag{4.80}$$

and hyperparameters are composed of

$$\boldsymbol{\Psi}^{(n)} = \{\boldsymbol{\rho}^{(n)}, \mathbf{V}^{(n)}, \mathbf{u}^{(n)}, \boldsymbol{\omega}^{(n)}, \boldsymbol{\Psi}_\pi^{(n)}, \boldsymbol{\Psi}_m^{(n)}, \boldsymbol{\Psi}_r^{(n)}\}. \tag{4.81}$$

In general, the exact solution to model inference based on posterior distribution $p(\boldsymbol{\Theta}^{(n)}|\mathbf{X}^{(n)}, \boldsymbol{\Psi}^{(n-1)})$ does not exist since latent variables are coupled and cannot be factorized in posterior manner. Exact Bayesian inference is not possible. It becomes crucial to realize an approximate Bayesian solution to ICA learning. The VB-EM algorithm (Chan et al., 2002, Lawrence and Bishop, 2000) is feasible to implement an approximate inference where the *negative free energy* or the variational lower bound for the logarithm of marginal likelihood is maximized. Here, we integrate the likelihood function with respect to uncertainties of various latent variables in $\boldsymbol{\Theta}^{(n)}$ and calculate the marginal likelihood $p(\mathbf{X}^{(n)}|\boldsymbol{\Psi}^{(n-1)})$ by using minibatch data with L training samples $\mathbf{X}^{(n)} = \{\mathbf{x}_t^{(n)}\}_{t=1}^L$ as

$$\prod_{t=1}^L \int p(\mathbf{x}_t^{(n)}|\mathbf{A}^{(n)}, \mathbf{s}_t^{(n)}, \boldsymbol{\varepsilon}_t^{(n)}) p(\mathbf{A}^{(n)}|\boldsymbol{\alpha}^{(n)}, \mathbf{H}^{(n)})$$

$$\times p(\boldsymbol{\alpha}^{(n)}\mathbf{H}^{(n)}|\boldsymbol{\rho}^{(n-1)}, \mathbf{V}^{(n-1)}) p(\boldsymbol{\varepsilon}_t^{(n)}|\mathbf{B}^{(n)})$$

$$\times p(\mathbf{B}^{(n)}|\mathbf{u}^{(n-1)}, \boldsymbol{\omega}^{(n-1)}) p(\mathbf{s}_t^{(n)}|\boldsymbol{\Pi}^{(n)}, \mathbf{M}^{(n)}, \mathbf{R}^{(n)}) \tag{4.82}$$

$$\times p(\boldsymbol{\Pi}^{(n)}|\boldsymbol{\Psi}_\pi^{(n-1)}) p(\mathbf{M}^{(n)}|\boldsymbol{\Psi}_m^{(n-1)}) p(\mathbf{R}^{(n)}|\boldsymbol{\Psi}_r^{(n-1)})$$

$$\times d\mathbf{A}^{(n)} d(\boldsymbol{\alpha}^{(n)}\mathbf{H}^{(n)}) d\boldsymbol{\varepsilon}_t^{(n)} d\mathbf{B}^{(n)} d\mathbf{s}_t^{(n)} d\boldsymbol{\Pi}^{(n)} d\mathbf{M}^{(n)} d\mathbf{R}^{(n)}.$$

The updated hyperparameters $\mathbf{\Psi}^{(n-1)}$ from previous minibatches $\mathcal{X}^{(n-1)}$ are given in the VB-EM procedure. A variational distribution $q(\mathbf{\Theta}^{(n)})$ is used to approximate the true posterior distribution $p(\mathbf{\Theta}^{(n)}|\mathbf{X}^{(n)}, \mathbf{\Psi}^{(n-1)})$ using each minibatch data $\mathbf{X}^{(n)}$. In online Bayesian learning, we maximize the lower bound of the log-likelihood $\log p(\mathbf{X}^{(n)}|\mathbf{\Psi}^{(n-1)})$ or, equivalently, maximize the expectation function $\mathbb{E}_q[\log p(\mathbf{X}^{(n)}|\mathbf{\Theta}^{(n)}, \mathbf{\Psi}^{(n-1)})]$ over variational distribution $q(\mathbf{\Theta}^{(n)})$ and minimize the Kullback–Leibler divergence between $q(\mathbf{\Theta}^{(n)})$ and $p(\mathbf{\Theta}^{(n)}|\mathbf{X}^{(n)}, \mathbf{\Psi}^{(n-1)})$. The factorized variational inference is run by using the factorized variational distribution

$$
\begin{aligned}
q(\mathbf{\Theta}^{(n)}) = \prod_l q(\mathbf{\Theta}_l^{(n)}) &= q(\mathbf{A}^{(n)})q(\alpha^{(n)}\mathbf{H}^{(n)}) \\
&\times q(\mathbf{E}^{(n)})q(\mathbf{B}^{(n)})q(\mathbf{S}^{(n)})q(\mathbf{\Pi}^{(n)})q(\mathbf{M}^{(n)})q(\mathbf{R}^{(n)}).
\end{aligned}
\tag{4.83}
$$

We then expand the variational lower bound of $\log p(\mathbf{X}^{(n)}|\mathbf{\Psi}^{(n-1)})$ as

$$
\begin{aligned}
&\mathbb{E}_q[\log p(\mathbf{X}^{(n)}|\mathbf{A}^{(n)}, \mathbf{S}^{(n)}, \mathbf{E}^{(n)})] \\
&+ \mathbb{E}_q[\log p(\mathbf{A}^{(n)}|\alpha^{(n)}, \mathbf{H}^{(n)})] + \mathbb{H}[q(\mathbf{A}^{(n)})] \\
&+ \mathbb{E}_q[\log p(\alpha^{(n)}\mathbf{H}^{(n)}|\rho^{(n-1)}, \mathbf{V}^{(n-1)})] + \mathbb{H}[q(\alpha^{(n)}\mathbf{H}^{(n)})] \\
&+ \mathbb{E}_q[\log p(\mathbf{E}^{(n)}|\mathbf{B}^{(n)})] + \mathbb{H}[q(\mathbf{E}^{(n)})] \\
&+ \mathbb{E}_q[\log p(\mathbf{B}^{(n)}|\mathbf{u}^{(n-1)}, \omega^{(n-1)})] + \mathbb{H}[q(\mathbf{B}^{(n)})] \\
&+ \mathbb{E}_q[\log p(\mathbf{S}^{(n)}|\mathbf{\Pi}^{(n)}, \mathbf{M}^{(n)}, \mathbf{R}^{(n)})] + \mathbb{H}[q(\mathbf{S}^{(n)})] \\
&+ \mathbb{E}_q[\log p(\mathbf{\Pi}^{(n)}|\mathbf{\Psi}_\pi^{(n-1)})] + \mathbb{H}[q(\mathbf{\Pi}^{(n)})] \\
&+ \mathbb{E}_q[\log p(\mathbf{M}^{(n)}|\mathbf{\Psi}_m^{(n-1)})] + \mathbb{H}[q(\mathbf{M}^{(n)})] \\
&+ \mathbb{E}_q[\log p(\mathbf{R}^{(n)}|\mathbf{\Psi}_r^{(n-1)})] + \mathbb{H}[q(\mathbf{R}^{(n)})].
\end{aligned}
\tag{4.84}
$$

In Eq. (4.84), $\mathbb{H}[q(\cdot)]$ expresses the entropy of a variational distribution. The variational distribution can be derived and estimated by (Bishop, 2006, Lawrence and Bishop, 2000)

$$
\log \widehat{q}(\mathbf{\Theta}_l^{(n)}) \propto \mathbb{E}_{q(\mathbf{\Theta} \neq \mathbf{\Theta}_l)}[\log p(\mathbf{X}^{(n)}, \mathbf{\Theta}^{(n)}|\mathbf{\Psi}^{(n-1)})],
\tag{4.85}
$$

which is obtained by taking the derivative of Eq. (4.84) with respect to the lth variational distribution $q(\mathbf{\Theta}_l^{(n)})$ and equating the result to zero. In Eq. (4.85), the expectation is calculated over all the previous distributions except that of l, i.e., $q(\mathbf{\Theta} \neq \mathbf{\Theta}_l)$. In the VB-E step, the hyperparameters are updated

$$
\mathbf{\Psi}^{(n-1)} \rightarrow \mathbf{\Psi}^{(n)}
\tag{4.86}
$$

through estimation of variational distributions $\widehat{q}(\mathbf{\Theta}^{(n)})$. In Choudrey et al. (2000), the hyperparameters

$$
\{\mathbf{u}^{(n)}, \omega^{(n)}, \mathbf{\Psi}_\pi^{(n)}, \mathbf{\Psi}_m^{(n)}, \mathbf{\Psi}_r^{(n)}\}
\tag{4.87}
$$

have been obtained. We now emphasize the derivation of optimal hyperparameters $\{\rho^{(n)}, \mathbf{V}^{(n)}\}$. This is accomplished by introducing prior densities of Eqs. (4.68) and (4.70) in Eq. (4.85). The estimation of variational distribution is given by

$$
\begin{aligned}
\log \prod_{j=1}^{m} \widehat{q}(\alpha_j^{(n)} \mathbf{H}_j^{(n)}) &\propto \mathbb{E}_{q(\mathbf{A}^{(n)})}[\log p(\mathbf{A}^{(n)}|\boldsymbol{\alpha}^{(n)}, \mathbf{H}^{(n)})] \\
&\quad + \log p(\boldsymbol{\alpha}^{(n)} \mathbf{H}^{(n)}|\boldsymbol{\rho}^{(n-1)}, \mathbf{V}^{(n-1)}) \\
&= \sum_{j=1}^{m} \left\{ \mathbb{E}_{q(\mathbf{a}_j^{(n)})}[\log p(\mathbf{a}_j^{(n)}|\alpha_j^{(n)}, \mathbf{H}_j^{(n)})] \right. \\
&\quad \left. + \log p(\alpha_j^{(n)} \mathbf{H}_j^{(n)}|\rho_j^{(n-1)}, \mathbf{V}_j^{(n-1)}) \right\} \\
&\propto \sum_{j=1}^{m} \left\{ \frac{\rho_j^{(n-1)} - m}{2} \log|\alpha_j^{(n)} \mathbf{H}_j^{(n)}| \right. \\
&\quad \left. - \frac{1}{2}\left[\mathrm{Tr}\left[\left(\mathbb{E}_{q(\mathbf{a}_j^{(n)})}[\mathbf{a}_j^{(n)}(\mathbf{a}_j^{(n)})^\top] + (\mathbf{V}_j^{(n-1)})^{-1} \right) \alpha_j^{(n)} \mathbf{H}_j^{(n)} \right] \right] \right\}.
\end{aligned}
\tag{4.88}
$$

The variational distribution is seen as a new Wishart distribution

$$
\widehat{q}(\alpha_j^{(n)} \mathbf{H}_j^{(n)}|\rho_j^{(n)}, V_j^{(n)}) = \mathcal{W}(\alpha_j^{(n)} \mathbf{H}_j^{(n)}|\rho_j^{(n)}, V_j^{(n)})
\tag{4.89}
$$

with the updated hyperparameters

$$
\rho_j^{(n)} = \rho_j^{(n-1)} + 1,
\tag{4.90}
$$

$$
\mathbf{V}_j^{(n)} = \left(\mathbb{E}_{q(\mathbf{a}_j^{(n)})}[\mathbf{a}_j^{(n)}(\mathbf{a}_j^{(n)})^\top] + (\mathbf{V}_j^{(n-1)})^{-1} \right)^{-1}.
\tag{4.91}
$$

The updating of Wishart distribution in VB-EM algorithm provides a means to sequential learning of individual mixing parameter

$$
\mathbf{a}_j^{(n-1)} \rightarrow \mathbf{a}_j^{(n)}.
\tag{4.92}
$$

As a result, nonstationary Bayesian learning is developed in the NB-ICA procedure by incrementally using minibatch data $\mathbf{X}^{(n)}$ to update hyperparameters

$$
\boldsymbol{\Psi}^{(n-1)} = \{\rho_j^{(n-1)}, \mathbf{V}_j^{(n-1)}\} \rightarrow \boldsymbol{\Psi}^{(n)} = \{\rho_j^{(n)}, \mathbf{V}_j^{(n)}\}
\tag{4.93}
$$

and applying the newest hyperparameters for future updating based on newly-enrolled minibatch data $\mathbf{X}^{(n+1)}$. Next, in the VB-M step, we use the updated hyperparameters $\boldsymbol{\Psi}^{(n)}$ or find the modes of variational distribution to derive the updating of model parameters

$$
\boldsymbol{\Theta}^{(n)} \rightarrow \boldsymbol{\Theta}^{(n+1)}.
\tag{4.94}
$$

The VB-EM algorithm is accordingly implemented to fulfill nonstationary source separation based on NB-ICA. A key idea of this method is to introduce online learning, Bayesian learning and sparse learning to build an ICA procedure which is driven by the incrementally estimated ARD parameters $\alpha^{(n)} = \{\alpha_j^{(n)}\}$ for individual sources j. Evaluation of the effects of ARD parameters and mixing matrix is conducted in the next section.

4.4.4 SYSTEM EVALUATION

Source separation under nonstationary mixing condition was simulated and evaluated by using real-world speech and music signals. The audio signals were again collected from ICA'99 BSS datasets accessible at http://sound.media.mit.edu/ica-bench/. A number of scenarios were simulated by using time-varying mixing coefficients and nonstationary source signals. Fig. 4.17(A) shows two channels of waveforms of nonstationary source signals containing 5 seconds of audio signals in the presence of two different speakers and one music source. Source signals 1 and 2 were shown by blue (dark gray in print version) and red (light gray in print version) waveforms, respectively. The first channel simulated a male speaker speaking in the beginning, leaving at 1.5 s and then replaced by a background music at 2.5 s. The scenario of changing distribution of the source signal was simulated. The moving sources and the source change were investigated. In addition, the second channel simulated a different male speaker speaking in the beginning, leaving at 2.5 s and coming back to speak again at 3.5 s. The scenario of leaving and coming back for the same speaker was examined. Notably, the nonstationary mixing condition was reflected by introducing a time-varying mixing matrix

$$\mathbf{A}(t) = \begin{bmatrix} \cos(2\pi f_1 t) & \sin(2\pi f_2 t) \\ -\sin(2\pi f_1 t) & \cos(2\pi f_2 t) \end{bmatrix} \tag{4.95}$$

where f_1 and f_2 indicate the varying frequencies of mixing coefficients. Each mixing coefficient reflected the spatial correlation or relevance between a source and a sensor. The relevance of two sources to different sensors was revealed by two columns in $\mathbf{A}(t)$. The settings of changing frequencies were $f_1 = 1/20$ Hz and $f_2 = 1/10$ Hz.

Using NB-ICA, the adaptation or compensation of ARD parameter using

$$\alpha_j^{(n)} \mathbf{I}_m \to \alpha_j^{(n)} \mathbf{H}_j^{(n)} \tag{4.96}$$

is performed. This adaptation aims to improve Bayesian modeling of the mixing matrix $\mathbf{A}^{(n)}$ for nonstationary source separation by using sophisticated priors for mixing coefficients. Fig. 4.18 compares the variational lower bounds of using NB-ICA with and without adaptation or transformation of ARD parameter. This result is obtained by using the same mixed speech and music signals as provided in Fig. 4.17(A). The higher the measured variational lower bound or the free energy, the better the achieved learning representation of demixed signals. This figure confirms the increase of variational lower bound by applying adaptation to a full precision matrix of mixing coefficients. Sequential learning of a desirable mixing matrix is assured for prediction and learning in the next minibatch $n + 1$. Moreover, Fig. 4.19 displays two diagonal elements of $\alpha_j^{(l)} \mathbf{H}_j^{(n)}$ (namely ARD parameters) estimated by NB-ICA using the same mixed signals. These ARD parameters truly reflect the activity of source signals as shown in Fig. 4.17 at different time frames. Irrelevant source signals are attained and

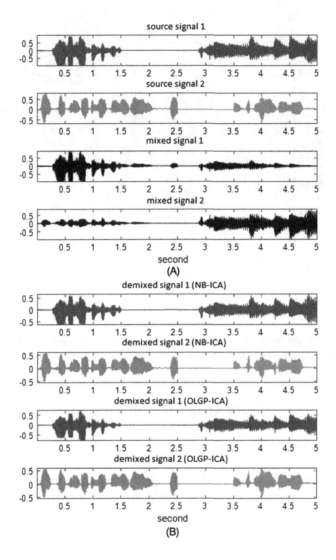

FIGURE 4.17

Comparison of speech waveforms for (A) source signal 1 (blue in web version or dark gray in print version), source signal 2 (red in web version or light gray in print version) and two mixed signals (black) and (B) two demixed signals 1 (blue in web version or dark gray in print version) and two demixed signals 2 (red in web version or light gray in print version) by using NB-ICA and OLGP-ICA algorithms.

deemphasized. Time-domain activities of sources are substantially reflected by ARD parameters. For example, the first speaker stops talking during the interval from 1.5 to 2.5 s, which obviously matches the active value in the ARD curve of the first demixed signal. The second speaker stops talking in the

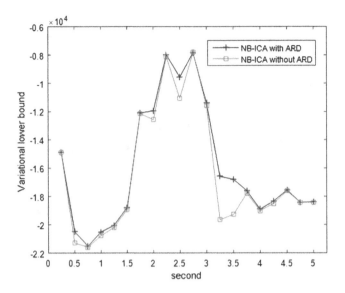

FIGURE 4.18

Comparison of variational lower bounds by using NB-ICA with (blue in web version or dark gray in print version) and without (red in web version or light gray in print version) adaptation of ARD parameter.

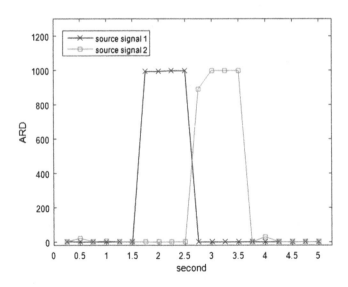

FIGURE 4.19

Comparison of the estimated ARD parameters of source signal 1 (blue in web version or dark gray in print version) and source signal 2 (red in web version or light gray in print version) using NB-ICA.

period from 2.5 to 3.5 s where ARD values of the second demixed signal are also active in that period. ARD accurately detects the dynamics of sources.

ARD parameters of NB-ICA have sufficiently reflected the activity of source signals. Such an advanced work will be compared with other ICA methods and will be evaluated by the experiments on source separation in terms of other metrics including signal-to-interference ratios. A more comparative study will be discussed in Section 4.5 where we will focus on another nonstationary source separation based on an online Gaussian process. Sequential learning and precise temporal modeling are both performed via Bayesian framework for nonstationary source separation.

4.5 ONLINE GAUSSIAN PROCESS ICA

Following the nonstationary source separation using NB-ICA in Section 4.4, this section introduces an advanced topic which captures the temporal information in audio signals in each minibatch training based on a Gaussian process. Again, a minibatch set of mixed signals $\mathbf{X}^{(n)}$ is sequentially observed from $\mathcal{X}^{(n)} = \{\mathbf{X}^{(1)}, \mathbf{X}^{(2)}, \ldots, \mathbf{X}^{(n)}\}$. An online learning for a Gaussian process (GP) is presented to build an online Gaussian process ICA (OLGP-ICA) for source separation. The challenging condition is based on nonstationary and *temporally-correlated* source signals $\mathbf{S} = \{\mathbf{s}_t\}_{t=1}^{T}$ and mixing coefficients $\mathbf{A} = [a_{ij}] \in \mathbb{R}^{m \times m}$. There are two components. One is online learning while the other is the Gaussian process. Online learning is performed to continuously identify the status of source signals so as to estimate the corresponding distributions. GP is an infinite-dimensional generalization of a multivariate Gaussian distribution. GP provides a nonparametric solution to represent the temporal structure of a time-varying mixing system. The temporally-correlated mixing coefficients and source signals are characterized by GP where the posterior statistics are continuously accumulated and propagated using individual minibatches according to a recursive Bayesian algorithm. We describe how GP prior is employed to represent temporal information in nonstationary source signals as well as mixing coefficients. In Ahmed et al. (2000), Costagli and Kuruoğlu (2007), a nonstationary ICA algorithm, called the sequential Monte Carlo ICA (SMC-ICA), was proposed and implemented for source separation based on the sequential importance sampling procedure where the implementation cost was demanding due to slow convergence in model inference.

The evolution of ICA methods for nonstationary source separation is now surveyed. Traditionally, ICA models were developed for static source separation by using a batch collection of T training samples. To tackle nonstationary mixing conditions, a Markov chain (Choudrey and Roberts, 2003, Hirayama et al., 2007, Ichir and Mohammad-Djafari, 2006, Snoussi and Mohammad-Djafari, 2004a), autoregressive model (Huang et al., 2007) and Gaussian process (Park and Choi, 2008) were proposed to characterize the temporally-correlated source signals. However, a fixed set of model parameters is insufficient to represent the complicated nonstationary environments. It is meaningful to detect the activity of source signals and adapt the system distributions to newest environments by using a varying set of model parameters which are dynamically learned from different minibatch collections. As addressed in Sections 3.5 and 4.4, the scenarios of nonstationary mixing conditions may contain moving sources, moving sensors, active/inactive sources and switching sources where sources may be speakers or music. It is important to elaborately characterize the temporally-correlated source signals and mixing coefficients for real-world blind source separation. Fig. 4.20 shows the evolution of nonstationary source separation methods from the nonstationary Bayesian ICA, which was addressed in Section 4.4,

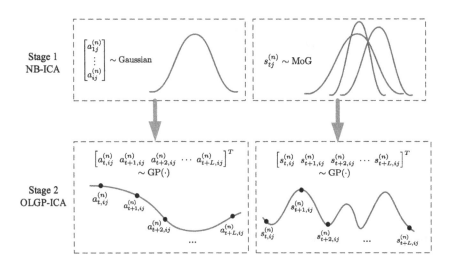

FIGURE 4.20

Evolution from NB-ICA to OLGP-ICA for nonstationary source separation.

to the online Gaussian process ICA, which will be the focus of this section. In this evolution, we first assume that the mixing coefficients are drawn from a Gaussian distribution within a minibatch but those between different minibatches are sampled by different Gaussians. A mixture of Gaussian distributions is employed in delicate modeling of source signals. Sequential and variational learning is developed. The variational and model parameters are continuously updated. The nonstationary Bayesian ICA (NB-ICA) is established as addressed in Section 4.4.3. Secondly, we are addressing an advanced solution where the mixing coefficients and source signals are distributed differently between different minibatches and at the same time are temporally-correlated within a minibatch. The sequential updating and temporally-correlated modeling are implemented for nonstationary source separation. A Gaussian process is applied to represent the temporal structures of mixing coefficients and source signals. Variational inference is used to obtain the resulting OLGP-ICA algorithm.

This section is organized as follows. We first introduce the modeling of a temporal structure and formulate the solution to the source separation problem in Section 4.5.1. The online Bayesian learning is then presented in Section 4.5.2 with an online learning procedure for a Gaussian process in the presence of a noisy ICA model. The variational Bayesian inference is addressed in Section 4.5.3. In Section 4.5.4, the sequential Monte Carlo ICA is presented. The differences of sequential ICA algorithms using an online Gaussian process and sequential Monte Carlo procedure are discussed. In Section 4.5.5, a set of experiments are reported to compare different ICA methods for nonstationary source separation.

4.5.1 TEMPORAL STRUCTURE FOR SEPARATION

Audio source signals are temporally correlated. The temporal correlation is regarded as a crucial information for signal representation for nonstationary source separation. A dynamic model is required to capture temporal correlation for source separation (Smaragdis et al., 2014). Traditionally, the autoregressive (AR) process (Gençağa et al., 2010, Huang et al., 2007) and Gaussian process (Park and Choi, 2008) were applied to characterize temporal correlation for blind source separation. An AR process was used to predict a future sample s_{tj} at time t based on the previous p samples

$$\mathbf{s}_{t-1,j} = [s_{t-1,j}, \ldots, s_{t-p,j}]^\top \tag{4.97}$$

according to a latent function

$$f(\mathbf{s}_{t-1,j}) = \sum_{\tau=1}^{p} h_{\tau,j} s_{t-\tau,j} \tag{4.98}$$

using the AR coefficients $\{h_{\tau,j}\}_{\tau=1}^{p}$. Different from the AR model using linear parametric function for prediction, GP adopted a *nonlinear nonparametric* function as a regression model for prediction. A source sample s_{tj} was drawn and predicted by a Gaussian prior with a zero-mean vector and a kernel-function-based covariance matrix (Park and Choi, 2008)

$$f(\mathbf{s}_{t-1,j}) \sim \mathcal{N}\left(f(\mathbf{s}_{t-1,j}) \Big| 0, \kappa(\mathbf{s}_{t-1,j}, \mathbf{s}_{\tau-1,j}) \right). \tag{4.99}$$

In Eq. (4.99), $f(\cdot)$ is a latent function and $\kappa(\mathbf{s}_{t-1,j}, \mathbf{s}_{\tau-1,j})$ is a kernel function of previous p samples at different times $t-1$ and $\tau-1$. Using GP, different source samples are jointly Gaussian. A regression function was used to represent the temporal information in the jth source by

$$s_{tj} = f(\mathbf{s}_{t-1,j}) + \varepsilon_{tj} \tag{4.100}$$

where ε_{tj} denotes a white Gaussian noise with zero mean and unit variance, i.e., $\mathcal{N}(\varepsilon_{tj}|0, 1)$. GP is a *nonparametric* model which serves as a prior distribution for regression function rather than assumes the parametric form of an unknown regression function. GP is a flexible and powerful model through implementing a high dimensional kernel space using an $L \times L$ positive definite Gram matrix $\mathbf{K}_{s_j}$ (Rasmussen and Williams, 2006)

$$[\mathbf{K}_{s_j}]_{t\tau} = \kappa(\mathbf{s}_{t-1,j}, \mathbf{s}_{\tau-1,j}) + \delta_{t\tau} \tag{4.101}$$

where $\delta_{t\tau} = 1$ at $t = \tau$, and $\delta_{t\tau} = 0$ otherwise. GP is a *Bayesian kernel* method where the GP prior in Eq. (4.99) is employed to drive Bayesian prediction for regression problem in a separation system. GP was applied to model the source signals for source separation in Park and Choi (2008). However, GP is not only useful for characterizing the source signals but also feasible to represent the mixing coefficients as addressed in what follows.

4.5.2 ONLINE LEARNING AND GAUSSIAN PROCESS

The advanced ICA model based on an online Gaussian process (Hsieh and Chien, 2011) is developed with two extensions. One is the online learning for nonstationary source separation while the other is the representation of temporally-correlated mixing coefficients using a Gaussian process. The mixing matrix is time-varying

$$\mathbf{A}^{(n)} \to \mathbf{A}_t^{(n)} \tag{4.102}$$

to reflect the scenario of moving sources or sensors where the relevance between any source j and sensor i using the mixing coefficient $a_{ij}^{(n)}$ is changing in time. Accordingly, we represent the temporal correlation for L samples of mixing coefficients $\{a_{t,ij}^{(n)}\}$ within a minibatch n and simultaneously characterize the temporal structure for those samples across different minibatches. Online learning, combined with temporal modeling, is performed for mixing coefficients and source signals under a noisy ICA model as given in Eq. (4.64). An ICA algorithm is undertaken by combining online learning and a Gaussian process. Sequential and variational learning is performed to estimate model parameters $\mathbf{\Theta}^{(n)}$ and variational parameters $\mathbf{\Psi}^{(n)}$ at each individual minibatch. The time-varying mixing matrix $\mathbf{A}_t^{(n)}$ is employed in OLGP-ICA, which is different from NB-ICA in Section 4.4 by using a fixed mixing matrix $\mathbf{A}^{(n)}$. Online learning is performed to capture the temporal structures in $\{a_{t,ij}^{(n)}\}_{t=1}^L$ and $\{s_{tj}^{(n)}\}_{t=1}^L$ based on GP.

The details of OLGP-ICA procedure are addressed. First of all, we consider a noisy ICA model by using the time-varying mixing matrix $\mathbf{A}_t^{(n)}$ and assume the generation of the temporally correlated mixing coefficients and source signals based on the distributions of nonparametric latent functions. A Gaussian process is introduced to elaborately express the temporal correlation of mixing coefficient $a_{t,ij}^{(n)}$ where the latent prediction function of current coefficient $a_{t,ij}^{(n)}$ based on the previous p coefficients

$$\mathbf{a}_{t-1,ij}^{(n)} = [a_{t-1,ij}^{(n)}, \ldots, a_{t-p,ij}^{(n)}]^\top \tag{4.103}$$

is calculated by

$$a_{t,ij}^{(n)} = f(\mathbf{a}_{t-1,ij}^{(n)}) + \varepsilon_{t,ij}^{(n)} \tag{4.104}$$

where $\varepsilon_{t,ij}^{(n)}$ is a white noise. Since we use a GP, this latent function is represented by a Gaussian distribution

$$f(\mathbf{a}_{t-1,ij}^{(n)}) \sim \mathcal{N}\left(f(\mathbf{a}_{t-1,ij}^{(n)})\middle|0, \kappa(\mathbf{a}_{t-1,ij}^{(n)}, \mathbf{a}_{\tau-1,ij}^{(n)})\right) \tag{4.105}$$

with zero mean and a kernel-function-based variance

$$\kappa(\mathbf{a}_{t-1,ij}^{(n)}, \mathbf{a}_{\tau-1,ij}^{(n)}) \triangleq \xi_{a_{ij}}^{(n-1)} \exp\left[-\frac{\lambda_{a_{ij}}^{(n-1)}}{2}\left\|\mathbf{a}_{t-1,ij}^{(n)} - \mathbf{a}_{\tau-1,ij}^{(n)}\right\|^2\right]. \tag{4.106}$$

An exponential-quadratic kernel function with parameters

$$\{\lambda_{a_{ij}}^{(n-1)}, \xi_{a_{ij}}^{(n-1)}\} \tag{4.107}$$

is considered. Importantly, a GP prior distribution for mixing coefficients

$$\mathbf{a}_{ij}^{(n)} = [a_{1,ij}^{(n)}, \ldots, a_{L,ij}^{(n)}]^\top \tag{4.108}$$

at minibatch n is obtained by

$$p(\mathbf{a}_{ij}^{(n)}|\boldsymbol{\mu}_{a_{ij}}^{(n-1)}, \mathbf{R}_{a_{ij}}^{(n-1)}) \tag{4.109}$$

which is derived by using a Gaussian prior for a set of latent functions $\{f(\mathbf{a}_{t-1,ij}^{(n)})\}$ given by

$$f(\mathbf{a}_{ij}^{(n)}) \sim \mathcal{N}\left(\mathbf{a}_{ij}^{(n)}\middle|\boldsymbol{\mu}_{a_{ij}}^{(n-1)} = 0, (\mathbf{R}_{a_{ij}}^{(n-1)})^{-1} = \mathbf{K}_{a_{ij}}^{(n-1)}\right). \tag{4.110}$$

This Gaussian prior has a zero-mean vector and an $L \times L$ covariance matrix $\mathbf{K}_{a_{ij}}^{(n-1)}$ where the (t, τ)th entry is given by

$$[\mathbf{K}_{a_{ij}}^{(n-1)}]_{t\tau} = \kappa(\mathbf{a}_{t-1,ij}^{(n)}, \mathbf{a}_{\tau-1,ij}^{(n)}) + \delta_{t\tau}. \tag{4.111}$$

On the other hand, a GP prior is derived to express the temporal correlation in the time series of source signal $\{s_{tj}^{(n)}\}$ at each minibatch n similar to the GP prior for the mixing coefficient in Eq. (4.110). A latent function $f(\mathbf{s}_{t-1,j}^{(n)})$ using previous p source samples

$$\mathbf{s}_{t-1,j}^{(n)} = [s_{t-1,j}^{(n)}, \ldots, s_{t-p,j}^{(n)}]^\top \tag{4.112}$$

is adopted to determine the value of the current sample $s_{tj}^{(n)}$. This function is driven by a Gaussian prior

$$\mathcal{N}\left(f(\mathbf{s}_{t-1,j}^{(n)})\middle|0, \kappa(\mathbf{s}_{t-1,j}^{(n)}, \mathbf{s}_{\tau-1,j}^{(n)})\right) \tag{4.113}$$

with zero mean and $\kappa(\mathbf{s}_{t-1,j}^{(n)}, \mathbf{s}_{\tau-1,j}^{(n)})$ as the variance parameter, which is controlled by an exponential-quadratic kernel function in Eq. (4.106) but using the kernel parameters $\{\lambda_{s_j}^{(n-1)}, \xi_{s_j}^{(n-1)}\}$. We accordingly derive the GP prior $p(\mathbf{s}_j^{(n)}|\boldsymbol{\mu}_{s_j}^{(n-1)}, \mathbf{R}_{s_j}^{(n-1)})$ for L samples $\mathbf{s}_j^{(n)} = [s_{1j}^{(n)}, \ldots, s_{Lj}^{(n)}]^\top$ as

$$\mathcal{N}\left(\mathbf{s}_j^{(n)}\middle|\boldsymbol{\mu}_{s_j}^{(n-1)} = 0, (\mathbf{R}_{s_j}^{(n-1)})^{-1} = \mathbf{K}_{s_j}^{(n-1)}\right) \tag{4.114}$$

where the (t, τ)th element of the covariance matrix is expressed by

$$[\mathbf{K}_{s_j}^{(n-1)}]_{t\tau} = \kappa(\mathbf{s}_{t-1,j}^{(n)}, \mathbf{s}_{\tau-1,j}^{(n)}) + \delta_{t\tau}. \tag{4.115}$$

The graphical model of the online Gaussian process ICA is depicted in Fig. 4.21. Similar to NB-ICA, a noisy ICA model is adopted in the signal representation using OLGP-ICA where the noise signal $\boldsymbol{\varepsilon}_t^{(n)}$ is modeled by a zero-mean Gaussian distribution

$$\boldsymbol{\varepsilon}_t^{(n)} \sim \mathcal{N}(\boldsymbol{\varepsilon}_t^{(n)}|0, (\mathbf{B}^{(n)})^{-1}) \tag{4.116}$$

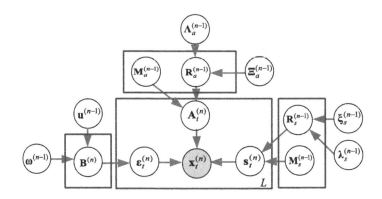

FIGURE 4.21

Graphical representation for online Gaussian process ICA model.

with a diagonal matrix for precision $\mathbf{B}^{(n)} = \mathrm{diag}\{\beta_i^{(n)}\}$. The diagonal element $\beta_i^{(n)}$ is distributed according to a conjugate prior based on a Gamma distribution

$$\beta_i^{(n)} \sim \mathrm{Gam}(\beta_i^{(n)} | u_{\beta_i}^{(l-1)}, \omega_{\beta_i}^{(l-1)}). \tag{4.117}$$

Under this model construction, OLGP-ICA parameters consist of

$$\Theta^{(n)} = \{\mathbf{A}^{(n)}, \mathbf{E}^{(n)}, \mathbf{B}^{(n)}, \mathbf{S}^{(n)}\} \tag{4.118}$$

and hyperparameters are composed of

$$\Psi^{(n)} = \{\mathbf{M}_a^{(n)}, \mathbf{R}_a^{(n)}, \mathbf{\Lambda}_a^{(n)}, \mathbf{\Xi}_a^{(n)}, \mathbf{u}^{(n)}, \boldsymbol{\omega}^{(n)}, \mathbf{M}_s^{(n)}, \mathbf{R}_s^{(n)}, \boldsymbol{\lambda}_s^{(n)}, \boldsymbol{\xi}_s^{(n)}\}, \tag{4.119}$$

which are Gaussian parameters of mixing coefficients

$$\{\mathbf{M}_a^{(n)} = \{\boldsymbol{\mu}_{a_{ij}}^{(n)}\}, \mathbf{R}_a^{(n)} = \{\mathbf{R}_{a_{ij}}^{(n)}\}\} \tag{4.120}$$

and source signals $\{\mathbf{M}_s^{(n)} = \{\boldsymbol{\mu}_{s_j}^{(n)}\}, \mathbf{R}_s^{(n)} = \{\mathbf{R}_{s_j}^{(n)}\}\}$, Gamma parameters of noise signals

$$\{\mathbf{u}^{(n)} = \{u_{\beta_i}^{(n)}\}, \boldsymbol{\omega}^{(n)} = \{\omega_{\beta_i}^{(n)}\}\} \tag{4.121}$$

as well as kernel parameters of mixing coefficients

$$\{\mathbf{\Lambda}_a^{(n)} = \{\lambda_{a_{ij}}^{(n)}\}, \mathbf{\Xi}_a^{(n)} = \{\xi_{a_{ij}}^{(n)}\}\} \tag{4.122}$$

and source signals

$$\{\boldsymbol{\lambda}_s^{(n)} = \{\lambda_{s_j}^{(n)}\}, \boldsymbol{\xi}_s^{(n)} = \{\xi_{s_j}^{(n)}\}\}. \tag{4.123}$$

As a result, an online Gaussian process is employed for blind source separation in nonstationary and temporally correlated environments. Interestingly, the scaling hyperparameter $\lambda_{a_{ij}}^{(n)}$ in OLGP-ICA is similar to the automatic relevance determination (ARD) in NB-ICA as addressed in Sections 3.4.1 and 4.4.2. This hyperparameter sufficiently reflects the temporal information of the mixing coefficient $a_{ij}^{(n)}$, which indicates the activity of source signals. When the value of $\lambda_{a_{ij}}^{(n)}$ is small, the corresponding previous coefficient $a_{t-1,ij}^{(n)}$ has a slight effect on prediction of the new coefficient $a_{t,ij}^{(n)}$. If the predicted coefficient $a_{t,ij}^{(n)}$ approaches zero, the source signal becomes inactive at that moment. For this case of small $\lambda_{a_{ij}}^{(n)}$, the previous samples of mixing coefficients affect prediction of new coefficient very little. Similarly, the status of source signals at different minibatches is reflected by the estimated scaling parameter $\lambda_{s_j}^{(n)}$.

4.5.3 SEQUENTIAL AND VARIATIONAL LEARNING

As addressed in Section 4.4, NB-ICA characterizes the mixing coefficients and source signals by using Gaussian and GMM (or MoG) distributions, respectively. This section presents a Gaussian process approach to discover temporal structures in the mixing matrix $\mathbf{A}_t^{(n)}$ as well as source signals $\mathbf{s}_t^{(n)}$. Basically, GP is a generalization of a Gaussian distribution for a time-series random variables. The nonparametric latent functions for ICA parameters $\{\mathbf{A}_t^{(n)}, \mathbf{s}_t^{(n)}\}$ are driven by the corresponding Gaussian priors controlled by kernel parameters. The OLGP-ICA algorithm is developed by implementing a VB-EM procedure as seen in Algorithm 4.1 where model parameters $\Theta^{(n)}$ and hyperparameters $\Psi^{(n)}$ are estimated at each minibatch n. In this implementation, we maximize the variational lower bound of the logarithm of marginal likelihood $p(\mathbf{X}^{(n)}|\Psi^{(n-1)})$ so as to estimate the variational distribution $\widehat{q}(\Theta^{(n)})$ and update the variational parameters or hyperparameters as

$$\Psi^{(n-1)} \to \Psi^{(n)}. \tag{4.124}$$

The VB-E step is run. In Choudrey et al. (2000), the variational parameters of the noise signal $\{\mathbf{u}^{(n)}, \omega^{(n)}\}$ have been formulated. Here, we focus on the derivation of the variational distributions of mixing coefficients $\mathbf{a}_{ij}^{(n)}$ and source signals $\mathbf{s}_j^{(n)} = \{s_{tj}^{(n)}\}_{t=1}^L$ at each minibatch n where a Gaussian process is applied. To do so, we realize the general VB solution in Eq. (4.85) and derive the variational distribution of mixing coefficients as follows:

$$\begin{aligned}
\log \widehat{q}(\mathbf{a}_{ij}^{(n)}) \propto\ & E_{q(\Theta \neq \mathbf{a}_{ij})}[\log p(\mathbf{X}^{(n)}|\mathbf{a}_{ij}^{(n)}, \mathbf{s}_j^{(n)}, \boldsymbol{\varepsilon}_i^{(n)})] \\
& + \log p(\mathbf{a}_{ij}^{(n)}|\boldsymbol{\mu}_{a_{ij}}^{(n-1)}, \mathbf{R}_{a_{ij}}^{(n-1)})
\end{aligned} \tag{4.125}$$

where

$$\boldsymbol{\varepsilon}_i^{(n)} = [\varepsilon_{i,1}^{(n)}, \dots, \varepsilon_{i,L}^{(n)}] \tag{4.126}$$

denotes the error signals in the ith microphone. The optimal variational distribution is interpreted as a new kind of posterior distribution, which is combined by integrating the expectation of the likelihood function (the first term in Eq. (4.125)) operated over all variational distributions $q(\Theta \neq \mathbf{a}_{ij})$ except that of $\mathbf{a}_{ij}$ and the GP prior (the second term in Eq. (4.125)) as shown in Eq. (4.110). An updated Gaussian with new variational parameters

$$\{\boldsymbol{\mu}_{a_{ij}}^{(n-1)}, \mathbf{R}_{a_{ij}}^{(n-1)}\} \to \{\boldsymbol{\mu}_{a_{ij}}^{(n)}, \mathbf{R}_{a_{ij}}^{(n)}\} \tag{4.127}$$

is obtained by combining two quadratic functions of $\mathbf{a}_{ij}^{(n)}$ from two RHS terms in Eq. (4.125). A new expression of multivariate Gaussian distribution is arranged as

$$\exp\left\{\mathbb{E}_{q(\Theta\neq\mathbf{a}_{ij})}\left[\log p(\mathbf{X}^{(n)}|\mathbf{a}_{ij}^{(n)},\mathbf{s}_j^{(n)},\boldsymbol{\varepsilon}_i^{(n)})\right]\right\}$$
$$\propto \mathcal{N}\left((\boldsymbol{\Psi}_{a_{ij}}^{(n)})^{-1}\widetilde{\mathbf{x}}_{a_{ij}}^{(n)}\Big|\mathbf{a}_{ij}^{(n)},(\boldsymbol{\Psi}_{a_{ij}}^{(n)})^{-1}\right) \tag{4.128}$$

where the L dimensional vector $\widetilde{\mathbf{x}}_{a_{ij}}^{(n)}$ has the tth element

$$\widetilde{x}_{t,a_{ij}}^{(n)} = \mathbb{E}_{q(\Theta\neq\mathbf{a}_{ij})}[\beta_{ti}^{(n)}]\mathbb{E}_{q(\Theta\neq\mathbf{a}_{ij})}[s_{tj}^{(n)}]$$
$$\times\left(x_{ti}^{(n)} - \sum_{k\neq j}^{m}\mathbb{E}_{q(\Theta\neq\mathbf{a}_{ij})}[a_{t,ik}^{(n)}]\mathbb{E}_{q(\Theta\neq\mathbf{a}_{ij})}[s_{tk}^{(n)}]\right) \tag{4.129}$$

and the $L \times L$ diagonal matrix $\boldsymbol{\Psi}_{a_{ij}}^{(n)}$ has the tth diagonal element

$$[\boldsymbol{\Psi}_{a_{ij}}^{(n)}]_{tt} = \mathbb{E}_{q(\Theta\neq\mathbf{a}_{ij})}[\beta_{ti}^{(n)}]\mathbb{E}_{q(\Theta\neq\mathbf{a}_{ij})}[(s_{tj}^{(n)})^2]. \tag{4.130}$$

Attractively, we arrange Eq. (4.128) as a Gaussian distribution for the transformed observation vector due to mixing coefficient $\mathbf{a}_{ij}^{(n)}$, i.e.,

$$(\boldsymbol{\Psi}_{a_{ij}}^{(n)})^{-1}\widetilde{\mathbf{x}}_{a_{ij}}^{(n)},$$

which can be manipulated as a quadratic function of $\mathbf{a}_{ij}^{(n)}$. By incorporating the likelihood function of Eq. (4.128) and the prior density of Eq. (4.110) into Eq. (4.125), the two exponents of quadratic functions of $\mathbf{a}_{ij}^{(n)}$ are summed up to estimate an optimal variational distribution

$$\widehat{q}(\mathbf{a}_{ij}^{(n)}|\boldsymbol{\mu}_{a_{ij}}^{(n)},\mathbf{R}_{a_{ij}}^{(n)}).$$

This distribution is viewed as an approximate posterior, which is yielded as a Gaussian with new variational parameters

$$\boldsymbol{\mu}_{a_{ij}}^{(n)} = \left(\mathbf{R}_{a_{ij}}^{(n-1)}\right)^{-1}\left(\boldsymbol{\Psi}_{a_{ij}}^{(n)}\right)^{-1}\widetilde{\mathbf{x}}_{a_{ij}}^{(n)}, \tag{4.131}$$

$$\mathbf{R}_{a_{ij}}^{(n)} = \left(\boldsymbol{\Psi}_{a_{ij}}^{(n)}\left(\mathbf{R}_{a_{ij}}^{(n-1)}\right)^{-1}\right)^{-1}. \tag{4.132}$$

This online Gaussian process ICA is constructed to carry out the perspective of reproducible prior/posterior pair via the VB-EM algorithm where a Gaussian likelihood is combined with a GP prior to reproduce a Gaussian variational posterior $\widehat{q}(\mathbf{a}_{ij}^{(n)}|\boldsymbol{\mu}_{a_{ij}}^{(n)},\mathbf{R}_{a_{ij}}^{(n)})$. The variational posterior drives the updating of hyperparameters by

$$\{\mathbf{M}_a^{(n-1)},\mathbf{R}_a^{(n-1)}\} \rightarrow \{\mathbf{M}_a^{(n)},\mathbf{R}_a^{(n)}\}. \tag{4.133}$$

The detailed updating formulas for the other hyperparameters $\{\mathbf{M}_s^{(n)}, \mathbf{R}_s^{(n)}\}$ and $\{\boldsymbol{\Lambda}_a^{(n)}, \boldsymbol{\Xi}_a^{(n)}, \boldsymbol{\lambda}_s^{(n)}, \boldsymbol{\xi}_s^{(n)}\}$ can be referred in Chien and Hsieh (2013b). Sequential and variational learning algorithm is executed by sequentially using the current minibatch $\mathbf{X}^{(n)}$ to estimate the variational parameters $\boldsymbol{\Psi}^{(n-1)} \to \boldsymbol{\Psi}^{(n)}$ and continuously propagating the posterior statistics to run for the next iteration by using the new minibatch $\mathbf{X}^{(n+1)}$. In the implementation, the hyperparameters are initialized from $\boldsymbol{\Psi}^{(0)}$ and then iteratively updated to new ones $\boldsymbol{\Psi}^{(n-1)} \to \boldsymbol{\Psi}^{(n)}$ at each minibatch n. As explained in Section 3.7.3, the continuous increase of variational lower bound is assured before saturated at a maximum point by running the sequential VB-EM iterations. Current hyperparameters $\boldsymbol{\Psi}^{(n)}$ at minibatch n are propagated to a new minibatch $n + 1$ and act as new statistics of the updated priors for online Bayesian learning. Once the hyperparameters are updated, the model parameters can be calculated in the VB-M step for OLGP-ICA. VB inference is implemented for nonstationary source separation. However, in addition to variational Bayesian (VB) learning, the approximate inference based on Markov chain Monte Carlo (MCMC) is also feasible to carry out for nonstationary source separation.

4.5.4 SEQUENTIAL MONTE CARLO ICA

As discussed in Sections 3.7.3 and 3.7.4, VB inference is seen as a greedy algorithm while MCMC inference is run as a local algorithm. Both algorithms perform similar strategy of local updating in Bayesian manner which is prone to trap in a local mode of posterior, in particular when the initial condition is far from optimum. VB and MCMC are algorithmically related and theoretically complimentary (Cemgil et al., 2007). VB has fast convergence towards a nearby local optimum, but MCMC has desirable robustness and generality. There were several MCMC solutions to ICA model. In Snoussi and Mohammad-Djafari (2004b), an image separation method was implemented by developing MCMC inference for ICA based on importance sampling. In Cemgil et al. (2007), the approximate Bayesian using Student's t distribution was introduced. Source separation using VB outperformed than that using MCMC. The previous MCMC methods conducted batch separation which was invalid for source separation in nonstationary environments. In Doucet et al. (2001), a particle filter with recursively-updated state parameters was designed to establish a jump Markov linear system where the sequential importance sampling was developed for a general solution to both continuous and discrete state processes. In Gençağa et al. (2010), a particle filter based on AR process was designed. This process did not involve an ICA problem. In Rowe (2002), an MCMC method was presented by introducing conjugate priors, but the authors did not investigate the temporally-correlated condition in blind source separation. In Ahmed et al. (2000), Costagli and Kuruoğlu (2007), a sequential Monte Carlo ICA (SMC-ICA) was proposed as an ICA solution to continuous-state process. SMC-ICA is addressed as follows. First of all, the noisy ICA model, as given in Eq. (4.64), is reshaped as a continuous-state space model with an expression of observed mixture signal

$$a^{(n)} = a^{(n-1)} + \mathbf{v}^{(n)}, \tag{4.134}$$

$$\mathbf{x}_t^{(n)} = \mathbf{C}_t^{(n)} a^{(n)} + \boldsymbol{\varepsilon}_t^{(n)}. \tag{4.135}$$

Here, a matrix of mixing coefficients is rewritten as an $m^2 \times 1$ vectorized parameter $a^{(n)} = \text{vec}\{\mathbf{A}^{(n)}\}$ with a relation

$$[a^{(n)}]_{m(j-1)+i} = a_{ij}^{(n)}. \tag{4.136}$$

In Eqs. (4.134)–(4.135), $\mathbf{v}^{(n)}$ is a Gaussian noise vector with zero mean, and

$$\mathbf{C}_t^{(n)} = (\mathbf{s}_t^{(n)})^\top \otimes \mathbf{I}_m \tag{4.137}$$

is a reexpression of source signals as an $m \times m^2$ matrix. The Gaussian mixture model in Eq. (4.61) is used for representation of source signals where an additional transition probability of different mixture components from $z_{tj}^{(n-1)} = l$ to $z_{tj}^{(n)} = k$, for $1 \le l, k \le K$, is incorporated and defined by

$$p(z_{tj}^{(n)} = k | z_{tj}^{(n-1)} = l) = \tau_{jlk}^{(n)}. \tag{4.138}$$

This transition probability also reveals the activity of the Gaussian component for source j at time moment t. The parameters of SMC-ICA consist of

$$\boldsymbol{\Theta}^{(n)} = \{\{\mathbf{s}_t^{(n)}\}, \{z_{tj}^{(n)}\}, \{\mu_{jk}^{(n)}\}, \{\gamma_{jk}^{(n)}\}, \{\tau_{jlk}^{(n)}\}\}. \tag{4.139}$$

This particle filter is constructed from those particles $\{a_q^{(n)}, \boldsymbol{\Theta}_q^{(n)}, 1 \le q \le Q\}$ which are drawn from the corresponding posterior distribution

$$\begin{aligned} &p(a^{(n)}, \boldsymbol{\Theta}^{(0:n)} | \mathbf{X}^{(1:n)}) \\ &= p(a^{(n)} | \boldsymbol{\Theta}^{(0:n)}, \mathbf{X}^{(1:n)}) p(\boldsymbol{\Theta}^{(0:n)} | \mathbf{X}^{(1:n)}) \end{aligned} \tag{4.140}$$

by using training samples

$$\mathbf{X}^{(1:n)} = \mathcal{X}^{(n)} = \{\mathbf{X}^{(1)}, \dots, \mathbf{X}^{(n)}\}. \tag{4.141}$$

In the implementation, the mixing coefficients are sampled by a derived Gaussian $p(a^{(n)} | \boldsymbol{\Theta}^{(0:n)}, \mathbf{X}^{(1:n)})$ given by an approximate distribution $p(\boldsymbol{\Theta}^{(0:n)} | \mathbf{X}^{(1:n)})$. These coefficients are recursively calculated in closed-form based on the *Kalman filter* in Eqs. (4.134)–(4.135). The posterior distribution $p(\boldsymbol{\Theta}^{(0:n)} | \mathbf{X}^{(1:n)})$ is recursively updated to fulfill particle filtering based on a suboptimal solution by applying an importance distribution (Ahmed et al., 2000, Costagli and Kuruoğlu, 2007)

$$\begin{aligned} &\pi(\boldsymbol{\Theta}^{(0:n)} | \mathbf{X}^{(1:n)}) \\ &= \pi(\boldsymbol{\Theta}^{(0)}) \prod_{l=1}^{n} \pi(\boldsymbol{\Theta}^{(l)} | \boldsymbol{\Theta}^{(0:n-1)}, \mathbf{X}^{(1:n)}). \end{aligned} \tag{4.142}$$

The prior importance function $\pi(\boldsymbol{\Theta}^{(n)} | \boldsymbol{\Theta}^{(n-1)})$ is expressed by

$$\begin{aligned} &p(\{\mathbf{s}_t^{(n)}\} | \{z_{tj}^{(n)}\}, \{\mu_{jk}^{(n)}\}, \{\gamma_{jk}^{(n)}\}) p(\{\mu_{jk}^{(n)}\} | \{\mu_{jk}^{(n-1)}\}) \\ &\times p(\{\gamma_{jk}^{(n)}\} | \{\gamma_{jk}^{(n-1)}\}) p(\{z_{tj}^{(n)}\} | \{z_{tj}^{(n-1)}\}, \{\tau_{jlk}^{(n)}\}) \\ &\times p(\{\tau_{jlk}^{(n)}\} | \{\tau_{jlk}^{(n-1)}\}). \end{aligned} \tag{4.143}$$

This function is determined by drawing $\{\mu_{jk}^{(n)}\}$ and $\{\log \gamma_{jk}^{(n)}\}$ at each frame or minibatch n based on the corresponding Gaussians where the means are given by the particles at the previous minibatch $n-1$ and the variance is obtained by the method in Ahmed et al. (2000), Costagli and Kuruoğlu (2007). There are two steps implemented for particle filtering (Ahmed et al., 2000, Costagli and Kuruoğlu, 2007, Doucet et al., 2001)

Step 1: sequential importance sampling

- For $q = 1, \ldots, Q$
  - draw $\widetilde{\Theta}_q^{(n)}$ by using the distribution $\pi(\Theta^{(n)} | \Theta_q^{(0:n-1)}, \mathbf{X}^{(1:n)})$ and define $\widetilde{\Theta}_q^{(0:n)} = \{\Theta_q^{(0:n-1)}, \widetilde{\Theta}_q^{(n)}\}$.
  - calculate the importance weights up to a normalization constant

$$w_q^{(n)} \propto \frac{p(\mathbf{X}^{(n)} | \widetilde{\Theta}_q^{(0:n)}, \mathbf{X}^{(1:n-1)}) p(\widetilde{\Theta}_q^{(n)} | \widetilde{\Theta}_q^{(n-1)})}{\pi(\widetilde{\Theta}_q^{(n)} | \widetilde{\Theta}_q^{(0:n-1)}, \mathbf{X}^{(1:n)})}. \tag{4.144}$$

  - calculate the normalized importance weights by

$$\widetilde{w}_q^{(n)} \propto \left[\sum_{l=1}^{Q} w_l^{(n)} \right]^{-1} w_q^{(n)}. \tag{4.145}$$

Step 2: particle selection

- Remove/enhance particles with small/large values of normalized importance weights to obtain particles $\{\Theta_q^{(0:n)}, q = 1, \ldots, Q\}$.

SMC parameters or latent sources $\Theta^{(n)} = \{\mathbf{s}_t^{(n)}\}$ are finally obtained via the expectation

$$\mathbb{E}_{p_Q(\Theta^{(0:n)} | \mathbf{X}^{(1:n)})}[f^{(n)}(\Theta^{(0:n)})] = \sum_{q=1}^{Q} f^{(n)}(\Theta_q^{(n)}) \widetilde{w}_q^{(n)} \tag{4.146}$$

by using those particles $\{\Theta_q^{(0:n)}, q = 1, \ldots, Q\}$ and importance weights $\{\widetilde{w}_q^{(n)}\}$.

When we treat $f^{(n)}$ as an identify function, the result of the minimum mean square estimate of $\Theta^{(n)}$ is implemented. Basically, SMC-ICA is feasible to capture the dynamics of mixing coefficients and source signals for nonstationary source separation.

4.5.5 SYSTEM EVALUATION

The experimental setup for online Gaussian process ICA (OLGP-ICA) is consistent with that for nonstationary Bayesian ICA (NB-ICA), which has been mentioned in Section 4.4.4. Several other ICA procedures, including VB-ICA (Lawrence and Bishop, 2000), BICA-HMM (Choudrey and Roberts, 2003), switching ICA (Hirayama et al., 2007), GP-ICA (Park and Choi, 2008), online VB-ICA (Honkela and Valpola, 2003), NS-ICA (Everson and Roberts, 2000) and SMC-ICA (Ahmed et al., 2000, Costagli and Kuruoğlu, 2007), were included for comparison. The variational Bayesian ICA (VB-ICA) (Lawrence and Bishop, 2000) conducted the batch learning without compensating the nonstationary mixing condition. The Bayesian ICA with hidden Markov sources (BICA-HMM) (Choudrey and Roberts, 2003) introduced an HMM to characterize dynamic sources but with batch training mode. The switching ICA (S-ICA) (Hirayama et al., 2007) treated the situation of a sudden presence and absence of sources but still in batch training way. The condition of moving sensors or sources was disregarded in Choudrey and Roberts (2003), Hirayama et al. (2007). The GP-ICA (Park and Choi, 2008) represented the temporal correlation of source signals rather than that of mixing coefficients. Batch training was performed. The online VB-ICA (OVB-ICA) (Honkela and Valpola, 2003) ran sequential and variational learning for nonstationary source signals while the mixing coefficients were kept unchanged. The nonstationary ICA (NS-ICA) (Everson and Roberts, 2000) implemented a sequential learning algorithm to compensate nonstationary mixing condition for mixing coefficients rather than that for source signals. The SMC-ICA (Ahmed et al., 2000, Costagli and Kuruoğlu, 2007) constructed a particle filter as a sequential Bayesian approach to source separation based on sequential importance sampling where temporally-correlated mixing condition was disregarded. This section presents the sequential and variational learning based on NB-ICA and OLGP-ICA where online Bayesian learning and temporally-correlated modeling are both performed to learn parameters of source signals and the mixing matrix. In the implementation, the minibatch size in online learning algorithms using NS-ICA, SMC-ICA, NB-ICA and OLGP-ICA was consistently specified as 0.25 s. The prediction order in temporally-correlated modeling based on GP-ICA and OLGP-ICA was fixed to be $p = 6$. Using variational Bayesian learning, the source signals and mixing coefficients were estimated and obtained from the variational parameters, sufficient statistics, or modes of variational distributions.

Fig. 4.22 shows the square error between true and the estimated mixing coefficients in $\mathbf{A}_t$ and $\mathbf{A}_t^{(n)}$ at different minibatches of the experimental data as displayed in Fig. 4.17. Different ICA methods are investigated and shown by different colors. The measure of square error was calculated from four estimated coefficients in the 2×2 matrix $\mathbf{A}_t^{(n)}$ and L samples in $\mathbf{X}^{(n)} = \{\mathbf{x}_t^{(n)}\}_{t=1}^L$ at each minibatch n. OLGP-ICA attains a lower square error curve than NB-ICA for most time frames. These two ICA methods are more accurate than NS-ICA and SMC-ICA at various minibatches. The improvement of SMC-ICA relative to NS-ICA is desirable. Nevertheless, SMC-ICA does not work as well as NB-ICA and OLGP-ICA in this set of experiments. In addition to the set of mixed signals adopted in Section 4.4.4, the other five sets of mixed signals (5 seconds on average) were collected to conduct a statistically meaningful evaluation. The same scenarios, addressed in Section 4.4.4, were applied but simulated with *different* speakers and music sources, which were sampled from the same dataset. Also, different mixing matrices with varying frequencies f_1 and f_2 were adopted. System evaluation was performed by measuring the absolute errors of predictability and the SIRs using different ICAs. The evaluation measure is shown over a total of six sets of experimental data.

Speech and music source signals are temporally correlated in nature. Mixing such highly correlated source signals results in a very complex signal which is more complicated than individual source signals. The problem of blind source separation turns out to decompose and identify the minimally complex source signals. Following (Stone, 2001), the performance of source separation is assessed in terms of the complexity of demixed signals. This metric is coincident with the ICA objective functions based on independence or non-Gaussianity. This complexity is measured by the temporal predictability of a signal. The higher the predictability of a signal sample from its previous samples, the *lower* the measured signal complexity. The temporal predictability of a source or demixed signal $\mathbf{s}_m = \{s_{m,t}\}$ is

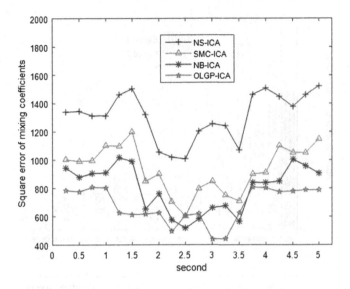

FIGURE 4.22

Comparison of square errors of the estimated mixing coefficients by using NS-ICA (black), SMC-ICA (pink in web version or light gray in print version), NB-ICA (blue in web version or dark gray in print version) and OLGP-ICA (red in web version or mid gray in print version).

calculated by (Stone, 2001)

$$F(\mathbf{s}_j) = \log \frac{\sum_t (s_{tj} - \bar{s}_j)^2}{\sum_t (s_{tj} - \tilde{s}_{tj})^2} \tag{4.147}$$

where

$$\bar{s}_j = \bar{\eta} \bar{s}_j + (1 - \bar{\eta}) s_{tj} \tag{4.148}$$

with $\bar{\eta} = 0.99$ and

$$\tilde{s}_{tj} = \tilde{\eta} \tilde{s}_{tj} + (1 - \tilde{\eta}) s_{tj} \tag{4.149}$$

with $\tilde{\eta} = 0.5$. In Eq. (4.147), the numerator is interpreted as an overall variance of a demixed signal while the denominator measures the extent to which a source or demixed sample s_{tj} is predicted by a short-term moving average $\tilde{s}_{tj}$ of previous samples in $\mathbf{s}_j$. High predictability happens in case of a high value of the overall signal variance in the numerator and a low value of prediction error of a smooth signal in the denominator. Fig. 4.23 compares the performance of different ICA methods by evaluating the absolute error of temporal predictabilities $|F(\mathbf{s}_j) - F(\widehat{\mathbf{s}}_j)|$ between true and demixed source signals $\{\mathbf{s}_j, \widehat{\mathbf{s}}_j\}$ which is calculated and averaged over six test examples. Two channel signals, as given in Fig. 4.17, are examined. The lower the error an ICA method achieves, the better the performance this method obtains regarding predictability or complexity in demixed signals, which is as accurate

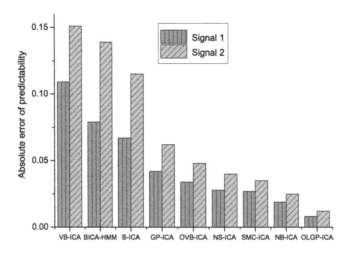

FIGURE 4.23

Comparison of absolute errors of temporal predictabilities between true and demixed source signals where different ICA methods are evaluated.

as that in true source signals. In this comparison, OLGP-ICA attains the lowest error in temporal predictability. In general, the ICAs with online learning (OVB-ICA, NS-ICA, SMC-ICA, NB-ICA and OLGP-ICA) outperform those with batch learning (VB-ICA, BICA-HMM, S-ICA and GP-ICA). Among different online learning methods, OLGP-ICA has the lowest predictability error because the nonstationary mixing condition is deliberately tackled by characterizing temporal information in source signals as well as mixing coefficients at each individual minibatch.

On the other hand, different ICA methods are evaluated in terms of signal-to-interference ratios (SIRs) in decibels which are calculated over all signal samples of true $\mathbf{s}_j = \{s_{tj}\}$ and estimated source signals $\widehat{\mathbf{s}}_j = \{\widehat{s}_{tj}\}$ at various minibatches. SIRs are calculated according to Eq. (2.10). Fig. 4.24 compares the SIRs of two channel signals by using different ICAs which are averaged over six test examples. Among these ICA methods, the lowest SIRs and the highest SIRs are obtained by VB-ICA and OLGP-ICA, respectively. The performance of VB-ICA is poor because VB-ICA does not handle the nonstationary condition for source separation. The reason why S-ICA obtains higher SIRs than BICA-HMM is because switching ICA properly estimates the source signals with sudden presence and absence via switching indicators. GP-ICA conducts temporally-correlated modeling and accordingly performs better than S-ICA. Again, the ICAs with online learning are better than those with batch learning in terms of SIRs. Basically, GP-ICA and OLGP-ICA suffer from high computation cost.

4.6 SUMMARY

We have presented a number of advanced ICA models for unsupervised learning and multi-channel source separation. Unsupervised learning solutions to build independent voices for speaker adaptation

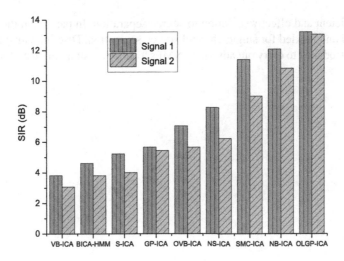

FIGURE 4.24

Comparison of signal-to-interference ratios of demixed signals where different ICA methods are evaluated.

and to construct multiple acoustic models for speech recognition were presented. The solutions to independent components with minimum redundancy reduction were illustrated. The ICA contrast function measured by hypothesis testing was presented. The nonparametric ICA procedure was developed by avoiding Gaussian assumption and realized by using nonparametric distribution for demixed signals. Moreover, the information-theoretic learning was conducted to carry out an ICA procedure based on a series of information-theoretic contrast functions. A general function, called convex divergence, was presented and incorporated in the implementation of the ICA procedure. A convexity parameter was found to achieve faster convergence in training and higher signal-to-interference ratio in the demixed signals when compared with standard ICA. The investigations on blind source separation in the presence of instantaneous, noisy and convolutive mixtures of speech and music signals were addressed. In addition to information-theoretic learning, the Bayesian learning, online learning and sparse learning were further explored to build an ICA procedure under the conditions of moving sources, missing sources and/or changing sources. Nonstationary source separation was implemented to deal with such complicated mixing scenarios. Recursive Bayesian algorithm was presented to continuously adapt ICA parameters using individual minibatches based on a sequential and noisy ICA model. The sparse learning based on an automatic relevance determination scheme was incorporated in detection of the activity for different sources. For those temporal structure in each minibatch, the ICA procedure combining online learning and Gaussian process (GP) was presented to further improve the SIRs of the demixed signals. The temporally-correlated time signals were delicately characterized by GP. The ICA solutions based on approximate inference methods using variational Bayesian (VB) inference and Markov chain Monte Carlo (MCMC) algorithm were presented and compared. The unknown sources were treated as the latent variables which were inferred from the parameters of variational distribution using VB or the samples by sampling methods using MCMC.

In the next section, we will present advanced solutions to source separation based on another paradigm called nonnegative matrix factorization (NMF). NMF has been attracting many researchers who pursue an efficient and effective solution to source separation. In particular, the underdetermined system is modeled and tackled for single-channel source separation. Discriminative learning and deep learning will be introduced to carry out advanced NMFs to achieve competitive performance for source separation.

NONNEGATIVE MATRIX FACTORIZATION

5

Nonnegative matrix factorization (NMF) was first proposed in Lee and Seung (1999, 2000) with multiplicative updating solution to decompose a nonnegative temporal-frequency data matrix $\mathbf{X} \in \mathbb{R}^{M \times N}$ into the sum of individual components or sources which are calculated from a nonnegative basis matrix $\mathbf{B} \in \mathbb{R}_+^{M \times K}$ and a nonnegative weight matrix $\mathbf{W} \in \mathbb{R}_+^{K \times N}$. Due to the nonnegative weights, NMF conducts the parts-based representation of a mixed signal so that the individual source signals are decomposed by using the corresponding sets of basis vectors $\{\mathbf{B}^s, \mathbf{B}^m\}$. NMF is attractive because of *good performance*, *easy interpretation* and *fast convergence*. The basics of NMF have been addressed in Section 2.2 with an example of separating a mixed spectrogram $\mathbf{X} \approx [\mathbf{B}^s \ \mathbf{B}^m]\mathbf{W}$ into a speech spectrogram $\mathbf{X}^s \approx \mathbf{B}^s \mathbf{W}^s$ and a music spectrogram $\mathbf{X}^m \approx \mathbf{B}^m \mathbf{W}^m$. Such a baseline NMF can be specialized to improve source separation by considering the following five issues. First, the temporal structure in separation system is missing. This information is important for signal reconstruction. Section 5.1 addresses the convolutive NMF which deals with audio enhancement and speech dereverberation. Second, the randomness in model representation is disregarded. The uncertainty in separation task due to ill-posed conditions may seriously deteriorate system performance. In Section 5.2, the probabilistic nonnegative factorization is described to deal with this issue. Third, in Section 5.3, a full Bayesian NMF is presented to address this problem with investigation of Bayesian NMF for singing voice separation. Fourth, the relevance of basis vectors for representation of target source is *not* explicitly characterized in standard NMF. Such a weakness may degrade the representation capability. Section 5.4 addresses a group sparse NMF to handle this issue. Finally, shallow NMF may not sufficiently capture the fine basis structure for reconstruction of a mixed signal. Section 5.5 presents a deep NMF with multiplicative updating formulas. This chapter addresses five advanced NMFs to deal with single-channel source separation under different conditions governed by different regularization solutions. Latent variable models in different NMFs are constructed by variational Bayesian inference or a sampling method.

5.1 CONVOLUTIVE NMF

Basis decomposition and convolutive operation have been the important tools in machine learning and signal processing, respectively. NMF is known as a powerful solution to basis decomposition. Combining signal processing and machine learning provides an avenue to many complicated tasks and applications. In particular, the convolutive basis decomposition is beneficial for source separation in reverberation environments where potential dependencies across successive columns or time frames of an input matrix $\mathbf{X}$ influence signal representation. This section first addressed the formulation of convolutive NMF in Section 5.1.1 based on nonnegative matrix factor deconvolution (NMFD) and nonnegative matrix factor two-dimensional deconvolution (NMF2D). Next, we address how convolutive NMF is considered to construct a speech dereverberation model in Section 5.1.2. In Section 5.1.3, Bayesian learning is introduced to carry out speech dereverberation using convolutive NMF. Variational Bayesian inference is applied to construct Bayesian speech dereverberation.

5.1.1 NONNEGATIVE MATRIX FACTOR DECONVOLUTION

Standard NMF is a well-known technique for data analysis, but it does not consider the temporal information of input data. In Smaragdis (2007), the nonnegative matrix factor deconvolution (NMFD) was proposed to extract the bases which took into account the dependencies across successive columns of input spectrogram for supervised single-channel speech separation. An extended NMF model leads to a solution of the problem:

$$\mathbf{X} \approx \widehat{\mathbf{X}} = \sum_{\tau=0}^{T-1} \mathbf{B}^\tau \overset{\tau \rightarrow}{\mathbf{W}} \tag{5.1}$$

where $\tau \rightarrow$ denotes the right shift operator which moves each column or time stamp of a matrix by τ positions or time delays, and similarly $\leftarrow \tau$ denotes the left shift operator. According to the learning objectives of square Euclidean distance and Kullback–Leibler (KL) divergence introduced in Section 2.2.2, the multiplicative updating rules for EU-NMFD and KL-NMFD are obtained by

$$\mathbf{B}^\tau \leftarrow \mathbf{B}^\tau \odot \frac{\mathbf{X} \overset{\tau \rightarrow}{\mathbf{W}}^\top}{\widehat{\mathbf{X}} \overset{\tau \rightarrow}{\mathbf{W}}^\top}, \tag{5.2}$$

$$\mathbf{W} \leftarrow \mathbf{W} \odot \frac{(\mathbf{B}^\tau)^\top \overset{\leftarrow \tau}{\mathbf{X}}}{(\mathbf{B}^\tau)^\top \overset{\leftarrow \tau}{\widehat{\mathbf{X}}}} \tag{5.3}$$

and

$$\mathbf{B}^\tau \leftarrow \mathbf{B}^\tau \odot \frac{\frac{\mathbf{X}}{\widehat{\mathbf{X}}} \overset{\tau \rightarrow}{\mathbf{W}}^\top}{\mathbf{1} \overset{\tau \rightarrow}{\mathbf{W}}^\top}, \tag{5.4}$$

$$\mathbf{W} \leftarrow \mathbf{W} \odot \frac{(\mathbf{B}^\tau)^\top \left[\overset{\leftarrow \tau}{\frac{\mathbf{X}}{\widehat{\mathbf{X}}}}\right]}{(\mathbf{B}^\tau)^\top \mathbf{1}}, \tag{5.5}$$

respectively. Here, the basis matrix $\mathbf{B}^\tau$ depends on each time τ, and the weight matrix is shared for all times τ. If we set $T = 1$, then no convolution operation is done. NMFD is accordingly reduced to the standard NMF. Nevertheless, we can generalize the solution to convolutive NMF in accordance with β divergence, given in Eq. (2.34), as the learning objective. The β-NMFD is derived accordingly. Then, the special realizations of EU-NMFD, KL-NMFD and Itakura–Saito (IS)-NMFD can be obtained by using $\beta = 2$, $\beta = 1$ and $\beta = 0$. The multiplicative updating rule of β-NMFD is provided by

$$\mathbf{B}^\tau \leftarrow \mathbf{B}^\tau \odot \frac{\left(\widehat{\mathbf{X}}^{\cdot[\beta-2]} \odot \mathbf{X}\right) \overset{\tau\rightarrow}{\mathbf{W}}^\mathsf{T}}{\widehat{\mathbf{X}}^{\cdot[\beta-1]} \overset{\tau\rightarrow}{\mathbf{W}}^\mathsf{T}}, \tag{5.6}$$

$$\mathbf{W} \leftarrow \mathbf{W} \odot \frac{(\mathbf{B}^\tau)^\mathsf{T} \left(\overset{\leftarrow\tau}{\widehat{\mathbf{X}}}{}^{\cdot[\beta-2]} \odot \mathbf{X}\right)}{(\mathbf{B}^\tau)^\mathsf{T} \overset{\leftarrow\tau}{\widehat{\mathbf{X}}}{}^{\cdot[\beta-1]}}. \tag{5.7}$$

In Schmidt and Morup (2006), the nonnegative matrix factor 2-D deconvolution (NMF2D) was exploited to discover the fundamental bases or notes for blind musical instrument separation in the presence of harmonic variations from a piano and trumpet with shift-invariance along the log-frequency. In addition to considering temporal information, NMF2D also represents the pitch shifting of different notes performed by an instrument. NMF2D attempts to use single component to shift up or down along the log-frequency spectrogram to learn all notes for each instrument instead of learning single component for each note of each instrument. The representation using NMF2D is expressed by

$$\mathbf{X} \approx \widehat{\mathbf{X}} = \sum_{\tau,\phi} \overset{\downarrow\phi}{\mathbf{B}^\tau} \overset{\rightarrow\tau}{\mathbf{W}^\phi} \tag{5.8}$$

where $\downarrow \phi$ denotes the downward shift operator which moves each row in the matrix by ϕ positions, and similarly $\uparrow \phi$ denotes the upward shift operator. If we set $\phi = 0$, NMF2D is reduced to the NMFD.

Considering the learning objectives based on square Euclidean distance in Eq. (2.21) and KL divergence in Eq. (2.26), the multiplicative updating rules for EU-NMF2D and KL-NMF2D are given by

$$\mathbf{B}^\tau \leftarrow \mathbf{B}^\tau \odot \frac{\sum_\phi \overset{\uparrow\phi}{\mathbf{X}} \overset{\tau\rightarrow}{\mathbf{W}^\phi}{}^\mathsf{T}}{\sum_\phi \overset{\uparrow\phi}{\widehat{\mathbf{X}}} \overset{\tau\rightarrow}{\mathbf{W}^\phi}{}^\mathsf{T}}, \tag{5.9}$$

$$\mathbf{W}^\phi \leftarrow \mathbf{W}^\phi \odot \frac{\sum_\tau \overset{\downarrow\phi}{\mathbf{B}^\tau}{}^\mathsf{T} \overset{\leftarrow\tau}{\mathbf{X}}}{\sum_\tau \overset{\downarrow\phi}{\mathbf{B}^\tau}{}^\mathsf{T} \overset{\leftarrow\tau}{\widehat{\mathbf{X}}}} \tag{5.10}$$

and

$$\mathbf{B}^\tau \leftarrow \mathbf{B}^\tau \odot \frac{\sum_\phi \left[\overset{\uparrow\phi}{\frac{\mathbf{X}}{\widehat{\mathbf{X}}}}\right] \overset{\tau\rightarrow}{\mathbf{W}^\phi}{}^\mathsf{T}}{\sum_\phi \mathbf{1} \overset{\tau\rightarrow}{\mathbf{W}^\phi}{}^\mathsf{T}}, \tag{5.11}$$

$$\mathbf{W}^\phi \leftarrow \mathbf{W}^\phi \odot \frac{\sum_\tau \overset{\downarrow\phi}{\mathbf{B}^\tau}{}^\mathsf{T} \left[\overset{\leftarrow\tau}{\frac{\mathbf{X}}{\widehat{\mathbf{X}}}}\right]}{\sum_\tau \overset{\downarrow\phi}{\mathbf{B}^\tau}{}^\mathsf{T} \mathbf{1}}, \tag{5.12}$$

respectively.

Table 5.1 Comparison of multiplicative updating rules of standard NMFD and sparse NMFD based on the objective functions of squared Euclidean distance and KL divergence

	NMFD	Sparse NMFD
EU	$\mathbf{B}^\tau \leftarrow \mathbf{B}^\tau \odot \dfrac{\mathbf{X}\overset{\tau\rightarrow}{\mathbf{W}}^\top}{\widehat{\mathbf{X}}\overset{\tau\rightarrow}{\mathbf{W}}^\top}$	$\mathbf{B}^\tau \leftarrow \mathbf{B}^\tau \odot \dfrac{\mathbf{X}\overset{\tau\rightarrow}{\mathbf{W}}^\top + \mathbf{B}^\tau \odot (1(\widehat{\mathbf{X}}\overset{\tau\rightarrow}{\mathbf{W}}^\top \odot \mathbf{B}^\tau))}{\widehat{\mathbf{X}}\overset{\tau\rightarrow}{\mathbf{W}}^\top + \mathbf{B}^\tau \odot (1(\mathbf{X}\overset{\tau\rightarrow}{\mathbf{W}}^\top \odot \mathbf{B}^\tau))}$
	$\mathbf{W} \leftarrow \mathbf{W} \odot \dfrac{(\mathbf{B}^\tau)^\top \overset{\leftarrow\tau}{\mathbf{X}}}{(\mathbf{B}^\tau)^\top \overset{\leftarrow\tau}{\widehat{\mathbf{X}}}}$	$\mathbf{W} \leftarrow \mathbf{W} \odot \dfrac{(\mathbf{B}^\tau)^\top \overset{\leftarrow\tau}{\mathbf{X}}}{(\mathbf{B}^\tau)^\top \overset{\leftarrow\tau}{\widehat{\mathbf{X}}} + \lambda}$
KL	$\mathbf{B}^\tau \leftarrow \mathbf{B}^\tau \odot \dfrac{\frac{\mathbf{X}}{\widehat{\mathbf{X}}}\overset{\tau\rightarrow}{\mathbf{W}}^\top}{1\overset{\tau\rightarrow}{\mathbf{W}}^\top}$	$\mathbf{B}^\tau \leftarrow \mathbf{B}^\tau \odot \dfrac{\frac{\mathbf{X}}{\widehat{\mathbf{X}}}\overset{\tau\rightarrow}{\mathbf{W}}^\top + \mathbf{B}^\tau \odot (1(1\overset{\tau\rightarrow}{\mathbf{W}}^\top \odot \mathbf{B}^\tau))}{1\overset{\tau\rightarrow}{\mathbf{W}}^\top + \mathbf{B}^\tau \odot (1(\frac{\mathbf{X}}{\widehat{\mathbf{X}}}\overset{\tau\rightarrow}{\mathbf{W}}^\top \odot \mathbf{B}^\tau))}$
	$\mathbf{W} \leftarrow \mathbf{W} \odot \dfrac{(\mathbf{B}^\tau)^\top \left[\frac{\mathbf{X}}{\widehat{\mathbf{X}}}\right]^{\leftarrow\tau}}{(\mathbf{B}^\tau)^\top 1}$	$\mathbf{W} \leftarrow \mathbf{W} \odot \dfrac{(\mathbf{B}^\tau)^\top \left[\frac{\mathbf{X}}{\widehat{\mathbf{X}}}\right]^{\leftarrow\tau}}{(\mathbf{B}^\tau)^\top 1 + \lambda}$

Table 5.2 Comparison of multiplicative updating rules of standard NMF2D and sparse NMF2D based on the objective functions of squared Euclidean distance and KL divergence

	NMF2D	Sparse NMF2D
EU	$\mathbf{B}^\tau \leftarrow \mathbf{B}^\tau \odot \dfrac{\sum_\phi \overset{\uparrow\phi}{\mathbf{X}}\overset{\tau\rightarrow}{\mathbf{W}^\phi}^\top}{\sum_\phi \overset{\uparrow\phi}{\widehat{\mathbf{X}}}\overset{\tau\rightarrow}{\mathbf{W}^\phi}^\top}$	$\mathbf{B}^\tau \leftarrow \mathbf{B}^\tau \odot \dfrac{\sum_\phi \overset{\uparrow\phi}{\mathbf{X}}\overset{\tau\rightarrow}{\mathbf{W}^\phi}^\top + \mathbf{B}^\tau \operatorname{diag}(\sum_\tau 1((\overset{\uparrow\phi}{\widehat{\mathbf{X}}}\overset{\tau\rightarrow}{\mathbf{W}^\phi}^\top) \odot \mathbf{B}^\tau))}{\sum_\phi \overset{\uparrow\phi}{\widehat{\mathbf{X}}}\overset{\tau\rightarrow}{\mathbf{W}^\phi}^\top + \mathbf{B}^\tau \operatorname{diag}(\sum_\tau 1((\overset{\uparrow\phi}{\mathbf{X}}\overset{\tau\rightarrow}{\mathbf{W}^\phi}^\top) \odot \mathbf{B}^\tau))}$
	$\mathbf{W}^\phi \leftarrow \mathbf{W}^\phi \odot \dfrac{\sum_\tau \overset{\downarrow\phi}{\mathbf{B}^\tau}^\top \overset{\leftarrow\tau}{\mathbf{X}}}{\sum_\tau \overset{\downarrow\phi}{\mathbf{B}^\tau}^\top \overset{\leftarrow\tau}{\widehat{\mathbf{X}}}}$	$\mathbf{W} \leftarrow \mathbf{W} \odot \dfrac{\sum_\tau \overset{\downarrow\phi}{\mathbf{B}^\tau}^\top \overset{\leftarrow\tau}{\mathbf{X}}}{\sum_\tau \overset{\downarrow\phi}{\mathbf{B}^\tau}^\top \overset{\leftarrow\tau}{\widehat{\mathbf{X}}} + \lambda}$
KL	$\mathbf{B}^\tau \leftarrow \mathbf{B}^\tau \odot \dfrac{\sum_\phi \overset{\uparrow\phi}{\left[\frac{\mathbf{X}}{\widehat{\mathbf{X}}}\right]}\overset{\tau\rightarrow}{\mathbf{W}^\phi}^\top}{\sum_\phi 1\overset{\tau\rightarrow}{\mathbf{W}^\phi}^\top}$	$\mathbf{B}^\tau \leftarrow \mathbf{B}^\tau \odot \dfrac{\sum_\phi \overset{\uparrow\phi}{\left[\frac{\mathbf{X}}{\widehat{\mathbf{X}}}\right]}\overset{\tau\rightarrow}{\mathbf{W}^\phi}^\top + \mathbf{B}^\tau \operatorname{diag}(\sum_\tau 1((1\overset{\tau\rightarrow}{\mathbf{W}^\phi}^\top) \odot \mathbf{B}^\tau))}{\sum_\phi 1\overset{\tau\rightarrow}{\mathbf{W}^\phi}^\top + \mathbf{B}^\tau \operatorname{diag}(\sum_\tau 1((\overset{\uparrow\phi}{\left[\frac{\mathbf{X}}{\widehat{\mathbf{X}}}\right]}\overset{\tau\rightarrow}{\mathbf{W}^\phi}^\top) \odot \mathbf{B}^\tau))}$
	$\mathbf{W}^\phi \leftarrow \mathbf{W}^\phi \odot \dfrac{\sum_\tau \overset{\downarrow\phi}{\mathbf{B}^\tau}^\top \left[\frac{\mathbf{X}}{\widehat{\mathbf{X}}}\right]^{\leftarrow\tau}}{\sum_\tau \overset{\downarrow\phi}{\mathbf{B}^\tau}^\top 1}$	$\mathbf{W}^\phi \leftarrow \mathbf{W}^\phi \odot \dfrac{\sum_\tau \overset{\downarrow\phi}{\mathbf{B}^\tau}^\top \left[\frac{\mathbf{X}}{\widehat{\mathbf{X}}}\right]^{\leftarrow\tau}}{\sum_\tau \overset{\downarrow\phi}{\mathbf{B}^\tau}^\top 1 + \lambda}$

In addition, the sparse NMFD (OGrady and Pearlmutter, 2006) and sparse NMF2D (Mørup and Schmidt, 2006a) were derived by imposing ℓ_1-norm, or lasso, regularization on either the basis matrix $\mathbf{B}$ or weight matrix $\mathbf{W}$ in the learning procedure. The ℓ_1-regularized learning objective for NMF was addressed in Eq. (2.37) given in Section 2.2.3. By combining sparse learning and convolutive processing, we can carry out sparse convolutive NMFs for source separation. Table 5.1 compares the multiplicative updating rules of NMFD and sparse NMFD based on the learning objectives of squared Euclidean distance (denoted by EU) and Kullback–Leibler divergence (denoted by KL). Furthermore, Table 5.2 shows the multiplicative updating rules for estimating NMF2D and sparse NMF2D parameters based on EU and KL learning objectives. In these formulas, $\odot$ denotes the element-wise multiplication and $\mathbf{A}^{\cdot[n]}$ denotes the matrix with entries $[\mathbf{A}]_{ij}^n$. Basically, the multiplicative updating rules guarantee the nonnegative parameters in the basis matrix $\mathbf{B}$ and weight matrix $\mathbf{W}$ during the iterative procedure. All these formulas are derived according to the gradient descent algorithm with a calculation of derivatives and a special manipulation of learning rate as illustrated in the descriptions of Kullback–Leibler divergence for NMF learning in Section 2.2.2. In general, convolutive NMFs with one- and two-dimensional deconvolution learn the temporal and frequency dependencies in successive time frames and frequency bins of a mixed signal, which are beneficial for single-channel source separation. Sparsity constraint is imposed to improve model regularization. In what follows, we address an advanced study which develops a speech dereverberation model based on convolutive NMF. Bayesian learning is merged in this convolutive NMF.

5.1.2 SPEECH DEREVERBERATION MODEL

As we know, the quality and intelligibility of speech signals recorded using a far-placed active microphone in a room is prone to be degraded due to reverberation, i.e., the reflections of sound from surrounding objects and walls. The effect of reverberation or signal mixing significantly changes the quality of hearing aids and the performance of automatic speech recognition. Recovering the reverberant speech signals from single-channel recordings has been impacting speech community for real-world applications in adverse conditions. As introduced in Section 1.1.3, a reverberant speech signal $x(t)$ at time t is represented by a linear convolution of a clean speech $s(t)$ and a room impulse response (RIR) $r(t)$ as given in Eq. (1.3). RIR length L is known beforehand. Such a speech reverberation problem can be solved through a statistical model (Mohammadiha et al., 2015) which integrated a nonnegative convolutive transfer function (NCTF) (Kameoka et al., 2009) and a nonnegative matrix factorization (NMF) where the room acoustics and the speech spectra were jointly modeled in the magnitude spectrum domain where the signals, noises and reverberation parameters are all nonnegative. Different from standard NMF, factorizing a magnitude spectral matrix $\mathbf{X}$ into a basis matrix $\mathbf{B}$ and a weight matrix $\mathbf{W}$ for estimation of multiple sources, NCTF-NMF carries out a speech dereverberation model which not only characterizes the clean speech $\mathbf{S} = \{S_{ft}\} \in \mathbb{R}_{+}^{F \times T}$ using NMF

$$S_{ft} \approx [\mathbf{BW}]_{ft} = \sum_{k} B_{fk} W_{kt} \tag{5.13}$$

but also represents the reverberant speech $\mathbf{X} = \{X_{ft}\} \in \mathbb{R}_{+}^{F \times T}$ using NCTF

$$X_{ft} \approx \sum_{l=0}^{L-1} R_{fl} S_{f,t-l} \tag{5.14}$$

as addressed in Section 1.1.3. There are T time steps and F frequency bins. Notably, the notation $\mathbf{X} = \{X_{ft}\} \in \mathbb{R}_{+}^{F \times T}$ for NCTF-NMF corresponds to $\mathbf{X} = \{X_{mn}\} \in \mathbb{R}_{+}^{M \times N}$ for standard NMF which was used in Section 2.2. In NCTF-NMF, the magnitude spectra of reverberant speech $X_{ft} = |x_c(f,t)|$, RIR $R_{fl} = |r_c(f,l)|$ and clean speech $S_{ft} = |s_c(f,t)|$ are all nonnegative. Here, c reflects the complex value in the spectra calculated by short-time Fourier transform (STFT); $x_c(f,t)$, $r_c(f,l)$ and $s_c(f,t)$ denote the complex spectra of reverberant signal $x(t)$, RIR $r(t)$ and clean signal $s(t)$, respectively. In the implementation, a set of pretrained bases $\mathbf{B}$ for clean speech is required. NCTF-NMF model parameters are composed of RIR parameters and weight parameters

$$\mathbf{\Theta} = \{\mathbf{R} = \{R_{fl}\}, \mathbf{W} = \{W_{kt}\}\}, \tag{5.15}$$

which are estimated by minimizing the reconstruction error function. In Eggert and Körner (2004), Chien and Yang (2016), the ℓ_1-norm of weight parameters was incorporated to handle an overcomplete problem in basis representation and carry out a sparse NCTF-NMF by solving the optimization problem

$$
\{\widehat{R}_{fl}, \widehat{W}_{kt}\} = \arg \min_{\{R_{fl}, W_{kt}\}} \sum_{f,t} \mathcal{D}_{\mathrm{KL}} \left(X_{ft} \left\| \sum_l R_{fl} \sum_k B_{fk} W_{k,t-l} \right. \right)
$$
$$
+ \lambda \sum_{k,t} W_{kt}
$$

(5.16)

where λ is the regularization parameter for a tradeoff between reconstruction error and weight sparseness. The reconstruction error function is measured by KL divergence between the reverberant speech **X** and the reconstructed speech. NMF is adopted in speech dereverberation so that the prior information of speech spectral bases can be employed to capture the dedicated structure in speech spectral signal.

However, the performance of dereverberation is considerably affected by the variations of reverberant signals in adverse environments, the uncertainty of the assumed model and the bias of the estimated parameters. Reverberation condition may be varied due to moving speakers. RIR parameter or RIR kernel R_{ft} is random in nature. The order of K basis vectors $\mathbf{B} = \{\mathbf{b}_k\}$ in NMF and the type of noise interference $\mathbf{N} = \{N_f\}$ in reverberation may be incorrectly determined. Therefore, uncertainty modeling via probabilistic framework is helpful to compensate for these unknown conditions and improve model regularization and generalization (Liang et al., 2015). From the Bayesian perspective, the randomness of spectral signals and model parameters is accommodated in a probabilistic model by introducing prior distributions for different signals and parameters (Watanabe and Chien, 2015). The dereverberated signals can be obtained by reflecting the nature of model randomness. This section introduces an advanced speech dereverberation model where Bayesian learning is merged into construction of NCTF-NMF. This framework is constructed under a probabilistic integration of NCTF and NMF where the Poisson likelihood is assumed to generate the reverberant signal and the underlying clean signal is represented by NMF governed by the exponential distribution. A full Bayesian solution to speech dereverberation is presented by characterizing the randomness of individual parameters including reverberation kernel, additive noise and NMF weight matrix by exponential prior, exponential prior and Gamma prior, respectively. NMF basis parameters were pretrained from development data. Model complexity can be controlled. Solution can be derived according to variational Bayesian expectation maximization (VB-EM) procedure where the lower bound of log marginal likelihood over clean signals and dereverberation parameters is maximized. The model-based speech dereverberation is developed without the need of prior knowledge about room configuration and speaker characteristics. This model can be realized with the existing methods, which did not consider or partially considered the randomness of dereverberation model (Mohammadiha et al., 2015, Liang et al., 2015). Detailed model construction and inference are shown in what follows.

5.1.3 BAYESIAN SPEECH DEREVERBERATION

Fig. 5.1 shows the graphical representation of a Bayesian speech reverberation model (Chien and Chang, 2016) where Bayesian learning is employed in single-channel source separation in a reverberant environment. Reverberant speech Y_{ft} is corresponding to X_{ft}. The likelihood function for reconstructed speech is modeled by a Poisson distribution where a sentence-based additive noise N_f

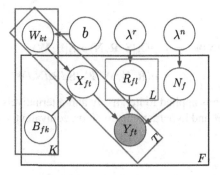

FIGURE 5.1

Graphical representation for Bayesian speech dereverberation.

is introduced. A noisy NCTF model is considered by merging the frequency-dependent noise signal as follows:

$$X_{ft} \approx \sum_l R_{fl} \sum_k B_{fk} W_{k,t-l} + N_f. \tag{5.17}$$

The exponential distribution is selected as the conjugate prior for latent variables including clean speech signal $\mathbf{S} = \{S_{ft}\}$, reverberation kernel $\mathbf{R} = \{R_{fl}\}$ and additive noise $\mathbf{N} = \{N_f\}$. Hyperparameters λ^r and λ^n are introduced as the hyperparameters of exponential priors of R_{fl} and N_f. The weight parameters in $\mathbf{W} = \{W_{kt}\}$ are Gamma distributed with hyperparameter b. The basis matrix $\mathbf{B} = \{B_{fk}\}$ is deterministic and pretrained from held-out speech signals. A probabilistic framework is constructed by using

$$X_{ft} \sim \text{Pois}\left(X_{ft}\Big|\sum_l R_{fl} S_{f,t-l} + N_f\right), \tag{5.18}$$

$$S_{ft} \sim \text{Exp}\left(S_{ft}\Big|c\sum_k B_{fk} W_{kt}\right), \tag{5.19}$$

$$R_{fl} \sim \text{Exp}(R_{fl}|\lambda^r), \tag{5.20}$$

$$N_f \sim \text{Exp}(N_f|\lambda^n), \tag{5.21}$$

$$W_{kt} \sim \text{Gam}(W_{kt}|b,b) \tag{5.22}$$

where c is a scaling parameter and

$$\mathbf{\Psi} = \{\lambda^r, \lambda^n, b\} \tag{5.23}$$

denote the hyperparameters. This model is an extension of Liang et al. (2015) by fully considering the randomness of all latent variables. Using the Bayesian approach as addressed in Section 3.7.2, we maximize the likelihood of reverberant speech $p(\mathbf{X}|\mathbf{\Psi})$, which is marginalized over latent variables

$$\mathbf{\Theta} = \{\mathbf{S}, \mathbf{R}, \mathbf{N}, \mathbf{W}\}, \tag{5.24}$$

which is yielded by

$$p(\mathbf{X}|\boldsymbol{\Psi}) = \int p(\mathbf{X}|\mathbf{S}, \mathbf{R}, \mathbf{N}) p(\mathbf{S}|\mathbf{B}, \mathbf{W})$$
$$\times p(\mathbf{R}, \mathbf{N}, \mathbf{W}|\boldsymbol{\Psi}) d\mathbf{S} d\mathbf{R} d\mathbf{N} d\mathbf{W}. \tag{5.25}$$

Variational Bayesian inference is applied to iteratively and alternatively perform the VB-E and VB-M steps to estimate parameters $\boldsymbol{\Theta}$ and hyperparameters $\boldsymbol{\Psi}$ or, equivalently, find the dereverberated speech spectra $\mathbf{S}$.

Variational Inference

As interpreted in Section 3.7.3, the exact solution to this optimization does not exist because of the couplings of latent variables $\{\mathbf{S}, \mathbf{R}, \mathbf{N}, \mathbf{W}\}$ in the posterior distribution $p(\mathbf{S}, \mathbf{R}, \mathbf{N}, \mathbf{W}|\mathbf{Y}, \boldsymbol{\Psi})$. To tackle this problem, variational inference procedure is performed by introducing a *factorized* variational distribution

$$q(\mathbf{S}, \mathbf{R}, \mathbf{N}, \mathbf{W}) = \prod_{t,f,k,l} q(S_{ft})q(R_{fl})q(N_f)q(W_{kt}), \tag{5.26}$$

which is used to approximate the true posterior $p(\mathbf{S}, \mathbf{R}, \mathbf{N}, \mathbf{W}|\mathbf{X}, \boldsymbol{\Phi})$. The variational distributions for individual latent variables are specified by

$$q(S_{ft}) = \text{Gam}(S_{ft}|v_{ft}^s, \rho_{ft}^s), \tag{5.27}$$
$$q(R_{fl}) = \text{Gam}(R_{fl}|v_{fl}^r, \rho_{fl}^r), \tag{5.28}$$
$$q(N_f) = \text{Gam}(N_f|v_f^n, \rho_f^n), \tag{5.29}$$
$$q(W_{kt}) = \text{GIG}(W_{kt}|v_{kt}^w, \rho_{kt}^w, \tau_{kt}^w) \tag{5.30}$$

where Gamma distributions are assumed for S_{ft}, R_{fl} and N_f with variational parameters $\{v_{ft}^s, \rho_{ft}^s, v_{fl}^r, \rho_{fl}^r, v_f^n, \rho_f^n\}$ and the generalized inverse-Gaussian (GIG) distribution (Liang et al., 2015)

$$\text{GIG}(w; v, \rho, \tau)$$
$$= \frac{\exp\{(v-1)\log w - \rho w - \tau/w\}\rho^{v/2}}{2\tau^{v/2}\mathcal{K}_v(2\sqrt{\rho\tau})} \tag{5.31}$$

is assumed for W_{kt}. In Eq. (5.31), $w \geq 0$, $\rho \geq 0$, and $\tau \geq 0$ and $\mathcal{K}_v(\cdot)$ denotes a modified Bessel function of the second kind. By Jensen's inequality, we have

$$\log p(\mathbf{X}|\boldsymbol{\Psi}) \geq \int q(\mathbf{S}, \mathbf{R}, \mathbf{N}, \mathbf{W}) \log \frac{p(\mathbf{X}, \mathbf{S}, \mathbf{R}, \mathbf{N}, \mathbf{W}|\boldsymbol{\Psi})}{q(\mathbf{S}, \mathbf{R}, \mathbf{N}, \mathbf{W})}$$
$$\times d\mathbf{S} d\mathbf{R} d\mathbf{N} d\mathbf{W} = \mathbb{E}_q[\log p(\mathbf{X}|\mathbf{S}, \mathbf{R}, \mathbf{N})]$$
$$+ \mathbb{E}_q[\log p(\mathbf{S}|\mathbf{B}, \mathbf{W})] - \mathbb{E}_q[\log q(\mathbf{S})]$$
$$+ \mathbb{E}_q[\log p(\mathbf{R}|\lambda^r)] - \mathbb{E}_q[\log q(\mathbf{R})]$$
$$+ \mathbb{E}_q[\log p(\mathbf{N}|\lambda^n)] - \mathbb{E}_q[\log q(\mathbf{N})]$$
$$+ \mathbb{E}_q[\log p(\mathbf{W}|b)] - \mathbb{E}_q[\log q(\mathbf{W})] \triangleq \mathcal{L}. \tag{5.32}$$

Instead of directly maximizing the marginal likelihood $p(\mathbf{X}|\boldsymbol{\Psi})$, the variational Bayesian (VB) inference is implemented to estimate the hyperparameters

$$\boldsymbol{\Psi} = \{\lambda^r, \lambda^n, b\}$$

and the variational parameters

$$\{v_{ft}^s, \rho_{ft}^s, v_{fl}^r, \rho_{fl}^r, v_f^n, \rho_f^n, v_{kt}^w, \rho_{kt}^w, \tau_{kt}^w\}$$

of speech dereverberation model by maximizing the lower bound of the logarithm of marginal likelihood $\mathcal{L}$. Variational lower bound is yielded by

$$
\begin{aligned}
\mathcal{L} = \sum_{f,t} &\left\{ X_{ft}\left[\phi_{ft}^n\left(\mathbb{E}_q[\log N_f] - \log \phi_{ft}^n \right) + \sum_l \phi_{ftl}^r\left(\mathbb{E}_q[\log S_{f,t-l}] \right.\right.\right.\\
&\left.\left. + \mathbb{E}_q[\log R_{fl}] - \log \phi_{ftl}^r \right) \right] - \mathbb{E}_q[N_f] - \sum_l \mathbb{E}_q[S_{f,t-l}]\mathbb{E}_q[R_{fl}] \right\}\\
&+ \sum_{f,t}\left\{ \left(\rho_{ft}^x - \frac{1}{c}\sum_k (\phi_{ftk}^s)^2 \mathbb{E}_q\left[\frac{1}{B_{fk}W_{kt}} \right] \right)\mathbb{E}_q[S_{ft}] - \log(c\omega_{ft}) \right.\\
&\left. + (1 - v_{ft}^x)\mathbb{E}_q[\log S_{ft}] - \frac{1}{\omega_{ft}}\sum_k \mathbb{E}_q[B_{fk}W_{kt}] - A^\Gamma(v_{ft}^s, \rho_{ft}^s) \right\}\\
&+ \sum_{f,l}\left\{ \left(\rho_{fl}^r - \frac{1}{\lambda^r} \right)\mathbb{E}_q[R_{fl}] + (1 - v_{fl}^r)\mathbb{E}_q[\log R_{fl}] - A^\Gamma(v_{fl}^r, \rho_{fl}^r) \right\}\\
&+ \sum_f \left\{ \left(\rho_f^n - \frac{1}{\lambda^n} \right)\mathbb{E}_q[N_f] + (1 - v_f^n)\mathbb{E}_q[\log N_f] - A^\Gamma(v_f^n, \rho_f^n) \right\}\\
&+ \sum_{k,t}\left\{ (b - v_{kt}^w)\mathbb{E}_q[\log W_{kt}] - (b - \rho_{kt}^w)\mathbb{E}_q[W_{kt}] + \tau_{kt}^w\mathbb{E}_q\left[\frac{1}{W_{kt}} \right] \right.\\
&\left. - A^{\text{GIG}}(v_{kt}^w, \rho_{kt}^w, \tau_{kt}^w) \right\} + \text{const.}
\end{aligned}
$$
$$(5.33)$$

In Eq. (5.33), the auxiliary variables

$$\{\phi_{ftk}^s, \phi_{ftl}^r, \phi_{ft}^n, \omega_{ft}\}$$

are introduced during the derivation subject to

$$\sum_k \phi_{ftk}^s = 1, \ \phi_{ftk}^s \geq 0, \ \omega_{ft} > 0, \tag{5.34}$$

$$\sum_l \phi_{ftl}^r + \phi_{ft}^n = 1, \ \phi_{ftl}^r \geq 0, \ \phi_{ft}^n \geq 0 \tag{5.35}$$

where $A^\Gamma(\cdot)$ and $A^{\text{GIG}}(\cdot)$ denote the log-partition functions for Gamma and GIG distributions, respectively.

Optimization Procedure

An iterative procedure is implemented for the VB-EM algorithm. In the VB E-step, we tighten the lower bound $\mathcal{L}$ with respect to auxiliary variables $\{\phi^s_{ftk}, \phi^r_{ftl}, \phi^n_{ft}, \omega_{ft}\}$ given by the constraints in Eq. (5.34). Lagrangian multiplies are introduced into this constrained optimization problem. The auxiliary variables are derived as

$$\phi^s_{ftk} \propto \left(\mathbb{E}_q \left[\frac{1}{B_{fk} W_{kt}} \right] \right)^{-1}, \tag{5.36}$$

$$\omega_{ft} = \sum_k \mathbb{E}_q [B_{fk} W_{kt}], \tag{5.37}$$

$$\phi^r_{ftl} = \frac{\exp\{\mathbb{E}_q[\log S_{f,t-l}] + \mathbb{E}_q[\log R_{fl}]\}}{\exp\{\mathbb{E}_q[\log N_f]\} + \sum_j \exp\{\mathbb{E}_q[\log S_{f,t-j}] + \mathbb{E}_q[\log R_{fj}]\}}, \tag{5.38}$$

$$\phi^n_{ft} = \frac{\exp\{\mathbb{E}_q[\log N_f]\}}{\exp\{\mathbb{E}_q[\log N_f]\} + \sum_j \exp\{\mathbb{E}_q[\log S_{f,t-j}] + \mathbb{E}_q[\log R_{fj}]\}}. \tag{5.39}$$

After updating these auxiliary variables, we estimate the variational parameters by maximizing the resulting lower bound $\mathcal{L}$ using coordinate ascent. The variational parameters corresponding to three Gamma distributions and one GIG distribution are estimated by

$$v^s_{ft} = 1 + \sum_l X_{f,t+l} \phi^r_{f,t+l,l}, \tag{5.40}$$

$$\rho^s_{ft} = \frac{1}{c} \left(\sum_k \mathbb{E}_q \left[\frac{1}{B_{fk} W_{kt}} \right]^{-1} \right)^{-1} + \sum_l \mathbb{E}_q[R_{fl}], \tag{5.41}$$

$$v^r_{fl} = 1 + \sum_t X_{ft} \phi^r_{ftl}, \tag{5.42}$$

$$\rho^r_{fl} = \frac{1}{\lambda^r} + \sum_t \mathbb{E}_q[S_{f,t-l}], \tag{5.43}$$

$$v^n_f = 1 + \sum_t X_{ft} \phi^n_{ft}, \tag{5.44}$$

$$\rho^n_f = \frac{1}{\lambda^n} + T, \tag{5.45}$$

$$v^w_{kt} = b, \tag{5.46}$$

$$\rho^w_{kt} = b + \sum_f \frac{\mathbb{E}_q[B_{fk}]}{\omega_{ft}}, \tag{5.47}$$

$$\tau^w_{kt} = \sum_f \frac{(\phi^s_{ftk})^2}{c} \mathbb{E}_q \left[\frac{1}{B_{fk}} \right] \mathbb{E}_q[S_{ft}]. \tag{5.48}$$

The scaling parameter c is updated by

$$c = \frac{1}{FT} \sum_{f,t} \mathbb{E}_q[S_{ft}] \left(\sum_k \mathbb{E}_q \left[\frac{1}{B_{fk}W_{kt}} \right]^{-1} \right)^{-1}. \tag{5.49}$$

After updating the auxiliary variables in Eqs. (5.36)–(5.39) and variational parameters in Eqs. (5.40)–(5.48) in the VB E-step, the VB M-step is performed by further tightening the lower bound $\mathcal{L}$ to find hyperparameters $\boldsymbol{\Psi} = \{\lambda^r, \lambda^n, b\}$. Auxiliary parameters, variational parameters and hyperparameters are iteratively estimated with convergence by the VB-EM steps. To obtain the enhanced or dereverberated spectra, we simply take the expectation of $\mathbf{S}$ with respect to the variational distribution to obtain

$$\mathbb{E}_q[S_{ft}] = \frac{v_{ft}^s}{\rho_{ft}^s} \tag{5.50}$$

for F frequencies and T frames. To recover the time-domain signals, we further apply the standard Wiener filter as a postprocessor based on the estimated dereverberated spectra $\mathbb{E}[\mathbf{S}]$.

5.1.4 SYSTEM EVALUATION

2014 REVERB Challenge dataset (Kinoshita et al., 2013) is used to feature the reverberation and stationary noise in the evaluation for reverberant source separation. There are eight different acoustic conditions, six of which are simulated by convolving the WSJCAM0 corpus with three measured RIRs at near (50 cm) and far microphone distances (200 cm), and adding the stationary noise recordings from the same rooms at SNR of 20 dB. There are 2176 utterances from 28 speakers collected as the evaluation data. The other two conditions are real recordings in a reverberant meeting room at two microphone distances (near at 100 cm and far at 250 cm) with stationary noise, taken from the MC-WSJ-AV corpus (Lincoln et al., 2005). Moreover, 372 utterances from ten speakers are collected for evaluation. For simulation data, reverberation time $T_{60} = 0.7$ s was considered. In real recordings, the meeting room has a measured T_{60} of 0.7 s. Reverberation time reflects the RIR condition. Sampling frequency in these data collections is 16 kHz. In general, this section addresses a full Bayesian approach (denoted by FB-NCTF-NMF) which deals with the random speech dereverberation problem and can be simplified to the NCTF-NMF in Mohammadiha et al. (2015) where no probabilistic framework is considered and the partial Bayesian NCTF-NMF (denoted by PB-NCTF-NMF) in Liang et al. (2015) where the noise signal $\mathbf{N} = \{N_f\}$ and the reverberation kernel $\mathbf{R} = \{R_{fl}\}$ are assumed to be deterministic. The results of speech dereverberation are evaluated by four metrics (Hu and Loizou, 2008) including cepstrum distance (CD) (lower is better), log-likelihood ratio (LLR) (lower is better), frequency-weighted segmental SNR (FWSegSNR) (higher is better), and speech-to-reverberation modulation energy ratio (SRMR) (Falk et al., 2010) (higher is better). For real recordings, only the non-intrusive SRMR can be used. These qualify measures are displayed in dB and individually measured by

- Cepstrum distance (CD): CD represents a distance between cepstra calculated from the target (i.e., enhanced or processed) and clean reference signals, and is an estimate of a smoothed spectral distance between the target and reference (Hu and Loizou, 2008). Cepstrum distance provides an estimate of the log spectral distance between two spectra. The cepstrum coefficients can be obtained recursively from linear prediction coefficients (LPCs) $\{a_m\}$ using the following expression:

$$c(m) = a_m + \sum_{k=1}^{m-1} \frac{k}{m} c(k) a_{m-k} \quad 1 \le m \le p \tag{5.51}$$

where p is the order of LPC. Cepstrum distance is computed by

$$d_{\text{CD}}(\mathbf{c}_c, \mathbf{c}_e) = \frac{10}{\log 10} \sqrt{2 \sum_{k=1}^{p} [c_c(k) - c_e(k)]^2} \tag{5.52}$$

where $\mathbf{c}_c$ and $\mathbf{c}_e$ are the cepstrum vectors of clean and enhanced signals, respectively.
- Log-likelihood ratio (LLR): LLR is an LPC-based measure, which represents the degree of discrepancy between the smoothed spectra of target and reference signals as defined (see Hu and Loizou, 2008) by

$$d_{\text{LLR}}(\mathbf{a}_e, \mathbf{a}_c) = \log \left(\frac{\mathbf{a}_e \mathbf{R}_c \mathbf{a}_c^T}{\mathbf{a}_c \mathbf{R}_c \mathbf{a}_c^T} \right) \tag{5.53}$$

where $\mathbf{a}_c$ is the LPC vector of clean speech, $\mathbf{a}_e$ is the LPC vector of enhanced speech frame, and $\mathbf{R}_c$ is the autocorrelation matrix of clean speech.
- Frequency-weighted segmental SNR (FWSegSNR): FWSegSNR is a reference-based speech quality measure which sufficiently reflects the perceptual quality of enhanced speech and is calculated (see Hu and Loizou, 2008) by

$$\begin{aligned} \text{FWSeqSNR} = &\frac{10}{T} \\ &\times \sum_{t=0}^{T-1} \frac{\sum_{j=1}^{F} W(f,t) \log_{10} \frac{|X(f,t)|^2}{(|X(j,m)| - |\hat{X}(f,t)|)^2}}{\sum_{j=1}^{F} W(f,t)} \end{aligned} \tag{5.54}$$

where $W(f,t)$ is the weight placed on the fth frequency band, F is the number of bands, T is the total number of frames in the signal, $|X(f,t)|$ is the weighted (by a Gaussian-shaped window) clean signal spectrum in the fth frequency band at the tth frame, and $|\hat{X}(f,t)|$ in the weighted enhanced signal spectrum in the same band.
- Speech-to-reverberation modulation energy ratio (SRMR): SRMR scores are used for both simulated data and real recordings. SRMR measure is defined as follows (Falk et al., 2010):

$$\text{SRMR} = \frac{\sum_{f=1}^{4} \bar{\mathcal{E}}_f}{\sum_{f=5}^{K^*} \bar{\mathcal{E}}_f} \tag{5.55}$$

where $\bar{\mathcal{E}}_f$ denotes the average per-modulation energy of fth band and the upper summation bound K^* in denominator is adapted to speech signal under test.

Table 5.3 Comparison of using different methods for speech dereverberation under various test conditions (near, near microphone; far, far microphone; sim, simulated data; real, real recording) in terms of evaluation metrics of CD, LLR, FWSegSNR and SRMR (dB)

Measure/condition	Unprocessed	Baseline	Partial-Bayes	Full-Bayes
CD (near-sim)	4.38	3.71	3.10	3.07
CD (far-sim)	4.96	4.13	3.97	3.71
LLR (near-sim)	0.65	0.90	0.82	0.75
LLR (far-sim)	0.84	0.82	0.80	0.72
FWSegSNR (near-sim)	2.27	6.02	7.61	9.01
FWSegSNR (far-sim)	0.24	4.58	6.01	7.10
SRMR (near-sim)	3.57	3.72	3.84	3.98
SRMR (far-sim)	2.73	3.04	3.29	3.40
SRMR (near-real)	3.17	3.53	3.69	3.81
SRMR (far-real)	3.19	3.49	3.61	3.73

In the implementation, the clean training utterances are randomly chosen from ten speakers of WSJCAM0 corpus to train basis parameters $\mathbf{B} = \{B_{fk}\}$ for clean speech signals with dictionary size $K = 50$ according to the sparse NMF (Eggert and Körner, 2004) with regularization parameter 0.1. These ten speakers are different from those in evaluation data. The 1024-point STFT was performed with zero-padded to 2048 points and 512-point overlap. Magnitude spectra of reverberant speech $\mathbf{X}$ were calculated. FB-NCTF-NMF was implemented per utterance without prior knowledge about test conditions. Initial hyperparameters $\lambda^r = 0.8$, $\lambda^n = 0.008$ and $b = 0.1$ were chosen and updated in the VB M-step according to the lower bound $\mathcal{L}$. The reverberation kernel length $L = 20$, corresponding to 640 ms, was selected. Variational parameters were initialized by $v_{ft}^s = v_{fl}^r = v_f^n = v_{kt}^w = 1$ and $\{\rho_{ft}^s, \rho_{fl}^r, \rho_f^n, \rho_{kt}^w, \tau_{kt}^w\}$ by a random draw from Gamma distributions or random function. Twenty iterations were run to estimate the auxiliary variables and variational parameters and 10 iterations were run to estimate hyperparameters. VB-EM steps were converged. The dereverberated spectra $\mathbb{E}[\mathbf{S}]$ were calculated and smoothed with an attenuation level 0.1.

Table 5.3 compares the speech dereverberation on a whole set of evaluation data under different test conditions including the simulated data and the real recording data by using near and far placed active microphones. The performance of unprocessed speech and dereverberated speech using NCTF-NMF (denoted by baseline), PB-NCTF-NMF (denoted by partial-Bayes) and FB-NCTF-NMF (denoted by full-Bayes) is compared in terms of CD, LLR, FWSegSNR and SRMR. All measures are expressed in decibel (dB). From these results, we find that NCTF-NMF improves the performance and compensates for reverberation in most cases. Bayesian learning using PB-NCTF-NMF and FB-NCTF-NMF consistently perform better than NCTF-NMF without Bayesian treatment. FB-NCTF-NMF is better than PB-NCTF-NMF in a number of conditions. The improvement of FB-NCTF-NMF over other methods is significant in terms of FWSegSNR and SRMR. The margin of improvement in real recordings is large by applying speech dereverberation using three NCTF-NMFs.

In addition to different extensions of convolutive NMF for reverberant source separation, in what follows, we aim at another advanced topic based on the probabilistic factorization framework which is useful for extracting the latent components of acoustic or music spectra for source separation.

5.2 PROBABILISTIC NONNEGATIVE FACTORIZATION

In general, uncertainty modeling via probabilistic framework is helpful to improve model regularization. The uncertainties in source separation may come from improper model assumptions, incorrect model order, nonstationary environment, reverberant distortion and possible noise interference. Under the probabilistic framework, nonnegative spectral signals are drawn from probability distributions. Section 5.1.3 has addressed Bayesian learning for speech dereverberation, which is based on a probabilistic extension and combination over a nonnegative convolutive transfer function and a nonnegative matrix factorization (NMF). However, such a solution is only specialized to deal with the reverberant source separation. In the following, Section 5.2.1 introduces a probabilistic solution to a general latent variable model with wide applications, which is called the probabilistic latent component analysis (PLCA) (Shashanka et al., 2008). This probabilistic nonnegative factorization model is developed for audio spectra analysis (Smaragdis et al., 2006) and singing voice separation (Raj et al., 2007). Section 5.2.3 presents the shift-invariant PLCA, which is feasible to handle complex data. Section 5.2.3 illustrates how PLCA is theoretically related with a generalization of NMF.

5.2.1 PROBABILISTIC LATENT COMPONENT ANALYSIS

Probabilistic latent component analysis (PLCA) (Shashanka et al., 2008, Smaragdis et al., 2006) is seen as a probabilistic latent variable model for analysis of nonnegative data via factorization. The basic idea of PLCA is closely related to the probabilistic latent semantic indexing (Hofmann, 1999), which has been popularly explored to extract latent semantic topics from discrete-valued text data for information retrieval. Here, PLCA is introduced to learn latent topics or latent mixture components z from the real-valued time–frequency audio observations which are applied for source separation. The probabilistic model is defined by

$$p(\mathbf{x}) = \sum_z p(\mathbf{x}, z) = \sum_z p(z) \prod_{i=1}^{J} p(x_i|z) \tag{5.56}$$

where $p(\mathbf{x})$ is a J-dimensional probability distribution with random variables $\mathbf{x} = \{x_1, x_2, \ldots, x_J\}$, and z denotes the latent variables. Basically, there is a one-to-one mapping between the variables in PLCA and NMF. The NMF model is expressed by $\mathbf{X} \approx \mathbf{BW}$ where $\mathbf{X} \in \mathbb{R}_+^{M \times N}$, $\mathbf{B} \in \mathbb{R}_+^{M \times K}$ and $\mathbf{W} \in \mathbb{R}_+^{K \times N}$ where

$$X_{mn} = \sum_k B_{mk} W_{kn}. \tag{5.57}$$

In case of $J = 2$, $p(x_1, x_2)$, $p(x_1|z)$, $p(x_2|z)p(z)$ (or $p(x_2, z)$) in PLCA corresponds to the entries X_{mn}, B_{mk} and W_{kn} of the matrices $\mathbf{X}$, $\mathbf{B}$ and $\mathbf{W}$ in NMF, respectively. Detailed theoretical equivalence between PLCA and NMF will be demonstrated in Section 5.2.3. In case of $J > 2$, the factorization in Eq. (5.56) can be further illustrated to relate with the nonnegative tensor factorization. This model expresses a J-dimensional distribution as a mixture where each J-dimensional component of the mixture is a product of one-dimensional marginal distributions. The motivation of this expression is to figure out how the underlying probability distributions in the latent variable model are related.

Since there is only one latent variable z in PLCA, the expectation-maximization (EM) algorithm (Dempster et al., 1977) as addressed in Section 3.7.1 is introduced here to estimate the PLCA parameters

$$\Theta = \{p(z), p(x_i|z)\} \tag{5.58}$$

according to the maximum likelihood (ML) criterion. In the E-step, the auxiliary function or the expectation function of the log-likelihood of observation data $\mathbf{x}$ with latent variable z is calculated by

$$Q(\Theta|\Theta^{(\tau)}) = \mathbb{E}_z[\log p(\mathbf{x}, z|\Theta)|\mathbf{x}, \Theta^{(\tau)}]$$
$$= \mathbb{E}_z\left[q(z)\log\left(p(z)\prod_{i=1}^{J} p(x_i|z)\right)\right] \tag{5.59}$$

where the posterior of latent variable z given current estimate of parameters

$$\Theta^{(\tau)} = \{(p(z))^{(\tau)}, (p(x_i|z))^{(\tau)}\}$$

is obtained by

$$q(z) \triangleq p(z|\mathbf{x}, \Theta^{(\tau)}) = \frac{(p(z))^{(\tau)}\prod_{i=1}^{J}(p(x_i|z))^{(\tau)}}{\sum_{z'}(p(z'))^{(\tau)}\prod_{i=1}^{J}(p(x_i|z'))^{(\tau)}}, \tag{5.60}$$

see Eq. (3.72). Notably, τ denotes the current EM iteration. Similar to the auxiliary function in the standard EM algorithm in Eq. (3.54), the expectation in Eq. (5.59) is calculated by using the whole training data $\mathbf{X} = \{\mathbf{x}\}$. Since the latent topic z of audio spectra $\mathbf{x}$ is a discrete variable and the observation x_i given latent variable z is a continuous variable, we need to impose the constraints

$$\sum_z p(z) = 1 \tag{5.61}$$

and

$$\int p(x_i|z)dx_i = 1, \quad 1 \le i \le J, \tag{5.62}$$

into ML estimation of two parameters $p(z)$ and $p(x_i|z)$. Constrained optimization problem is formulated. Accordingly, the Lagrange multipliers $\{\lambda, \lambda_{zi}\}$ are incorporated into the calculation of the extended auxiliary function

$$\tilde{Q}(\Theta, \Theta^{(\tau)}) = \mathbb{E}_z\left[\log p(\mathbf{x}, z|\Theta)|\mathbf{x}, \Theta^{(\tau)}\right]$$
$$- \lambda\sum_z p(z) - \sum_{z,i}\left(\lambda_{zi}\int p(x_i|z)dx_i\right). \tag{5.63}$$

In the M-step, new estimates

$$\Theta^{(\tau+1)} = \{(p(z))^{(\tau+1)}, (p(x_i|z))^{(\tau+1)}\}$$

of PLCA parameters are obtained by maximizing the extended auxiliary function in Eq. (5.63) with respect to $p(z)$ and $p(x_i|z)$, i.e.,

$$\Theta^{(\tau+1)} = \arg \max_{\Theta} \tilde{Q}(\Theta|\Theta^{(\tau)}). \tag{5.64}$$

As a result, the updated PLCA parameters based on ML estimation are derived as

$$(p(z))^{(\tau+1)} = \int p(\mathbf{x})p(z|\mathbf{x}|\Theta^{(\tau)})d\mathbf{x} = \int p(\mathbf{x})q(z)d\mathbf{x} \tag{5.65}$$

and

$$(p(x_i|z))^{(\tau+1)} = \frac{\int\int \cdots \int p(\mathbf{x})p(z|\mathbf{x}, \Theta^{(\tau)})dx_j}{p(z)}$$

$$= \frac{\int\int \cdots \int p(\mathbf{x})q(z)dx_j}{p(z)}, \quad \forall j \neq i. \tag{5.66}$$

If the observation $\mathbf{x}$ is a discrete random variable, the updating rules are replaced by

$$(p(z))^{(\tau+1)} = \sum_i \sum_{x_i} p(\mathbf{x})p(z|\mathbf{x}, \Theta^{(\tau)}) \tag{5.67}$$

and

$$(p(x_i|z))^{(\tau+1)} = \frac{\sum_j \sum_{x_j} p(\mathbf{x})p(z|\mathbf{x}, \Theta^{(\tau)})}{p(z)}, \quad \forall j \neq i. \tag{5.68}$$

Iteratively and alternatively performing E- and M-steps is required to assure convergence in the EM procedure so as to find the ML parameters Θ_{ML} for source separation. It is noted that the updating rules in Eqs. (5.65)–(5.68) are comparable with the multiplicative updating rules in Eqs. (2.27)–(2.28) for KL-NMF parameters B_{mk} and W_{kn} which are estimated by minimizing the Kullback–Leibler (KL) divergence for data reconstruction based on NMF. A detailed connection between PLCA and KL-NMF will be described in Section 5.2.3.

5.2.2 SHIFT-INVARIANT PLCA

In general, a simple PLCA model with a single set of frequency observations x does not take into account additional consideration such as temporal information y when the joint distribution $p(x, y)$ is complex with shifting variations. When image data or audio spectrograms with time–frequency observations are encountered, the shift-invariant or convolutive property becomes crucial for identifying the latent topics in the PLCA model, which are robust to perturbation of signals due to shifting in the time and frequency spaces. In Smaragdis and Raj (2008), the shift invariant PLDA is correspondingly developed to tackle this weakness. This method can be also realized as a kind of convolutive NMF, which was addressed in Section 5.1. In what follows, shift-invariant PLCA is presented by learning for shift invariance across either one or two dimensions. These two dimensions x and y can be either time t or frequency f which are observed for audio signal analysis and separation.

Shift-Invariant 1-D PLCA

First, the distribution of a two-dimensional (2-D) model $p(x, y)$ with one-dimensional (1-D) left–right shifting ($\rightarrow$ or $\leftarrow$) is expressed by

$$p(x, y) = \sum_z p(z) \int p(x, \tau|z) p(y - \tau|z) d\tau \qquad (5.69)$$

where τ denotes the shifting size, $p(x, \tau|z)$ and $p(y - \tau|z)$ are seen as the kernel and impulse distributions, respectively. The parameters of shift-invariant 1-D PLCA

$$\Theta = \{p(z), p(x, \tau|z), p(y|z)\} \qquad (5.70)$$

are estimated by the maximum likelihood theory using the EM algorithm. Similarly, the posterior distribution of latent variables, including z and τ, is calculated in the E-step by

$$q(z, \tau) \triangleq p(z, \tau|x, y, \Theta^{(\tau)})$$
$$= \frac{(p(z))^{(\tau)} (p(x, \tau|z))^{(\tau)} (p(y - \tau|z))^{(\tau)}}{\sum_{z'} (p(z'))^{(\tau)} \int (p(x, \tau|z'))^{(\tau)} (p(y - \tau|z'))^{(\tau)} d\tau}. \qquad (5.71)$$

Then, the extended auxiliary function $\tilde{Q}(\Theta|\Theta^{(\tau)})$, similar to Eq. (5.63), is constructed by imposing the constraints due to three probability parameters including Eq. (5.61) and

$$\int \int p(x, \tau|z) dx d\tau = 1, \qquad (5.72)$$

$$\int p(y|z) dy = 1. \qquad (5.73)$$

In the M-step, the extended auxiliary function is maximized to find the updating formulas for the new estimates of the three parameters:

$$(p(z))^{(\tau+1)} = \int \int \int p(x, y) p(z, \tau|x, y, \Theta^{(\tau)}) dx dy d\tau, \qquad (5.74)$$

$$(p(x, \tau|z))^{(\tau+1)} = \frac{\int p(x, y) p(z, \tau|x, y, \Theta^{(\tau)}) dy}{p(z)}, \qquad (5.75)$$

$$(p(y|z))^{(\tau+1)} = \frac{\int \int p(x, y + \tau) p(z, \tau|x, y + \tau, \Theta^{(\tau)}) dx d\tau}{\int \int \int p(x, y + \tau) p(z, \tau|x, y + \tau, \Theta^{(\tau)}) dx dy d\tau}. \qquad (5.76)$$

If the observations x and y and the shifting τ are discrete variables, the updating rules of the M-step are modified by replacing the integral operator with the summation operator. Note that if

$$p(\tau|z) = \delta(\tau), \qquad (5.77)$$

the shift-invariant 1-D PLCA is equivalent to the original PLCA. Basically, the solutions to shift-invariant 1-D PLCA in Eqs. (5.74)–(5.76) are seen in a style of multiplicative updating, which is comparable with that for the nonnegative matrix factor deconvolution (NMFD) based on KL divergence as shown in Eqs. (5.4) and (5.5). The shifting property is incorporated into PLCA to implement the probabilistic convolutive NMF (Smaragdis and Raj, 2008).

Shift-Invariant 2-D PLCA

Next, we simultaneously consider the left–right shifting ($\rightarrow$ or $\leftarrow$) and the other shifting direction, i.e., up–down shift ($\uparrow$ or $\downarrow$). The observed two-dimensional (2-D) data such as audio spectrograms are represented in a better way. This 2-D model with left–right and up–down shifting is defined as

$$p(x, y) = \sum_z p(z) \iint p(\tau_x, \tau_y | z) p(x - \tau_x, y - \tau_y | z) d\tau_x d\tau_y \tag{5.78}$$

where τ_x and τ_y denote the shifting sizes of the two dimensions x and y, respectively. There are three latent variables, z, τ_x and τ_y. Again, the parameters

$$\Theta = \{p(z), p(\tau_x, \tau_y | z), p(x, y | z)\} \tag{5.79}$$

for shift-invariant 2-D PLCA are estimated by the maximum likelihood method based on an EM procedure. Following the same estimation procedure of standard PLCA and shift-invariant 1-D PLCA, the new estimates $\Theta^{(\tau+1)}$ of shift-invariant 2-D PLCA at the EM iteration $\tau + 1$ are obtained by

$$(p(z))^{(\tau+1)} = \iiiint p(x, y) p(z, \tau_x, \tau_y | x, y, \Theta^{(\tau)}) dx \, dy \, d\tau_x \, d\tau_y, \tag{5.80}$$

$$(p(\tau_x, \tau_y | z))^{(\tau+1)} = \frac{\iint p(x, y) p(z, \tau_x, \tau_y | x, y, \Theta^{(\tau)}) dx \, dy}{P(z)}, \tag{5.81}$$

$$(p(y|z))^{(\tau+1)}$$
$$= \frac{\iint p(x + \tau_x, y + \tau_y) p(z, \tau_x, \tau_y | x + \tau_x, y + \tau_y, \Theta^{(\tau)}) dx \, dy}{\iiiint p(x + \tau_x, y + \tau_y) p(z, \tau_x, \tau_y | x + \tau_x, y + \tau_y, \Theta^{(\tau)}) dx \, dy \, d\tau_x \, d\tau_y} \tag{5.82}$$

where the posterior distribution is calculated by

$$q(z, \tau_x, \tau_y) \triangleq p(z, \tau_x, \tau_y | x, y, \Theta^{(\tau)})$$
$$= \frac{(p(z))^{(\tau)} (p(\tau_x, \tau_y | z))^{(\tau)} (p(x - \tau_x, y - \tau_y | z))^{(\tau)}}{\sum_{z'} (p(z'))^{(\tau)} \iint (p(\tau_x, \tau_y | z'))^{(\tau)} (p(x - \tau_x, y - \tau_y | z'))^{(\tau)} d\tau_x d\tau_y}. \tag{5.83}$$

Notably, new estimates in Eqs. (5.80)–(5.82) are represented in a style of multiplicative updating, which is in accordance with the form of different NMFs. Without loss of generality, the shifting property in the shift-invariant 2-D PLCA corresponds to that in the nonnegative matrix factor 2-D deconvolution (NMF2D) based on KL divergence, which has been addressed in Eqs. (5.11) and (5.12). In particular, the equivalence between the basic PLCA model and KL-NMF model is illustrated in what follows.

5.2.3 PLCA VERSUS NMF

To bridge the relation between PLCA and NMF, we first express the basic PLCA for two-dimensional data (x, y) or time–frequency data (t, f) in a form of

$$p(x, y) = \sum_z p(z)p(x|z)p(y|z) \tag{5.84}$$

where the property of shifting invariance is disregarded. Here, the discrete-valued data x and y are collected or the digital audio signals of time step t and frequency bin f are observed. By referring to the previous subsection, the updating rules for 2-D PLCA parameters $p(z)$, $p(x|z)$ and $p(y|z)$ are obtained by

$$(p(z))^{(\tau+1)} = \sum_x \sum_y p(x, y)p(z|x, y, \Theta^{(\tau)}) = \sum_y (p(y, z))^{(\tau+1)}, \tag{5.85}$$

$$(p(y|z))^{(\tau+1)} = \frac{\sum_x p(x, y)p(z|x, y, \Theta^{(\tau)})}{\sum_x \sum_y p(x, y)p(z|x, y, \Theta^{(\tau)})} = \frac{(p(y, z))^{(\tau+1)}}{\sum_y p(y, z))^{(\tau+1)}}, \tag{5.86}$$

$$(p(x|z))^{(\tau+1)} = \frac{\sum_y p(x, y)p(z|x, y, \Theta^{(\tau)})}{\sum_x \sum_y p(x, y)p(z|x, y, \Theta^{(\tau)})} = \frac{\sum_y p(x, y)p(z|x, y, \Theta^{(\tau)})}{\sum_y (p(y, z))^{(\tau+1)}} \tag{5.87}$$

where the posterior probability is calculated by

$$
\begin{aligned}
q(z) \triangleq p(z|x, y, \Theta^{(\tau)}) &= \frac{(p(z))^{(\tau)}(p(x|z))^{(\tau)}(p(y|z))^{(\tau)}}{\sum_z (p(z))^{(\tau)}(p(x|z))^{(\tau)}(p(y|z))^{(\tau)}} \\
&= \frac{(p(x|z))^{(\tau)}(p(y, z))^{(\tau)}}{(p(x, y))^{(\tau)}}.
\end{aligned} \tag{5.88}
$$

In Eqs. (5.85)–(5.87), we use an intermediate variable $(p(y, z))^{(\tau+1)}$ defined by

$$(p(y, z))^{(\tau+1)} \triangleq \sum_x p(x, y)p(z|x, y, \Theta^{(\tau)}). \tag{5.89}$$

From the above equations, we find that the updating rules for $p(x|z)$ in Eq. (5.87) and $p(y, z)$ in Eq. (5.89) are essential to carry out the estimation of the three PLCA parameters

$$\Theta = \{p(z), p(x|z), p(y|z)\}.$$

The PLCA updating rules are all expressed because $(p(z))^{(\tau+1)}$ and $(p(y|z))^{(\tau+1)}$ can be obtained by updating the intermediate variable $(p(y, z))^{(\tau+1)}$. Consequently, the rearrangement of $(p(x|z))^{(\tau+1)}$

and $(p(y, z))^{(\tau+1)}$ is performed as follows:

$$
\begin{aligned}
(p(x|z))^{(\tau+1)} &= \frac{\sum_y p(x, y) p(z|x, y, \Theta^{(\tau)})}{\sum_y (p(y, z))^{(\tau+1)}} \\
&= \frac{\sum_y p(x, y) \frac{(p(x|z))^{(\tau)}(p(y,z))^{(\tau)}}{(p(x,y))^{(\tau)}}}{\sum_y (p(y, z))^{(\tau+1)}} \\
&= (p(x|z))^{(\tau)} \frac{\sum_y \frac{p(x,y)}{\sum_z p(x|z))^{(\tau)}(p(y,z))^{(\tau)}}(p(y, z))^{(\tau)}}{\sum_y (p(y, z))^{(\tau+1)}},
\end{aligned}
\tag{5.90}
$$

$$
\begin{aligned}
(p(y, z))^{(\tau+1)} &= \sum_x p(x, y) p(z|x, y, \Theta^{(\tau)}) \\
&= \sum_x p(x, y) \frac{(p(x|z))^{(\tau)}(p(y, z))^{(\tau)}}{(p(x, y))^{(\tau)}} \\
&= (p(y, z))^{(\tau)} \sum_x \frac{(p(x|z))^{(\tau)} \frac{p(x,y)}{\sum_z p(x|z))^{(\tau)}(p(y,z))^{(\tau)}}}{\sum_x (p(x|z))^{(\tau)}}.
\end{aligned}
\tag{5.91}
$$

The right- and left-hand sides (RHS and LHS) of Eqs. (5.90) and (5.91) show the parameter updating from $(p(x|z))^{(\tau)}$ to $(p(x|z))^{(\tau+1)}$ and from $(p(y, z))^{(\tau)}$ to $(p(y, z))^{(\tau+1)}$. In this manipulation, we use the properties

$$
(p(x, y))^{(\tau)} = \sum_z p(x|z))^{(\tau)}(p(y, z))^{(\tau)},
\tag{5.92}
$$

$$
\sum_x (p(x|z))^{(\tau)} = 1.
\tag{5.93}
$$

Let $\mathbf{X}$ denote the matrix with entries $p(x, y)$, $\mathbf{B}^{(\tau)}$ denote the matrix with entries $(p(x|z))^{(\tau)}$, and $\mathbf{W}^{(\tau)}$ denote the matrix with entries $(p(y, z))^{(\tau)}$. Therefore, Eqs. (5.91) and (5.90) can be rewritten in the matrix form as

$$
\mathbf{B}^{(\tau+1)} = \mathbf{B}^{(\tau)} \odot \frac{\frac{\mathbf{X}}{\mathbf{B}^{(\tau)}\mathbf{W}^{(\tau)}}\mathbf{W}^\top}{\mathbf{1}(\mathbf{W}^{(\tau+1)})^\top},
\tag{5.94}
$$

$$
\mathbf{W}^{(\tau+1)} = \mathbf{W}^{(\tau)} \odot \frac{(\mathbf{B}^{(\tau)})^\top \frac{\mathbf{X}}{\mathbf{B}^{(\tau)}\mathbf{W}^{(\tau)}}}{(\mathbf{B}^{(\tau)})^\top \mathbf{1}},
\tag{5.95}
$$

which are identical to the multiplicative updating for basis matrix $\mathbf{B}$ and weight matrix W of KL-NMF as shown in Table 2.1. The terms in last RHSs of Eqs. (5.90) and (5.91) based on PLCA exactly correspond to those in RHSs of Eqs. (5.94) and (5.95) based on KL-NMF, respectively.

5.2.4 INTERPRETATION AND APPLICATION

In general, the two-dimensional PLCA model is numerically identical to the nonnegative matrix factorization (NMF) model in the presence of two-dimensional samples $\{x, y\}$ or with time–frequency observations $\{t, f\}$. Higher-dimensional PLCA corresponds to the nonnegative tensor factorization (NTF) with multidimensional data $\{x_1, \ldots, x_J\}$. Different from the decomposition techniques using NMF, PLCA was developed for analysis of acoustic spectra, which provided explicit model and probabilistic interpretation for the observed spectra based on the learned distribution parameters $\{p(z), p(x|z), p(y|z)\}$ with semantically meaningful components z (Smaragdis et al., 2006). This model facilitates a variety of applications, ranging from source separation and denoising to music transcription and sound recognition. For the application of source separation, we deal with separation of a mixture signal in spectrogram domain with time–frequency data $\{t, f\}$. If we know the frequency marginals $p(f|z)$, which correspond to each source in a mixture, we can reconstruct the sources by using these marginals. However, we assume that we know the types of sounds in the mixture and that we have pretrained frequency marginals for describing those sounds. Once we estimate the corresponding time marginals $p(t|z)$ that best describe an input, we can perform reconstruction by appropriately multiplying and summing all marginals. A selective reconstruction using the marginals from a single source will result in a reconstruction that contains only one source. PLCA was applied to obtain reasonable results for separation of speech and chimes based on their corresponding frequency marginals $p(f|z)$ in Smaragdis et al. (2006). In Raj et al. (2007), a statistical model for signal spectra of time t and frequency bin f based on PLCA was employed in separating vocals from background music. The parameters of the underlying distributions $\{p(z), p(t|z), p(f|z)\}$ for the singing voice (indexed by v) and accompanied music (indexed by m) were learned from the observed magnitude spectrogram of the song (index by s) according to

$$
p_s(t, f) = p(m) \sum_z p_m(z) p_m(f|z) p_m(t|z)
$$
$$
+ p(v) \sum_z p_v(z) p_v(f|z) p_v(t|z). \tag{5.96}
$$

In the implementation, only the frequency marginal $p_m(f|z)$ is pretrained and known, all the other parameters $p(m)$, $p(v)$, $p_m(z)$, $p_m(t|z)$, $p_v(z)$, $p_v(t|z)$ and $p_v(f|z)$ are unknown and are to be estimated. Such a solution allows the personalization of songs by separating out the vocals, processing them to one's own taste, and then remixing them.

In Smaragdis and Raj (2008), shift invariant PLCA was applied to discover a shift-invariant structure in a music signal, as well as extract multiple kernel distributions from speech signals. The distribution or count data of time–frequency audio signals was characterized and decomposed into shift-invariant components. In case of a music application, PLCA was able to decompose the time–frequency distribution of a piano notes passage into a kernel distribution and an impulse distribution. The kernel function was seen as a harmonic series while the impulse function was used to place the harmonic series in the time–frequency plane. The impulse function correctly identified the time and frequency placement of all the notes played. Furthermore, PLCA was applied for collaborative audio enhancement (CAS) (Kim and Smaragdis, 2013), which aims to recover the common audio sources from multiple recordings of a specific audio scene. An extended model, called the probabilistic latent component sharing (PLCS), was proposed by adopting the common component sharing concept.

In general, there are many overlapping recordings of an audio scene from different sensors. Each recording is uniquely corrupted. Corruption may be caused by frequency band limitation or unwanted inference. CAS is developed to fully utilize low cost noisy data by extracting common audio sources from them so as to produce a higher quality rendering of the recorded event. To do so, PLCS was applied to simultaneously represent these synchronized recording data. After decomposition, some of the parameters were fixed to be the same during and after the learning process to capture the shared audio content while the remaining parameters represented those undesirable recording-specific interferences and artifacts. PLCS allows the incorporation of prior knowledge about the model parameters or, equivalently, the representative spectra of the components.

Although PLCA presents a probabilistic latent variable model for source separation, the limitations of PLCA lie at the simple multinomial parameters and learning objective without treating model regularization and controlling model complexity. In what follows, we present a direct model for probabilistic nonnegative matrix factorization (NMF), which opens an avenue to a wide range of solutions to flexible parts-based representation. In particular, Bayesian learning is implemented to deal with the uncertainty in a learning representation based on NMF and carry out the solution to supervised and unsupervised source separation with applications for speech and music as well as singing voice separation as mentioned in the next section.

5.3 BAYESIAN NMF

Under Bayesian framework, nonnegative spectral signals are drawn from probability distributions. The nonnegative parameters are represented by prior distributions. Bayesian learning is introduced to deal with uncertainty modeling by maximizing the marginal likelihood over the randomness of model parameters. Model complexity is controlled to improve a generalization of NMF for unknown variations in mixing environments. Section 3.3 has addressed the importance of Bayesian learning for source separation in adverse environments. This section will present an advanced study on a number of Bayesian factorization and learning approaches to monaural source separation. Different settings of the likelihood function and prior distributions are investigated to find different solutions with different inference procedures as addressed in Sections 5.3.1, 5.3.2 and 5.3.3. We will illustrate the pros and cons of different solutions to Bayesian nonnegative matrix factorization (BNMF) in terms of theoretical and practical concerns. Then, an attractive BNMF method, where the modeling error is drawn from a Poisson distribution and the model parameters are characterized by the exponential distributions, is introduced in Section 5.3.3. A closed-form solution to hyperparameters based on the VB-EM algorithm will be derived for rapid implementation. The regularization parameters are optimally estimated from training data without empirical selection from validation data. This BNMF will be connected to the standard NMF with a *sparseness* constraint in Section 2.2.3. The dependencies of the variational objective on hyperparameters are sufficiently characterized with an efficient implementation for single-channel source separation. Using BNMF, the number of bases K is adaptively determined from the mixed signal according to the variational lower bound of the logarithm of the marginal likelihood over NMF basis and weight matrices. System variations are taken into account for source separation. The variations come from improper model assumptions, incorrect model order, possible noise interference, nonstationary environment, reverberant distortion, as well as a number of variations from different source signals, including singers, instruments, speakers and room acoustics. In Section 5.3.4, system evaluation for speech and music separation is reported while that for singing voice separation is mentioned in Section 5.3.5 where the adaptive supervised and unsupervised source separation will be also examined.

5.3.1 GAUSSIAN–EXPONENTIAL BAYESIAN NMF

As mentioned in Section 3.7.1, the maximum likelihood (ML) estimation of NMF parameters is seen as an optimization problem where the log-likelihood function based on the Poisson distribution, given in Eq. (3.79), is maximized. Such a problem is resolved for KL-NMF where Kullback–Leibler divergence for data reconstruction is minimized. However, ML estimation is prone to find an overtrained model, which is sensitive to unknown test conditions (Bishop, 2006, Watanabe and Chien, 2015). To improve model regularization, Bayesian approach is introduced to establish NMF for monaural source separation. ML-based NMF was improved by considering the priors of basis matrix **B** and weight matrix **W** when constructing BNMF. Different specifications of the likelihood function and prior distribution result in different solutions with different inference procedures. In what follows, we introduce a couple of BNMFs with different combinations of the likelihood function and prior distribution.

In Schmidt et al. (2009), the approximation error of X_{mn} using $\sum_k B_{mk} W_{kn}$ was modeled by a zero-mean Gaussian distribution

$$X_{mn} \sim \mathcal{N}\left(X_{mn} \middle| \sum_k B_{mk} W_{kn}, \sigma^2\right) \tag{5.97}$$

with the variance parameter σ^2, which was characterized by an inverse Gamma distribution

$$
\begin{aligned}
p(\sigma^2) &= \text{Inv-Gam}(\sigma^2|\alpha, \beta) \\
&= \frac{\beta^\alpha}{\Gamma(\alpha)} (\sigma^2)^{-\alpha-1} \exp\left(-\frac{\beta}{\sigma^2}\right).
\end{aligned} \tag{5.98}
$$

The priors of nonnegative parameters B_{mk} and W_{kn} were represented by the exponential distributions with means $(\lambda_{mk}^b)^{-1}$ and $(\lambda_{kn}^w)^{-1}$, namely

$$B_{mk} \sim \text{Exp}(B_{mk}|\lambda_{mk}^b), \tag{5.99}$$

$$W_{kn} \sim \text{Exp}(W_{kn}|\lambda_{kn}^w), \tag{5.100}$$

where

$$\text{Exp}(x; \theta) = \theta \exp(-\theta x). \tag{5.101}$$

Without loss of generality, this method is here named the *Gaussian–Exponential BNMF*, owing to its usage of the particular likelihood function and prior density in BNMF. Graphical representation of this BNMF is shown in Fig. 5.2. The right-bottom corner of the panel shows the number of variables. The numbers of two-dimensional variables are shown in the corresponding two overlapped panels.

Using this BNMF, the marginal likelihood is computed by integrating and applying the posterior distribution, which is a product of the likelihood function in Eq. (5.97) and prior functions in Eqs. (5.98), (5.99) and (5.100). Since the integral is analytically intractable, the MCMC sampling

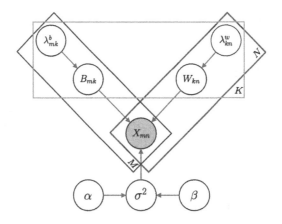

FIGURE 5.2

Graphical representation for Gaussian–Exponential Bayesian nonnegative matrix factorization.

method, as addressed in Section 3.7.4, is introduced to conduct approximate inference. Using Gibbs sampling, the conditional posteriors of model parameters

$$\boldsymbol{\Theta} = \{\mathbf{B}, \mathbf{W}, \sigma^2\} \tag{5.102}$$

are necessary. Consequently, the conditional distribution of basis B_{mk} is obtained by

$$p(B_{mk}|\mathbf{X}, \mathbf{B}_{\backslash(mk)}, \mathbf{W}, \sigma^2) = \mathcal{R}(B_{mk}|\mu_{B_{mk}}, \sigma^2_{B_{mk}}, \lambda^b_{mk}) \tag{5.103}$$

where $\mathbf{B}_{\backslash(mk)}$ denotes all the entries in $\mathbf{B}$ except for B_{mk} and

$$\mu_{B_{mk}} = \frac{\sum_n (X_{mn} - \sum_{k' \neq k} B_{mk'} W_{k'n}) W_{kn}}{\sum_n W^2_{kn}}, \tag{5.104}$$

$$\sigma^2_{B_{mk}} = \frac{\sigma^2}{\sum_n W^2_{kn}}. \tag{5.105}$$

Note that $\mathcal{R}$ denotes the rectified normal distribution, which is defined as

$$\mathcal{R}(x|\mu, \sigma^2, \lambda) \propto \mathcal{N}(x|\mu, \sigma^2) \text{Exp}(x|\lambda). \tag{5.106}$$

Similarly, the posterior distribution for the sampling weight W_{kn} is obtained by

$$p(W_{kn}|\mathbf{X}, \mathbf{W}_{\backslash(kn)}, \mathbf{B}, \sigma^2) = \mathcal{R}(W_{kn}|\mu_{W_{kn}}, \sigma^2_{W_{kn}}, \lambda^w_{kn}) \tag{5.107}$$

where $\mathbf{W}_{\backslash(kn)}$ denotes all the entries in $\mathbf{W}$ except for W_{kn} and

$$\mu_{W_{kn}} = \frac{\sum_m (X_{mn} - \sum_{k' \neq k} B_{mk'} W_{k'n}) B_{mk}}{\sum_m B^2_{mk}}, \tag{5.108}$$

$$\sigma_{W_{kn}}^2 = \frac{\sigma^2}{\sum_m B_{mk}^2}.$$ (5.109)

The posterior distribution for sampling noise variance σ^2 is obtained by

$$p(\sigma^2|\mathbf{X}, \mathbf{B}, \mathbf{W}) = \text{Inv-Gam}(\sigma^2|\alpha_{\sigma^2}, \beta_{\sigma^2})$$ (5.110)

where

$$\alpha_{\sigma^2} = \frac{MN}{2} + 1 + \alpha,$$ (5.111)

$$\beta_{\sigma^2} = \frac{1}{2}\sum_m\sum_n(\mathbf{X} - \mathbf{BW})_{mn}^2 + \beta.$$ (5.112)

The Gibbs sampling algorithm for Gaussian–Exponential BNMF is implemented in Algorithm 5.1. L samples of BNMF parameters

$$\{\mathbf{B}^{(l)}, \mathbf{W}^{(l)}\}_{l=1}^L$$

are therefore obtained. Typically, the larger the applied exponential parameter θ, the sparser shaped the exponential distribution. The sparsity of basis B_{mk} and weight W_{kn} parameters is controlled by hyperparameters λ_{mk}^b and λ_{kn}^w, respectively. As addressed in Section 3.4.1, a kind of sparse Bayesian learning is performed in Gaussian–Exponential BNMF. However, there are two weaknesses in this BNMF. First, the hyperparameters $\{\lambda_{mk}^b, \lambda_{kn}^w\}$ were assumed to be fixed and selected empirically. Some validation data were collected for this selection. The second weakness is that the exponential distribution is not a conjugate prior to the Gaussian likelihood for NMF. There is no closed-form solution to NMF parameters $\Theta = \{\mathbf{B}, \mathbf{W}, \sigma^2\}$. These parameters were estimated by Gibbs sampling procedure where a sequence of posterior samples of Θ were drawn from the corresponding conditional posterior probabilities. The Gibbs sampling procedure is computationally demanding.

Algorithm 5.1 Gibbs Sampling for Gaussian–Exponential BNMF.

Initialize with $\mathbf{B}^{(0)}$ and $\mathbf{W}^{(0)}$
For each sampling iteration l
   For each basis component k
      Sample $\mathbf{B}_{:k} = \{B_{mk}\}_{m=1}^M$ using Eq. (5.103)
      Sample $\mathbf{W}_{k:} = \{W_{kn}\}_{n=1}^N$ using Eq. (5.107)
      $k \leftarrow k + 1$
   Sample σ^2 using Eq. (5.110)
   Check convergence
   $l \leftarrow l + 1$
Return $\{\mathbf{B}^{(l)}, \mathbf{W}^{(l)}\}_{l=1}^L$

5.3.2 POISSON–GAMMA BAYESIAN NMF

In Cemgil (2009), another type of Bayesian NMF was proposed for image reconstruction, which was based on the likelihood function using the Poisson distribution as shown in Eq. (3.79) and the prior densities of basis and weight matrices using the Gamma distributions:

$$B_{mk} \sim \text{Gam}(B_{mk}|\alpha^b_{mk}, \beta^b_{mk}), \tag{5.113}$$

$$W_{kn} \sim \text{Gam}(W_{kn}|\alpha^w_{kn}, \beta^w_{kn}). \tag{5.114}$$

Gamma distribution is represented by a shape parameter a and a scale parameter b in a form of

$$\begin{aligned} &\text{Gam}(x; a, b) \\ &= \exp\left((a-1)\log x - \frac{x}{b} - \log \Gamma(a) - a \log b\right). \end{aligned} \tag{5.115}$$

The entropy of Gamma distribution is yielded by

$$\mathbb{H}[x] = -(a-1)\Psi(a) + \log b + a + \log \Gamma(a) \tag{5.116}$$

where

$$\Psi(x) \triangleq \frac{d \log \Gamma(x)}{dx} \tag{5.117}$$

denotes the digamma function. This method is therefore called the *Poisson–Gamma BNMF* (Cemgil, 2009, Dikmen and Fevotte, 2012). Notably, Gamma distribution is known as the *conjugate prior* for Poisson likelihood function.

Optimization Criteria

Fig. 5.3 depicts the graphical model of Poisson–Gamma BNMF where the model parameters $\Theta = \{\mathbf{B}, \mathbf{W}\}$ are treated as latent variables for estimation of the hyperparameters

$$\Psi = \{\alpha^b_{mk}, \beta^b_{mk}, \alpha^w_{kn}, \beta^w_{kn}\}. \tag{5.118}$$

As addressed in Section 3.7.3, variational Bayesian (VB) inference procedure can be introduced to estimate the hyperparameters Ψ by maximizing the marginal likelihood $p(\mathbf{X}|\Psi)$. This likelihood is marginalized with respect to the latent variables including $\mathbf{Z}$ (as defined in Eq. (3.52)) and model parameters $\{\mathbf{B}, \mathbf{W}\}$ of NMF model

$$\begin{aligned} p(\mathbf{X}|\Phi) = \int \int \sum_{\mathbf{Z}} p(\mathbf{X}|\mathbf{Z}) p(\mathbf{Z}|\mathbf{B}, \mathbf{W}) \\ \times\, p(\mathbf{B}, \mathbf{W}|\Phi) d\mathbf{B} d\mathbf{W}. \end{aligned} \tag{5.119}$$

However, the exact solution to the optimization of Eq. (5.119) does not exist because the posterior $p(\mathbf{Z}, \mathbf{B}, \mathbf{W}|\mathbf{X}, \Phi)$ with three latent variables $\{\mathbf{Z}, \mathbf{B}, \mathbf{W}\}$ can not be factorized. To deal with this issue,

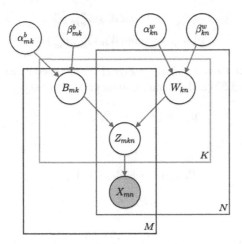

FIGURE 5.3

Graphical representation for Poisson–Gamma Bayesian nonnegative matrix factorization.

VB inference is accordingly performed by maximizing the lower bound of log marginal likelihood $\mathcal{L}(q)$, which derived as

$$
\begin{aligned}
\log p(\mathbf{X}|\mathbf{\Phi}) \geq & \int \int \sum_{\mathbf{Z}} q(\mathbf{Z}, \mathbf{B}, \mathbf{W}) \\
& \times \log \frac{p(\mathbf{X}, \mathbf{Z}, \mathbf{B}, \mathbf{W}|\mathbf{\Phi})}{q(\mathbf{Z}, \mathbf{B}, \mathbf{W})} d\mathbf{B} d\mathbf{W} \\
= & \, \mathbb{E}_q[\log p(\mathbf{X}, \mathbf{Z}, \mathbf{B}, \mathbf{W}|\mathbf{\Phi})] \\
& + \mathbb{H}[q(\mathbf{Z}, \mathbf{B}, \mathbf{W})] \triangleq \mathcal{L}(q),
\end{aligned}
\tag{5.120}
$$

where $\mathbb{H}[\cdot]$ denotes the entropy function and the factorized variational distribution

$$
\begin{aligned}
q(\mathbf{Z}, \mathbf{B}, \mathbf{W}) &= q(\mathbf{Z}) q(\mathbf{B}) q(\mathbf{W}) \\
&= \prod_m \prod_n \prod_k q(Z_{mkn}) q(B_{mk}) q(W_{kn})
\end{aligned}
\tag{5.121}
$$

is used to approximate the true posterior $p(\mathbf{Z}, \mathbf{B}, \mathbf{W}|\mathbf{X}, \mathbf{\Phi})$.

Variational Bayesian Learning

Variational Bayesian EM (VB-EM) algorithm, as addressed in Section 3.7.3, is applied here to derive the approximate solution to the Poisson–Gamma Bayesian NMF. This algorithm is implemented by performing the E and M steps of VB. The variational distributions of three latent variations are represented by

$$
q(\mathbf{z}_{mn}) = \text{Mult}(\mathbf{z}_{mn}|X_{mn}, \mathbf{p}_{mn}),
\tag{5.122}
$$

$$q(B_{mk}) = \mathrm{Gam}(B_{mk}|\widetilde{\alpha}^b_{mk}, \widetilde{\beta}^b_{mk}), \tag{5.123}$$

$$q(W_{kn}) = \mathrm{Gam}(W_{kn}|\widetilde{\alpha}^w_{kn}, \widetilde{\beta}^w_{kn}) \tag{5.124}$$

where the latent variable vector $\mathbf{z}_{mn} = [Z_{m1n} \cdots Z_{mKn}]^\top$ is modeled by a multinomial distribution with the observation X_{mn} in Eq. (3.52) and the multinomial parameter vector

$$\mathbf{p}_{mn} = [P_{m1n} \cdots P_{mKn}]^T \tag{5.125}$$

subject to

$$P_{m1n} + \cdots + P_{mKn} = 1, \tag{5.126}$$

given by

$$\mathrm{Mult}(\mathbf{s}; x, \boldsymbol{\theta}) = \delta\left(x - \sum_{k=1}^{K} s_k\right) x! \prod_{k=1}^{K} \frac{\theta_k^{s_k}}{s_k!}. \tag{5.127}$$

The entropy of multinomial distribution is given by

$$\mathbb{H}[\mathbf{s}] = -\log\Gamma(x+1) - \sum_k \langle s_k \rangle \log\theta_k$$
$$+ \sum_k \langle \log\Gamma(s_k+1) \rangle - \left\langle \log\delta\left(x - \sum_k s_k\right) \right\rangle. \tag{5.128}$$

Notably, the variational parameters

$$\{\widetilde{\alpha}^b_{mk}, \widetilde{\beta}^b_{mk}, \widetilde{\alpha}^w_{kn}, \widetilde{\beta}^w_{kn}\}$$

in Eqs. (5.123) and (5.124) are different from the hyperparameters

$$\boldsymbol{\Psi} = \{\alpha^b_{mk}, \beta^b_{mk}, \alpha^w_{kn}, \beta^w_{kn}\} \tag{5.129}$$

in Eqs. (5.113) and (5.114). Comparable with Eq. (3.110), a general solution to the variational distribution q_j of an individual latent variable $j \in \{\mathbf{Z}, \mathbf{B}, \mathbf{W}\}$ is given by

$$\log\widehat{q}_j \propto \mathbb{E}_{q_{(i\neq j)}}[\log p(\mathbf{X}, \mathbf{Z}, \mathbf{B}, \mathbf{W}|\boldsymbol{\Phi})], \tag{5.130}$$

which is obtained by maximizing the variational lower bound $\mathcal{L}(q)$. The variational parameters $\{P_{mkn}, \widetilde{\alpha}^b_{mk}, \widetilde{\beta}^b_{mk}, \widetilde{\alpha}^w_{kn}, \widetilde{\beta}^w_{kn}\}$ of multinomial Z_{mkn}, Gamma B_{mk} and Gamma W_{kn} are individually estimated as

$$\widehat{P}_{mkn} = \frac{\exp(\langle\log B_{mk}\rangle + \langle\log W_{kn}\rangle)}{\sum_i \exp(\langle\log B_{mi}\rangle + \langle\log W_{in}\rangle)} \tag{5.131}$$

$$\widehat{\alpha}^b_{mk} = \alpha^b_{mk} + \sum_n \langle Z_{mkn} \rangle, \tag{5.132}$$

$$\widehat{\beta}_{mk}^b = \left(\frac{\alpha_{mk}^b}{\beta_{mk}^b} + \sum_n \langle W_{kn} \rangle \right)^{-1},$$

(5.133)

$$\widehat{\alpha}_{kn}^w = \alpha_{kn}^w + \sum_m \langle Z_{mkn} \rangle,$$

(5.134)

$$\widehat{\beta}_{kn}^w = \left(\frac{\alpha_{kn}^w}{\beta_{kn}^w} + \sum_m \langle B_{mk} \rangle \right)^{-1}$$

(5.135)

where the expectation function $\mathbb{E}_q[\cdot]$ is replaced by the notation $\langle \cdot \rangle$ for simplicity. In Eqs. (5.131)–(5.135), the sufficient statistics or the means of multinomial Z_{mkn}, Gamma B_{mk} and Gamma W_{kn} are given by

$$\langle Z_{mkn} \rangle = X_{mn} \widehat{P}_{mkn},$$

(5.136)

$$\langle B_{mk} \rangle = \widehat{\alpha}_{mk}^b \widehat{\beta}_{mk}^b,$$

(5.137)

$$\langle \log B_{mk} \rangle = \Psi(\widehat{\alpha}_{mk}^b) + \log(\widehat{\beta}_{mk}^b),$$

(5.138)

$$\langle W_{kn} \rangle = \widehat{\alpha}_{kn}^w \widehat{\beta}_{kn}^w,$$

(5.139)

$$\langle \log W_{kn} \rangle = \Psi(\widehat{\alpha}_{kn}^w) + \log(\widehat{\beta}_{kn}^w).$$

(5.140)

After having these variational distributions and sufficient statistics in the VB-E step, the VB-M step is performed by substituting them into the variational lower bound $\mathcal{L}(q)$ in Eq. (5.120) and maximizing this bound to find the updated hyperparameters $\Psi = \{\alpha_{mk}^b, \beta_{mk}^b, \alpha_{kn}^w, \beta_{kn}^w\}$. In the implementation, the solution to $\{\beta^b, \beta^w\}$ is obtained by means of basis and weight matrices, respectively. However, a closed-form solution to the shape parameters $\{\alpha_{mk}^b, \alpha_{kn}^w\}$ of Gamma distributions does not exist. These two parameters are estimated by Newton's method. The computation cost is increased. In addition, we find that some dependency of the variational lower bound on model parameter was ignored in Cemgil (2009). The resulting parameters did *not* reach a true optimum of the variational objective.

Algorithm 5.2 Variational Bayesian Learning for Poisson–Gamma BNMF.

Initialize with $\mathbf{B}^{(0)}$ and $\mathbf{W}^{(0)}$
For each VB-EM iteration τ
   VB-E step: update variational distributions
      Accumulate sufficient statistics $\{\langle Z_{mkn} \rangle, \langle B_{mk} \rangle, \langle W_{kn} \rangle\}$
      Update variational parameters $\{P_{mkn}, \widetilde{\alpha}_{mk}^b, \widetilde{\beta}_{mk}^b, \widetilde{\alpha}_{kn}^w, \widetilde{\beta}_{kn}^w\}$
   VB-M step: estimate model parameters
      Update hyperparameters $\Psi = \{\alpha_{mk}^b, \beta_{mk}^b, \alpha_{kn}^w, \beta_{kn}^w\}$
   Check if $\| \Psi^{(\tau)} - \Psi^{(\tau-1)} \|$ is small enough
   $\tau \leftarrow \tau + 1$
Return Ψ

The variational Bayesian learning of Poisson–Gamma BNMF is expressed in Algorithm 5.2. Initial parameters $\mathbf{B}^{(0)}$ and $\mathbf{W}^{(0)}$ are randomly sampled from a Gamma distribution. In the VB-E step, the sufficient statistics in Eqs. (5.136)–(5.140) are calculated before finding the variational parameters

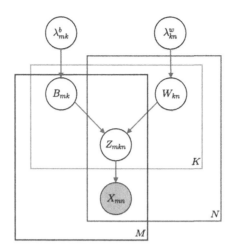

FIGURE 5.4

Graphical representation for Poisson–Exponential Bayesian nonnegative matrix factorization.

in Eqs. (5.131)–(5.135). Given the updated variational distributions in Eqs. (5.122)–(5.124), in the VB-M step, we update the hyperparameters $\mathbf{\Psi} = \{\alpha_{mk}^b, \beta_{mk}^b, \alpha_{kn}^w, \beta_{kn}^w\}$ where $\{\alpha_{mk}^b, \alpha_{kn}^w\}$ are estimated by Newton's method. Given the variational Gamma parameters $\{\widetilde{\alpha}_{mk}^b, \widetilde{\beta}_{mk}^b\}$ and $\{\widetilde{\alpha}_{kn}^w, \widetilde{\beta}_{kn}^w\}$, the final estimates of $\mathbf{B}$ and $\mathbf{W}$ are realized by finding the corresponding Gamma means $\langle B_{mk} \rangle$ and $\langle W_{kn} \rangle$ using Eqs. (5.137) and (5.139), respectively.

5.3.3 POISSON–EXPONENTIAL BAYESIAN NMF

Next, we would like to present a closed-form solution to full Bayesian NMF where all dependencies of variational lower bound on variational parameters and model hyperparameters are modeled. In the Bayesian framework, it is important to select probabilistic distributions for the likelihood function and prior density so that a meaningful solution could be obtained to meet the requirements of applications. Considering the spirit of NMF based on Kullback–Leibler (KL) divergence, we adopt the Poisson likelihood, as given in Eq. (3.79), and the exponential distribution as the *conjugate prior* for nonnegative parameters B_{mk} and W_{kn} with hyperparameters λ_{mk}^b and λ_{kn}^w, as given in Eqs. (5.99) and (5.100), respectively. The resulting algorithm is called the *Poisson–Exponential BNMF* (Chien and Yang, 2016). Fig. 5.4 displays the graphical representation for Poisson–Exponential BNMF with parameters

$$\mathbf{\Phi} = \{\lambda_{mk}^b, \lambda_{kn}^w\} \tag{5.141}$$

where the latent variables inside the green panel (light gray in print version)

$$\{Z_{mkn}, B_{mk}, W_{kn}\}$$

are indexed by $1 \le k \le K$, and the observation variable X_{mn} is from an $M \times N$ nonnegative matrix. Using this approach, the maximum *a posteriori* (MAP) estimates of parameters $\mathbf{\Theta} = \{\mathbf{B}, \mathbf{W}\}$ are obtained by maximizing the posterior distribution or minimizing a regularized KL divergence between $\mathbf{X}$ and $\mathbf{BW}$, that is,

$$
\begin{aligned}
& -\log p(\mathbf{B}, \mathbf{W}|\mathbf{X}) \\
& \propto \mathcal{D}_{\mathrm{KL}}(\mathbf{X}\|\mathbf{BW}) + \sum_m \sum_k \lambda_{mk}^b B_{mk} + \sum_k \sum_n \lambda_{kn}^w W_{kn}.
\end{aligned}
\tag{5.142}
$$

Notably, the regularization terms (the second and third terms on the RHS of Eq. (5.142)) are seen as the ℓ_1 or Lasso regularizers (Tibshirani, 1996), which are controlled by hyperparameters $\boldsymbol{\Phi} = \{\lambda_{mk}^b, \lambda_{kn}^w\}$. These regularizers impose the *sparseness* in the estimated MAP parameters $\boldsymbol{\Theta}_{\mathrm{MAP}}$ as addressed in Sections 2.2.3 and 3.4.1. Interestingly, Laplace distribution is thought of as two exponential distributions spliced together back-to-back, and accordingly called a double exponential distribution.

There are three issues regarding the MAP solution. First, it is difficult to empirically determine so many hyperparameters $\{\lambda_{mk}^b, \lambda_{kn}^w\}$ from validation data. Second, the model complexity is fixed and cannot be adapted. Third, MAP estimates are seen as point estimates. The randomness of parameters is neglected in model construction. Similar to Poisson–Gamma BNMF, Poisson–Exponential BNMF also deals with these issues by developing the full Bayesian solution based on maximization of marginal likelihood $p(\mathbf{X}|\boldsymbol{\Phi})$ over latent variables and parameters $\{\mathbf{Z}, \mathbf{B}, \mathbf{W}\}$ as expressed in Eq. (5.119) with a variational lower bound given in Eq. (5.120). The sparsity-controlled regularizers $\boldsymbol{\Phi} = \{\lambda^b, \lambda^w\} = \{\lambda_{mk}^b, \lambda_{kn}^w\}$ are estimated by variational Bayesian learning. No validation data is required. The resulting evidence function $p(\mathbf{X}|\boldsymbol{\Phi})$ can be applied for model selection which is feasible to judge how many bases K should be selected or to identify how many sources exist in a single channel system. The selected number of bases or sources is adaptive to fit different experimental conditions with varying lengths and contents of experimental data.

Variational Expectation Step

Similar to VB-EM steps for Poisson–Gamma BNMF in Section 5.3.2, Poisson–Exponential BMNF is implemented by first estimating the variational distributions $q(\mathbf{z}_{mn})$, $q(B_{mk})$ and $q(W_{kn})$ (defined in Eqs. (5.122)–(5.124)) in the VB E-step. To do so, we calculate the logarithm of the joint distribution as

$$
\begin{aligned}
\log p(\mathbf{X}, \mathbf{Z}, \mathbf{B}, \mathbf{W}|\boldsymbol{\Phi}) &= \log p(\mathbf{X}|\mathbf{Z}) + \log p(\mathbf{Z}|\mathbf{B}, \mathbf{W}) \\
&+ \log p(\mathbf{B}|\lambda^b) + \log p(\mathbf{W}|\lambda^w) \\
&\propto \sum_m \sum_n \log \delta \left(X_{mn} - \sum_k Z_{mkn} \right) \\
&+ \sum_m \sum_k \sum_n \left(Z_{mkn} \log(B_{mk} W_{kn}) - B_{mk} W_{kn} \right. \\
&- \left. \log \Gamma(Z_{mkn}+1) \right) + \sum_m \sum_k \left(\log \lambda_{mk}^b - \lambda_{mk}^b B_{mk} \right) \\
&+ \sum_k \sum_n \left(\log \lambda_{kn}^w - \lambda_{kn}^w W_{kn} \right).
\end{aligned}
\tag{5.143}
$$

The variational distribution is estimated by using the terms in Eq. (5.143), which are related to the target latent variable, and treating all the other terms as constants. Then, the variational multinomial distribution for $\mathbf{z}_{mn}$ is derived as

$$\widehat{q}(\mathbf{z}_{mn}) \propto \exp\left(\mathbb{E}_{q_{(i \neq \mathbf{z}_{mn})}}[\log p(\mathbf{X}, \mathbf{Z}, \mathbf{B}, \mathbf{W}|\Phi)]\right)$$

$$\propto \exp\left(\log \delta\left(X_{mn} - \sum_k Z_{mkn}\right) + \sum_k Z_{mkn}\right.$$

$$\times \left(\langle\log B_{mk}\rangle + \langle\log W_{kn}\rangle\right) - \sum_k \log \Gamma(Z_{mkn} + 1)\bigg)$$

$$\propto \exp\left(\log \delta\left(X_{mn} - \sum_k Z_{mkn}\right) + \log(X_{mn}!)\right. \tag{5.144}$$

$$\left. + \sum_k \left(Z_{mkn} \log \widehat{P}_{mkn} - \log(Z_{mkn}!)\right)\right)$$

$$\propto \delta\left(X_{mn} - \sum_k Z_{mkn}\right) X_{mn}! \prod_k \frac{\widehat{P}_{mkn}^{Z_{mkn}}}{Z_{mkn}!}$$

$$\propto \text{Mult}(\mathbf{z}_{mn}|X_{mn}, \widehat{\mathbf{p}}).$$

The optimal variational parameters

$$\widehat{\mathbf{p}} = [\widehat{P}_{m1n} \cdots \widehat{P}_{mKn}]^T$$

are derived as given in Eq. (5.131). This parameter set is the same for Poisson–Gamma and Poisson–Exponential BNMFs.

Different from Poisson–Gamma BNMF, the variational distributions for individual latent variables B_{mk} and W_{kn} in Poisson–Exponential BNMF are obtained by combining the statistics from the Poisson likelihood and exponential prior which are manipulated as a new Gamma distribution as follows:

$$\widehat{q}(B_{mk}) \propto \exp\left(\mathbb{E}_{q_{(i \neq B_{mk})}}[\log p(\mathbf{X}, \mathbf{Z}, \mathbf{B}, \mathbf{W}|\Phi)]\right)$$

$$\propto \exp\left(\sum_n \langle Z_{mkn}\rangle \log B_{mk} - \left(\sum_n \langle W_{kn}\rangle + \lambda_{mk}^b\right) B_{mk}\right) \tag{5.145}$$

$$\propto \text{Gam}(B_{mk}|\widehat{\alpha}_{mk}^b, \widehat{\beta}_{mk}^b),$$

$$\widehat{q}(W_{kn}) \propto \exp\left(\mathbb{E}_{q_{(i \neq W_{kn})}}[\log p(\mathbf{X}, \mathbf{Z}, \mathbf{B}, \mathbf{W}|\Phi)]\right)$$

$$\propto \exp\left(\sum_m \langle Z_{mkn}\rangle \log W_{kn} - \left(\sum_m \langle B_{mk}\rangle + \lambda_{kn}^w\right) W_{kn}\right) \tag{5.146}$$

$$\propto \text{Gam}(W_{kn}|\widehat{\alpha}_{kn}^w, \widehat{\beta}_{kn}^w).$$

In calculation of variational distributions, the sufficient statistics $\langle Z_{mkn}\rangle$, $\langle B_{mk}\rangle$, $\langle\log B_{mk}\rangle$, $\langle W_{kn}\rangle$ and $\langle\log W_{kn}\rangle$ are computed by Eqs. (5.136)–(5.140). The variational Gamma parameters are then obtained by

$$\widehat{\alpha}_{mk}^b = 1 + \sum_n \langle Z_{mkn} \rangle, \tag{5.147}$$

$$\widehat{\beta}_{mk}^b = \left(\sum_n \langle W_{kn} \rangle + \lambda_{mk}^b \right)^{-1}, \tag{5.148}$$

$$\widehat{\alpha}_{kn}^w = 1 + \sum_m \langle Z_{mkn} \rangle, \tag{5.149}$$

$$\widehat{\beta}_{kn}^w = \left(\sum_m \langle B_{mk} \rangle + \lambda_{kn}^w \right)^{-1}. \tag{5.150}$$

Variational Maximization Step

To estimate the regularization parameters $\mathbf{\Phi} = \{ \lambda_{mk}^b, \lambda_{kn}^w \}$, we need to substitute the optimal variational distributions

$$\{ \widehat{q}(\mathbf{z}_{mn}), \widehat{q}(B_{mk}), \widehat{q}(W_{kn}) \}$$

with the updated parameters

$$\{ \widehat{P}_{mkn}, \widehat{\alpha}_{mk}^b, \widehat{\beta}_{mk}^b, \widehat{\alpha}_{kn}^w, \widehat{\beta}_{kn}^w \}$$

into the variational lower bound $\mathcal{L}(q)$ in Eq. (5.120). The first term is calculated as

$$
\begin{aligned}
\mathbb{E}_{\widehat{q}}[\log p(\mathbf{X}, \mathbf{Z}, \mathbf{B}, \mathbf{W} | \mathbf{\Phi})] &\propto \sum_m \sum_n \left\langle \log \delta \left(X_{mn} - \sum_k Z_{mkn} \right) \right\rangle \\
&+ \sum_m \sum_k \langle \log B_{mk} \rangle \sum_n \langle Z_{mkn} \rangle + \sum_k \sum_n \langle \log W_{kn} \rangle \sum_m \langle Z_{mkn} \rangle \\
&- \sum_m \sum_k \sum_n \langle B_{mk} \rangle \langle W_{kn} \rangle - \sum_m \sum_k \sum_n \langle \log \Gamma(Z_{mkn} + 1) \rangle \\
&+ \sum_m \sum_k (\log \lambda_{mk}^b - \lambda_{mk}^b \langle B_{mk} \rangle) + \sum_k \sum_n (\log \lambda_{kn}^w - \lambda_{kn}^w \langle W_{kn} \rangle)
\end{aligned}
\tag{5.151}
$$

and the second term is obtained as the entropy of the variational distributions

$$
\begin{aligned}
\mathbb{H}[\widehat{q}(\mathbf{Z}, \mathbf{B}, \mathbf{W})] &= -\langle \log \widehat{q}(\mathbf{Z}, \mathbf{B}, \mathbf{W}) \rangle_{\widehat{q}(\mathbf{Z}, \mathbf{B}, \mathbf{W})} \\
&= \sum_m \sum_n \mathbb{H}[\mathbf{z}_{mn}] + \sum_m \sum_k \mathbb{H}[B_{mk}] + \sum_k \sum_n \mathbb{H}[W_{kn}] \\
&= \sum_m \sum_n \left(-\log \Gamma(X_{mn} + 1) - \sum_k \langle Z_{mkn} \rangle \log \widehat{P}_{mkn} \right) \\
&+ \sum_m \sum_k \sum_n \langle \log \Gamma(Z_{mkn} + 1) \rangle
\end{aligned}
\tag{5.152}
$$

$$-\sum_m \sum_n \left\langle \log \delta \left(X_{mn} - \sum_k Z_{mkn} \right) \right\rangle$$

$$+\sum_m \sum_k \left(-(\widehat{\alpha}_{mk}^b - 1)\Psi(\widehat{\alpha}_{mk}^b) + \log \widehat{\beta}_{mk}^b + \widehat{\alpha}_{mk}^b + \log \Gamma(\widehat{\alpha}_{mk}^b) \right)$$

$$+\sum_k \sum_n \left(-(\widehat{\alpha}_{kn}^w - 1)\Psi(\widehat{\alpha}_{kn}^w) + \log \widehat{\beta}_{kn}^w + \widehat{\alpha}_{kn}^w + \log \Gamma(\widehat{\alpha}_{kn}^w) \right)$$

where the entropy functions of the multinomial and Gamma distributions are expressed in Eqs. (5.128) and (5.116), respectively. The updated variational lower bound is obtained by adding Eqs. (5.151) and (5.152) to yield

$$
\begin{aligned}
\mathcal{L}(q) = \sum_m \sum_n &\left(-\log \Gamma(X_{mn} + 1) - \sum_k \langle Z_{mkn} \rangle \log \widehat{P}_{mkn} \right) \\
&+ \sum_m \sum_k \langle \log B_{mk} \rangle \sum_n \langle Z_{mkn} \rangle + \sum_k \sum_n \langle \log W_{kn} \rangle \sum_m \langle Z_{mkn} \rangle \\
&- \sum_m \sum_n \sum_k \langle B_{mk} \rangle \langle W_{kn} \rangle + \sum_m \sum_k (\log \lambda_{mk}^b - \lambda_{mk}^b \langle B_{mk} \rangle) \\
&+ \sum_k \sum_n (\log \lambda_{kn}^w - \lambda_{kn}^w \langle W_{kn} \rangle) + \sum_m \sum_k \left(-(\widehat{\alpha}_{mk}^b - 1)\Psi(\widehat{\alpha}_{mk}^b) \right. \\
&+ \log \widehat{\beta}_{mk}^b + \widehat{\alpha}_{mk}^b + \log \Gamma(\widehat{\alpha}_{mk}^b) \Bigg) + \sum_k \sum_n \left(-(\widehat{\alpha}_{kn}^w - 1) \right. \\
&\times \Psi(\widehat{\alpha}_{kn}^w) + \log \widehat{\beta}_{kn}^w + \widehat{\alpha}_{kn}^w + \log \Gamma(\widehat{\alpha}_{kn}^w) \Bigg).
\end{aligned}
\tag{5.153}
$$

The optimal regularization parameters $\widehat{\Phi} = \{\widehat{\lambda}_{mk}^b, \widehat{\lambda}_{kn}^w\}$ are derived by maximizing Eq. (5.153) with respect to individual parameters to yield

$$\frac{\partial \mathcal{L}(q)}{\partial \lambda_{mk}^b} = \frac{1}{\lambda_{mk}^b} - \langle B_{mk} \rangle + \frac{\partial \log \widehat{\beta}_{mk}^b}{\partial \lambda_{mk}^b} = 0, \tag{5.154}$$

$$\frac{\partial \mathcal{L}(q)}{\partial \lambda_{kn}^w} = \frac{1}{\lambda_{kn}^w} - \langle W_{kn} \rangle + \frac{\partial \log \widehat{\beta}_{kn}^w}{\partial \lambda_{kn}^w} = 0. \tag{5.155}$$

Eq. (5.154) turns out to solve the following quadratic equations:

$$(\lambda_{mk}^b)^2 + \sum_n \langle W_{kn} \rangle \lambda_{mk}^b - \frac{\sum_n \langle W_{kn} \rangle}{\langle B_{mk} \rangle} = 0, \tag{5.156}$$

$$(\lambda_{kn}^w)^2 + \sum_m \langle B_{mk} \rangle \lambda_{kn}^w - \frac{\sum_m \langle B_{mk} \rangle}{\langle W_{kn} \rangle} = 0. \tag{5.157}$$

Table 5.4 Comparison of different Bayesian NMFs in terms of inference algorithm, closed-form solution and optimization theory

Likelihood–Prior	Inference	Closed-Form	Optimization
Gaussian–Exponential	Gibbs	No	–
Poisson–Gamma	VB	Partial	Simplified
Poisson–Exponential	VB	Yes	Full

Considering the property of nonnegative regularization parameters, we can find the positive and unique roots

$$\widehat{\lambda}_{mk}^{b} = \frac{1}{2}\left(-\sum_{n}\langle W_{kn}\rangle + \sqrt{\left(\sum_{n}\langle W_{kn}\rangle\right)^{2} + 4\frac{\sum_{n}\langle W_{kn}\rangle}{\langle B_{mk}\rangle}} \right), \tag{5.158}$$

$$\widehat{\lambda}_{kn}^{w} = \frac{1}{2}\left(-\sum_{m}\langle B_{mk}\rangle + \sqrt{\left(\sum_{m}\langle B_{mk}\rangle\right)^{2} + 4\frac{\sum_{m}\langle B_{mk}\rangle}{\langle W_{kn}\rangle}} \right). \tag{5.159}$$

The E and M steps of VB are alternatively and iteratively performed to estimate the BNMF hyperparameters $\widehat{\Phi}$ with convergence. In the previously described Poisson–Gamma BNMF (Cemgil, 2009), the dependency of the variational parameter on the model hyperparameter Φ was ignored in the optimization procedure. The true optimum of the variational lower bound was not assured. In contrast, this work thoroughly characterizes the dependencies of variational parameters $\{\widehat{\beta}_{mk}^{b}, \widehat{\beta}_{kn}^{w}\}$ on model parameters $\{\widehat{\lambda}_{mk}^{b}, \widehat{\lambda}_{kn}^{w}\}$ and incorporates the derivatives

$$\left\{ \frac{\partial \log \widehat{\beta}_{mk}^{b}}{\partial \lambda_{mk}^{b}}, \frac{\partial \log \widehat{\beta}_{kn}^{w}}{\partial \lambda_{kn}^{w}} \right\}$$

into the solution $\widehat{\Phi} = \{\widehat{\lambda}_{mk}^{b}, \widehat{\lambda}_{kn}^{w}\}$ for the hyperparameters of Poisson–Exponential BNMF.

Properties of Bayesian NMFs

Table 5.4 compares the Gaussian–Exponential BNMF (also denoted as GE-BNMF) in Schmidt et al. (2009), the Poisson–Gamma BNMF (denoted as PG-BNMF) in Cemgil (2009) and the Poisson–Exponential BNMF (denoted as PE-BNMF) in terms of the likelihood–prior pair, inference algorithm, closed-form solution and optimization condition. Basically, PE-BNMF is superior to GE-BNMF and PG-BNMF for two reasons. First, using the exponential priors in BNMF provides a tractable solution as given in Eqs. (5.158)–(5.159). BNMF with exponential priors is equivalent to the sparse NMF, as addressed in Section 2.2.3, which is beneficial for robust source separation. The Gibbs sampling in GE-BNMF and the Newton's solution in PG-BNMF are computationally demanding. Second, the dependencies of three terms of the variational lower bound on the exponential hyperparameters λ_{mk}^{b} and λ_{kn}^{w} are fully characterized in finding an optimum solution to PE-NMF while some dependencies of $\mathcal{L}(q)$ on Gamma hyperparameters were disregarded in the simplified solution to PG-BNMF (Cemgil, 2009). Besides, the observations in GE-BNMF (Schmidt et al., 2009) were not constrained to be nonnegative and could not reflect the nonnegativity condition in NMF. Although Gibbs sampling and VB inference procedures are used to implement full Bayesian solutions, these solutions are *approximate* since the exact posterior distribution over $\{\mathbf{Z}, \mathbf{B}, \mathbf{W}\}$ is not analytic in these procedures (Bishop, 2006, Watanabe and Chien, 2015).

The uncertainties in monaural source separation due to an ill-posed model and heterogeneous data could be characterized by a prior distribution of model parameters $\boldsymbol{\Theta}$ and compensated by integrating the prior $p(\boldsymbol{\Theta}|\boldsymbol{\Phi})$ into the marginal likelihood $p(\mathbf{X}|\boldsymbol{\Phi})$. BNMF maximizes the marginal likelihood to estimate the hyperparameters or regularization parameters $\boldsymbol{\Phi}$ as the *distribution estimates* of the NMF model. Such a BNMF could improve the robustness of singing voice separation to a variety of singers, songs and instruments. In the implementation, the variational lower bound $\mathcal{L}(q)$ of log marginal likelihood is treated as the objective to find the optimal hyperparameters $\boldsymbol{\Phi}$, as well as to select the best number of basis vectors K. The model order K is adaptively selected for different mixed signal $\mathbf{X}$ according to $\mathcal{L}(q)$ in Eq. (5.153) with the converged variational parameters

$$\{\widehat{P}_{mkn}, \widehat{\alpha}_{mk}^b, \widehat{\beta}_{mk}^b, \widehat{\alpha}_{kn}^w, \widehat{\beta}_{kn}^w\}$$

and hyperparameters

$$\{\widehat{\lambda}_{mk}^b, \widehat{\lambda}_{kn}^w\}$$

using

$$\widehat{K} = \arg\max_K \mathcal{L}(q, K). \tag{5.160}$$

The best model order $\widehat{K}$ with the highest variational lower bound $\mathcal{L}(q, K)$ is determined (Yang et al., 2014a). This scheme can be extended for different BNMFs based on VB inference.

In system evaluation, Bayesian NMFs were implemented for two single-channel source separation tasks. The first task is to conduct a supervised speech and music separation where the labeled training data are collected to estimate basis matrix $\mathbf{B}$ for two sources in advance. This matrix is known for separation of a test mixed signal. The second task is to implement an unsupervised singing voice separation without any labeled data given beforehand. The basis matrix is estimated in an online unsupervised manner.

5.3.4 EVALUATION FOR SUPERVISED SEPARATION

In the evaluation of speech and music separation, a set of 600 speech sentences, randomly selected from TIMIT corpus, was collected as the training utterances. Each sentence had a length of 2–3 seconds. TIMIT core test set consisting of 192 utterances from unseen speakers was used to mixed with the music signals. A set of high quality music recordings were sampled from Saarland Music Data (SMD) (Muller et al., 2011). SMD dataset consisted of two music collections. The first collection contained MIDI-Audio pairs of piano music and the second collection contained various Western classical music repertoire. We selected one piano clip and one violin clip composed by Bach from the second collection and down-sampled the signals to 16 kHz sampling frequency. The mixed signals were generated by corrupting with a randomly selected music segments at the speech-to-music ratio (SMR) of 0 dB. The speech bases $\mathbf{B}^s$ were learned from the spectrogram of training utterances. The music bases $\mathbf{B}^m$ were

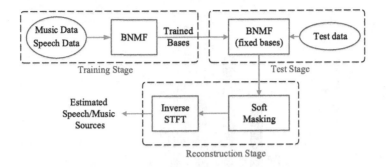

FIGURE 5.5

Implementation procedure for supervised speech and music separation.

computed from a disjoint music piece of the same track which was used for generating the test sample. The 1024-point STFT was calculated to obtain the Fourier magnitude spectrogram with frame duration of 40 ms and frame shift of 10 ms. Fig. 5.5 depicts the implementation procedure for supervised speech and music separation. A standard supervised NMF learning procedure has been addressed in Section 2.2.1 and plotted in Fig. 2.6. The BSS Eval toolbox http://bass-db.gforge.inria.fr/bss_eval/ was used to measure the signal-to-distortion ratio (SDR) (Vincent et al., 2006) to evaluate the system performance. Baseline NMF (KL-NMF) (Lee and Seung, 2000), GE-BNMF (Schmidt et al., 2009), PG-BNMF (Cemgil, 2009) and PE-BNMF (Chien and Yang, 2016) were implemented. NMF was run with 100 iterations. In the implementation of GE-BNMF, 500 Gibbs sampling iterations were performed while the first 100 iterations were treated as burn-in samples. A small set of validation data was collected to find hyperparameters in GE-BNMF. Using PG-BNMF and PE-BNMF, 200 VB-EM iterations were run.

First of all, the number of bases for speech and music separation is adaptively selected by Eq. (5.160) based on PE-BNMF. Fig. 5.6 displays the histogram of the selected model order or number of bases $\widehat{K}$ from 1200 samples of speech and music source signals where the music signals of piano and violin are investigated. It is found that the estimated model order is changed sample by sample. Most of model orders are selected and distributed in the range between 20 and 50. On average, the estimated number ($\widehat{K}$) for the speech source signal is smaller than those for the piano and violin source signals. This adaptive scheme is required to estimate the model complexity for BNMF source separation.

Fig. 5.7 further investigates the effect of the number of bases K of NMF on the SDR of the separated speech and music signals. Separation results for the mixed speech/piano signals (upper) and the mixed speech/violin signals (lower) are shown. PG-BNMF and PE-BNMF with adaptive number of bases $\widehat{K}$ are compared. The SDRs of NMF with fixed number of bases ($K = 20, 30, 40, 50$ and 60) are shown in the first five pairs of bars while those of PG-BNMF with fixed $K = 40$ (for mixed speech/piano signals) or 50 (for mixed speech/violin signals) are shown in the sixth pair of bars, those of PG-BNMF with adaptive K are shown in the seventh pair of bars, those of PE-BNMF with fixed $K = 50$ (for mixed speech/piano signals) or 40 (for mixed speech/violin signals) are shown in the eighth pair of bars and those of PE-BNMF with adaptive K are shown in the last pair of bars. SDR is here averaged over 600 mixed signals. The averaged SDRs are changed for NMF with different K. Using the fixed K, PG-BNMF and PE-BNMF perform better than NMF. Bayesian NMF is more robust

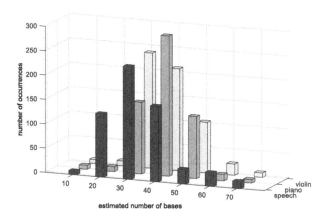

FIGURE 5.6

Histogram of the estimated number of bases $(\widehat{K})$ using PE-BNMF for source signals of speech, piano and violin.

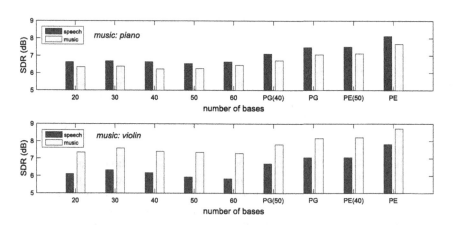

FIGURE 5.7

Comparison of the averaged SDR using NMF with a fixed number of bases (1–5 pairs of bars), PG-BNMF with the best fixed number of bases (sixth pair of bars), PG-BNMF with adaptive number of bases (seventh pair of bars), PE-BNMF with the best fixed number of bases (eighth pair of bars) and PE-BNMF with adaptive number of bases (ninth pair of bars).

than ML-based NMF over different mixed signals from different speakers, genders and instruments. When comparing the PG-BNMF and PE-BNMF with fixed and adaptive model complexity, it is found that adaptive K increases the SDR in the presence of different music sources. Adaptive basis selection is crucial for speech and music separation. PE-BNMF performs better than PG-BNMF in terms of SDR.

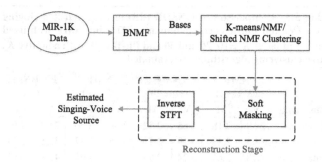

FIGURE 5.8

Implementation procedure for unsupervised singing voice separation.

5.3.5 EVALUATION FOR UNSUPERVISED SEPARATION

MIR-1K dataset (https://sites.google.com/site/unvoicedsoundseparation/mir-1k) (Hsu and Jang, 2010, Yang et al., 2014b) is introduced for evaluation of unsupervised singing voice separation. This dataset was composed of 1000 song clips extracted from 110 Chinese karaoke pop songs performed by 8 female and 11 male amateurs. Each clip had a duration ranging from 4 to 13 seconds. Three different sets of monaural mixtures at SMRs of 5, 0 and −5 dB were collected. A general unsupervised source separation was addressed in Section 2.2.1 and depicted in Fig. 2.6. Here, Fig. 5.8 illustrates the implementation procedure of singing voice separation. When implementing the BNMF, baseline NMF was adopted as the initialization and 200 iterations were run to find the posterior means of basis and weight parameters. The total number of bases K and the grouping of these bases into vocal source and music source are both unknown and should be learned from the test singing voice signal in an unsupervised way. No training data was used. The model complexity in BNMF based on K was determined in accordance with Eq. (5.160) using the test signal while the grouping of bases for two sources was simply performed via the clustering algorithm by using the estimated basis vectors in $\mathbf{B}$ or, equivalently, from the estimated hyperparameters $\boldsymbol{\lambda}^b = \{\lambda^b_{mk}\}$. Three clustering algorithms were investigated. First, the K-means clustering algorithm was run over the basis vectors $\mathbf{B}$ in the MFCC domain (Spiertz and Gnann, 2009). Second, the NMF clustering was applied according to the ML criterion. The Mel-scaled basis matrix was factorized into two matrices $\widetilde{\mathbf{B}}$ of size M-by-2 and $\widetilde{\mathbf{W}}$ of size 2-by-K. The third clustering method was based on the shifted NMF as detailed in Jaiswal et al. (2011). The same soft mask and reconstruction stages were performed. The soft mask scheme based on Wiener gain was applied to smooth the separation of $\mathbf{B}$ into two groups of basis vectors. The separated singing voice and music accompaniment signals in time domain were obtained by the overlap-and-add method and the inverse STFT. For system evaluation, the normalized SDR (NSDR) and the global NSDR (GNSDR) are measured by

$$\mathrm{NSDR}(\widehat{\mathbf{V}}, \mathbf{V}, \mathbf{X}) = \mathrm{SDR}(\widehat{\mathbf{V}}, \mathbf{V}) - \mathrm{SDR}(\mathbf{X}, \mathbf{V}), \qquad (5.161)$$

$$\mathrm{GNSDR}(\widehat{\mathbf{V}}, \mathbf{V}, \mathbf{X}) = \frac{\sum_{n=1}^{\widetilde{N}} l_n \mathrm{NSDR}(\widehat{\mathbf{V}}_n, \mathbf{V}_n, \mathbf{X}_n)}{\sum_{n=1}^{\widetilde{N}} l_n} \qquad (5.162)$$

Table 5.5 Comparison of GNSDR (dB) of the separated singing voices (V) and music accompaniments (M) using NMF with fixed number of bases $K = 10, 20$ and 30 and PE-BNMF with adaptive $\widehat{K}$. Three clustering algorithms are evaluated

Clustering	NMF ($K = 10$)		NMF ($K = 20$)		NMF ($K = 30$)		PE-BNMF ($\widehat{K}$)	
	V	M	V	M	V	M	V	M
K-means	2.59	2.18	2.65	2.52	2.69	2.50	2.92	2.92
NMF	3.05	2.48	3.19	2.75	3.15	2.77	3.25	3.21
Shifted NMF	3.18	2.61	3.24	2.93	3.26	2.91	4.01	3.95

where $\widehat{V}, V, X$ denote the estimated singing voice, the original clean singing voice, and the mixture signal, respectively, $\widetilde{N}$ is the total number of the clips, and l_n is the length of the nth clip. NSDR is used to measure the improvement of SDR between the estimated singing voice $\widehat{V}$ and the mixture signal X. GNSDR is used to calculate the overall separation performance by taking the weighted mean of the NSDRs.

Table 5.5 reports a comparison of GNSDR of the separated singing voices and music accompaniments under SMR of 0 dB. The baseline NMF with fixed number of bases $K = 10, 20$ and 30 and the PE-BNMF with adaptive number of bases $\widehat{K}$ are compared. K-means, NMF and shifted NMF clustering are examined. NMF has the highest GNSDR for the case of $K = 20$, which is very close to the mean value (18.4) of the estimated number of bases $\widehat{K}$ by using PE-BNMF. NMF clustering performs better than the K-means clustering. However, the highest GNSDR is achieved by using the shifted NMF clustering. In this evaluation, the best performance of the separated singing voices and the separated music accompaniments is obtained by combining the shifted NMF clustering into PE-BNMF. PE-BNMF helps in the uncertainty modeling of NMF parameters and the adaptive number of NMF bases.

In addition, GNSDR of PE-NMF are compared with those of five competitive methods, including Hsu (Hsu and Jang, 2010), Huang (Huang et al., 2012), Yang (Yang, 2012), Rafii1 (Rafii and Pardo, 2011) and Rafii2 (Rafii and Pardo, 2013). GNSDR was averaged over 1000 test mixed signals. The PE-BNMFs with three clustering algorithms (BNMF1, BNMF2 and BNMF3) are compared. Fig. 5.9 depicts the results of GNSDR of the separated singing voices for different methods under SMRs of -5, 0 and 5 dB. When using K-means clustering in the MFCC domain for PE-NMF, the resulting BNMF1 outperforms the other five methods under SMRs of -5 and 0 dB while the results using Huang (Huang et al., 2012) and Yang (Yang, 2012) are better than that using BNMF1 under SMR of 5 dB. Nevertheless, using the PE-BNMF with NMF clustering (BNMF2), the overall evaluation consistently achieves 0.33–0.57 dB absolute improvement in GNSDR when compared with BNMF1 including the SMR condition at 5 dB. Among these methods, the highest GNSDR is obtained by using PE-BNMF with the shifted NMF clustering (BNMF3). The absolute improvement in GNSDR using BNMF3 is 0.44–0.81 dB over different SMRs when compared with the second best method (BNMF2). In the following experiments, we fix the shifted NMF clustering in implementation of different Bayesian NMFs.

Table 5.6 further compares system performance of baseline NMF, GE-BNMF, PG-BNMF and PE-BNMF for two monaural source separation tasks. The NMF with $K = 30$ was run. NMF converges much faster than GE-BNMF, PG-BNMF and PE-BNMF. In task 1 on speech and music separation, we report the SDR which was averaged over 1200 test samples with speech/piano and speech/violin mixed

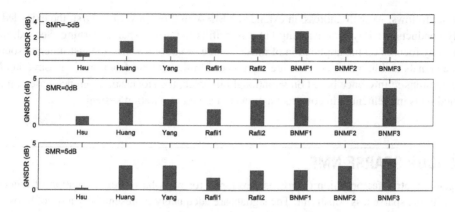

FIGURE 5.9

Comparison of GNSDR of the separated singing voices at different SMRs using PE-BNMFs with K-means clustering (denoted by BNMF1), NMF clustering (denoted by BNMF2) and shifted NMF clustering (denoted by BNMF3). Five competitive methods are included for comparison.

Table 5.6 Comparison of NMF and different BNMFs in terms of SDR and GNSDR for two separation tasks. Standard deviation is given in the parentheses

	Separation task 1 SDR (dB)	Separation task 2 GNSDR (dB)
NMF	6.75 (0.12)	3.03 (0.09)
GE-BNMF	7.31 (0.13)	3.51 (0.12)
PG-BNMF	7.43 (0.09)	3.59 (0.10)
PE-BNMF	8.08 (0.10)	3.74 (0.09)

signals. In task 2 on singing voice separation, we report the GNSDR of the separated singing voices which was averaged over 1000 test samples under different SMR conditions. NMF obtains the lowest SDR or GNSDR in two tasks. Three BNMFs performs better than NMF. In terms of SDR in task 1 and GNSDR in task 2, the best performance among different BNMFs is achieved by using PE-BNMF. Basically, the PE-BNMF solution $\Phi = \{\lambda_{mk}^b, \lambda_{kn}^w\}$ in Eqs. (5.158) and (5.159) based on VB-EM algorithm is computationally demanding when compared with the baseline NMF solution $\Theta = \{B_{mk}, W_{kn}\}$ which was addressed in Section 3.7.1. The key reason is due to the multiple latent variables $\{\mathbf{Z}, \mathbf{B}, \mathbf{W}\}$ when performing the variational inference for PE-BNMF. The computation overhead due to Bayesian learning is mainly caused by the updating of variational parameters in Eqs. (5.147)–(5.150) with calculation of sufficient statistics in Eqs. (5.136)–(5.140) which are scaled up by the number of training samples N, the number of bases K, the dimension of observations M. The additional memory is required by the variational parameters

$$\{P_{mkn}, \alpha_{mk}^b, \beta_{mk}^b, \alpha_{kn}^w, \beta_{kn}^w\}$$

of latent variables $\{\mathbf{Z}, \mathbf{B}, \mathbf{W}\}$.

In what follows, we will address an extended NMF with three novelties. First, the extended NMF not only conducts the Bayesian learning but also fulfills the group sparse learning. Second, this advanced study focuses on the application of music source separation where the instrumental sources of rhythmic signals and harmonic signals are separated. Third, different from previous Bayesian NMFs using deterministic inference based on variational inference, the stochastic inference based on a sampling method is implemented to construct a monaural source separation system.

5.4 GROUP SPARSE NMF

Nonnegative matrix factorization (NMF) is developed for parts-based representation of nonnegative signals with the sparseness constraint. The signals are adequately represented by a set of basis vectors and the corresponding weight parameters. Typically, controlling the degree of sparseness and characterizing the uncertainty of model parameters are two critical issues for model regularization using NMF. This section presents the *Bayesian group sparse (BGS) learning* for NMF and applies it for single-channel music source separation. Grouping of basis vectors in NMF provides different subspaces for signal reconstruction. Considering the characteristics of music signals, this method reconstructs the rhythmic or repetitive signal from a *common subspace* spanned by the shared bases for the whole signal, and simultaneously decodes the harmonic or residual signal from an *individual subspace* consisting of separate bases for different signal segments. Importantly, a group basis representation is introduced to deal with this learning representation in Section 5.4.1. In particular, Section 5.4.2 addresses a specialized BGS model where the *Laplacian scale mixture* distribution is implemented for sparse coding given a sparseness control parameter. The relevance of basis vectors for reconstructing two groups of music signals is automatically determined. Sparser priors identify fewer but more relevant bases. After describing the construction of group sparse NMF, a Markov chain Monte Carlo procedure is presented to infer two sets of model parameters and hyperparameters through a Gibbs sampling procedure based on the conditional posterior distributions as illustrated in Section 5.4.3. Another type of full Bayesian theory is implemented. In Section 5.4.4, a number of experiments are conducted to show how group basis representation and sparse signal reconstruction are developed for monaural audio source separation in presence of music environments with rhythmic and harmonic source signals.

5.4.1 GROUP BASIS REPRESENTATION

In the literature, the group basis representation has been developed for monaural source separation in presence of rhythmic signal as well as harmonic signal. In Yoo et al. (2010), the nonnegative matrix partial co-factorization (NMPCF) was proposed for rhythmic source separation. Given the magnitude spectrogram as input data matrix $\mathbf{X}$, NMPCF decomposes the music signal into a drum or rhythmic part and a residual or harmonic part

$$\mathbf{X} \approx \mathbf{B}_r \mathbf{W}_r + \mathbf{B}_h \mathbf{W}_h \tag{5.163}$$

with the factorized matrices, including the basis and weight matrices for rhythmic source $\{\mathbf{B}_r, \mathbf{W}_r\}$ and for harmonic source $\{\mathbf{B}_h, \mathbf{W}_h\}$. The prior knowledge from drum-only signal $\mathbf{Y} \approx \mathbf{B}_r \mathbf{W}_r$ given the same rhythmic bases $\mathbf{B}_r$ is incorporated in joint minimization of two Euclidean error functions

$$\|\mathbf{X} - \mathbf{B}_r\mathbf{W}_r - \mathbf{B}_h\mathbf{W}_h\|^2 + \lambda \|\mathbf{Y} - \mathbf{B}_r\mathbf{W}_r\|^2 \qquad (5.164)$$

where λ is a regularization parameter for a tradeoff between the first and the second reconstruction errors due to $\mathbf{X}$ and $\mathbf{Y}$, respectively. In Kim et al. (2011), the mixed signals were divided into L segments. Each segment $\mathbf{X}^{(l)}$ is decomposed into common and individual parts which reflect the rhythmic and harmonic sources, respectively. Notably, the segment index l is comparable with the minibatch index n in online learning as described in Section 3.5. We use this index l to avoid conflict with the time index n of spectral observation data $\mathbf{X}^{(l)} \in \mathbb{R}_+^{M \times N}$ in the NMF model. The common bases $\mathbf{B}_r$ are shared for different segments due to high temporal repeatability in rhythmic sources. The individual bases $\mathbf{B}_h^{(l)}$ are separate for individual segment l due to the changing frequency and low temporal repeatability. The resulting objective function consists of a weighted Euclidean error function and the regularization terms due to bases $\mathbf{B}_r$ and $\mathbf{B}_h^{(l)}$, which are expressed by

$$\sum_{l=1}^{L} w^{(l)} \|\mathbf{X}^{(l)} - \mathbf{B}_r\mathbf{W}_r^{(l)} - \mathbf{B}_h^{(l)}\mathbf{W}_h^{(l)}\|^2$$
$$+ \lambda L \|\mathbf{B}_r\|^2 + \lambda \sum_{l=1}^{L} \|\mathbf{B}_h^{(l)}\|^2 \qquad (5.165)$$

where $\{w^{(l)}, \mathbf{W}_r^{(l)}, \mathbf{W}_h^{(l)}\}$ denotes the segment-dependent weights and weight matrices for common basis and individual basis, respectively. This is an NMPCF for L segments.

The signal reconstruction methods in Eqs. (5.164)–(5.165) correspond to the group basis representation where two groups of bases $\mathbf{B}_r$ and $\mathbf{B}_h^{(l)}$ are applied. Separation of single-channel mixed signal into two source signals is implemented. In Lee and Choi (2009), the group-based NMF (GNMF) was developed by conducting group analysis and constructing two groups of bases. The intrasubject variations for a subject in different trials and the intersubject variations for different subjects could be compensated. Given L subjects or segments, the lth segment is generated by

$$\mathbf{X}^{(l)} \approx \mathbf{B}_r^{(l)}\mathbf{W}_r^{(l)} + \mathbf{B}_h^{(l)}\mathbf{W}_h^{(l)} \qquad (5.166)$$

where $\mathbf{B}_r^{(l)}$ denotes the common bases which capture the intra- and intersubject variations and $\mathbf{B}_h^{(l)}$ denotes the individual bases which reflect the residual information. In general, different common bases $\mathbf{B}_r^{(l)}$ should be close together since these bases represent the shared information in mixed signal. In contrast, individual bases $\mathbf{B}_h^{(l)}$ characterize individual features which should be discriminated and mutually far apart. The object function of GNMF is formed by

$$\sum_{l=1}^{L} \|\mathbf{X}^{(l)} - \mathbf{B}_r^{(l)}\mathbf{W}_r^{(l)} - \mathbf{B}_h^{(l)}\mathbf{W}_h^{(l)}\|^2 + \lambda_b \sum_{l=1}^{L} \|\mathbf{B}_r^{(l)}\|^2 + \lambda_b \sum_{l=1}^{L} \|\mathbf{B}_h^{(l)}\|^2$$
$$+ \lambda_{b_r} \sum_{l=1}^{L}\sum_{j=1}^{L} \|\mathbf{B}_r^{(l)} - \mathbf{B}_r^{(j)}\|^2 - \lambda_{b_h} \sum_{l=1}^{L}\sum_{j=1}^{L} \|\mathbf{B}_h^{(l)} - \mathbf{B}_h^{(j)}\|^2 \qquad (5.167)$$

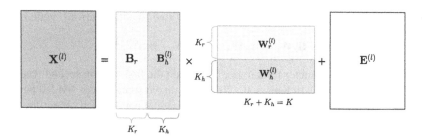

FIGURE 5.10

Illustration for group basis representation.

where the second and third terms are seen as the ℓ_2- or weight-decay regularization functions, the fourth term enforces the distance between different common bases to be small, and the fifth term enforces the distance between different individual bases to be large. Regularization parameters

$$\{\lambda_b, \lambda_{b_r}, \lambda_{b_h}\}$$

are used. The NMPCFs in Yoo et al. (2010), Kim et al. (2011) and GNMF in Lee and Choi (2009) did not consider sparsity in group basis representation.

More generally, a group sparse coding algorithm (Bengio et al., 2009) was proposed for basis representation of group instances $\{\mathbf{x}_n\}_{n \in \mathcal{G}}$ where the objective function is defined by

$$\sum_{n \in \mathcal{G}} \left\| \mathbf{x}_n - \sum_{k=1}^{|D|} w_k^n \mathbf{b}_k \right\|^2 + \lambda \sum_{k=1}^{|D|} \|\mathbf{w}_k\|. \tag{5.168}$$

All the instances within a group $\mathcal{G}$ share the same dictionary D with weight vectors $\{\mathbf{w}_k\}_{k=1}^{|D|}$. The weight parameters $\{w_k^n\}$ are estimated from different group instances $n \in \mathcal{G}$ using different bases $k \in D$. Here, ℓ_1-, or Lasso, regularization term is incorporated to carry out group sparse coding. The group sparsity was further extended to structural sparsity for dictionary learning and basis representation. Basically, all the above-mentioned methods (Yoo et al., 2010, Kim et al., 2011, Lee and Choi, 2009) did not apply a probabilistic framework.

We are presenting the BGS learning for an NMF model where the group basis representation is performed. To implement this representation, the magnitude spectrogram $\mathbf{X} = \{\mathbf{X}^{(l)}\}$ of a mixed audio signal is calculated and then chopped it into L segments. The audio signal is assumed to be mixed from two kinds of source signals. One is rhythmic, or repetitive, source signal and the other is harmonic, or residual, source signal. As illustrated in Fig. 5.10, BGS-NMF aims to decompose a nonnegative matrix $\mathbf{X}^{(l)} \in \mathbb{R}_+^{M \times N}$ of the lth segment into a product of two nonnegative matrices $\mathbf{B}^{(l)}$ and $\mathbf{W}^{(l)}$. A linear decomposition model is constructed in a form of

$$\mathbf{X}^{(l)} = \mathbf{B}_r \mathbf{W}_r^{(l)} + \mathbf{B}_h^{(l)} \mathbf{W}_h^{(l)} + \mathbf{E}^{(l)} \tag{5.169}$$

where $\mathbf{B}_r \in \mathbb{R}_+^{M \times K_r}$ denotes the shared basis matrix for all segments $\{\mathbf{X}^{(l)}\}_{l=1}^L$, while $\mathbf{B}_h^{(l)} \in \mathbb{R}_+^{M \times K_h}$ and $\mathbf{E}^{(l)}$ denote the individual and noise matrices for a given segment l. Typically, common bases capture the repetitive patterns, which continuously happen in different segments of a whole signal. Individual bases are used to compensate the residual information that common bases cannot handle. Basically, common bases and individual bases are applied to recover the rhythmic and harmonic signals from a mixed audio signal, respectively. Such a signal recovery problem could be interpreted from the perspective of a subspace approach. An observed signal is demixed into one signal from the *principal subspace* spanned by common bases and the other signal from the *minor subspace* spanned by individual bases (Chien and Ting, 2008). Moreover, the sparseness constraint is imposed on two groups of reconstruction weights $\mathbf{W}_r^{(l)} \in \mathbb{R}_+^{K_r \times N}$ and $\mathbf{W}_h^{(l)} \in \mathbb{R}_+^{K_h \times N}$. It is assumed that the reconstruction weights of rhythmic sources $\mathbf{W}_r^{(l)}$ and harmonic sources $\mathbf{W}_h^{(l)}$ are independent, but the dependencies between reconstruction weights within each group are allowed. Assuming that the nth noise vector $\mathbf{E}_{:n}^{(l)}$ is Gaussian with zero mean and $M \times M$ diagonal covariance matrix

$$\mathbf{\Sigma}^{(l)} = \mathrm{diag}\{[\mathbf{\Sigma}^{(l)}]_{mm}\}, \tag{5.170}$$

which is shared for all samples n within a segment l, the likelihood function of an audio signal segment $\mathbf{X}^{(l)} = \{X_{mn}^{(l)}\}$ is expressed by

$$
\begin{aligned}
&p(\mathbf{X}^{(l)}|\mathbf{\Theta}^{(l)}) \\
&= \prod_{m=1}^M \prod_{n=1}^N \mathcal{N}\left(X_{mn}^{(l)}|[\mathbf{B}_r \mathbf{W}_r^{(l)} + \mathbf{B}_h^{(l)} \mathbf{W}_h^{(l)}]_{mn}, [\mathbf{\Sigma}^{(l)}]_{mm}\right).
\end{aligned}
\tag{5.171}
$$

This NMF is constructed with parameters

$$\mathbf{\Theta}^{(l)} = \{\mathbf{B}_r, \mathbf{B}_h^{(l)}, \mathbf{W}_r^{(l)}, \mathbf{W}_h^{(l)}, \mathbf{\Sigma}^{(l)}\} \tag{5.172}$$

for each segment l where only the rhythmic bases $\mathbf{B}_r$ are shared for different segments.

5.4.2 BAYESIAN GROUP SPARSE LEARNING

To enhance the robustness of NMF-based source separation in heterogeneous environments, we present an advanced NMF, called the BGS-NMF (Chien and Hsieh, 2013a), which simultaneously considers the *uncertainty* of model parameters and controls the *sparsity* of weight parameters in group basis representation for instrumental music separation. The common bases $\mathbf{B}_r$ are constructed to represent the characteristics of repetitive patterns for different data segments while the individual bases $\mathbf{B}_h^{(l)}$ are estimated to reflect the unique information in each segment l. Sparsity control is enforced in the corresponding reconstruction weights $\mathbf{W}_r^{(l)}$ and $\mathbf{W}_h^{(l)}$ so that relevant bases are retrieved for group basis representation. To implement BGS-NMF, it is necessary to define the prior densities for model parameters $\mathbf{\Theta}^{(l)}$.

Sparse Prior

In accordance with Moussaoui et al. (2006), the nonnegative basis parameters are assumed to be Gamma distributed by

$$p(\mathbf{B}_r) = \prod_{m=1}^{M} \prod_{k=1}^{K_r} \mathrm{Gam}\left([\mathbf{B}_r]_{mk}|\alpha_{rk}, \beta_{rk}\right), \tag{5.173}$$

$$p(\mathbf{B}_h^{(l)}) = \prod_{m=1}^{M} \prod_{k=1}^{K_h} \mathrm{Gam}\left([\mathbf{B}_h^{(l)}]_{mk}|\alpha_{hk}^{(l)}, \beta_{hk}^{(l)}\right) \tag{5.174}$$

where

$$\boldsymbol{\Phi}_b^{(l)} = \{\alpha_{rk}, \beta_{rk}, \alpha_{hk}^{(l)}, \beta_{hk}^{(l)}\} \tag{5.175}$$

denote the hyperparameters of Gamma distributions and $\{K_r, K_h\}$ denote the numbers of common bases and individual bases, respectively. Gamma distribution is fitted to characterize *nonnegative data*. Its two parameters $\{\alpha, \beta\}$ are adjusted to shape the distribution. Here, all entries in matrices $\mathbf{B}_r$ and $\mathbf{B}_h^{(l)}$ are assumed to be independent.

Importantly, the sparsity of reconstruction weights is controlled by using a prior density based on the *Laplacian scale mixture* (LSM) distribution (Garrigues and Olshausen, 2010). The LSM of a reconstruction weight of common basis is constructed by

$$[\mathbf{W}_r^{(l)}]_{kn} = (\lambda_{rk}^{(l)})^{-1} u_{rk}^{(l)} \tag{5.176}$$

where $u_{rk}^{(l)}$ is a Laplace distribution

$$p(u_{rk}^{(l)}) = \frac{1}{2} \exp\left\{-|u_{rk}^{(l)}|\right\} \tag{5.177}$$

with scale 1 and $\lambda_{rk}^{(l)}$ is an inverse scale parameter. The nonnegative parameter $[\mathbf{W}_r^{(l)}]_{kn}$ becomes a Laplace distribution

$$p([\mathbf{W}_r^{(l)}]_{kn}|\lambda_{rk}^{(l)}) = \frac{\lambda_{rk}^{(l)}}{2} \exp\left\{-\lambda_{rk}^{(l)}[\mathbf{W}_r^{(l)}]_{kn}\right\}, \tag{5.178}$$

which is controlled by a nonnegative continuous mixture parameter $\lambda_{rk}^{(l)} \geq 0$. Considering a Gamma distribution

$$p(\lambda_{rk}^{(l)}) = \mathrm{Gam}(\lambda_{rk}^{(l)}|\gamma_{rk}^{(l)}, \delta_{rk}^{(l)}) \tag{5.179}$$

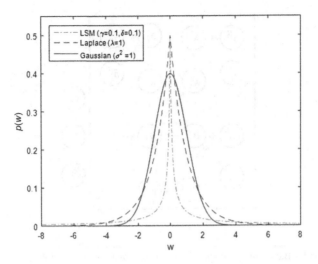

FIGURE 5.11

Comparison of Gaussian, Laplace and LSM distributions centered at zero.

for the mixture parameter, the marginal distribution of a nonnegative reconstruction weight can be calculated by

$$
\begin{aligned}
p([\mathbf{W}_r^{(l)}]_{kn}) &= \int_0^{\infty} p([\mathbf{W}_r^{(l)}]_{kn}|\lambda_{rk}^{(l)}) p(\lambda_{rk}^{(l)}) d\lambda_{rk}^{(l)} \\
&= \frac{\gamma_{rk}^{(l)}(\delta_{rk}^{(l)})^{\gamma_{rk}^{(l)}}}{2\left(\delta_{rk}^{(l)} + [\mathbf{W}_r^{(l)}]_{kn}\right)^{\gamma_{rk}^{(l)}+1}}.
\end{aligned}
\tag{5.180}
$$

This LSM distribution is obtained by adopting the property that Gamma distribution is the *conjugate prior* for Laplace distribution. LSM distribution was measured to be sparser than Laplace distribution by approximately a factor of 2 (Garrigues and Olshausen, 2010). Fig. 5.11 shows that LSM distribution is the sharpest distribution when compared with Gaussian and Laplace distributions. In addition, a truncated LSM prior for nonnegative parameter $[\mathbf{W}_r^{(l)}]_{jk} \in \mathbb{R}_+$ is adopted, namely, the distribution of negative parameter is forced to be zero. The sparse prior for reconstruction weight for individual basis $[\mathbf{W}_h^{(l)}]_{kn}$ is also expressed by LSM distribution with hyperparameter $\{\gamma_{hk}^{(l)}, \delta_{hk}^{(l)}\}$. The hyperparameters of BGS-NMF are formed by

$$
\Phi^{(l)} = \{\Phi_b^{(l)}, \Phi_w^{(l)} = \{\gamma_{rk}^{(l)}, \delta_{rk}^{(l)}, \gamma_{hk}^{(l)}, \delta_{hk}^{(l)}\}\}.
\tag{5.181}
$$

Fig. 5.12 displays a graphical representation for the construction of BGS-NMF with different parameters $\Theta^{(l)}$ and hyperparameters $\Phi^{(l)}$.

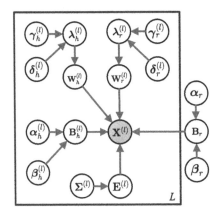

FIGURE 5.12

Graphical representation for Bayesian group sparse nonnegative matrix factorization.

Learning Objective

By combining the likelihood function in Eq. (5.171) and the prior densities in Eqs. (5.173)–(5.178), the new learning objective $-\log p(\mathbf{B}_r, \mathbf{B}_h^{(l)}, \mathbf{W}_r^{(l)}, \mathbf{W}_h^{(l)} | \mathbf{X})$ for source separation, which is seen as the negative logarithm of the posterior distribution, is calculated by

$$
\sum_{l=1}^{L} \sum_{m=1}^{M} \sum_{n=1}^{N} (X_{mn}^{(l)} - [\mathbf{B}_r \mathbf{W}_r^{(l)}]_{mn} - [\mathbf{B}_h^{(l)} \mathbf{W}_h^{(l)}]_{mn})^2
$$

$$
+ \lambda_b L \sum_{m=1}^{M} \sum_{k=1}^{K_r} ((1 - \alpha_{rk}) \log[\mathbf{B}_r]_{mk} + \beta_{rk} [\mathbf{B}_r]_{mk})
$$

$$
+ \lambda_b \sum_{l=1}^{L} \sum_{m=1}^{M} \sum_{k=1}^{K_h} ((1 - \alpha_{hk}^{(l)}) \log[\mathbf{B}_h^{(l)}]_{mk} + \beta_{hk}^{(l)} [\mathbf{B}_h^{(l)}]_{mk})
$$

$$
+ \lambda_{w_r} \sum_{l=1}^{L} \sum_{k=1}^{K_r} \sum_{n=1}^{N} [\mathbf{W}_r^{(l)}]_{kn} + \lambda_{w_h} \sum_{l=1}^{L} \sum_{k=1}^{K_h} \sum_{n=1}^{N} [\mathbf{W}_h^{(l)}]_{kn}
$$

(5.182)

where $\{\lambda_b, \lambda_{w_r}, \lambda_{w_h}\}$ denote the regularization parameters for two groups of bases and reconstruction weights. Some BGS-NMF parameters or hyperparameters have been absorbed in these regularization parameters. Comparing with the learning objectives Eq. (5.165) for NMPCF and Eq. (5.167) for GNMF, the optimization of Eq. (5.182) for BGS-NMF will lead to two groups of signals, which are reconstructed from the sparse common bases $\mathbf{B}_r$ and sparse individual bases $\mathbf{B}_h^{(l)}$. The regularization terms due to two Gamma bases are naturally considered. Typically, BGS-NMF is implemented by a general learning objective. After some manipulation and simplification, BGS-NMF can be realized into the solutions to NMF (Lee and Seung, 1999), NMPCF (Yoo et al., 2010), GNMF (Lee and Choi, 2009), the probabilistic matrix factorization in Salakhutdinov and Mnih (2008) and the Bayesian NMF

(BNMF) in (Moussaoui et al., 2006). Notably, the objective function in Eq. (5.182) only considers the BGS-NMF based on Laplace prior. Nevertheless, in what follows, the model inference procedure for BGS-NMF with LSM prior is addressed.

5.4.3 MARKOV CHAIN MONTE CARLO SAMPLING

Full Bayesian solution to BGS-NMF model based on the posterior distribution of parameters and hyperparameters $p(\boldsymbol{\Theta}, \boldsymbol{\Phi}|\mathbf{X})$ is not analytically tractable. A stochastic optimization scheme is adopted according to the Markov chain Monte Carlo (MCMC) sampling algorithm which is realized by iteratively generating samples of parameters $\boldsymbol{\Theta}$ and hyperparameters $\boldsymbol{\Phi}$ using the corresponding posterior distributions. This algorithm converges by using those samples as addressed in Section 3.7.4. MCMC sampling is based on a stationary ergodic Markov chain whose samples asymptotically follow the posterior distribution $p(\boldsymbol{\Theta}, \boldsymbol{\Phi}|\mathbf{X})$. Estimates of parameters $\boldsymbol{\Theta}$ and hyperparameters $\boldsymbol{\Phi}$ are then computed via Monte Carlo integrations on the simulated Markov chains. For simplicity, the segment index l is neglected in derivation of the MCMC algorithm for BGS-NMF. At each new iteration $\tau + 1$, the BGS-NMF parameters $\boldsymbol{\Theta}^{(\tau+1)}$ and hyperparameters $\boldsymbol{\Phi}^{(\tau+1)}$ are sequentially sampled by

$$\{\mathbf{B}_r, \mathbf{W}_r, \mathbf{B}_h, \mathbf{W}_h, \boldsymbol{\Sigma}, \alpha_r, \beta_r, \alpha_h, \beta_h, \lambda_r, \lambda_h, \gamma_r, \delta_r, \gamma_h, \delta_h\}$$

according to their corresponding conditional posterior distributions. Nonnegativity constraint is imposed on $\{\mathbf{B}_r, \mathbf{B}_h, \mathbf{W}_r, \mathbf{W}_h\}$ during the sampling procedure. In this subsection, we describe the calculation of conditional posterior distributions for individual eight parameters and hyperparameters, which is described in the following:

1. Sampling of $[\mathbf{B}_r]_{mk}$: First of all, the common basis parameter $[\mathbf{B}_r^{(\tau+1)}]_{mk}$ is sampled by the conditional posterior distribution

$$
\begin{aligned}
&p([\mathbf{B}_r]_{mk}|\mathbf{X}_{m:}^{\top}, \boldsymbol{\Theta}_{B_{rmk}}^{(\tau)}, \boldsymbol{\Phi}_{B_{rmk}}^{(\tau)}) \\
&\propto p(\mathbf{X}_{m:}^{\top}|\boldsymbol{\Theta}_{B_{rmk}}^{(\tau)}) p([\mathbf{B}_r]_{mk}|\boldsymbol{\Phi}_{B_{rmk}}^{(\tau)})
\end{aligned}
\tag{5.183}
$$

where

$$
\boldsymbol{\Theta}_{B_{rmk}}^{(\tau)} = \{[\mathbf{B}_r^{(\tau+1)}]_{m(1:k-1)}, [\mathbf{B}_r^{(\tau)}]_{m(k+1:K_r)}, \mathbf{W}_r^{(\tau)}, \mathbf{B}_h^{(\tau)}, \mathbf{W}_h^{(\tau)}, \boldsymbol{\Sigma}^{(\tau)}\}
\tag{5.184}
$$

and

$$
\boldsymbol{\Phi}_{B_{rmk}}^{(\tau)} = \{\alpha_{rk}^{(\tau)}, \beta_{rk}^{(\tau)}\}.
\tag{5.185}
$$

Here, $\mathbf{X}_{m:}$ denotes the mth row vector of $\mathbf{X}$. Notably, for each sampling, we use the preceding bases $[\mathbf{B}_r^{(\tau+1)}]_{m(1:k-1)}$ at the new iteration $\tau + 1$ and subsequent bases $[\mathbf{B}_r]_{m(k+1:K_r)}^{(\tau)}$ at the current iteration τ. Basically, the conditional posterior distribution in Eq. (5.183) is derived by combining the likelihood function $p(\mathbf{X}_{m:}^{\top}|\boldsymbol{\Theta}_{B_{rmk}}^{(\tau)})$ and Gamma prior $p([\mathbf{B}_r]_{mk}|\boldsymbol{\Phi}_{B_{rmk}}^{(\tau)})$ of Eq. (5.173). For this arrangement, we first express the exponent of the likelihood function

$$p(\mathbf{X}_{m:}^{\top}|[\mathbf{B}_r^{(\tau+1)}]_{m(1:k-1)}, [\mathbf{B}_r^{(\tau)}]_{m(k+1:K_r)}, \mathbf{W}_r^{(\tau)}, \mathbf{B}_h^{(\tau)}, \mathbf{W}_h^{(\tau)}, \boldsymbol{\Sigma}^{(\tau)})$$

as

$$
-\frac{1}{2[\mathbf{\Sigma}^{(\tau)}]_{mm}} \sum_{n=1}^{N} \left[X_{mn} - \sum_{j=1}^{k-1} [\mathbf{B}_r^{(\tau+1)}]_{mj} [\mathbf{W}_r^{(\tau)}]_{mk} - [\mathbf{B}_r]_{mk} [\mathbf{W}_r^{(\tau)}]_{kn} \right.
$$

$$
\left. - \sum_{j=k+1}^{K_r} [\mathbf{B}_r^{(\tau)}]_{mj} [\mathbf{W}_r^{(\tau)}]_{mk} - \sum_{j=1}^{K_h} [\mathbf{B}_h^{(\tau)}]_{mj} [\mathbf{W}_h^{(\tau)}]_{mk} \right]^2, \tag{5.186}
$$

which can be manipulated as a quadratic function of parameter $[\mathbf{B}_r]_{mk}$ and leads to a Gaussian likelihood of $[\mathbf{B}_r]_{mk}$

$$
p(\mathbf{X}_{m:}^{\top} | \mathbf{\Theta}_{B_{rmk}}^{(\tau)}) \propto \exp \left\{ -\frac{([\mathbf{B}_r]_{mk} - \mu_{B_{rmk}}^{\text{likel}})^2}{2[\sigma_{B_{rmk}}^{\text{likel}}]^2} \right\} \tag{5.187}
$$

where

$$
\mu_{B_{rmk}}^{\text{likel}} = [\sigma_{B_{rmk}}^{\text{likel}}]^{-2} \sum_{n=1}^{N} [\mathbf{W}_r^{(\tau)}]_{kn} (\varepsilon_{mn})_{\setminus k}, \tag{5.188}
$$

$$
(\varepsilon_{mn})_{\setminus k} = X_{mn} - \left(\sum_{j=1}^{k-1} [\mathbf{B}_r^{(\tau+1)}]_{mj} [\mathbf{W}_r^{(\tau)}]_{mk} \right.
$$

$$
\left. + \sum_{j=k+1}^{K_r} [\mathbf{B}_r^{(\tau)}]_{mj} [\mathbf{W}_r^{(\tau)}]_{mk} \right) - \sum_{j=1}^{K_h} [\mathbf{B}_h^{(\tau)}]_{mj} [\mathbf{W}_h^{(\tau)}]_{mk} \tag{5.189}
$$

$$
[\sigma_{B_{rmk}}^{\text{likel}}]^2 = [\mathbf{\Sigma}^{(\tau)}]_{mm} \left(\sum_{n=1}^{N} [\mathbf{W}_r^{(\tau)}]_{kn} \right)^{-1}. \tag{5.190}
$$

The conditional posterior distribution

$$
p([\mathbf{B}_r]_{mk} | \mathbf{X}_{m:}^{\top}, \mathbf{\Theta}_{B_{rmk}}^{(\tau)}, \mathbf{\Phi}_{B_{rmk}}^{(\tau)})
$$

turns out to be

$$
[\mathbf{B}_r]_{mk}^{\alpha_{rk}^{(\tau)}-1} \exp \left\{ -\frac{[\mathbf{B}_r]_{mk}^2 - 2(\mu_{B_{rmk}}^{\text{likel}} - \beta_{rk}^{(\tau)}[\sigma_{B_{rmk}}^{\text{likel}}]^2)[\mathbf{B}_r]_{mk} + [\mu_{B_{rmk}}^{\text{likel}}]^2)}{2[\sigma_{B_{rmk}}^{\text{likel}}]^2} \right\}
$$

$$
\times \mathbb{I}_{[0,+\infty[}([\mathbf{B}_r]_{mk}), \tag{5.191}
$$

which is proportional to

$$
[\mathbf{B}_r]_{mk}^{\alpha_{rk}^{(\tau)}-1} \exp \left\{ -\frac{([\mathbf{B}_r]_{mk} - \mu_{B_{rmk}}^{\text{post}})^2}{2[\sigma_{B_{rmk}}^{\text{post}}]^2} \right\} \mathbb{I}_{[0,+\infty[}([\mathbf{B}_r]_{mk}) \tag{5.192}
$$

where

$$\mu_{B_{rmk}}^{\text{post}} = \mu_{B_{rmk}}^{\text{likel}} - \beta_{rk}^{(\tau)} [\sigma_{B_{rmk}}^{\text{likel}}]^2, \tag{5.193}$$

$$[\sigma_{B_{rmk}}^{\text{post}}]^2 = [\sigma_{B_{rmk}}^{\text{likel}}]^2 \tag{5.194}$$

and $\mathbb{I}_{[0,+\infty[}(z)$ denotes an indicator function, which has value either 1 if $z \in [0, +\infty[$, or 0 otherwise. In Eq. (5.192), the posterior distribution for negative $[\mathbf{B}_r]_{mk}$ is forced to be zero.

However, Eq. (5.192) is not an usual distribution, therefore its sampling requires the use of a *rejection sampling* method, such as the Metropolis–Hastings algorithm as addressed in Section 3.7.4. Using this algorithm, an *instrumental distribution* $q([\mathbf{B}_r]_{mk})$ is chosen to fit the best target distribution in Eq. (5.192) so that high rejection condition is avoided or, equivalently, rapid convergence toward the true parameter could be achieved. In case of rejection, the previous parameter sample is used, namely

$$[\mathbf{B}_r^{(\tau+1)}]_{mk} \leftarrow [\mathbf{B}_r^{(\tau)}]_{mk}. \tag{5.195}$$

Generally, the shape of the target distribution is characterized by its mode and width. The instrumental distribution is constructed as a truncated Gaussian distribution, which is calculated by

$$q([\mathbf{B}_r]_{mk}) = \mathcal{N}_+ \left([\mathbf{B}_r]_{mk} \middle| \mu_{B_{rmk}}^{\text{inst}}, [\sigma_{B_{rmk}}^{\text{inst}}]^2 \right) \tag{5.196}$$

where the mode $\mu_{B_{rmk}}^{\text{inst}}$ is obtained by finding the roots of a quadratic equation of $[\mathbf{B}_r]_{mk}$, which appears in the exponent of the posterior distribution in Eq. (5.192). The subscript $+$ in Eq. (5.196) reflects the truncation for nonnegative $[\mathbf{B}_r]_{mk}$. For mode finding, we first take the logarithm of Eq. (5.192) and solve the corresponding quadratic equation of $[\mathbf{B}_r]_{mk}$

$$\frac{\partial}{\partial [\mathbf{B}_r]_{mk}} \left\{ (\alpha_{rk}^{(\tau)} - 1) \log[\mathbf{B}_r]_{mk} - \frac{([\mathbf{B}_r]_{mk} - \mu_{B_{rmk}}^{\text{post}})^2}{2[\sigma_{B_{rmk}}^{\text{post}}]^2} \right\} = 0 \tag{5.197}$$

or, equivalently, find the root of

$$[\mathbf{B}_r]_{mk}^2 - \mu_{B_{rmk}}^{\text{post}} [\mathbf{B}_r]_{mk} - (\alpha_{rk}^{(\tau)} - 1)[\sigma_{B_{rmk}}^{\text{post}}]^2 = 0. \tag{5.198}$$

By defining

$$\triangle \triangleq (\mu_{B_{rmk}}^{\text{post}})^2 + 4(\alpha_{rk}^{(\tau)} - 1)[\sigma_{B_{rmk}}^{\text{post}}]^2, \tag{5.199}$$

the mode is then determined by

$$\mu_{B_{rmk}}^{\text{inst}} = \begin{cases} 0, & \text{if } \triangle < 0, \\ \max\{\frac{1}{2}(\mu_{B_{rmk}}^{\text{post}} + \sqrt{\triangle}), 0\}, & \text{else.} \end{cases} \tag{5.200}$$

In case of complex- or negative-valued root, the mode is forced by $\mu_{B_{rmk}}^{\text{inst}} = 0$. The width of instrumental distribution is controlled by

$$[\sigma_{B_{rmk}}^{\text{inst}}]^2 = [\sigma_{B_{rmk}}^{\text{post}}]^2. \tag{5.201}$$

2. Sampling of $[\mathbf{W}_r]_{kn}$: The sampling of reconstruction weight of common basis $[\mathbf{W}_r^{(\tau+1)}]_{kn}$ depends on the conditional posterior distribution

$$
\begin{aligned}
&p([\mathbf{W}_r]_{kn}|\mathbf{X}_{:n}, \mathbf{\Theta}_{W_{rkn}}^{(\tau)}, \mathbf{\Phi}_{W_{rkn}}^{(\tau)}) \\
&\propto p(\mathbf{X}_{:n}|[\mathbf{W}_r]_{kn}, \mathbf{\Theta}_{W_{rkn}}^{(\tau)})p([\mathbf{W}_r]_{kn}|\mathbf{\Phi}_{W_{rkn}}^{(\tau)}).
\end{aligned}
\tag{5.202}
$$

Here,

$$
\mathbf{\Theta}_{W_{rkn}}^{(\tau)} = \{\mathbf{B}_r^{(\tau+1)}, [\mathbf{W}_r^{(\tau+1)}]_{(1:k-1)n}, [\mathbf{W}_r^{(\tau)}]_{(k+1:K_r)k}, \mathbf{B}_h^{(\tau)}, \mathbf{W}_h^{(\tau)}, \mathbf{\Sigma}^{(\tau)}\}
\tag{5.203}
$$

and

$$
\mathbf{\Phi}_{W_{rkn}}^{(\tau)} = \lambda_{rk}^{(\tau)}.
\tag{5.204}
$$

In the above formula $\mathbf{X}_{:n}$ denotes the nth column of $\mathbf{X}$. Again, the preceding weights $[\mathbf{W}_r^{(\tau+1)}]_{(1:k-1)n}$ at the new iteration $\tau + 1$ and subsequent weights $[\mathbf{W}_r^{(\tau)}]_{(k+1:K_r)n}$ at the current iteration τ are used. The likelihood function is rewritten as a Gaussian distribution of $[\mathbf{W}_r]_{kn}$ given by

$$
p(\mathbf{X}_{:n}|[\mathbf{W}_r]_{kn}, \mathbf{\Theta}_{W_{rkn}}^{(\tau)}) \propto \exp\left\{-\frac{([\mathbf{W}_r]_{kn} - \mu_{W_{rkn}}^{\text{likel}})^2}{2[\sigma_{W_{rkn}}^{\text{likel}}]^2}\right\}.
\tag{5.205}
$$

The Gaussian parameters are obtained by

$$
\mu_{W_{rkn}}^{\text{likel}} = [\sigma_{W_{rkn}}^{\text{likel}}]^{-2}\sum_{i=1}^{N}[\mathbf{\Sigma}^{(\tau)}]_{mm}^{-1}[\mathbf{B}_r^{(\tau+1)}]_{mk}(\varepsilon_{mn})_{\backslash k},
\tag{5.206}
$$

$$
\begin{aligned}
(\varepsilon_{mn})_{\backslash k} = X_{mn} - &\left(\sum_{j=1}^{k-1}[\mathbf{B}_r^{(\tau+1)}]_{mj}[\mathbf{W}_r^{(\tau+1)}]_{mk}\right. \\
&\left. + \sum_{j=k+1}^{K_r}[\mathbf{B}_r^{(\tau+1)}]_{mj}[\mathbf{W}_r^{(\tau)}]_{mk}\right) - \sum_{j=1}^{K_h}[\mathbf{B}_h^{(\tau)}]_{mj}[\mathbf{W}_h^{(\tau)}]_{mk}
\end{aligned}
\tag{5.207}
$$

$$
[\sigma_{W_{rkn}}^{\text{likel}}]^2 = \left(\sum_{i=1}^{N}[\mathbf{\Sigma}^{(\tau)}]_{mm}^{-1}([\mathbf{B}_r^{(\tau+1)}]_{mk})^2\right)^{-1}.
\tag{5.208}
$$

Given the Gaussian likelihood and Laplace prior, the conditional posterior distribution is calculated by

$$
\lambda_{rk}^{(\tau)}\exp\left\{-\frac{([\mathbf{W}_r]_{kn} - \mu_{W_{rkn}}^{\text{post}})^2}{2[\sigma_{W_{rkn}}^{\text{post}}]^2}\right\}\mathbb{I}_{[0,+\infty[}([\mathbf{W}_r]_{kn})
\tag{5.209}
$$

where

$$\mu_{W_{rkn}}^{\text{post}} = \mu_{W_{rkn}}^{\text{likel}} - \lambda_{rk}^{(\tau)} [\sigma_{W_{rkn}}^{\text{likel}}]^2, \tag{5.210}$$

$$[\sigma_{W_{rkn}}^{\text{post}}]^2 = [\sigma_{W_{rkn}}^{\text{likel}}]^2. \tag{5.211}$$

Notably, the hyperparameters $\{\gamma_{rk}^{(\tau+1)}, \delta_{rk}^{(\tau+1)}\}$ in LSM prior are also sampled and used to sample LSM parameter $\lambda_{rk}^{(\tau+1)}$ based on a Gamma distribution. Here, Metropolis–Hastings algorithm is applied again. The best instrumental distribution $q([\mathbf{W}_r]_{kn})$ is selected to fit Eq. (5.209). This distribution is derived as a truncated Gaussian distribution

$$\mathcal{N}_+([\mathbf{W}_r]_{kn}|\mu_{W_{rkn}}^{\text{inst}}, [\sigma_{W_{rkn}}^{\text{inst}}]^2)$$

where the mode $\mu_{W_{rkn}}^{\text{inst}}$ is derived by finding the root of a quadratic equation of $[\mathbf{W}_r]_{kn}$ and the width is obtained by

$$[\sigma_{W_{rkn}}^{\text{inst}}]^2 = [\sigma_{W_{rkn}}^{\text{post}}]^2. \tag{5.212}$$

In addition, the conditional posterior distributions for sampling the individual basis parameter $[\mathbf{B}_h^{(\tau+1)}]_{mk}$ and its reconstruction weight $[\mathbf{W}_h^{(\tau+1)}]_{kn}$ are similar to those for sampling $[\mathbf{B}_r^{(\tau+1)}]_{mk}$ and $[\mathbf{W}_r^{(\tau+1)}]_{kn}$, respectively. We don't address these two distributions.

3. Sampling of $[\mathbf{\Sigma}]_{mm}^{-1}$: The sampling of the inverse of noise variance $([\mathbf{\Sigma}]_{mm}^{(\tau+1)})^{-1}$ is performed according to the conditional posterior distribution

$$p\left([\mathbf{\Sigma}]_{mm}^{-1}\middle|\mathbf{X}_{m:}^{\mathsf{T}}, \mathbf{\Theta}_{\Sigma_{mm}}^{(\tau)}, \mathbf{\Phi}_{\Sigma_{mm}}^{(\tau)}\right)$$

$$\propto p\left(\mathbf{X}_{m:}^{\mathsf{T}}\middle|[\mathbf{\Sigma}]_{mm}^{-1}, \mathbf{\Theta}_{\Sigma_{mm}}^{(\tau)}\right) p\left([\mathbf{\Sigma}]_{mm}^{-1}\middle|\mathbf{\Phi}_{\Sigma_{mm}}^{(\tau)}\right) \tag{5.213}$$

where

$$\mathbf{\Theta}_{\Sigma_{mm}}^{(\tau)} = \{\mathbf{B}_r^{(\tau+1)}, \mathbf{W}_r^{(\tau+1)}, \mathbf{B}_h^{(\tau+1)}, \mathbf{W}_h^{(\tau+1)}\} \tag{5.214}$$

and a Gamma prior

$$p([\mathbf{\Sigma}]_{mm}^{-1}|\mathbf{\Phi}_{\Sigma_{mm}}^{(\tau)}) = \text{Gam}([\mathbf{\Sigma}]_{mm}^{-1}|\alpha_{\Sigma_{mm}}, \beta_{\Sigma_{mm}}) \tag{5.215}$$

with hyperparameters

$$\mathbf{\Phi}_{\Sigma_{mm}}^{(\tau)} = \{\alpha_{\Sigma_{mm}}, \beta_{\Sigma_{mm}}\} \tag{5.216}$$

is assumed. The resulting posterior distribution can be derived as a new Gamma distribution with updated hyperparameters

$$\alpha_{\Sigma_{mm}}^{\text{post}} = \frac{N}{2} + \alpha_{\Sigma_{mm}}, \tag{5.217}$$

$$\beta_{\Sigma_{mm}}^{\text{post}} = \frac{1}{2} \sum_{n=1}^{N} \left(X_{mn} - \sum_{j=1}^{K_r} [\mathbf{B}_r^{(\tau+1)}]_{mj} [\mathbf{W}_r^{(\tau+1)}]_{mk} \right.$$
$$\left. - \sum_{j=1}^{K_h} [\mathbf{B}_h^{(\tau+1)}]_{mj} [\mathbf{W}_h^{(\tau+1)}]_{mk} \right)^2 + \beta_{\Sigma_{mm}}. \qquad (5.218)$$

4. Sampling of α_{rk}: The hyperparameter $\alpha_{rk}^{(\tau+1)}$ is sampled according to a conditional posterior distribution which is obtained by combining the likelihood function of $[\mathbf{B}_r]_{mk}$ and the exponential prior density of α_{rk} with parameter $\lambda_{\alpha_{rk}}$. The resulting distribution is written as

$$p(\alpha_{rk} \mid [\mathbf{B}_r^{(\tau+1)}]_{mk}, \beta_{rk}^{(\tau)})$$
$$\propto \left(\frac{1}{\Gamma(\alpha_{rk})} \exp\{\lambda_{\alpha_{rk}}^{\text{post}} \alpha_{rk}\} \right)^{K_r} \mathbb{I}_{[0,+\infty[}(\alpha_{rk}) \qquad (5.219)$$

where

$$\lambda_{\alpha_{rk}}^{\text{post}} = \log \beta_{rk}^{(\tau)} + \frac{1}{K_r} \sum_{j=1}^{K_r} \log [\mathbf{B}_r^{(\tau+1)}]_{mj} - \frac{1}{K_r} \lambda_{\alpha_{rk}}. \qquad (5.220)$$

This distribution does not belong to a known family, so the Metropolis–Hastings algorithm is applied. An instrumental distribution $q(\alpha_{rk})$ is obtained by fitting the term within the parentheses of Eq. (5.219) through a Gamma distribution as detailed in Moussaoui et al. (2006).

5. Sampling of β_{rk}: The hyperparameter $\beta_{rk}^{(\tau+1)}$ is sampled according to a conditional posterior distribution, which is obtained by combining the likelihood function of $[\mathbf{B}_r]_{mk}$ and the Gamma prior density of β_{rk} with parameters $\{\alpha_{\beta_{rk}}, \beta_{\beta_{rk}}\}$, i.e.,

$$p(\beta_{rk} \mid [\mathbf{B}_r^{(\tau+1)}]_{mk}, \alpha_{rk}^{(\tau+1)}) \propto (\beta_{rk})^{K_r \alpha_{rk}^{(\tau+1)}}$$
$$\times \exp \left\{ -\beta_{rk} \sum_{j=1}^{K_r} [\mathbf{B}_r^{(\tau+1)}]_{mj} \right\} \text{Gam}(\beta_{rk} \mid \alpha_{\beta_{rk}}, \beta_{\beta_{rk}}) \qquad (5.221)$$
$$\propto \text{Gam}(\beta_{rk} \mid \alpha_{\beta_{rk}}^{\text{post}}, \beta_{\beta_{rk}}^{\text{post}}),$$

which is arranged as a new Gamma distribution where

$$\alpha_{\beta_{rk}}^{\text{post}} = 1 + K_r \alpha_{rk}^{(\tau+1)} + \alpha_{\beta_{rk}}, \qquad (5.222)$$

$$\beta_{\beta_{rk}}^{\text{post}} = \sum_{j=1}^{K_r} [\mathbf{B}_r^{(\tau+1)}]_{mj} + \beta_{\beta_{rk}}. \qquad (5.223)$$

Here, we don't describe the sampling of $\alpha_{hj}^{(\tau+1)}$ and $\beta_{hj}^{(\tau+1)}$ since the conditional posterior distributions for sampling these two hyperparameters are similar to those for sampling $\alpha_{rk}^{(\tau+1)}$ and $\beta_{rk}^{(\tau+1)}$.

6. Sampling of λ_{rk} or λ_{hj}: For sampling of scaling parameter $\lambda_{rk}^{(\tau+1)}$, the conditional posterior distribution is obtained by

$$p(\lambda_{rk}|[\mathbf{W}_r^{(\tau+1)}]_{k(n=1:N)}, \gamma_{rk}^{(\tau)}, \delta_{rk}^{(\tau)})$$

$$\propto \prod_{n=1}^{N} p([\mathbf{W}_r^{(\tau+1)}]_{kn}|\lambda_{rk}) p(\lambda_{rk}|\gamma_{rk}^{(\tau)}, \delta_{rk}^{(\tau)}) \tag{5.224}$$

$$\propto (\lambda_{rk})^{N\gamma_{rk}^{(\tau)}} \exp\left\{-N\lambda_{rk}\left(\delta_{rk}^{(\tau)} + \sum_{n=1}^{N}[\mathbf{W}_r^{(\tau+1)}]_{kn}\right)\right\}.$$

7. Sampling of γ_{rk}: The sampling of LSM parameter $\gamma_{rk}^{(\tau+1)}$ is performed by using the conditional posterior distribution, which is derived by combining the likelihood function of λ_{rk} and the exponential prior density of γ_{rk} with parameter $\lambda_{\gamma_{rk}}$. The resulting distribution is expressed as

$$p(\gamma_{rk}|\lambda_{rk}^{(\tau+1)}, \delta_{rk}^{(\tau)})$$

$$\propto \frac{1}{\Gamma(\gamma_{rk})} \exp\{\lambda_{\gamma_{rk}}^{\text{post}}\gamma_{rk}\}\mathbb{I}_{[0,+\infty[}(\gamma_{rk}) \tag{5.225}$$

where

$$\lambda_{\gamma_{rk}}^{\text{post}} = \log\delta_{rk}^{(\tau)} + \frac{\gamma_{rk}-1}{\gamma_{rk}}\log\lambda_{rk}^{(\tau+1)} - \lambda_{\gamma_{rk}}. \tag{5.226}$$

Again, we need to find an instrumental distribution $q(\gamma_{rk})$, which optimally fits the conditional posterior distribution $p(\gamma_{rk}|\lambda_{rk}^{(\tau+1)}, \delta_{rk}^{(\tau)})$. An approximate Gamma distribution is found accordingly. The Metropolis–Hastings algorithm is then applied.

8. Sampling of δ_{rk}: The sampling of the other LSM parameter $\delta_{rk}^{(\tau+1)}$ is performed by using the conditional posterior distribution, which is derived from the likelihood function of λ_{rk} and the Gamma prior density of δ_{rk} with parameters $\{\alpha_{\delta_{rk}}, \beta_{\delta_{rk}}\}$, namely,

$$p(\delta_{rk}|\lambda_{rk}^{(\tau+1)}, \gamma_{rk}^{(\tau+1)}) \propto (\delta_{rk})^{\gamma_{rk}^{(\tau+1)}}$$

$$\times \exp\{-\delta_{rk}\lambda_{rk}^{(\tau+1)}\}\text{Gam}(\delta_{rk}|\alpha_{\delta_{rk}}, \beta_{\delta_{rk}}) \tag{5.227}$$

$$\propto \text{Gam}(\delta_{rk}|\alpha_{\delta_{rk}}^{\text{post}}, \beta_{\delta_{rk}}^{\text{post}}),$$

which is also arranged as a new Gamma distribution where

$$\alpha_{\delta_{rk}}^{\text{post}} = K_r\gamma_{rk}^{(\tau+1)} + \alpha_{\delta_{rk}}, \tag{5.228}$$

$$\beta_{\delta_{rk}}^{\text{post}} = \lambda_{rk}^{(\tau+1)} + \beta_{\delta_{rk}}. \tag{5.229}$$

Similarly, the conditional posterior distributions for sampling $\gamma_{hj}^{(\tau+1)}$ and $\delta_{hj}^{(\tau+1)}$ could be formulated by referring to those for sampling $\gamma_{rk}^{(\tau+1)}$ and $\delta_{rk}^{(\tau+1)}$, respectively.

In the implementation, the MCMC sampling procedure is performed with τ_{max} iterations. However, the first τ_{min} iterations are not stable. These burn-in samples are ignored. The marginal posterior estimates of common basis $[\widehat{\mathbf{B}}_r]_{mk}$, individual basis $[\widehat{\mathbf{B}}_h]_{mk}$ and their reconstruction weights $[\widehat{\mathbf{W}}_r]_{kn}$ and $[\widehat{\mathbf{W}}_h]_{kn}$ are calculated by finding the following sample means, e.g.,

$$[\widehat{\mathbf{B}}_r]_{mk} = \frac{1}{\tau_{max} - \tau_{min}} \sum_{\tau=\tau_{min}+1}^{\tau_{max}} [\mathbf{B}_r]_{mk}^{(\tau)}. \tag{5.230}$$

With these posterior estimates, the rhythmic and harmonic sources are calculated by $\widehat{\mathbf{B}}_r \widehat{\mathbf{W}}_r$ and $\widehat{\mathbf{B}}_h \widehat{\mathbf{W}}_h$, respectively. The MCMC inference procedure is completed for BGS-NMF.

5.4.4 SYSTEM EVALUATION

Bayesian sparse learning is performed to conduct probabilistic reconstruction based on the relevant group bases for monaural music separation in the presence of rhythmic and harmonic sources. In system evaluation, six rhythmic and six harmonic signals, sampled from http://www.free-scores.com/index_uk.php3 and http://www.freesound.org/, respectively, were collected to generate six mixed music signals "music 1": bass+piano, "music 2": drum+guitar, "music 3": drum+violin, "music 4": cymbal+organ, "music 5": drum+saxophone, and "music 6": cymbal+singing, which contained combinations of different rhythmic and harmonic source signals. Three different drum signals and two different cymbal signals were included. These music clips were mixed by using different mixing matrices to simulate single-channel mixed signals. Each audio signal was 21 second long. In the implementation, the magnitude of STFT of audio signal was extracted every 1024 samples with 512 samples in the frame overlapping. Each mixed signal was equally chopped into L segments for music source separation. Each segment had a length of 3 seconds. Sufficient rhythmic signal existed within a segment. The numbers of common bases and individual bases were empirically set as $K_r = 15$ and $K_h = 10$, respectively. The common bases were sufficiently allocated so as to capture the shared base information from different segments. The initial common bases $\mathbf{B}_r^{(0)}$ and individual bases $\mathbf{B}_h^{(0)}$ were estimated by applying k-means clustering using the automatically-detected rhythmic and harmonic segments, respectively. The MCMC sampling iterations were specified by $\tau_{max} = 1000$ and $\tau_{min} = 200$. The separation performance was evaluated according to the signal-to-interference ratio (SIR) in decibels as shown in Eq. (2.10) which was measured by the Euclidean distance between original signal $\{\mathbf{X}_{:n}^{(l)}\}$ and reconstructed signal $\{\widehat{\mathbf{X}}_{:n}^{(l)}\}$ for different samples n in different segments l in the rhythmic and harmonic sources. In the initialization, two short segments with only rhythmic or harmonic signal were automatically detected and used to find rhythmic parameters $\{\mathbf{B}_r^{(0)}, \mathbf{W}_r^{(0)}\}$ and harmonic parameters $\{\mathbf{B}_h^{(0)}, \mathbf{W}_h^{(0)}\}$, respectively. This prior information was used to implement NMF methods for instrumental music separation. Unsupervised learning was performed without additional source signals as training data.

For comparison, the baseline NMF (Hoyer, 2004), Bayesian NMF (BNMF) (Moussaoui et al., 2006), group-based NMF (GNMF) (Lee and Choi, 2009) (or NMPCF (Kim et al., 2011)) and BGS-NMF (Chien and Hsieh, 2013a) were carried out under consistent experimental conditions. BGS-NMF was implemented by using LSM priors for $\mathbf{W}_r^{(l)}$ and $\mathbf{W}_h^{(l)}$. The case of using Laplace priors was realized by ignoring the sampling of LSM parameters $\{\gamma_{r_k}, \delta_{r_k}, \gamma_{h_k}, \delta_{h_k}\}$. The BGS-NMFs with Laplace and LSM distributions are denoted by BGS-NMF-LP and BGS-NMF-LSM, respectively. Using GNMF, the regularization parameters in Eq. (5.167) were empirically determined as

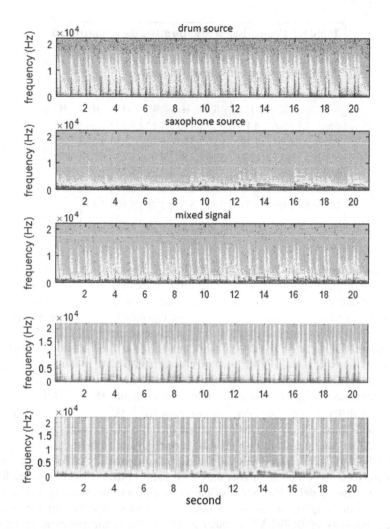

FIGURE 5.13

Spectrograms of "music 5" containing the drum signal (first panel), the saxophone signal (second panel), the mixed signal (third panel), the demixed drum signal (fourth panel) and the demixed saxophone signal (fifth panel).

$\{\lambda_b = 0.35, \lambda_{b_r} = 0.2, \lambda_{b_h} = 0.2\}$. Using BGS-NMF, the groups of common bases $\mathbf{B}_r$ and individual bases $\mathbf{B}_h$ were estimated to capture between-segment repetitive patterns and within-segment residual information, respectively. The relevant bases were detected via sparse priors. Parameters and hyperparameters by using different frames were automatically sampled from six music signals. The averaged values of regularization parameters were measured by $\{\lambda_b = 0.41, \lambda_{b_r} = 0.31, \lambda_{b_h} = 0.26\}$, which reflected different physical meaning when compared with those empirically chosen for GNMF.

The first three panels of Fig. 5.13 show the spectrograms of a drum, saxophone and the resulting mixed signals in "music 5". The corresponding spectrograms of the demixed drum and saxophone sig-

Table 5.7 Comparison of SIRs (in dB) of the reconstructed rhythmic signal (denoted by R) and harmonic signal (denoted by H) based on NMF, BNMF, GNMF and BGS-NMF. Six mixed music signals are investigated

	NMF		BNMF		GNMF		BGS-NMF	
	R	**H**	**R**	**H**	**R**	**H**	**R**	**H**
Music 1	6.47	4.17	6.33	4.29	9.19	6.10	9.86	8.63
Music 2	6.30	1.10	8.08	5.18	8.22	3.03	8.55	7.45
Music 3	3.89	−1.11	5.16	3.80	6.01	3.22	8.63	8.79
Music 4	2.66	6.03	3.28	6.28	3.59	8.36	8.20	9.78
Music 5	1.85	3.71	3.03	2.55	3.97	6.44	8.35	8.50
Music 6	1.06	6.37	3.34	5.56	2.78	7.10	5.19	7.23
Avg	**3.71**	**3.38**	**4.87**	**4.61**	**5.63**	**5.71**	**8.13**	**8.40**

nals are displayed in the last two panels. BGS-NMF-LSM algorithm is applied to find the demixed signals. In addition, a quantitative comparison over different NMFs is conducted by measuring SIRs of reconstructed rhythmic signal and reconstructed harmonic signal. Table 5.7 shows the experimental results on six mixed music signals. The averaged SIRs are reported in the last row. Comparing NMF and BNMF, we find that BNMF obtains higher SIRs on the reconstructed signals. BNMF is more robust to different combination of rhythmic signals and harmonic signals. The variation of SIRs using NMF is relatively high. On the other hand, GNMF performs better than BNMF in terms of averaged SIR of the reconstructed signals. The key difference between BNMF and GNMF is the reconstruction of rhythmic signal. BNMF estimates the rhythmic bases for individual segments while GNMF calculates the shared rhythmic bases for different segments. Prior information $\{\mathbf{B}_r^{(0)}, \mathbf{W}_r^{(0)}, \mathbf{B}_h^{(0)}, \mathbf{W}_h^{(0)}\}$ is applied for these methods. The importance of basis grouping in signal reconstruction based on NMF is illustrated. In particular, BGS-NMF (here means BGS-NMF-LSM) performs much better than other NMF methods. Using BGS-NMF-LSM, the averaged SIRs of reconstructed rhythmic and harmonic signals are measured by 8.13 and 8.40 dB, which are higher than 7.91 and 8.11 dB obtained by using BGS-NMF-LP. Reconstruction weights modeled by LSM distributions are better than those by Laplace distributions. Sparser reconstruction weights identify fewer but more relevant basis vectors for signal separation. Basically, the superiority of BGS-NMF to other NMFs is due to the Bayesian probabilistic modeling, group basis representation and sparse reconstruction weight. In addition to Bayesian learning and group sparsity, in what follows, we are introducing how deep learning and discriminative learning are developed to improve NMF-based monaural source separation.

5.5 DEEP LAYERED NMF

Deep learning has emerged as a powerful learning machine in recent years. This learning paradigm has achieved state-of-the-art performance in many tasks, including source separation. Basically, deep learning adopts a hierarchical architecture to grasp high-level information in data for various classification and regression tasks. There are attempts to incorporate a hierarchical architecture into the standard NMF, which was only built as a single-layer architecture with flat-layered basis vectors. In Roux et al.

(2015), a deep NMF was proposed to untie or unfold the multiplicative updating rule of NMF so as to facilitate a new form of backpropagation procedure to estimate deep layered parameters. Exploring deep structure and conducting deep factorization provide a sophisticated approach to monaural source separation in the presence of complicated mixing condition. Deep unfolding inference was also explored for other learning model (Chien and Lee, 2018). In addition to deep factorization, as addressed in Section 3.6, discriminative learning is crucial to improve separation performance of the demixed signals. For instance, in a supervised speech separation system, NMF is first applied to learn the nonnegative bases for each individual speaker. After learning, the speaker-dependent bases are used to separate the mixed spectrogram. However, the bases of each speaker are learned without considering the interfering effect from the other speakers during training. The mismatch accordingly happens between training and test conditions so that the separation performance is constrained. To address these issues, this section presents the deep layered NMF (LNMF) (Hsu et al., 2015), which can be trained either in a generative manner or in a discriminative way. First of all, Section 5.5.1 surveys a couple of works, which focus on deep structural factorization. Then, Section 5.5.2 develops the layered NMF, which is trained in a generative manner. Hierarchical bases are learned independently for each source. A tree of bases is estimated for multilevel or multiaspect decomposition of a complex mixed signal. In Section 5.5.3, such a generative LNMF is further extended to the discriminative layered NMF (DLNMF) (Hsu et al., 2016), which is constructed according to discriminative learning method. The bases are adaptive by enforcing an optimal recovery of the mixed from separated spectra and minimizing the reconstruction error between the separated and original spectra. In Section 5.5.4, a set of experiments are evaluated and analyzed to illustrate the merit of discriminative layered representation for monaural speech separation and decomposition.

5.5.1 DEEP STRUCTURAL FACTORIZATION

Structural information has been exploited for NMF-based data representation. For instance, the multilayer NMF was proposed as a sequential factorization for ill-conditioned and badly scaled data, which cause local minimum in an optimization procedure (Cichocki and Zdunek, 2006). For the case of an L-layer LNMF model, the hierarchical factorization is performed for signal reconstruction by

$$\mathbf{X} \approx \mathbf{X}_L = \mathbf{B}_1 \mathbf{B}_2 \cdots \mathbf{B}_L \mathbf{W}_L. \tag{5.231}$$

In Eq. (5.231), a set of hierarchical bases $\{\mathbf{B}_1, \ldots, \mathbf{B}_L\}$ are used for sequential factorization, which is recursively performed from the first layer using

$$\mathbf{X} \approx \mathbf{B}_1 \mathbf{W}_1 \tag{5.232}$$

with

$$\mathbf{W}_1 \approx \mathbf{B}_2 \mathbf{W}_2 \tag{5.233}$$

towards the Lth layer. The parameters of this multilayer NMF are estimated by sequentially solving the following two optimization steps (Cichocki and Zdunek, 2006):

$$\mathbf{W}_l^{(\tau+1)} = \arg\min_{\mathbf{W} \geq 0} \mathcal{D}_\alpha(\mathbf{X}_l \| \mathbf{B}_l^{(\tau)} \mathbf{W}), \qquad \forall l = 1, \ldots, L, \tag{5.234}$$

$$B_l^{(\tau+1)} = \arg\min_{B \geq 0} \mathcal{D}_\alpha(X_l \| BW_l^{(\tau+1)}), \qquad \forall\, l = 1, \ldots, L \tag{5.235}$$

where the α divergence $\mathcal{D}_\alpha(\cdot)$, as addressed in Section 3.2.1, was adopted in the optimization problem. Here, the entries of the basis matrix $B = [B_{mk}]$ are normalized at each layer l in each updating iteration $\tau + 1$ based on

$$B_l^{(\tau+1)} \leftarrow \left[\frac{B_{mk}}{\sum_{j=1}^{K} B_{mj}} \right]_l^{(\tau+1)}. \tag{5.236}$$

For each layer l, the first step of the optimization procedure is to estimate the weight matrix $W_l^{\tau+1}$ at iteration $\tau + 1$ by minimizing the α divergence between the converged signal X_l and the reconstructed signal $B_l^{(\tau)}W$ using the current basis matrix $B_l^{(\tau)}$ at iteration τ. Given the updated weight matrix $W_l^{(\tau+1)}$, the second step is to update the basis matrix $B_l^{(\tau+1)}$ by minimizing again the same α divergence. With sufficient number of iterations, this method was shown effective in terms of SIR when demixing four nonnegative badly scaled sources with a Hilbert mixing matrix (Cichocki and Zdunek, 2006). Nevertheless, in a regular source separation situation, the reconstruction error due to sequential factorization will increase with increasing number of layers due to the lack of an error correction procedure through different layers. The deep unfolding algorithm (Roux et al., 2015, Chien and Lee, 2018), which performs error backpropagation in a deep inference procedure, is helpful to deal with the learning problem in the multilayer NMF. However, deep unfolding only conducts deep inference and keeps the original model to be single-layered.

In Ding et al. (2010), the semi-NMF relaxes the nonnegativity constraint of NMF and allows the data matrix X and basis matrix B to have mixed signs

$$X^\pm \approx B^\pm W^+ \tag{5.237}$$

where $\pm$ denotes the sign relaxation. Using this method, only addition combination of bases is performed for signal reconstruction due to the nonnegative weight matrix W. However, data is often rather complex and has a collection of distinct, related, but unknown attributes. Single-level clustering could not physically interpret such a real-world situation. In Trigeorgis et al. (2014), a deep semi-NMF was proposed to deal with this situation by recursively factorizing X into $L + 1$ factors as

$$X^\pm \approx B_1^\pm B_2^\pm \cdots B_L^\pm W_L^+. \tag{5.238}$$

Similar to Cichocki and Zdunek (2006), a deep semi-NMF is recursively obtained through the following hierarchical factorization procedure:

$$\begin{aligned}
X^\pm &\approx B_1^\pm W_1^+, \\
X^\pm &\approx B_1^\pm B_2^\pm W_2^+, \\
X^\pm &\approx B_1^\pm B_2^\pm B_3^\pm W_3^+, \\
&\;\vdots \\
X^\pm &\approx B_1^\pm B_2^\pm \cdots B_L^\pm W_L^+.
\end{aligned} \tag{5.239}$$

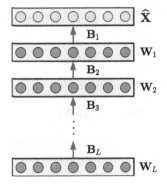

FIGURE 5.14

Illustration for layered nonnegative matrix factorization.

This method was proposed to learn a hierarchical representation of features from an image dataset for attribute-based clustering. For the application of audio source separation, the nonnegative constraint is imposed not only for the weight matrix **W** but also for the basis matrix **B**. Such a constraint for the basis matrix is required to reconstruct a nonnegative data matrix **X**, which reflects the essence of time–frequency observations in audio signals.

Basically, the above-mentioned works (Cichocki and Zdunek, 2006, Trigeorgis et al., 2014, Lyu and Wang, 2013) aim to generalize the standard NMF to decompose a nonnegative matrix into a product of a series of matrices. In what follows, we address a couple of layered NMFs for monaural speech separation, which are constructed either as a generative or as a discriminative model.

5.5.2 LAYERED NMF

The layered NMF (LNMF) with L layers is first introduced. Typically, a general form of factor analysis (FA) can be written as (Oh and Seung, 1998)

$$\mathbf{X} \approx \widehat{\mathbf{X}} = g\left(\mathbf{B}_1 \cdot g\left(\mathbf{B}_2 \cdot g\left(\cdots g\left(\mathbf{B}_L \mathbf{W}_L\right)\right)\right)\right) \tag{5.240}$$

where $g(\cdot)$ is a nonlinear function. The standard NMF is thought as a simple and shallow version of FA with the nonnegative constraints and the assumption of linear function $g(\cdot)$. Using LNMF, the signal reconstruction is performed by (Hsu et al., 2015)

$$\mathbf{X} \approx \widehat{\mathbf{X}} = \left(\prod_{l=1}^{L} \mathbf{B}_l\right) \mathbf{W}_L. \tag{5.241}$$

This approximation is performed based on a hierarchical architecture with L layers as shown in Fig. 5.14. In the training procedure, the factors $\{\mathbf{B}_l\}_{l=1}^{L}$ and $\mathbf{W}_L$ are initialized layer by layer. In the first step, the standard NMF with a single layer is performed to find a factorization

$$\mathbf{X} \approx \mathbf{B}_1 \mathbf{W}_1 \tag{5.242}$$

where $\mathbf{B}_1 \in \mathbb{R}_+^{M \times K_1}$ and $\mathbf{W}_1 \in \mathbb{R}_+^{K_1 \times N}$. Then the same factorization is performed on the result obtained from the first step as

$$\mathbf{W}_1 \approx \mathbf{B}_2 \mathbf{W}_2 \tag{5.243}$$

where $\mathbf{B}_2 \in \mathbb{R}_+^{K_1 \times K_2}$ and $\mathbf{W}_2 \in \mathbb{R}_+^{K_2 \times N}$. We continue the procedure to pretrain all layers until layer L

$$\mathbf{W}_{L-1} \approx \mathbf{B}_L \mathbf{W}_L \tag{5.244}$$

where $\mathbf{B}_L \in \mathbb{R}_+^{K_{L-1} \times K_L}$ and $\mathbf{W}_L \in \mathbb{R}_+^{K_L \times N}$. After the initialization, we fine-tune the parameters of all layers, $\{\mathbf{B}_l\}_{l=1}^L$ and $\mathbf{W}_L$, to reduce the total reconstruction error via minimizing the following divergence measure:

$$\min_{\{\mathbf{B}_l\}, \mathbf{W}_L \geq 0} \mathcal{D}\left(\mathbf{X} \left\| \left(\prod_{l=1}^L \mathbf{B}_l\right) \mathbf{W}_L\right.\right), \qquad \forall l = 1, \ldots, L. \tag{5.245}$$

According to Eq. (2.20), the multiplicative updating rules for parameters in all layers are derived as

$$\mathbf{B}_l \leftarrow \mathbf{B}_l \otimes \frac{[\nabla_{\mathbf{B}_l} \mathcal{D}]^-}{[\nabla_{\mathbf{B}_l} \mathcal{D}]^+}, \qquad \forall l = 1, \ldots, L, \tag{5.246}$$

$$\mathbf{W}_L \leftarrow \mathbf{W}_L \otimes \frac{[\nabla_{\mathbf{W}_L} \mathcal{D}]^-}{[\nabla_{\mathbf{W}_L} \mathcal{D}]^+} \tag{5.247}$$

where $[\nabla_{\mathbf{B}_l} \mathcal{D}]^+$ and $[\nabla_{\mathbf{B}_l} \mathcal{D}]^-$ denote the positive and negative parts of the gradient with respect to each layer $\mathbf{B}_l$, and $[\nabla_{\mathbf{W}_L} \mathcal{D}]^+$ and $[\nabla_{\mathbf{W}_L} \mathcal{D}]^-$ denote the positive and negative parts of the gradient with respect to $\mathbf{W}_L$, respectively. For instance, if the squared Euclidean distance $\mathcal{D}_{\text{EU}}(\cdot)$ is chosen as the cost function and the number of layers L is set to 2, the multiplicative updating rules are derived by

$$\mathbf{B}_1 \leftarrow \mathbf{B}_1 \otimes \frac{\mathbf{X}(\mathbf{W}_2)^\top (\mathbf{B}_2)^\top}{\mathbf{B}_1 \mathbf{B}_2 \mathbf{W}_2 (\mathbf{W}_2)^\top (\mathbf{B}_2)^\top}, \tag{5.248}$$

$$\mathbf{B}_2 \leftarrow \mathbf{B}_2 \otimes \frac{(\mathbf{B}_1)^\top \mathbf{X}(\mathbf{W}_2)^\top}{(\mathbf{B}_1)^\top \mathbf{B}_1 \mathbf{B}_2 \mathbf{W}_2 (\mathbf{W}_2)^\top}, \tag{5.249}$$

$$\mathbf{W}_2 \leftarrow \mathbf{W}_2 \otimes \frac{(\mathbf{B}_2)^\top (\mathbf{B}_1)^\top \mathbf{X}}{(\mathbf{B}_2)^\top (\mathbf{B}_1)^\top \mathbf{B}_1 \mathbf{B}_2 \mathbf{W}_2}. \tag{5.250}$$

When compared with standard NMF, this LNMF can realize more complex bases via the structural learning by combining sparse parts-based bases extracted by the single-layer NMF to interpret the data differently, hence improve separation performance. Algorithm 5.3 shows the learning procedure, which implements the inference of LNMF in two stages. Parameter initialization aims to find the initial factorized parameters

$$\Theta^{(0)} = \{\{\mathbf{B}_l^{(\tau)}\}_{l=1}^L, \mathbf{W}_L^{(0)}\} \tag{5.251}$$

while parameter fine-tuning is performed through L layers for updating

$$\Theta^{(\tau+1)} \leftarrow \Theta^{(\tau)}. \tag{5.252}$$

Algorithm 5.3 Learning Procedure for Layered Nonnegative Matrix Factorization.

Require $\mathbf{X} \in \mathbb{R}_+^{M \times N}$, L and $\{K_1, \ldots, K_L\}$

Parameter initialization

For each layer l

  If $l = 1$ then

    $\{\widehat{\mathbf{B}}_l, \widehat{\mathbf{W}}_l\} = \arg\min_{\{\mathbf{B}_l \geq 0, \mathbf{W}_l \geq 0\}} \mathcal{D}\left(\mathbf{X} \parallel \mathbf{B}_l \mathbf{W}_l\right)$

  Else

    $\{\widehat{\mathbf{B}}_{l+1}, \widehat{\mathbf{W}}_{l+1}\} = \arg\min_{\{\mathbf{B}_{l+1} \geq 0, \mathbf{W}_{l+1} \geq 0\}} \mathcal{D}\left(\widehat{\mathbf{W}}_l \parallel \mathbf{B}_{l+1} \mathbf{W}_{l+1}\right)$

Parameter fine-tuning

For each iteration τ

  For each layer l

    Updating basis matrix $\mathbf{B}_l^{(\tau)}$ via Eq. (5.246)

    $l \leftarrow l + 1$

  Updating weight matrix $\mathbf{W}_L^{(\tau)}$ via Eq. (5.247)

  Check convergence

  $\tau \leftarrow \tau + 1$

Return $\{\{\mathbf{B}_l\}_{l=1}^L, \mathbf{W}_L\}$

5.5.3 DISCRIMINATIVE LAYERED NMF

Basically, both standard and layered NMFs are seen as generative models, which provide a way to represent mixed audio signals so that decomposed source signals can be obtained. However, a sophisticated generative model may not guarantee desirable performance in source separation if the discrimination between source signals are disregarded. Section 3.6.1 has addressed several approaches to enhance the discrimination for NMF modeling. In what follows, an advanced study on discriminative learning for the layered NMF (LNMF) in Section 5.5.2 is presented. Considering the task of supervised speech separation in the presence two different speakers s_1 and s_2, L hierarchical bases of individual speakers $\{\mathbf{B}_{s_1 1}, \mathbf{B}_{s_1 2}, \ldots, \mathbf{B}_{s_1 L}\}$ and $\{\mathbf{B}_{s_2 1}, \mathbf{B}_{s_2 2}, \cdots, \mathbf{B}_{s_2 L}\}$ are separately learned from his/her clean sentences using the LNMF algorithm. Fig. 5.15 illustrates the topology of discriminative layered NMF (DLNMF) (Hsu et al., 2016), which is extended from Fig. 5.14 for LNMF with individual sets of bases for different sources. Given the input mixed spectrogram $\mathbf{X}$, the weight parameters $\mathbf{W}_{s_1 L}$ and $\mathbf{W}_{s_2 L}$ of two sources in layer L are estimated by minimizing the divergence measure where the basis parameters $\{\mathbf{B}_{s_1 1}, \mathbf{B}_{s_1 2}, \ldots, \mathbf{B}_{s_1 L}\}$ and $\{\mathbf{B}_{s_2 1}, \mathbf{B}_{s_2 2}, \ldots, \mathbf{B}_{s_2 L}\}$ of two sources in L layers are provided as follows:

$$\min_{\{\mathbf{W}_{s_1 L}, \mathbf{W}_{s_2 L}\} \geq 0} \mathcal{D}\left(\mathbf{X} \left\| [\mathbf{I}, \mathbf{I}] \begin{bmatrix} \mathbf{B}_{s_1 1} & \mathbf{0} \\ \mathbf{0} & \mathbf{B}_{s_2 1} \end{bmatrix} \right.\right.$$
$$\left.\left. \times \begin{bmatrix} \mathbf{B}_{s_1 2} & \mathbf{0} \\ \mathbf{0} & \mathbf{B}_{s_2 2} \end{bmatrix} \cdots \begin{bmatrix} \mathbf{B}_{s_1 L} & \mathbf{0} \\ \mathbf{0} & \mathbf{B}_{s_2 L} \end{bmatrix} \begin{bmatrix} \mathbf{W}_{s_1 L} \\ \mathbf{W}_{s_2 L} \end{bmatrix} \right) \right. \tag{5.253}$$

where $\mathbf{I}$ and $\mathbf{0}$ are identity and zero matrices of proper sizes, respectively. For ease of expression, the learning objective in Eq. (5.253) is rewritten by using the compound matrices for each individual matrix

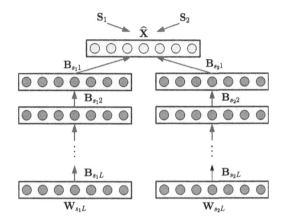

FIGURE 5.15

Illustration for discriminative layered nonnegative matrix factorization.

in a form of

$$\min_{\mathbb{W}_L \geq 0} \mathcal{D}\left(\mathbf{X} \middle\| \mathbb{I} \left(\prod_{l=1}^{L} \mathbb{B}_l\right) \mathbb{W}_L\right) \tag{5.254}$$

where compound matrices are defined by

$$\mathbb{I} \triangleq [\mathbf{I}, \mathbf{I}], \tag{5.255}$$

$$\mathbb{B}_l \triangleq \begin{bmatrix} \mathbf{B}_{s_1 l} & \mathbf{0} \\ \mathbf{0} & \mathbf{B}_{s_2 l} \end{bmatrix}, \ \forall \, l = 1, \ldots, L, \tag{5.256}$$

$$\mathbb{W}_L \triangleq \begin{bmatrix} \mathbf{W}_{s_1 L} \\ \mathbf{W}_{s_2 L} \end{bmatrix}. \tag{5.257}$$

By adopting the Kullback–Leibler divergence $\mathcal{D}_{\text{KL}}(\cdot)$ with additional consideration of matrix $\mathbb{I}$, the multiplicative updating rule of weight parameters $\mathbb{W}_L$ in layer L is formulated by

$$\mathbb{W}_L \leftarrow \mathbb{W}_L \otimes \frac{\left(\mathbb{I}\left(\prod_{l=1}^{L} \mathbb{B}_l\right)\right)^{\top} \left(\frac{\mathbf{X}}{\widehat{\mathbf{X}}}\right)}{\left(\mathbb{I}\left(\prod_{l=1}^{L} \mathbb{B}_l\right)\right)^{\top} \mathbf{1}} \tag{5.258}$$

where $\widehat{\mathbf{X}}$ is the reconstructed mixed spectra expressed by

$$\widehat{\mathbf{X}} = \mathbb{I}\left(\prod_{l=1}^{L} \mathbb{B}_l\right) \mathbb{W}_L \tag{5.259}$$

and $\mathbf{1}$ is a vector of ones of proper size. This step is actually the same as the parameter fine-tuning stage in LNMF (Eq. (5.247)), which is consistent with the updating rule based on KL divergence in the standard NMF provided in Eq. (2.28) and second row of Table 2.1. Given the estimated weight matrices of two sources $\{\mathbf{W}_{s_1 L}, \mathbf{W}_{s_2 L}\}$, the second step is to use the fixed weight matrices to estimate the basis matrices in different layers. To enhance the separation between speaker sources, the hierarchical basis matrices of different speakers $\{\mathbf{B}_{s_1 1}, \mathbf{B}_{s_1 2}, \ldots, \mathbf{B}_{s_1 L}\}$ and $\{\mathbf{B}_{s_2 1}, \mathbf{B}_{s_2 2}, \cdots, \mathbf{B}_{s_2 L}\}$ are estimated according to a discriminative criterion, which is performed by directly minimizing KL divergence between the original source signals and the reconstructed signals as

$$\min_{\{\mathbf{B}_{s_1 l}, \mathbf{B}_{s_2 l}\} \geq 0} \mathcal{D}_{\mathrm{KL}} \left(\begin{bmatrix} \mathbf{S}_1 \\ \mathbf{S}_2 \end{bmatrix} \middle\| \begin{bmatrix} \mathbf{B}_{s_1 1} & \mathbf{0} \\ \mathbf{0} & \mathbf{B}_{s_2 1} \end{bmatrix} \right.$$
$$\left. \times \begin{bmatrix} \mathbf{B}_{s_1 2} & \mathbf{0} \\ \mathbf{0} & \mathbf{B}_{s_2 2} \end{bmatrix} \cdots \begin{bmatrix} \mathbf{B}_{s_1 L} & \mathbf{0} \\ \mathbf{0} & \mathbf{B}_{s_2 L} \end{bmatrix} \begin{bmatrix} \mathbf{W}_{s_1 L} \\ \mathbf{W}_{s_2 L} \end{bmatrix} \right),$$
$$\forall l = 1, \ldots, L \tag{5.260}$$

where $\mathbf{S}_1$ and $\mathbf{S}_2$ are the original spectra, i.e., the target spectra of two speakers. Again, we rewrite this KL divergence by using compound matrices as

$$\min_{\{\mathbb{B}_l\} \geq 0} \mathcal{D}_{\mathrm{KL}} \left(\mathbb{S} \middle\| \left(\prod_{j=1}^{L} \mathbb{B}_j \right) \mathbb{W}_L \right), \quad \forall l = 1, \ldots, L \tag{5.261}$$

where the compound matrix of the original source signals is defined by

$$\mathbb{S} \triangleq \begin{bmatrix} \mathbf{S}_1 \\ \mathbf{S}_2 \end{bmatrix}. \tag{5.262}$$

By applying the formula of estimating basis matrix in Eq. (5.246), the multiplicative updating rules for basis matrices of DLNMF in different layers are derived by

$$\mathbb{B}_1 \leftarrow \mathbb{B}_1 \otimes \frac{\left(\frac{\mathbb{S}}{\mathbb{S}} \right) \left(\left(\prod_{l=2}^{L} \mathbb{W}_l \right) \mathbb{W}_L \right)^{\mathsf{T}}}{\mathbf{1} \left(\left(\prod_{l=2}^{L} \mathbb{W}_l \right) \mathbb{W}_L \right)^{\mathsf{T}}}, \tag{5.263}$$

$$\mathbb{B}_l \leftarrow \mathbb{B}_l \otimes \frac{\left(\prod_{j=1}^{l-1} \mathbb{B}_j \right)^{\mathsf{T}} \left(\frac{\mathbb{S}}{\mathbb{S}} \right) \left(\left(\prod_{j=l+1}^{L} \mathbb{W}_j \right) \mathbb{W}_L \right)^{\mathsf{T}}}{\left(\prod_{j=1}^{l-1} \mathbb{B}_j \right)^{\mathsf{T}} \mathbf{1} \left(\left(\prod_{j=l+1}^{L} \mathbb{W}_j \right) \mathbb{W}_L \right)^{\mathsf{T}}}, \tag{5.264}$$
$$\forall l = 2, \ldots, L - 1,$$

$$\mathbb{B}_L \leftarrow \mathbb{B}_L \otimes \frac{\left(\prod_{l=1}^{L-1} \mathbb{W}_l \right)^{\mathsf{T}} \left(\frac{\mathbb{S}}{\mathbb{S}} \right) (\mathbb{W}_L)^{\mathsf{T}}}{\left(\prod_{l=1}^{L-1} \mathbb{W}_l \right)^{\mathsf{T}} \mathbf{1} (\mathbb{W}_L)^{\mathsf{T}}} \tag{5.265}$$

where the compound matrix of the reconstructed source signals is given by

$$\widehat{\mathbb{S}} = \left(\prod_{l=1}^{L} \mathbb{B}_l \right) \mathbb{W}_L. \tag{5.266}$$

In summary, the model structure with the basis matrices of DLNMF in different layers is optimally constructed by

$$\{\widehat{\mathbb{B}}_l\} = \arg \min_{\{\mathbb{B}_l\} \geq 0} \mathcal{D}_{\mathrm{KL}} \left(\mathbb{S} \middle\| \left(\prod_{j=1}^{L} \mathbb{B}_j \right) \widehat{\mathbb{W}}_L \right), \tag{5.267}$$

$$\forall l = 1, \ldots, L$$

where the weight matrix of layer L is estimated by

$$\widehat{\mathbb{W}}_L = \arg \min_{\mathbb{W}_L \geq 0} \mathcal{D}_{\mathrm{KL}} \left(\mathbf{X} \middle\| \mathbb{I} \left(\prod_{l=1}^{L} \mathbb{B}_l \right) \mathbb{W}_L \right). \tag{5.268}$$

In practice, the procedure of updating $\widehat{\mathbb{W}}_L$ and $\{\widehat{\mathbb{B}}_l\}_{l=1}^{L}$ is repeated by several iterations τ.

The discriminative learning method in DLNMF differs from those in discriminative NMFs which were mentioned in Section 3.6.1. The differences are due to the discovery of a deep-layered structure and the way of imposing discrimination. LNMF estimates L layers of basis matrices $\{\mathbf{B}_l\}_{l=1}^{L}$ by disregarding the mutual correlation between two sources s_1 and s_2. On the contrary, there are two reasons that DLNMF enforces the discrimination between the demixed source signals. First, the basis matrices in L layers $\{\mathbf{B}_{s_1 1}, \mathbf{B}_{s_1 2}, \ldots, \mathbf{B}_{s_1 L}\}$ and $\{\mathbf{B}_{s_2 1}, \mathbf{B}_{s_2 2}, \cdots, \mathbf{B}_{s_2 L}\}$ are individually initialized and calculated by using the training data corresponding to individual sources s_1 and s_2, respectively. Importantly, these two sets of basis matrices in different layers $\{\mathbb{B}_l\}_{l=1}^{L}$ are jointly estimated by directly minimizing the reconstruction errors for target spectra $\mathbf{S}_1$ and $\mathbf{S}_2$. However, LNMF minimizes the reconstruction errors for mixed spectra $\mathbf{X}$ and indirectly uses the estimated parameters $\{\{\mathbf{B}_l\}_{l=1}^{L}, \mathbf{W}_L\}$ to find demixed signals $\widehat{\mathbf{S}}_1$ and $\widehat{\mathbf{S}}_2$. Secondly, when comparing Eq. (5.245) of LNMF and Eq. (5.254) of DLNMF, it is found that the weight matrices $\mathbf{W}_{s_1 L}$ and $\mathbf{W}_{s_2 L}$ in DLNMF are *jointly* estimated by considering the *correlation* between two sources through the contribution from the off-diagonal terms using compound matrices

$$\{\mathbb{I}, \{\mathbb{B}_l\}_{l=1}^{L}, \mathbb{W}_L\}$$

in Eq. (5.258). Similar situation is also observed when estimating the basis matrices $\{\mathbf{B}_{s_1 1}, \mathbf{B}_{s_1 2}, \ldots, \mathbf{B}_{s_1 L}\}$ and $\{\mathbf{B}_{s_2 1}, \mathbf{B}_{s_2 2}, \cdots, \mathbf{B}_{s_2 L}\}$ through the multiplicative updating rules for compound matrices $\mathbb{B}_1$, $\mathbb{B}_l$ and $\mathbb{B}_L$ in Eqs. (5.263), (5.264) and (5.265), respectively. The key difference between LNMF and DLNMF can be also reflected by the way of signal reconstruction when comparing Eq. (5.241) and Eq. (5.259). In general, the implementation procedure of DLNMF is similar to that of LNMF as shown in Algorithm 5.3 with the stages of parameter initialization and parameter fine-tuning except for replacement of multiplicative updating rules of parameters from

$$\Theta = \{\{\mathbf{B}_l\}_{l=1}^{L}, \mathbf{W}_L\} \tag{5.269}$$

for LNMF to

$$\Theta = \{\mathbb{W}_L, \mathbb{B}_1, \{\mathbb{B}_l\}_{l=2}^{L-1}, \mathbb{B}_L\} \tag{5.270}$$

for DLNMF by using Eqs. (5.258), (5.263), (5.264) and (5.265), respectively.

5.5.4 SYSTEM EVALUATION

Three variants of NMFs (standard NMF, LNMF and DLNMF) were investigated for supervised speaker-dependent source separation. These methods were evaluated by illustrating how the layered structure and the discriminative learning affected the performance of monaural speech separation. In the system setup, the speech mixtures were generated by combining sentences from one female and one male speaker from TSP corpus (Kabal, 2002, Hsu et al., 2016). There were 60 sentences spoken by each speaker. The 1024-point STFT with a 64-ms frame length and a 16-ms frame shift was calculated to obtain the Fourier magnitude spectrogram of a mixed signal $\mathbf{X}$. When implementing DLNMF, the initial hierarchical bases for the two speakers

$$\{\mathbf{B}_{s_1 1}^{(0)}, \mathbf{B}_{s_1 2}^{(0)}, \dots, \mathbf{B}_{s_1 L}^{(0)}\}$$

and

$$\{\mathbf{B}_{s_2 1}^{(0)}, \mathbf{B}_{s_2 2}^{(0)}, \dots, \mathbf{B}_{s_2 L}^{(0)}\}$$

were trained from clean training data using LNMF. Then, the discriminative criterion was applied to adapt the layered basis matrices according to Eqs. (5.263), (5.264) and (5.265) to obtain

$$\{\widehat{\mathbb{B}}_l\}_{l=1}^{L} = \{\widehat{\mathbf{B}}_{s_1 l}, \widehat{\mathbf{B}}_{s_2 l}\}_{l=1}^{L}. \tag{5.271}$$

After the adaptation, for the spectrogram of a new test mixture, the weight matrices

$$\widehat{\mathbb{W}}_L = [(\widehat{\mathbf{W}}_{s_1 L})^\top (\widehat{\mathbf{W}}_{s_2 L})^\top]^\top \tag{5.272}$$

were obtained by using Eq. (5.258) and the separated spectrograms were then calculated by applying the Wiener gain as follows:

$$\widetilde{\mathbf{X}}_{s_1} = \mathbf{X} \otimes \frac{\left(\prod_{l=1}^{L} \widehat{\mathbf{B}}_{s_1 l}\right) \widehat{\mathbf{W}}_{s_1 L}}{\mathbb{I}\left(\prod_{l=1}^{L} \widehat{\mathbb{B}}_l\right) \widehat{\mathbb{W}}_L}, \tag{5.273}$$

$$\widetilde{\mathbf{X}}_{s_2} = \mathbf{X} \otimes \frac{\left(\prod_{l=1}^{L} \widehat{\mathbf{B}}_{s_2 l}\right) \widehat{\mathbf{W}}_{s_2 L}}{\mathbb{I}\left(\prod_{l=1}^{L} \widehat{\mathbb{B}}_l\right) \widehat{\mathbb{W}}_L}, \tag{5.274}$$

which is comparable with the soft mask function of Eq. (2.15) used in standard NMF.

In the implementation, 70% of each speaker's sentences (42 sentences) were randomly selected for training his or her hierarchical bases. Then, another 20% (12 mixed utterances and their corresponding clean utterances) were used for discriminative learning in DLNMF. The remaining 10% (6 test mixed

Table 5.8 Performance of speech separation by using NMF, LNMF and DLNMF in terms of SDR, SIR and SAR (dB)

Method	Number of bases	Female			Male		
		SDR	SIR	SAR	SDR	SIR	SAR
NMF	$K = 10$	6.76	10.34	9.94	6.28	8.92	10.51
	$K = 20$	6.83	10.05	10.27	6.34	8.82	10.81
LNMF	$K_1 = 180, K_2 = 10$	7.26	11.09	10.14	6.75	9.46	10.78
	$K_1 = 180, K_2 = 20$	7.38	10.94	10.43	6.70	9.16	11.21
DLNMF	$K_1 = 180, K_2 = 10$	7.42	11.11	10.36	6.87	9.36	11.19
	$K_1 = 180, K_2 = 20$	7.87	11.33	11.04	7.48	10.30	11.38

utterances) were used for testing. In total, 90 test mixed signals were created from the female–male speaker pair. Then a 15-fold validation was performed, i.e., the results were averaged over 15 randomly selected training and test data. For both LNMF and DLNMF, the layer-related parameters were set as $L = 2$, $K_1 = 180$ and $K_2 = [10, 20]$. The performance of speech separation was assessed in terms of source-to-distortion ratio (SDR), signal-to-interference ratio (SIR) (defined in Eq. (2.10)) and source-to-artifacts ratio (SAR) (Vincent et al., 2006). Table 5.8 reports the separation results of using standard NMF, LNMF and DLNMF. The averaged SDR, SIR and SAR over 90 test mixed signals were measured. It is obvious that LNMF outperforms the standard NMF in both SDR and SIR measures since the layered structure is introduced. DLNMF further improves SDR, SIR and SAR by incorporating the discriminative cost function. Increasing the number of bases does not always improve the separation performance. DLNMF obtains more consistent improvement with more bases compared to LNMF. A further study on a deeper structure for DLNMF will be required.

5.6 SUMMARY

A series of advanced NMF approaches to various monaural source separation tasks have been presented and evaluated. These approaches provided different extensions to deal with different issues in source separation. A variety of learning algorithms were incorporated to carry out these extensions such that the NMF solutions are really flexible and efficient for many applications, including audio source separation. This section has described various applications of NMF models consisting of speech dereverberation, speech and music separation, singing voice separation, instrumental music separation and speech separation. Supervised and unsupervised learning was investigated under latent variable models. Convolutive processing, probabilistic modeling, Bayesian learning, dictionary learning, information-theoretical learning, online learning sparse learning, group representation, discriminative learning, hierarchical learning were merged into NMF in different ways from different perspectives. First of all, convolutive NMF paved a way to capture temporal and spatial correlation in mixed signals for signal reconstruction and deconvolution. The application of nonnegative convolution and factorization for reverberant source separation was demonstrated. Bayesian learning was applied to characterize the variations in speech dereverberation where a number of parameter distributions were used to carry out variational inference for model construction. Next, a probabilistic framework was introduced to perform nonnegative factorization for source separation. This framework established a kind of latent topic model and adopted the latent codes or topics to represent the bases of a dictionary for signal reconstruction and source separation. The one- and two-dimensional convolutive processing was presented to estimate the shift invariant probability parameters. A strong relation to NMF was theoretically illustrated. Moreover, the full Bayesian approaches to NMF were presented by using different types of distribution function. The inference algorithms using variational inference and Gibbs sampling were addressed. The pros and cons of using different distributions for NMF parameters and applying different inference algorithms for model training were presented. Supervised learning for speech and music separation, as well as unsupervised learning for singing voice separation, was evaluated. In addition, the group basis representation was presented for monaural source separation in the presence of rhythmic and harmonic sources where the shared and individual bases were estimated to represent repetitive and residual patterns in audio signals, respectively. Importantly, a Bayesian group sparse learning was introduced to learn the groups of relevant bases based on the Bayesian approach using a Laplace scale mixture distribution. The sampling method was addressed to infer a set of parameters and hyperparameters using the corresponding conditional posterior distributions. Finally, deep-layered information was explored for speech separation based on the layered NMF. The structural factorization was implemented through the learning stages for parameter initialization and fine-tuning. In particular, the mutual correlation between source signals was learned through estimation of structural bases by directly minimizing the reconstruction errors of source rather than mixed signals. Discriminative learning for individual sources was performed by using multiplicative updating rules.

The next section will further extend source separation models from nonnegative matrix factorization to nonnegative tensor factorization (NTF), which allows source separation in the presence of multiway or multichannel observation data. Again, the learning strategies or algorithms based on convolutive processing, probabilistic modeling and Bayesian learning will be incorporated into tensor factorization, which provides a variety of solutions to flexible source separation. Scalable modeling will be presented to carry out an advanced work for music source separation where the tensor factorization model adapts to different amounts of training data.

NONNEGATIVE TENSOR FACTORIZATION

6

Nonnegative tensor factorization (NTF) is seen as an extension of nonnegative matrix factorization (NMF) where the features from the third way in addition to the traditional ways of time and frequency are learned. NTF paves an avenue to conduct multiway decomposition to extract rich information from structural data. Multiways provide multiple features from different horizons which are helpful for source separation. Section 2.3 surveys background knowledge about the theoretical fundamentals and the practical solutions to NTF or generally the tensor factorization without nonnegativity constraint. This chapter is devoted to address a variety of NTF methods for audio source separation or related application. Section 6.1 will introduce a couple of advanced studies where the mixed tensors from modulation spectrogram or multiresolution spectrogram are decomposed. Section 6.2 addresses the extraction of temporal patterns based on the convolutive NTF by using log-frequency spectrograms from multichannel recording. Section 6.3 presents a probabilistic framework of NTF, which is realized by using Markov chain Monte Carlo (MCMC). The prediction performance can be improved. Furthermore, in Section 6.4, a Bayesian probabilistic tensor factorization is developed to carry out the full Bayesian solution where hyperparameter tuning can be avoided. Finally, Section 6.5 addresses an infinite positive semidefinite tensor factorization for source separation where the model complexity is autonomously controlled through a Bayesian nonparametric method based on a variational inference procedure. Fig. 6.1 illustrates the evolution of these advanced tensor factorization methods. The probabilistic assumption is beneficial to represent the model uncertainty and assure the improved generalization.

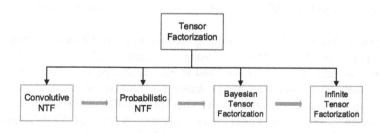

FIGURE 6.1

Categorization and evolution for different tensor factorization methods.

Source Separation and Machine Learning. https://doi.org/10.1016/B978-0-12-804566-4.00018-8

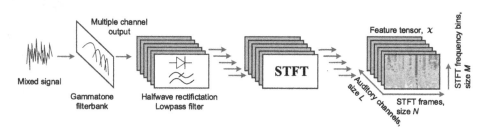

FIGURE 6.2

Procedure of producing the modulation spectrograms for tensor factorization.

6.1 NTF FOR SOURCE SEPARATION

There are several works, which are related to nonnegative tensor factorization (NTF) and audio signal processing. In Welling and Weber (2001), a positive tensor factorization was proposed for ease of interpretation in data analysis when compared with the singular value decomposition where negative components were contained. In Seki et al. (2016), NTF was developed for stereo channel music source separation where the mixed signals were represented as dual-channel signals in a form of tensor observations. NTF was performed with regularization by matching the cepstrum values in the separated signals. This section addresses a couple of examples showing how NTF is developed to factorize the mixed signals for monaural source separation in the presence of the modulation spectrograms (Barker and Virtanen, 2013) or the multiresolution spectrograms (Kırbız and Günsel, 2014), which are addressed in what follows.

6.1.1 MODULATION SPECTROGRAMS

Conventional nonnegative matrix factorization (NMF) operates on the magnitude or power spectrogram while in the human auditory system the sound is transduced to a representation based on the low frequency amplitude modulations within frequency bands. Basically, the generation of modulation spectrogram is based on a computation model of the cochlea, which follows the structure of the ear in a way of transduction of acoustic vibration into an electrical signal. It is crucial to conduct source separation using the spectrogram of a modulation envelope or simply the modulation spectrogram where the tensor factorization is able to identify the redundant patterns with similar features in an unsupervised way.

Fig. 6.2 illustrates the procedure of producing the tensors of modulation spectrograms for tensor factorization. The mixed signal is filtered by using a gammatone filterbank, which fits the cochlea model in the human auditory system. Each band is linearly spaced according to the corresponding rectangular bandwidth of the filter. Each band is half-wave rectified and low-pass filtered to obtain the modulation envelope using a single pole recursive filter with −3 dB bandwidth of approximately 26 Hz (Barker and Virtanen, 2013). Much of high frequency content is removed. Modulation spectrogram is obtained from the modulation envelope for each channel. These envelopes are segmented into a series of overlapping frames with Hamming window before calculating the short-time Fourier transform (STFT) for each channel. The magnitude outputs from STFTs are truncated into 150 frequency bins.

Finally, a three-way nonnegative tensor $\mathcal{X} = \{X_{lmn}\} \in \mathbb{R}_+^{L \times M \times N}$ is formed with the three-way dimensions for the number of modulated filterbank channels L, size of truncated STFT frequency bins M and number of observation frames N. Namely, the modulation spectrogram is formed by

$$\begin{aligned} &\text{modulation spectrogram} \\ &= \text{modulation channel} \times \text{frequency bin} \times \text{time frame.} \end{aligned} \tag{6.1}$$

The entries of the observed tensor $\{X_{lmn}\}$ can be written by using the approximated tensor $\widehat{\mathcal{X}} = \{\widehat{X}_{lmn}\}$ via

$$X_{lmn} \approx \widehat{X}_{lmn} = \sum_{k=1}^{K} C_{lk} B_{mk} W_{nk} \tag{6.2}$$

where

$$\mathbf{C} = \{\mathbf{c}_k\} = \{C_{lk}\} \in \mathbb{R}_+^{L \times K} \tag{6.3}$$

denotes the nonnegative factor matrix containing the auditory channel dependent gains,

$$\mathbf{B} = \{\mathbf{b}_k\} = \{B_{mk}\} \in \mathbb{R}_+^{M \times K} \tag{6.4}$$

denotes the nonnegative factor matrix containing the frequency basis functions for spectral content of modulation envelope features, and

$$\mathbf{W} = \{\mathbf{w}_k\} = \{W_{nk}\} \in \mathbb{R}_+^{N \times K} \tag{6.5}$$

denotes the nonnegative factor matrix containing the time-varying activations or weights of the components. Such a factorization model is therefore able to represent the modulation envelopes of the components k existing at different channels l at a particular frequency m and time n. There are K components or bases assumed in this tensor factorization. The parallel factor analysis (PARAFAC) structure allows representing a three-way tensor $\mathcal{X}$ by means of three factor matrices $\mathbf{C}$, $\mathbf{B}$ and $\mathbf{W}$. As addressed in Section 2.3.2, NTF in Eq. (6.2) is seen as a specialized CP decomposition based on

$$\begin{aligned} \mathcal{X} &\approx \mathcal{I} \times_1 \mathbf{C} \times_2 \mathbf{B} \times_3 \mathbf{W} \\ &= \sum_k \mathbf{c}_k \circ \mathbf{b}_k \circ \mathbf{w}_k. \end{aligned} \tag{6.6}$$

Similar to the learning procedure for nonnegative matrix factorization (NMF) addressed in Section 2.2.2, NTF can be formulated by minimizing the Kullback–Leibler (KL) divergence between real tensor $\mathcal{X}$ and approximate tensor $\widehat{\mathcal{X}}$:

$$\begin{aligned} &\mathcal{D}_{\text{KL}}(\mathcal{X} \| \widehat{\mathcal{X}}) \\ &= \sum_{l,m,n} \left(X_{lmn} \log \frac{X_{lmn}}{\widehat{X}_{lmn}} + \widehat{X}_{lmn} - X_{lmn} \right). \end{aligned} \tag{6.7}$$

The solution to this minimization problem can be derived as a formula for multiplicative updating where the entries of parameters $\mathbf{C}$, $\mathbf{B}$ and $\mathbf{W}$ are calculated by

$$C_{lk} \leftarrow C_{lk} \frac{\sum_{m,n} \mathcal{R}_{lmn} B_{mk} W_{nk}}{\sum_{m,n} B_{mk} W_{nk}}, \tag{6.8}$$

$$B_{mk} \leftarrow B_{mk} \frac{\sum_{l,n} \mathcal{R}_{lmn} C_{lk} W_{nk}}{\sum_{l,n} C_{lk} W_{nk}}, \tag{6.9}$$

$$W_{nk} \leftarrow W_{nk} \frac{\sum_{l,m} \mathcal{R}_{lmn} C_{lk} B_{mk}}{\sum_{l,m} C_{lk} B_{mk}} \tag{6.10}$$

where an auxiliary ratio variable of tensor $\mathcal{R} = \{\mathcal{R}_{lmn}\}$ for multiplicative updating is defined by

$$\mathcal{R}_{lmn} \triangleq \frac{\mathcal{X}_{lmn}}{\widehat{\mathcal{X}}_{lmn}}. \tag{6.11}$$

This variable should be reevaluated during each updating for $\mathbf{C}$, $\mathbf{B}$ and $\mathbf{W}$. The multiplicative updating rules in Eqs. (6.8)–(6.10) assure the nonnegative condition in the estimated factor matrices. The total number of entries in the factor matrices $\mathbf{C}$, $\mathbf{B}$ and $\mathbf{W}$ is determined by $K \times (L + M + N)$. Due to the multiway factorized structure, three-way factorization in NTF requires fewer parameters than two-way factorization based on NMF (Chien and Bao, 2018). Hence the overfitting problem is mitigated.

In Barker and Virtanen (2013), NTF was developed for single-channel source separation in the presence of speech and music sources, which were sampled at 16 kHz. NTF was implemented and compared with the standard NMF and the nonnegative matrix factor deconvolution (NMFD) with convolution over the frequency domain as addressed in Section 5.1.1 (Smaragdis, 2007, Schmidt and Morup, 2006). The sparseness constraint was imposed in different methods. A Hamming window with 1024 samples and 50% overlapping was used. Overlap-and-add reconstruction and soft mask function, as addressed in Section 2.2.1, were consistently performed. An auditory filterbank with 20 channels was used in NTF. Different factorization methods were run by 200 iterations with sufficient convergence. In the implementation of source separation, the blind clustering based on 13 Mel-frequency cepstral coefficients (MFCCs), as addressed in Section 5.3.5, was performed. A convolution filter length of 10 frequency bins was used in the implementation of NMFD. A different number of components K was evaluated. NTF was demonstrated to be better than NMF and NMFD with different K in the presence of two sources in terms of the averaged signal-to-distortion ratio (SDR) (Vincent et al., 2006).

6.1.2 MULTIRESOLUTION SPECTROGRAMS

On the other hand, nonnegative tensor factorization (NTF) was developed for single-channel audio source separation where each layer of the tensor represents the mixed signal at a different time–frequency resolution (Kırbız and Günsel, 2014). Fig. 6.3 shows an example of time resolution and frequency resolution of an mixed audio signal with the nonnegative tensor $\mathcal{X} \in \mathbb{R}_+^{L \times M \times N}$ and $L = 2$.

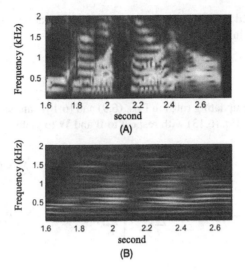

FIGURE 6.3

(A) Time resolution and (B) frequency resolution of a mixed audio signal.

Multiresolution spectrogram is constructed by

$$\text{multiresolution spectrogram} = \text{resolution} \times \text{frequency} \times \text{time.} \tag{6.12}$$

In order to fuse the information at different layers, the source separation was represented as a joint optimization problem where the KL divergence between the observed tensor $\mathcal{X}$ and the reconstructed tensor $\mathcal{X}$, as given in Eq. (6.7), was minimized. By eliminating the terms which are constant, the learning objective is yielded by

$$
\begin{aligned}
\mathcal{D}_{\mathrm{KL}}(\mathcal{X} \| \widehat{\mathcal{X}}) \\
= \sum_{l,m,n} \left(-\mathcal{X}_{lmn} \log \widehat{\mathcal{X}}_{lmn} + \widehat{\mathcal{X}}_{lmn} \right) \\
= \sum_{l,m,n} \left(-\mathcal{X}_{lmn} \log \left(\sum_{k} C_{lk} B_{mk} W_{nk} \right) + \sum_{k} C_{lk} B_{mk} W_{nk} \right).
\end{aligned}
\tag{6.13}
$$

Taking the gradient of learning objective with respect to $\mathbf{C}$ yields the following updating equation:

$$
C_{lk} \leftarrow C_{lk} + \eta_{lk} \sum_{m} B_{mk} \left[\sum_{n} \mathcal{R}_{lmn} W_{nk} - \sum_{n} W_{nk} \right] \tag{6.14}
$$

where $\mathcal{R}_{lmn}$ has been defined in Eq. (6.11). The multiplicative updating rule in Eq. (6.8) is obtained by setting the learning rate η_{lk} in Eq. (6.14) as

$$\eta_{lk} = \frac{C_{lk}}{\sum_{m,n} B_{mk} W_{nk}}. \tag{6.15}$$

Similarly, the multiplicative updating rules in Eqs. (6.9) and (6.10) can be derived by taking the gradients of learning objective in Eq. (6.13) with respect to $\mathbf{B}$ and $\mathbf{W}$ to yield

$$B_{mk} \leftarrow B_{mk} + \eta_{mk} \sum_{n} W_{nk} \left[\sum_{l} \mathcal{R}_{lmn} C_{lk} - \sum_{l} C_{lk} \right], \tag{6.16}$$

$$W_{nk} \leftarrow W_{nk} + \eta_{nk} \sum_{l} C_{lk} \left[\sum_{m} \mathcal{R}_{lmn} B_{mk} - \sum_{m} B_{mk} \right] \tag{6.17}$$

and using the learning rates

$$\eta_{mk} = \frac{B_{mk}}{\sum_{l,n} C_{lk} W_{nk}}, \tag{6.18}$$

$$\eta_{nk} = \frac{W_{nk}}{\sum_{l,m} C_{lk} B_{mk}}, \tag{6.19}$$

respectively.

In Kırbız and Günsel (2014), the multiresolution spectrograms of a mixed signal were formed as a nonnegative tensor $\mathcal{X} \in \mathbb{R}_+^{L \times M \times N}$ for single-channel source separation where nonnegative tensor factorization (NTF) was applied. The magnitude spectrograms at various resolutions were observed. The factor matrices $\mathbf{C} = \{C_{lk}\}$, $\mathbf{B} = \{B_{mk}\}$ and $\mathbf{W} = \{W_{nk}\}$ denote the nonnegative components for the resolution gains, frequency basis functions, and time-varying amplitude envelopes. In the implementation, the factor matrices for resolution gains $\mathbf{C}$ and time-varying activations $\mathbf{W}$ were learned from the multiplicative updating rules in Eqs. (6.8) and (6.10), respectively. The frequency basis matrix $\mathbf{B}$ was learned from the training signals via NMF at different resolutions. Multiple components were extracted for each source. A matching problem of which component belongs to which source appeared. To deal with this problem, a training set of original sources was prepared for dictionary learning. The basis vectors of the sources, either music or speech, were learned. Supervised separation of two sources from a single mixture was performed. The variable resolution of the filterbank had to be adequate in subspace learning to optimally reconstruct the frequency and temporal characteristics of the input signal. The synthesis of the separated sources in the time domain was performed by applying the inverse STFT and merging the phase information estimated from the mixed signal using a spectral filter based on an adaptive Wiener filter. The optimal compact representation was achieved by enhancing the sparsity in the magnitude spectrograms of the factorized sources. In the experiments, source separation was benefited from the fusion of the information gathered from representations extracted at various time–frequency resolutions. NTF performed better than NMF in terms of signal-to-distortion ratio (SDR), signal-to-interference ratio (SIR) and signal-to-artifacts ratio (SAR). The separation scenarios of two speech signals, two music signals, as well as one speech and one music signal, were investigated.

6.2 CONVOLUTIVE NTF

Similar to the convolutive nonnegative matrix factorization (NMF) in Section 5.1, the convolutive operation can be employed in nonnegative tensor factorization (NTF) to carry out the convolutive NTF. Our goal is to learn the temporal correlation and frequency correlation for single-channel source separation. In Mørup and Schmidt (2006b), the sparse nonnegative tensor factor two-dimensional deconvolution (NTF2D) was developed for source separation and music transcription. In what follows, we address the nonnegative tensor factor deconvolution in Section 6.2.1 and the sparse nonnegative tensor factor deconvolution in Section 6.2.2 based on the parallel factor analysis (PARAFAC) model. The learning objectives based on the squared Euclidean distance and the Kullback–Leibler divergence are investigated.

6.2.1 NONNEGATIVE TENSOR FACTOR DECONVOLUTION

As addressed in Section 2.3.2, a CP model is feasible to deal with single-channel source separation. In particular, the PARAFAC model in Eq. (6.6) can be applied to carry out NTF for source separation. In Mørup and Schmidt (2006b), the nonnegative tensor factor deconvolution (NTFD) and NTF2D were developed for multichannel time–frequency analysis, which was based on the PARAFAC model. In general, NTFD and NTF2D are seen as extensions of matrix factor deconvolution using NMFD and NMF2D, which have been described for convolutive NMF in Section 5.1.1. In the following, we derive the solution to NTF2D where convolution in both time and frequency domain is performed. The simplification of NTF2D to NTFD is straightforward and neglected in this section.

The signal reconstruction of an observed three-way nonnegative tensor $\mathcal{X} = \{\mathcal{X}_{lmn}\} \in \mathbb{R}_+^{L \times M \times N}$ based on the nonnegative tensor factor two-dimensional deconvolution (NTF2D) can be expressed by

$$\mathcal{X}_{lmn} \approx \widehat{\mathcal{X}}_{lmn} = \sum_{k,\phi,\tau} C_{lk} B_{m-\phi,k}^{\tau} W_{n-\tau,k}^{\phi} \tag{6.20}$$

where C_{lk} denotes that source k is present in channel l, B_{mk}^{τ} denotes the basis of source or component k at frequency m with delay τ, and W_{nk}^{ϕ} denotes the weight of source or component k at time n with note ϕ. If we set $\phi = 0$ and $\tau = 0$, NTF2D model is equivalent to the standard NTF model. Notably, the notations in NTF2D are consistent with those in NMF2D. In addition to the basis matrix $\mathbf{B}$ and weight matrix $\mathbf{W}$ for capturing the frequency and temporal information, the channel matrix $\mathbf{C}$ is incorporated to learn for the auxiliary information from different modulations or resolutions as addressed in Sections 6.1.1 and 6.1.2, respectively. The expression of NTF2D for each entry in Eq. (6.20) can be also rewritten in matrix form by defining the Khatri–Rao product $\star$ for two matrices $\mathbf{B}$ and $\mathbf{W}$

$$\mathbf{B} \star \mathbf{W} = [\mathbf{b}_1 \otimes \mathbf{w}_1 \cdots \mathbf{b}_K \otimes \mathbf{w}_K] \in \mathbb{R}^{MN \times K} \tag{6.21}$$

where $\otimes$ denotes the Kronecker product. The observed tensor

$$\mathcal{X} \in \mathbb{R}_+^{L \times M \times N}$$

is reexpressed by matrices in three forms:

$$\mathbf{X}_{(1)} = \mathbf{X}^{L \times MN} \in \mathbb{R}_+^{L \times MN}, \tag{6.22}$$

$$\mathbf{X}_{(2)} = \mathbf{X}^{M \times NL} \in \mathbb{R}_+^{M \times NL}, \tag{6.23}$$

$$\mathbf{X}_{(3)} = \mathbf{X}^{N \times LM} \in \mathbb{R}_+^{N \times LM}. \tag{6.24}$$

These three matrices are constructed by matricizing or flattening the original tensor in three different horizons. The approximation in Eq. (6.20) is thus equivalent to the following three approximations:

$$\mathbf{X}_{(1)} \approx \widehat{\mathbf{X}}_{(1)} = \mathbf{C} \left(\sum_{\phi,\tau} \overset{\downarrow\tau}{\mathbf{W}^{\phi}} \star \overset{\downarrow\phi}{\mathbf{B}^{\tau}} \right)^{\mathsf{T}}, \tag{6.25}$$

$$\mathbf{X}_{(2)} \approx \widehat{\mathbf{X}}_{(2)} = \sum_{\phi,\tau} \overset{\downarrow\phi}{\mathbf{B}^{\tau}} \left(\overset{\downarrow\tau}{\mathbf{W}^{\phi}} \star \mathbf{C} \right)^{\mathsf{T}}, \tag{6.26}$$

$$\mathbf{X}_{(3)} \approx \widehat{\mathbf{X}}_{(3)} = \sum_{\phi,\tau} \overset{\downarrow\tau}{\mathbf{W}^{\phi}} \left(\overset{\downarrow\phi}{\mathbf{B}^{\tau}} \star \mathbf{C} \right)^{\mathsf{T}} \tag{6.27}$$

where $\downarrow \phi$ denotes the downward shift operator which moves each row in the matrix by ϕ positions with zero-padding the rows. NTF is therefore solved by minimizing a divergence measure based on these matrices

$$\mathcal{D}(\mathbf{X}_{(p)} \| \widehat{\mathbf{X}}_{(p)}), \qquad p = 1, 2, 3, \tag{6.28}$$

where the squared Euclidean distance $\mathcal{D}_{\mathrm{EU}}(\mathbf{X}_{(p)} \| \widehat{\mathbf{X}}_{(p)})$ or the Kullback–Leibler divergence $\mathcal{D}_{\mathrm{KL}}(\mathbf{X}_{(p)} \| \widehat{\mathbf{X}}_{(p)})$ is applied.

Squared Euclidean Distance

When minimizing the squared Euclidean distance

$$\mathcal{D}_{\mathrm{EU}}(\mathbf{X}_{(p)} \| \widehat{\mathbf{X}}_{(p)})$$
$$= \| \mathbf{X}_{(p)} - \widehat{\mathbf{X}}_{(p)} \|^2 = \sum_{i,j} \left([\mathbf{X}_{(p)}]_{ij} - [\widehat{\mathbf{X}}_{(p)}]_{ij} \right)^2, \tag{6.29}$$

the multiplicative updating rules for three factorized matrices for operation channels, frequency bases and temporal weights are yielded by

$$\mathbf{C} \leftarrow \mathbf{C} \odot \frac{\mathbf{X}_{(1)} \mathbf{Z}}{\mathbf{C} \mathbf{Z}^{\mathsf{T}} \mathbf{Z}}, \tag{6.30}$$

$$\mathbf{B}^{\tau} \leftarrow \mathbf{B}^{\tau} \odot \frac{\sum_{\phi} \overset{\uparrow\phi}{\mathbf{X}}_{(2)} \left(\overset{\downarrow\tau}{\mathbf{W}^{\phi}} \star \mathbf{C} \right)}{\sum_{\phi} \overset{\uparrow\phi}{\widehat{\mathbf{X}}}_{(2)} \left(\overset{\downarrow\tau}{\mathbf{W}^{\phi}} \star \mathbf{C} \right)}, \tag{6.31}$$

$$
\mathbf{W}^{\phi} \leftarrow \mathbf{W}^{\phi} \odot \frac{\sum_{\tau} \overset{\uparrow\tau}{\mathbf{X}_{(3)}} \left(\overset{\downarrow\phi}{\mathbf{B}^{\tau}} \star \mathbf{C} \right)}{\sum_{\tau} \overset{\uparrow\tau}{\widehat{\mathbf{X}}_{(3)}} \left(\overset{\downarrow\phi}{\mathbf{B}^{\tau}} \star \mathbf{C} \right)},
\tag{6.32}
$$

respectively, where $\mathbf{A} \odot \mathbf{B}$ denotes the element-wise multiplication of matrices $\mathbf{A}$ and $\mathbf{B}$, $\frac{\mathbf{A}}{\mathbf{B}}$ denotes the element-wise division of matrices $\mathbf{A}$ and $\mathbf{B}$, and an auxiliary matrix is defined by

$$
\mathbf{Z} \triangleq \sum_{\phi,\tau} \overset{\downarrow\phi}{\mathbf{B}^{\tau}} \star \overset{\downarrow\tau}{\mathbf{W}^{\phi}}.
\tag{6.33}
$$

It is suggested that the matrices $\mathbf{C}$ and $\mathbf{B}^{\tau}$ are normalized to have unit norm via (Mørup and Schmidt, 2006b)

$$
C_{lk} = \frac{C_{lk}}{\|\mathbf{c}_k\|},
\tag{6.34}
$$

$$
B_{mk}^{\tau} = \frac{B_{mk}^{\tau}}{\|\mathbf{b}_k\|}
\tag{6.35}
$$

for each channel l, component k, frequency m and delay τ.

Kullback–Leibler Divergence

When minimizing the Kullback–Leibler divergence

$$
\begin{aligned}
&\mathcal{D}_{\mathrm{KL}}(\mathbf{X}_{(p)} \| \widehat{\mathbf{X}}_{(p)}) \\
&= \sum_{i,j} \left([\mathbf{X}_{(p)}]_{ij} \log \frac{[\mathbf{X}_{(p)}]_{ij}}{[\widehat{\mathbf{X}}_{(p)}]_{ij}} + [\widehat{\mathbf{X}}_{(p)}]_{ij} - [\mathbf{X}_{(p)}]_{ij} \right),
\end{aligned}
\tag{6.36}
$$

the multiplicative updating rules for three factorized matrices $\mathbf{C}$, $\mathbf{B}^{\tau}$ and $\mathbf{W}^{\phi}$ for different frequency note ϕ and time delay τ are derived as

$$
\mathbf{C} \leftarrow \mathbf{C} \odot \frac{\frac{\mathbf{X}_{(1)}}{\mathbf{C}\mathbf{Z}^{\top}}\mathbf{Z}}{\mathbf{1}\mathbf{Z}},
\tag{6.37}
$$

$$
\mathbf{B}^{\tau} \leftarrow \mathbf{B}^{\tau} \odot \frac{\sum_{\phi} \left[\overset{\uparrow\phi}{\frac{\mathbf{X}_{(2)}}{\widehat{\mathbf{X}}_{(2)}}} \right] \left(\overset{\downarrow\tau}{\mathbf{W}^{\phi}} \star \mathbf{C} \right)}{\sum_{\phi} \mathbf{1} \left(\overset{\downarrow\tau}{\mathbf{W}^{\phi}} \star \mathbf{C} \right)},
\tag{6.38}
$$

$$\mathbf{W}^{\phi} \leftarrow \mathbf{W}^{\phi} \odot \frac{\sum_{\tau}\left[\overset{\uparrow\tau}{\frac{\mathbf{X}_{(3)}}{\widehat{\mathbf{X}}_{(3)}}}\right]\left(\overset{\downarrow\phi}{\mathbf{B}^{\tau}} \star \mathbf{C}\right)}{\sum_{\tau} \mathbf{1}\left(\overset{\downarrow\phi}{\mathbf{B}^{\tau}} \star \mathbf{C}\right)}, \tag{6.39}$$

respectively, where $\mathbf{1}$ is a matrix of ones. Note that the matrices $\mathbf{C}$ and $\mathbf{B}^{\tau}$ should be normalized to have unit norm.

6.2.2 SPARSE NONNEGATIVE TENSOR FACTOR DECONVOLUTION

On the other hand, the sparsity constraint can be imposed in the weight parameters $\mathbf{W}$ to derive the sparse solution to nonnegative tensor factor deconvolution problem. Consequently, the sparse version of NTF2D is established by minimizing the regularized divergence measure

$$\mathcal{D}(\mathbf{X}_{(p)}\|\widehat{\mathbf{X}}_{(p)}) + \lambda\|\mathbf{W}\|_{1}, \qquad p = 1, 2, 3, \tag{6.40}$$

where the lasso regularization is enforced to construct the ℓ_1-regularized objective function. The issue of overcomplete representation can be resolved. Again, we consider two divergence measures as learning objectives for implementation of sparse NTF2D.

Squared Euclidean Distance

Considering the squared Euclidean distance as a learning objective, the multiplicative updating rules for three factorized matrices are derived by extending the solutions which have been obtained for NTF2D in Section 6.2.1. The solutions are expressed by

$$\mathbf{C} \leftarrow \mathbf{C} \odot \frac{\mathbf{X}_{(1)}\mathbf{Z} + \mathbf{C}\operatorname{diag}\left(\mathbf{1}((\mathbf{C}\mathbf{Z}^{T}\mathbf{Z})\odot\mathbf{C})\right)}{\mathbf{C}\mathbf{Z}^{T}\mathbf{Z} + \mathbf{C}\operatorname{diag}\left(\mathbf{1}((\mathbf{X}_{(1)}\mathbf{Z})\odot\mathbf{C})\right)}, \tag{6.41}$$

$$\mathbf{B}^{\tau} \leftarrow \mathbf{B}^{\tau} \odot \frac{\sum_{\phi}\overset{\uparrow\phi}{\mathbf{X}_{(2)}}\left(\overset{\downarrow\tau}{\mathbf{W}^{\phi}} \star \mathbf{C}\right) + \mathbf{B}^{\tau}\operatorname{diag}\left(\mathbf{1}\sum_{\tau}\left(\overset{\uparrow\phi}{\widehat{\mathbf{X}}_{(2)}}\left(\overset{\downarrow\tau}{\mathbf{W}^{\phi}} \star \mathbf{C}\right)\right)\odot\mathbf{B}^{\tau}\right)}{\sum_{\phi}\overset{\uparrow\phi}{\widehat{\mathbf{X}}_{(2)}}\left(\overset{\downarrow\tau}{\mathbf{W}^{\phi}} \star \mathbf{C}\right) + \mathbf{B}^{\tau}\operatorname{diag}\left(\mathbf{1}\sum_{\tau}\left(\overset{\uparrow\phi}{\mathbf{X}_{(2)}}\left(\overset{\downarrow\tau}{\mathbf{W}^{\phi}} \star \mathbf{C}\right)\right)\odot\mathbf{B}^{\tau}\right)}, \tag{6.42}$$

$$\mathbf{W}^{\phi} \leftarrow \mathbf{W}^{\phi} \odot \frac{\sum_{\tau}\overset{\uparrow\tau}{\mathbf{X}_{(3)}}\left(\overset{\downarrow\phi}{\mathbf{B}^{\tau}} \star \mathbf{C}\right)}{\sum_{\tau}\overset{\uparrow\tau}{\widehat{\mathbf{X}}_{(3)}}\left(\overset{\downarrow\phi}{\mathbf{B}^{\tau}} \star \mathbf{C}\right) + \lambda} \tag{6.43}$$

where $\operatorname{diag}(\mathbf{a})$ denotes a diagonal matrix containing the diagonal entries with the values in $\mathbf{a}$. Again, the matrices $\mathbf{C}$ and $\mathbf{B}^{\tau}$ are also normalized to have unit norm.

Kullback–Leibler Divergence

Considering the Kullback–Leibler divergence as a learning objective, the multiplicative updating rules are derived by extended the solutions obtained for NTF2D in Section 6.2.1. The sparse NTF2D is

Table 6.1 Comparison of multiplicative updating rules of NMF2D and NTF2D based on the objective functions of squared Euclidean distance and Kullback–Leibler divergence

	NMF2D	NTF2D
EU	$\mathbf{B}^{\tau} \leftarrow \mathbf{B}^{\tau} \odot \dfrac{\sum_{\phi} \overset{\uparrow\phi}{\mathbf{X}} \overset{\tau\rightarrow}{\mathbf{W}^{\phi}}{}^{T}}{\sum_{\phi} \overset{\uparrow\phi}{\mathbf{X}} \overset{\tau\rightarrow}{\mathbf{W}^{\phi}}{}^{T}}$	$\mathbf{B}^{\tau} \leftarrow \mathbf{B}^{\tau} \odot \dfrac{\sum_{\phi} \overset{\uparrow\phi}{\mathbf{X}}_{(2)} \left(\overset{\downarrow\tau}{\mathbf{W}^{\phi}} \star \mathbf{C}\right)}{\sum_{\phi} \overset{\uparrow\phi}{\mathbf{X}}_{(2)} \left(\overset{\downarrow\tau}{\mathbf{W}^{\phi}} \star \mathbf{C}\right)}$
	$\mathbf{W}^{\phi} \leftarrow \mathbf{W}^{\phi} \odot \dfrac{\sum_{\tau} \overset{\downarrow\phi}{\mathbf{B}^{\tau}}{}^{T} \overset{\leftarrow\tau}{\mathbf{X}}}{\sum_{\tau} \overset{\downarrow\phi}{\mathbf{B}^{\tau}}{}^{T} \overset{\leftarrow\tau}{\mathbf{X}}}$	$\mathbf{W}^{\phi} \leftarrow \mathbf{W}^{\phi} \odot \dfrac{\sum_{\tau} \overset{\uparrow\tau}{\mathbf{X}}_{(3)} \left(\overset{\downarrow\phi}{\mathbf{B}^{\tau}} \star \mathbf{C}\right)}{\sum_{\tau} \overset{\uparrow\tau}{\mathbf{X}}_{(3)} \left(\overset{\downarrow\phi}{\mathbf{B}^{\tau}} \star \mathbf{C}\right)}$
KL	$\mathbf{B}^{\tau} \leftarrow \mathbf{B}^{\tau} \odot \dfrac{\sum_{\phi} \left[\overset{\uparrow\phi}{\dfrac{\mathbf{X}}{\bar{\mathbf{X}}}}\right] \overset{\tau\rightarrow}{\mathbf{W}^{\phi}}{}^{T}}{\sum_{\phi} \mathbf{1} \overset{\tau\rightarrow}{\mathbf{W}^{\phi}}{}^{T}}$	$\mathbf{B}^{\tau} \leftarrow \mathbf{B}^{\tau} \odot \dfrac{\sum_{\phi} \left[\overset{\uparrow\phi}{\dfrac{\mathbf{X}_{(2)}}{\bar{\mathbf{X}}_{(2)}}}\right] \left(\overset{\downarrow\tau}{\mathbf{W}^{\phi}} \star \mathbf{C}\right)}{\sum_{\phi} \mathbf{1} \left(\overset{\downarrow\tau}{\mathbf{W}^{\phi}} \star \mathbf{C}\right)}$
	$\mathbf{W}^{\phi} \leftarrow \mathbf{W}^{\phi} \odot \dfrac{\sum_{\tau} \overset{\downarrow\phi}{\mathbf{B}^{\tau}}{}^{T} \left[\overset{\leftarrow\tau}{\dfrac{\mathbf{X}}{\bar{\mathbf{X}}}}\right]}{\sum_{\tau} \overset{\downarrow\phi}{\mathbf{B}^{\tau}}{}^{T} \mathbf{1}}$	$\mathbf{W}^{\phi} \leftarrow \mathbf{W}^{\phi} \odot \dfrac{\sum_{\tau} \left[\overset{\uparrow\tau}{\dfrac{\mathbf{X}_{(3)}}{\bar{\mathbf{X}}_{(3)}}}\right] \left(\overset{\downarrow\phi}{\mathbf{B}^{\tau}} \star \mathbf{C}\right)}{\sum_{\tau} \mathbf{1} \left(\overset{\downarrow\phi}{\mathbf{B}^{\tau}} \star \mathbf{C}\right)}$

formulated by

$$\mathbf{C} \leftarrow \mathbf{C} \odot \frac{\frac{\mathbf{X}_{(1)}}{\mathbf{C}\mathbf{Z}^{T}}\mathbf{Z} + \mathbf{C}\,\mathrm{diag}(\mathbf{1}((\mathbf{1}\mathbf{Z}) \odot \mathbf{C}))}{\mathbf{1}\mathbf{Z} + \mathbf{C}\,\mathrm{diag}\left(\mathbf{1}\left(\left(\frac{\mathbf{X}_{(1)}}{\mathbf{C}\mathbf{Z}^{T}}\mathbf{Z}\right) \odot \mathbf{C}\right)\right)}, \tag{6.44}$$

$$\mathbf{B}^{\tau} \leftarrow \mathbf{B}^{\tau} \odot \frac{\sum_{\phi} \left[\overset{\uparrow\phi}{\dfrac{\mathbf{X}_{(2)}}{\bar{\mathbf{X}}_{(2)}}}\right] \left(\overset{\downarrow\tau}{\mathbf{W}^{\phi}} \star \mathbf{C}\right) + \mathbf{B}^{\tau}\,\mathrm{diag}\left(\mathbf{1}\sum_{\tau}\left(\mathbf{1}(\overset{\downarrow\tau}{\mathbf{W}^{\phi}} \star \mathbf{C})\right) \odot \mathbf{B}^{\tau}\right)}{\sum_{\phi} \mathbf{1}\left(\overset{\downarrow\tau}{\mathbf{W}^{\phi}} \star \mathbf{C}\right) + \mathbf{B}^{\tau}\,\mathrm{diag}\left(\mathbf{1}\sum_{\tau}\left(\left[\overset{\uparrow\phi}{\dfrac{\mathbf{X}_{(2)}}{\bar{\mathbf{X}}_{(2)}}}\right]\left(\overset{\downarrow\tau}{\mathbf{W}^{\phi}} \star \mathbf{C}\right)\right) \odot \mathbf{B}^{\tau}\right)}, \tag{6.45}$$

$$\mathbf{W}^{\phi} \leftarrow \mathbf{W}^{\phi} \odot \frac{\sum_{\tau} \left[\overset{\uparrow\tau}{\dfrac{\mathbf{X}_{(3)}}{\bar{\mathbf{X}}_{(3)}}}\right] \left(\overset{\downarrow\phi}{\mathbf{B}^{\tau}} \star \mathbf{C}\right)}{\sum_{\tau} \mathbf{1}\left(\overset{\downarrow\phi}{\mathbf{B}^{\tau}} \star \mathbf{C}\right) + \lambda}. \tag{6.46}$$

Now tensor factorization is compared with matrix factorization. Table 6.1 shows a comparison of multiplicative updating rules based on NMF2D with NTF2D with different objective functions. For consistency, only the rules of basis matrix $\mathbf{B}^{\tau}$ and weight matrix $\mathbf{W}^{\phi}$ under different delay τ and note ϕ are shown. The additional channel matrix $\mathbf{C}$ in NTF2D is excluded in this table. We can see that, for the updating rule of $\mathbf{B}^{\tau}$ using the squared Euclidean distance, the variable $\mathbf{X}$ in the solution to NMF2D is replaced by $\mathbf{X}_{(2)}$ in the solution to NTF2D and also $\mathbf{W}^{\phi}$ in NMF2D is replaced by $\mathbf{W}^{\phi} \star \mathbf{C}$

in NTF2D. For the updating rule of $\mathbf{W}^\phi$ using the squared Euclidean distance, the variable $\mathbf{X}$ in the solution to NMF2D is replaced by $\mathbf{X}_{(3)}$ in the solution to NTF2D and also $\mathbf{B}^\tau$ in NMF2D is replaced by $\mathbf{B}^\tau \star \mathbf{C}$ in NTF2D.

Sparse nonnegative tensor factor deconvolution was proposed for multichannel time–frequency analysis and source separation (Mørup and Schmidt, 2006b). A two-dimensional convolutive model in the time, as well as log-frequency, axis was implemented. Log-frequency spectrograms of stereo music signals $\mathcal{X} \in \mathbb{R}_+^{L \times M \times N}$ under mixing conditions were collected as an observed tensor. The instrument signals based on harp and flute were mixed. Different delays and notes were investigated for matrix factorization using NMF2D and tensor factorization using NTF2D. It has been shown that sparse NTF2D performed better than sparse NMF2D. NTF2D was likely to be less overcomplete when compared with NMF2D. The framework used here is generalizable to a wide range of higher order data analysis where 2-dimensional convolutive NTF could be extended to 3-dimensional or 4-dimensional convolutive NTF. Nevertheless, the choice of regularization parameter λ in sparse solution to NTF2D shows influence and should be further evaluated.

6.3 PROBABILISTIC NTF

This section develops a probabilistic framework for nonnegative tensor factorization (NTF) (Schmidt and Mohamed, 2009), which will be further extended to Bayesian learning for tensor factorization as addressed in Section 6.4. This framework is constructed by starting from a probabilistic matrix factorization (PMF) (Salakhutdinov and Mnih, 2008) in Section 6.3.1 and then extending to the probabilistic tensor factorization (Xiong et al., 2010) in Section 6.3.2. Section 6.3.3 will address a solution to probabilistic nonnegative tensor factorization problem where Markov chain Monte Carlo inference will be implemented. These tensor factorization methods were developed for collaborative filtering and could be implemented for single-channel source separation.

6.3.1 PROBABILISTIC MATRIX FACTORIZATION

Collaborating filtering algorithms have been developed for recommendation systems, and applied in many real-world applications such as *Netflix*, *YouTube*, and *Amazon*. How to analyze such a large-scale data set is crucial. Matrix and tensor factorization with or without probabilistic treatment are all feasible to carry out efficient algorithms to deal with this issue. In general, the observed matrix $\mathbf{X} = \{X_{mn}\}$ contains the rating values of M users and N items (movies, videos or products), which are definitely nonnegative. Some entries in $\mathbf{X}$ may be missing. A sparse matrix may be collected. The goal of collaborative filtering is to analyze the past history of user preferences and ratings and use the analyzed information to make prediction of future rating for a user m over a specific item n. In Salakhutdinov and Mnih (2008), the probabilistic matrix factorization (PMF) was proposed to approximate an observed rating matrix $\mathbf{X} \in \mathbb{R}^{M \times N}$ using a product of a user matrix $\mathbf{B} \in \mathbb{R}^{M \times K}$ and an item matrix $\mathbf{W} \in \mathbb{R}^{K \times N}$. This is comparable with the standard source separation method by using NMF where the observed matrix $\mathbf{X} = \{X_{mn}\}$ is collected as the log-magnitude spectrograms over different frequency bins m and time frames n. The nonnegative matrix $\mathbf{X}$ is factorized as a product of basis matrix $\mathbf{B}$ and weight matrix $\mathbf{W}$. Let the matrices $\mathbf{B}$ and $\mathbf{W}$ be represented by their corresponding row and column vectors, i.e.,

$$\mathbf{B} = [\mathbf{B}_{1:}^\top \cdots \mathbf{B}_{M:}^\top]^\top, \tag{6.47}$$

$$\mathbf{W} = [\mathbf{W}_{:1} \cdots \mathbf{W}_{:N}]. \tag{6.48}$$

PMF is constructed according to probabilistic assumptions. The likelihood function of generating the rating matrix is assumed to follow a Gaussian distribution

$$
\begin{aligned}
&p(\mathbf{X}|\mathbf{B}, \mathbf{W}, \sigma^2) \\
&= \prod_{m,n} \left[\mathcal{N}(X_{mn}|\mathbf{B}_{m:}\mathbf{W}_{:n}, \sigma^2) \right]^{I_{mn}} \\
&= \prod_{m,n} \left[\mathcal{N}\left(X_{mn} \Big| \sum_k B_{mk} W_{kn}, \sigma^2 \right) \right]^{I_{mn}}
\end{aligned}
\tag{6.49}
$$

where $\mathbf{B}_{m:}$ and $\mathbf{W}_{:n}$ denote the K-dimensional user-specific and item-specific latent feature vectors, respectively, σ^2 is a shared variance parameter for all entries in $\mathbf{X}$ and I_{mn} denotes the indicator, which is one when X_{mn} is observed and zero when X_{mn} is missing. The prior densities of PMF parameters are assumed as zero-mean Gaussians with the shared precision parameters α_b and α_w for all entries in matrices $\mathbf{B}$ and $\mathbf{W}$ given by

$$p(\mathbf{B}|\sigma_b^2) = \prod_{m,k} \mathcal{N}\left(B_{mk}|0, \sigma_b^2 \right), \tag{6.50}$$

$$p(\mathbf{W}|\sigma_w^2) = \prod_{n,k} \mathcal{N}\left(W_{nk}|0, \sigma_w^2 \right). \tag{6.51}$$

The maximum *a posteriori* (MAP) estimates of $\mathbf{B}$ and $\mathbf{W}$ are calculated by maximizing the logarithm of the posterior distribution over the user and item matrices given by the fixed variances $\{\sigma^2, \sigma_b^2, \sigma_w^2\}$. Accordingly, maximizing the logarithm of the posterior distribution is equivalent to minimizing the following objective function:

$$
\begin{aligned}
&\frac{1}{2} \sum_{m,n} I_{mn} \left(X_{mn} - \sum_k B_{mk} W_{kn} \right)^2 \\
&+ \frac{\lambda_b}{2} \sum_{m,k} B_{mk}^2 + \frac{\lambda_w}{2} \sum_{n,k} W_{nk}^2
\end{aligned}
\tag{6.52}
$$

where the regularization parameters are determined by using different variance parameters,

$$\lambda_b = \frac{\sigma^2}{\sigma_b^2} \quad \text{and} \quad \lambda_w = \frac{\sigma^2}{\sigma_w^2}. \tag{6.53}$$

In the implementation, selecting proper hyperparameters $\{\sigma^2, \sigma_b^2, \sigma_w^2\}$ or regularization parameters $\{\lambda_b, \lambda_w\}$ is crucial for model regularization. Generalization for future data based on PMF depends on the complexity control. Particularly, the generalization in the presence of sparse and imbalanced datasets is critical in real-world applications. Gibbs sampling was developed to realize the Bayesian PMF (Salakhutdinov and Mnih, 2008) with the fixed hyperparameters or regularization parameters. In Ma et al. (2015), a related work was proposed for probabilistic matrix factorization where variational Bayesian inference was implemented.

6.3.2 PROBABILISTIC TENSOR FACTORIZATION

Traditionally, the collaborative filtering algorithms are not dynamic and cannot deal with the nonstationary models where the statistics of a model is changed with time. In Xiong et al. (2010), temporal collaborative filtering was proposed to handle this issue by incorporating the additional time horizon in the factorization model. Temporal relation data are included for the factorized representation. Probabilistic matrix factorization is then extended to the probabilistic tensor factorization (PTF), which allows collaborative filtering to be performed by using three-way tensor data $\mathcal{X} \in \mathbb{R}^{L \times M \times N}$ containing the rating values over L times, M users and N items. As a result, a three-way tensor is reconstructed according to the PARAFAC model or a specialized CP model

$$\mathcal{X} \approx \widehat{\mathcal{X}} = \sum_k \mathbf{c}_k \circ \mathbf{b}_k \circ \mathbf{w}_k \qquad (6.54)$$

where $\circ$ denotes the vector outer product, and the time matrix

$$\mathbf{C} = \{\mathbf{c}_k\} = \{C_{lk}\} \in \mathbb{R}^{L \times K}, \qquad (6.55)$$

user matrix

$$\mathbf{B} = \{\mathbf{b}_k\} = \{B_{mk}\} \in \mathbb{R}^{M \times K}, \qquad (6.56)$$

and item matrix

$$\mathbf{W} = \{\mathbf{w}_k\} = \{W_{nk}\} \in \mathbb{R}^{N \times K} \qquad (6.57)$$

are included in tensor factorization. Fig. 2.10 illustrates the reconstruction of the three-way tensor based on the summation of outer products. Similar to PMF, the likelihood function of generating the rating tensor is expressed by Gaussian distributions

$$p(\mathcal{X}|\mathbf{C}, \mathbf{B}, \mathbf{W}, \sigma^2)$$
$$= \prod_{l,m,n} \left[\mathcal{N} \left(\mathcal{X}_{lmn} \middle| \sum_k C_{lk} B_{mk} W_{nk}, \sigma^2 \right) \right]^{I_{lmn}}. \qquad (6.58)$$

Note that if $\mathbf{C}$ is a matrix of ones, this PTF model is then simplified to a PMF model. Again, the prior densities of the user matrix $\mathbf{B}$ and item matrix $\mathbf{W}$ are assumed by individual zero-mean Gaussians with variance parameters σ_b^2 and σ_w^2, respectively:

$$p(\mathbf{B}|\sigma_b^2) = \prod_{m,k} \mathcal{N} \left(B_{mk}|0, \sigma_b^2 \right), \qquad (6.59)$$

$$p(\mathbf{W}|\sigma_w^2) = \prod_{n,k} \mathcal{N}\left(W_{nk}|0, \sigma_w^2\right), \tag{6.60}$$

while the prior density of time matrix $\mathbf{C}$ is defined by a Gaussian distribution, which takes the continuity of time l into account as

$$p(\mathbf{C}|\sigma_c^2) = \sum_{l,k} \mathcal{N}\left(C_{lk}|C_{l-1,k}, \sigma_{ck}^2\right) \tag{6.61}$$

where the component-dependent variance parameters

$$\sigma_c^2 = \{\sigma_{ck}^2\}_{k=1}^K \tag{6.62}$$

are merged. The distribution of C_{lk} at the current time l is driven by a Gaussian mean $C_{l-1,k}$ at the previous time $l-1$. Starting from the beginning time $l=0$ with the assumption

$$p(\mathbf{C}_0|\sigma_0^2) = \mathcal{N}\left(\mathbf{C}_0|\boldsymbol{\mu}_c, \sigma_0^2 \mathbf{I}\right) \tag{6.63}$$

where $\mathbf{C}_0$ denotes an initial vector, $\boldsymbol{\mu}_c$ denotes the K-dimensional mean vector, σ_0^2 denotes the variance parameter, $\mathbf{I}$ denotes the $K \times K$ identity matrix, the MAP estimation of three factor matrices $\mathbf{C}$, $\mathbf{B}$ and $\mathbf{W}$ and one initial time vector $\mathbf{C}_0$ is performed by maximizing the logarithm of the posterior distribution

$$\begin{aligned}
\log p(\mathbf{C}, \mathbf{C}_0, \mathbf{B}, \mathbf{W}|\mathcal{X}) \\
\propto \log p(\mathcal{X}|\mathbf{C}, \mathbf{B}, \mathbf{W}, \sigma^2) + \log p(\mathbf{C}|\sigma_c^2) + \log p(\mathbf{C}_0|\sigma_0^2) \\
+ \log p(\mathbf{B}|\sigma_b^2) + \log p(\mathbf{W}|\sigma_w^2).
\end{aligned} \tag{6.64}$$

Maximizing the log-posterior distribution is equivalent to minimizing the following objective function:

$$\begin{aligned}
\frac{1}{2} \sum_{l,m,n} I_{lmn} \left(\mathcal{X}_{lmn} - \sum_k C_{lk} B_{mk} W_{nk}\right)^2 \\
+ \frac{\lambda_{ck}}{2} \sum_{l,k} (C_{lk} - C_{l-1,k})^2 + \frac{\lambda_0}{2} \|\mathbf{C}_0 - \boldsymbol{\mu}_c\|^2 \\
+ \frac{\lambda_b}{2} \sum_{m,k} B_{mk}^2 + \frac{\lambda_w}{2} \sum_{n,k} W_{nk}^2
\end{aligned} \tag{6.65}$$

where four regularization parameters are determined by

$$\lambda_{ck} = \frac{\sigma^2}{\sigma_{ck}^2}, \qquad \lambda_0 = \frac{\sigma^2}{\sigma_0^2}, \qquad \lambda_b = \frac{\sigma^2}{\sigma_b^2}, \qquad \lambda_w = \frac{\sigma^2}{\sigma_w^2}. \tag{6.66}$$

Consequently, MAP solutions to $\mathbf{C}$, $\mathbf{B}$ and $\mathbf{W}$ depend on the hyperparameters

$$\{\sigma_{ck}^2, \sigma_0^2, \sigma_b^2, \sigma_w^2\}$$

or regularization parameters

$$\{\lambda_{ck}, \lambda_0, \lambda_b, \lambda_w\},$$

which should be carefully selected to control the model complexity and avoid parameter overfitting so that a desirable prediction can be achieved. Section 6.4 will present a Bayesian approach to PTF where the hyperparameters or regularization parameters are automatically inferred in a Bayesian learning procedure instead of empirically selected from validation data.

6.3.3 PROBABILISTIC NONNEGATIVE TENSOR FACTORIZATION

Basically, the probabilistic tensor factorization addressed in Section 6.3.2 does not impose the nonnegativity constraint in factor matrices $\mathbf{C}$, $\mathbf{B}$ and $\mathbf{W}$ so that the property of parts-based representation is not satisfied. In Schmidt and Mohamed (2009), a probabilistic framework of nonnegative tensor factorization (NTF) was proposed to carry out the parts-based representation based on tensor observation. The probabilistic NTF was constructed according to the PARAFAC representation, which was provided in Eqs. (6.2) and (6.6). Nonnegative factor matrices $\mathbf{C} \in \mathbb{R}_+^{L \times K}$, $\mathbf{B} \in \mathbb{R}_+^{M \times K}$ and $\mathbf{W} \in \mathbb{R}_+^{N \times K}$ are used in signal reconstruction of nonnegative tensor $\mathcal{X} \in \mathbb{R}_+^{L \times M \times N}$. In the implementation, each entry in the observed tensor $\mathcal{X}$ is represented by a Gaussian distribution given by the mean using the reconstructed signal $\widehat{\mathcal{X}}_{lmn}$ and the shared variance σ^2 across different entries, namely

$$
p(\mathcal{X}_{lmn} | \widehat{\mathcal{X}}_{lmn}, \sigma^2)
$$
$$
= \mathcal{N}\left(\mathcal{X}_{lmn} \,\middle|\, \sum_k C_{lk} B_{mk} W_{nk}, \sigma^2\right)
$$
$$
= \frac{1}{\sqrt{2\pi\sigma^2}} \exp\left\{-\frac{1}{2\sigma^2}\left(\mathcal{X}_{lmn} - \sum_k C_{lk} B_{mk} W_{nk}\right)^2\right\}. \tag{6.67}
$$

Hierarchical Bayesian Model

For ease of Bayesian inference, a conjugate prior is introduced to characterize the data variance, namely an inverse Gamma distribution with shape parameter α and scale parameter β, that is,

$$
p(\sigma^2) = \text{Inv-Gam}(\sigma^2 | \alpha, \beta)
$$
$$
= \frac{\beta^\alpha}{\Gamma(\alpha)} (\sigma^2)^{-\alpha-1} \exp\left(-\frac{\beta}{\sigma^2}\right) \tag{6.68}
$$

is assumed. In particular, the latent factors in different ways are drawn by the rectified Gaussian priors:

$$
C_{lk} \sim \mathcal{R}\left(C_{lk} | \mu_{C_{lk}}, \sigma^2_{C_{lk}}\right), \tag{6.69}
$$

$$
B_{mk} \sim \mathcal{R}\left(B_{mk} | \mu_{B_{mk}}, \sigma^2_{B_{mk}}\right), \tag{6.70}
$$

$$
W_{nk} \sim \mathcal{R}\left(W_{nk} | \mu_{W_{nk}}, \sigma^2_{W_{nk}}\right) \tag{6.71}
$$

where the rectified Gaussian distribution is expressed by (Schmidt and Mohamed, 2009)

$$
\begin{aligned}
&\mathcal{R}\left(\theta | \mu_\theta, \sigma_\theta^2\right) \\
&= \frac{\sqrt{\frac{2}{\pi \sigma_\theta^2}}}{\text{erfc}\left(\frac{-\mu_\theta}{\sqrt{2\sigma_\theta^2}}\right)} \exp\left\{-\frac{(\theta - \mu_\theta)^2}{2\sigma_\theta^2}\right\} U(\theta).
\end{aligned}
\tag{6.72}
$$

In Eq. (6.72), $U(\theta)$ denotes the Heaviside unit step function, which is one for $\theta \geq 0$ and zero for $\theta < 0$, and erfc($\cdot$) denotes the error function defined by

$$
\text{erfc}(x) = \frac{1}{\sqrt{\pi}} \int_{-x}^{x} e^{-t^2} dt.
\tag{6.73}
$$

Notably, the rectified Gaussian distribution in Eq. (6.72) is different from the rectified normal distribution, which was adopted for implementation of Gaussian–Exponential Bayesian NMF as introduced in Section 5.3.1. The rectified normal distribution in Eq. (5.106) is constructed by a product of Gaussian and exponential distributions while the rectified Gaussian distribution in Eq. (6.72) is comparable with the truncated Gaussian distribution $\mathcal{N}_+(\theta | \mu_\theta, \sigma_\theta^2)$, which was adopted in Eq. (5.196) for MCMC inference of Bayesian group sparse NMF as addressed in Section 5.4.2.

Using the hierarchical Bayesian model, the conjugate prior for the mean μ_θ and variance σ_θ^2 of a rectified Gaussian distribution is conveniently formed as

$$
p(\mu_\theta, \sigma_\theta^2 | m_\theta, s_\theta, a, b) = \frac{1}{c}\sqrt{\sigma_\theta^2} \cdot \text{erfc}\left(\frac{-\mu_\theta}{\sqrt{2\sigma_\theta^2}}\right)
$$
$$
\times \mathcal{N}(\mu_\theta | m_\theta, s_\theta) \text{Inv-Gam}(\sigma_\theta^2 | a, b)
\tag{6.74}
$$

where c is a normalization constant. Based on this prior density, the hyperparameters μ_θ and σ_θ^2 are decoupled and the corresponding posterior distribution of μ_θ and σ_θ^2 is accumulated as the Gaussian and inverse Gamma distributions, respectively. Notably, Eq. (6.74) is applied for the three factor matrices using parameters $\{\mu_{C_{lk}}, \sigma_{C_{lk}}^2, \mu_{B_{mk}}, \sigma_{B_{mk}}^2, \mu_{W_{nk}}, \sigma_{W_{nk}}^2\}$ based on their shared hyperparameters $\{m_\theta, s_\theta, a, b\}$. A hierarchical Bayesian model is constructed accordingly.

Bayesian Inference

In general, there are three layers in this hierarchical Bayesian model, i.e., parameters $\{\mathbf{C}, \mathbf{B}, \mathbf{W}, \sigma^2\}$, hyperparameters $\{\mu_{C_{lk}}, \sigma_{C_{lk}}^2, \mu_{B_{mk}}, \sigma_{B_{mk}}^2, \mu_{W_{nk}}, \sigma_{W_{nk}}^2, \alpha, \beta\}$ and hyper-hyperparameters $\{m_\theta, s_\theta, a, b\}$. However, for ease of model inference, we simply treat the three layers as two layers $\boldsymbol{\Theta}$ and $\boldsymbol{\Psi}$ in Bayesian inference. To conduct Bayesian learning, the posterior distribution based on the parameters $\boldsymbol{\Theta} = \{\mathbf{C}, \mathbf{B}, \mathbf{W}, \sigma^2, \mu_{C_{lk}}, \sigma_{C_{lk}}^2, \mu_{B_{mk}}, \sigma_{B_{mk}}^2, \mu_{W_{nk}}, \sigma_{W_{nk}}^2\}$ and hyperparameters

$$
\boldsymbol{\Psi} = \{\alpha, \beta, m_\theta, s_\theta, a, b\}
\tag{6.75}
$$

is calculated by

$$p(\boldsymbol{\Theta}|\boldsymbol{\mathcal{X}}, \boldsymbol{\Psi}) \propto p(\boldsymbol{\mathcal{X}}|\boldsymbol{\Theta}) p(\boldsymbol{\Theta}|\boldsymbol{\Psi}) \propto (\sigma^2)^{-\frac{L+M+N}{2}-\alpha-1}$$

$$\times \prod_{l,m,n} \exp\left\{-\frac{1}{2\sigma^2}\left(\mathcal{X}_{lmn} - \sum_k C_{lk} B_{mk} W_{nk}\right)^2\right\} \exp\left(-\frac{\beta}{\sigma^2}\right)$$

$$\times \prod_k \left\{ \prod_l \exp\left[-\frac{(C_{lk} - \mu_{C_{lk}})^2}{2\sigma_{C_{lk}}^2}\right] U(C_{lk}) \exp\left[-\frac{(\mu_{C_{lk}} - m_\theta)^2}{2s_\theta}\right] \right.$$

$$\times (\sigma_{C_{lk}}^2)^{-a-1} \exp\left(-\frac{b}{\sigma_{C_{lk}}^2}\right) \prod_m \exp\left[-\frac{(B_{mk} - \mu_{B_{mk}})^2}{2\sigma_{B_{mk}}^2}\right] U(B_{mk}) \qquad (6.76)$$

$$\times \exp\left[-\frac{(\mu_{B_{mk}} - m_\theta)^2}{2s_\theta}\right] (\sigma_{B_{mk}}^2)^{-a-1} \exp\left(-\frac{b}{\sigma_{B_{mk}}^2}\right)$$

$$\times \prod_n \exp\left[-\frac{(W_{nk} - \mu_{W_{nk}})^2}{2\sigma_{W_{nk}}^2}\right] U(W_{nk}) \exp\left[-\frac{(\mu_{W_{nk}} - m_\theta)^2}{2s_\theta}\right]$$

$$\times (\sigma_{W_{nk}}^2)^{-a-1} \exp\left(-\frac{b}{\sigma_{W_{nk}}^2}\right) \right\}.$$

In Schmidt and Mohamed (2009), Gibbs sampling was introduced to infer J latent variables in model parameters

$$\boldsymbol{\Theta} = \{\mathbf{C}, \mathbf{B}, \mathbf{W}, \sigma^2, \mu_{C_{lk}}, \sigma_{C_{lk}}^2, \mu_{B_{mk}}, \sigma_{B_{mk}}^2, \mu_{W_{nk}}, \sigma_{W_{nk}}^2\}$$
$$\triangleq \{\Theta_j\}_{i=1}^J. \qquad (6.77)$$

The conditional distribution $p(\Theta_i|\boldsymbol{\Theta}_{\backslash i})$ or posterior distribution $p(\Theta_i|\boldsymbol{\mathcal{X}}, \boldsymbol{\Theta}_{\backslash i}, \boldsymbol{\Psi})$ of a target parameter Θ_i given by the observed tensor $\boldsymbol{\mathcal{X}}$ and all the other non-target parameters $\boldsymbol{\Theta}_{\backslash i}$ is calculated and then drawn to fulfill a Gibbs sampling procedure. In each learning epoch, the model parameters in $\boldsymbol{\Theta}$ are sampled sequentially from Θ_1 to Θ_J. A new sample of target parameter $\Theta_i^{(\tau+1)}$ is drawn by using the distribution

$$\Theta_i^{(\tau+1)} \sim p\left(\Theta_i \middle| \Theta_1^{(\tau+1)}, \ldots, \Theta_{i-1}^{(\tau+1)}, \Theta_{i+1}^{(\tau)}, \ldots, \Theta_J^{(\tau)}\right) \qquad (6.78)$$

given by the newest samples including the preceding parameters $\{\Theta_1, \ldots, \Theta_{i-1}\}$ in epoch $\tau + 1$ and the subsequent parameters $\{\Theta_{i+1}, \ldots, \Theta_J\}$ in epoch τ. In the implementation, the posterior distributions for sampling three-way factors

$$\{C_{lk}, B_{mk}, W_{nk}\}$$

are built as individual rectified Gaussian distributions. The posterior distributions for sampling the variance parameters

$$\{\sigma^2, \sigma_{C_{lk}}^2, \sigma_{B_{mk}}^2, \sigma_{W_{nk}}^2\}$$

are formed as the inverse Gamma distributions while the posterior distributions for sampling mean parameters

$$\{\mu_{C_{lk}}, \mu_{B_{mk}}, \mu_{W_{nk}}\}$$

are constructed as the Gaussian distributions. The updated parameters in different posterior distributions should be individually calculated, as detailed in Schmidt and Mohamed (2009), for iteratively sampling and updating in implementation of probabilistic NTF. Such a solution is seen as a Bayesian framework of NTF because of Gibbs sampling which plays a prominent role in modern Bayesian inference.

6.4 BAYESIAN TENSOR FACTORIZATION

In addition to the work Schmidt and Mohamed (2009), we are introducing another Bayesian tensor factorization, which is an extension of the probabilistic matrix factorization (PMF) in Section 6.3.1 and the probabilistic tensor factorization (PTF) in Section 6.3.2 as referred to Xiong et al. (2010). Bayesian tensor factorization was developed for collaborative filtering where the observed tensor $\mathcal{X} = \{\mathcal{X}_{lmn}\}$ was collected as the rating values along the horizons of L times, M users and N items. This Bayesian framework also follows a hierarchical Bayesian model where the distributions of variables in different layers are assumed. A full Bayesian inference is implemented.

6.4.1 HIERARCHICAL BAYESIAN MODEL

In model construction of Bayesian tensor factorization, the likelihood function $p(\mathcal{X}|\mathbf{C}, \mathbf{B}, \mathbf{W}, \sigma^2)$ is defined the same as in Eq. (6.58). The prior distributions of factor parameters $\mathbf{C} = \{C_l\}_{l=1}^{L}$, C_0, $\mathbf{B} = \{B_m\}_{m=1}^{M}$ and $\mathbf{W} = \{W_n\}_{n=1}^{N}$ are defined similar to Eqs. (6.61), (6.63), (6.59) and (6.60), respectively, but now represented by using their corresponding vectors $\{C_l, C_0, B_m, W_n\}$ along different channels l, frequency m and time n with nonzero mean vectors $\{\boldsymbol{\mu}_c, \boldsymbol{\mu}_b, \boldsymbol{\mu}_w\}$ and full precision matrices $\{\mathbf{R}_c, \mathbf{R}_b, \mathbf{R}_w\}$, that is,

$$p(C_l|\mathbf{R}_c^{-1}) = \mathcal{N}\left(C_l|C_{l-1}, \mathbf{R}_c^{-1}\right), \tag{6.79}$$

$$p(C_0|\boldsymbol{\mu}_c, \mathbf{R}_c^{-1}) = \mathcal{N}\left(C_0|\boldsymbol{\mu}_c, \mathbf{R}_c^{-1}\right), \tag{6.80}$$

$$p(B_m|\boldsymbol{\mu}_b, \mathbf{R}_b^{-1}) = \mathcal{N}\left(B_m|\boldsymbol{\mu}_b, \mathbf{R}_b^{-1}\right), \tag{6.81}$$

$$p(W_n|\boldsymbol{\mu}_w, \mathbf{R}_w^{-1}) = \mathcal{N}\left(W_n|\boldsymbol{\mu}_w, \mathbf{R}_w^{-1}\right). \tag{6.82}$$

The hierarchical Bayesian model is constructed for Bayesian tensor factorization. In this representation, model parameters consist of three-way factor matrices, as well as their Gaussian parameters, i.e.,

$$\begin{aligned}
\boldsymbol{\Theta} &= \{\mathbf{C}, \mathbf{B}, \mathbf{W}, \sigma^2, \boldsymbol{\Theta}_c = \{\boldsymbol{\mu}_c, \mathbf{R}_c\}, \\
\boldsymbol{\Theta}_b &= \{\boldsymbol{\mu}_b, \mathbf{R}_b\}, \boldsymbol{\Theta}_w = \{\boldsymbol{\mu}_w, \mathbf{R}_w\}\}.
\end{aligned} \tag{6.83}$$

The hyperparameters $\boldsymbol{\Psi}$ are composed of those parameters assumed in the distributions for $\mathbf{C}$, $\mathbf{B}$ and $\mathbf{W}$. Here, the mean vectors $\{\boldsymbol{\mu}_c, \boldsymbol{\mu}_b, \boldsymbol{\mu}_w\}$ and precision matrices $\{\mathbf{R}_c, \mathbf{R}_b, \mathbf{R}_w\}$ are represented by the conjugate priors which are constructed by the Gaussian–Wishart distribution:

$$p(\boldsymbol{\Theta}_c) = p(\boldsymbol{\mu}_c|\mathbf{R}_c)p(\mathbf{R}_c)$$
$$= \mathcal{N}\left(\boldsymbol{\mu}_c|\boldsymbol{\mu}_0, (\beta_0\mathbf{R}_c)^{-1}\right)\mathcal{W}\left(\mathbf{R}_c|\rho_0, \mathbf{V}_0\right), \tag{6.84}$$

$$p(\boldsymbol{\Theta}_b) = p(\boldsymbol{\mu}_b|\mathbf{R}_b)p(\mathbf{R}_b)$$
$$= \mathcal{N}\left(\boldsymbol{\mu}_b|\boldsymbol{\mu}_0, (\beta_0\mathbf{R}_b)^{-1}\right)\mathcal{W}\left(\mathbf{R}_b|\rho_0, \mathbf{V}_0\right), \tag{6.85}$$

$$p(\boldsymbol{\Theta}_w) = p(\boldsymbol{\mu}_w|\mathbf{R}_w)p(\mathbf{R}_w)$$
$$= \mathcal{N}\left(\boldsymbol{\mu}_w|\boldsymbol{\mu}_0, (\beta_0\mathbf{R}_w)^{-1}\right)\mathcal{W}\left(\mathbf{R}_w|\rho_0, \mathbf{V}_0\right), \tag{6.86}$$

and the variance parameter σ^2 is also modeled by a conjugate prior using the inverse Gamma distribution as given in Eq. (6.68) with hyperparameters $\{\alpha, \beta\}$. The complete set of hyperparameters consists of those shared parameters in Gaussian–Wishart and inverse Gamma distributions, namely,

$$\boldsymbol{\Psi} = \{\boldsymbol{\mu}_0, \beta_0, \rho_0, \mathbf{V}_0, \alpha, \beta\}. \tag{6.87}$$

Here, $\boldsymbol{\mu}_0$ is a mean vector and β_0 is a scaling factor of precision matrix $\mathbf{R}$ for a Gaussian distribution in $p(\boldsymbol{\mu}|\mathbf{R})$ while ρ_0 denotes the degree of freedom and $\mathbf{V}_0$ denotes a $K \times K$ scale matrix for a Wishart distribution in $p(\mathbf{R})$. Wishart distribution is defined in Appendix A.

6.4.2 BAYESIAN LEARNING ALGORITHM

According to full Bayesian inference, we need to calculate the predictive distribution $p(\mathcal{X}|\boldsymbol{\Psi})$ for data reconstruction. The marginal distribution is calculated by integrating out all parameters

$$\boldsymbol{\Theta} = \{\mathbf{C}, \mathbf{B}, \mathbf{W}, \sigma^2, \boldsymbol{\Theta}_c, \boldsymbol{\Theta}_b, \boldsymbol{\Theta}_w\} \tag{6.88}$$

by

$$p(\mathcal{X}|\boldsymbol{\Psi}) = \prod_{l,m,n} p(\mathcal{X}_{lmn}|\boldsymbol{\Psi})$$
$$= \prod_{l,m,n} \int p(\mathcal{X}_{lmn}, \boldsymbol{\Theta}|\boldsymbol{\Psi})d\boldsymbol{\Theta}$$
$$= \prod_{l,m,n} \int p\left(\mathcal{X}_{lmn}\left|\sum_k C_{lk}B_{mk}W_{nk}, \sigma^2\right.\right)$$
$$\times p(\mathbf{C}, \mathbf{B}, \mathbf{W}, \sigma^2, \boldsymbol{\Theta}_c, \boldsymbol{\Theta}_b, \boldsymbol{\Theta}_w|\boldsymbol{\Psi})d\{\mathbf{C}, \mathbf{B}, \mathbf{W}, \sigma^2, \boldsymbol{\Theta}_c, \boldsymbol{\Theta}_b, \boldsymbol{\Theta}_w\}. \tag{6.89}$$

However, since an exact solution of the above predictive distribution is analytically intractable, the Markov Chain Monte Carlo sampling method is introduced to approximate the integral by averaging the likelihood functions using those samples of latent variables

$$\{C_{l'}^{(l)}, B_m^{(l)}, W_n^{(l)}, (\sigma^2)^{(l)}, \Theta_c^{(l)}, \Theta_b^{(l)}, \Theta_w^{(l)}\}$$

drawn from the corresponding posterior distributions. Predictive distribution is accordingly yielded by

$$p(\mathcal{X}|\Psi) \approx \frac{1}{L} \sum_{l=1}^{L} \prod_{l',m,n} p\left(\mathcal{X}_{l'mn} \middle| C_{l'}^{(l)}, B_m^{(l)}, W_n^{(l)}, (\sigma^2)^{(l)}, \Theta_c^{(l)}, \Theta_b^{(l)}, \Theta_w^{(l)}, \Psi\right). \qquad (6.90)$$

In the implementation, there are L samples drawn to run the Gibbs sampling procedure. To avoid notation conflict, the channel index in tensor data has been changed to l', which differs from the sample index l in Gibbs sampling. The posterior distributions of model parameters and hyperparameters are detailed and formulated in Xiong et al. (2010). The Gibbs sampling algorithm for Bayesian tensor factorization is shown in Algorithm 6.1. Bayesian tensor factorization was successfully developed for sales prediction and movie recommendation.

6.5 INFINITE TENSOR FACTORIZATION

Bayesian learning is not only feasible to carry out uncertainty modeling to compensate the randomness in latent variable model for source separation but also helpful to improve the prediction capability via controlling the complexity of source separation model. In Yoshii et al. (2013), an infinite positive semidefinite tensor factorization was proposed for single-channel source separation in the presence of music mixture signals. The Bayesian nonparametric learning was introduced to control model complexity and find an appropriate number of bases for nonnegative tensor factorization. Learning model complexity based on Bayesian nonparametrics has been successfully developed for topic model (Blei et al., 2003) with applications for natural language processing (Chien, 2015a, 2016, 2018, Ma and Leijon, 2011). In what follows, we address model construction for nonnegative tensor factorization in Section 6.5.1 where Bayesian nonparametric model will be introduced. In Section 6.5.3, a solution to Bayesian nonparametric tensor factorization based on variational inference will be formulated.

6.5.1 POSITIVE SEMIDEFINITE TENSOR FACTORIZATION

In audio analysis, the nonnegative matrix factorization based on Itakura–Saito divergence (denoted by IS-NMF), as addressed in Section 2.2.2, is suitable for decomposing a power spectrogram $\mathbf{X}$ into a product of a sound source power spectra $\mathbf{B}$ and the corresponding temporal activations $\mathbf{W}$. IS-NMF is theoretically justified if the frequency bins of source spectra are independent. However, the short-time Fourier transform (STFT) is unable to perfectly decorrelate the frequency components with harmonic structure. It is not appropriate to incorporate the Gamma prior for basis matrix $\mathbf{B}$ as shown in Févotte et al. (2009) and minimize the divergence $\mathcal{D}_{\text{IS}}(\mathbf{X}\|\mathbf{BW})$ in an *element-wise manner*. This issue was tackled by using the positive semidefinite tensor factorization (PSDTF) in Yoshii et al. (2013). As seen in Fig. 6.4, each nonnegative vector at time n is embedded into a PSD matrix that represents the covariance structure of multivariate elements. A tensor $\mathcal{X} \in \mathbb{R}_+^{M \times M \times N}$ is accordingly constructed from the original power spectrogram matrix $\mathbf{X} \in \mathbb{R}_+^{M \times N}$. Tensor data consist of N slices of matrices

---

Algorithm 6.1 Gibbs Sampling for Bayesian Tensor Factorization.

Initialization with $\{\mathbf{C}^{(1)}, \mathbf{B}^{(1)}, \mathbf{W}^{(1)}\}$

For each sampling iteration l

    Sample the hyperparameters using individual posterior distributions

$$\Theta_c^{(l)} \sim p\left(\Theta_c \middle| \mathbf{C}^{(l)}\right)$$

$$\Theta_b^{(l)} \sim p\left(\Theta_b \middle| \mathbf{B}^{(l)}\right)$$

$$\Theta_w^{(l)} \sim p\left(\Theta_w \middle| \mathbf{W}^{(l)}\right)$$

$$(\sigma^2)^{(l)} \sim p\left(\sigma^2 \middle| \mathbf{C}^{(l)}, \mathbf{B}^{(l)}, \mathbf{W}^{(l)}, \mathcal{X}\right)$$

    For each user (or frequency bin) m

        Sample the user (or frequency) factors

$$B_m^{(l+1)} \sim p\left(B_m \middle| \mathbf{C}^{(l)}, \mathbf{W}^{(l)}, \Theta_b^{(l)}, (\sigma^2)^{(l)}, \mathcal{X}\right)$$

    For each item (or time frame) n

        Sample the item (or time) factors

$$W_n^{(l+1)} \sim p\left(W_n \middle| \mathbf{C}^{(l)}, \mathbf{B}^{(l+1)}, \Theta_w^{(l)}, (\sigma^2)^{(l)}, \mathcal{X}\right)$$

    For each time (or operation channel) l'

        Sample the time (or channel) factors

        For the case $l' = 0$

$$C_0^{(l+1)} \sim p\left(C_0 \middle| \mathbf{B}^{(l+1)}, \mathbf{W}^{(l+1)}, C_1^{(l)}, \Theta_c^{(l)}, (\sigma^2)^{(l)}, \mathcal{X}\right)$$

        For the case $l' = 2, \ldots, L' - 1$

$$C_{l'}^{(l+1)} \sim p\left(C_{l'} \middle| \mathbf{B}^{(l+1)}, \mathbf{W}^{(l)}, C_{l'-1}^{(l+1)}, C_{l'+1}^{(l+1)}, \Theta_c^{(l)}, (\sigma^2)^{(l)}, \mathcal{X}\right)$$

        For the case $l' = L'$

$$C_{L'}^{(l+1)} \sim p\left(C_{L'} \middle| \mathbf{B}^{(l+1)}, \mathbf{W}^{(l+1)}, C_{L'-1}^{(l+1)}, \Theta_c^{(l)}, (\sigma^2)^{(l)}, \mathcal{X}\right)$$

Return $\{C_{l'}^{(l)}, B_m^{(l)}, W_n^{(l)}, (\sigma^2)^{(l)}, \Theta_c^{(l)}, \Theta_b^{(l)}, \Theta_w^{(l)}\}$

$\mathbf{X}_n \in \mathbb{R}_+^{M \times M}$. The goal of PSDTF aims to decompose N positive semidefinite (PSD) matrices $\{\mathbf{X}_n\}$ as the convex combination of K PSD matrices or basis matrices $\mathbf{B}_k \in \mathbb{R}_+^{M \times M}$, that is,

$$\mathbf{X}_n \approx \sum_{k=1}^{K} w_k^g w_{kn}^l \mathbf{B}_k = \sum_{k=1}^{K} w_k^g \mathbf{w}_k^l \otimes \mathbf{B}_k \tag{6.91}$$

$$\triangleq \widehat{\mathbf{X}}_n,$$

where K is much smaller than original dimensions M and N, w_k^g is a global weight shared over N slices and w_{kn}^l is a local weight only for slice n. In Eq. (6.91), $\mathbf{w}_k^l = \{w_{kn}^l\}$, $w_k^g \geq 0$ and $w_{kn}^l \geq 0$ are

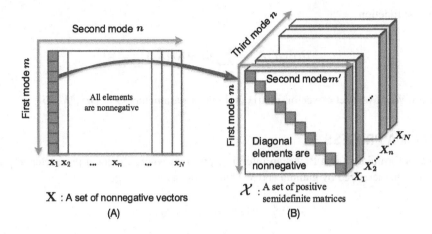

FIGURE 6.4

Comparison between (A) nonnegative matrix factorization and (B) positive semidefinite tensor factorization.

imposed, and Kronecker product is used in the reconstruction of tensor

$$\widehat{\mathcal{X}} = \{\widehat{\mathbf{X}}_n\} \in \mathbb{R}_+^{M \times M \times N}.$$ (6.92)

The learning objective in PSDTF was constructed by using the Bregman matrix divergence (Bregman, 1967, Yoshii et al., 2013)

$$
\begin{aligned}
\mathcal{D}_\phi(\mathbf{X}_n \| \widehat{\mathbf{X}}_n) \\
= \phi(\mathbf{X}_n) - \phi(\widehat{\mathbf{X}}_n) - \mathrm{tr}\left(\nabla\phi(\widehat{\mathbf{X}}_n)^\top(\mathbf{X}_n - \widehat{\mathbf{X}}_n)\right)
\end{aligned}
$$ (6.93)

where $\phi(\cdot)$ denotes a strictly convex matrix function. The log-determinant (LD) divergence was therefore introduced

$$
\begin{aligned}
\mathcal{D}_{\mathrm{LD}}(\mathbf{X}_n | \widehat{\mathbf{X}}_n) \\
= -\log |\mathbf{X}_n \widehat{\mathbf{X}}_n^{-1}| + \mathrm{tr}\left(\mathbf{X}_n \widehat{\mathbf{X}}_n^{-1}\right) - M
\end{aligned}
$$ (6.94)

by using the convex function $\phi(\mathbf{X}) = -\log |\mathbf{X}|$. Interestingly, referring to Section 2.2.2, Itakura–Saito (IS) divergence is seen as a special realization of log-determinant (LD) divergence when $M = 1$. LD-PSDTF is therefore viewed as an extension of IS-NMF. Here, LD-PSDTF is built by estimating the parameters

$$\mathbf{w}^g = \{w_k^g\} \in \mathbb{R}_+^K,$$ (6.95)

$$\mathbf{W}^l = \{\mathbf{w}_k^l\} \in \mathbb{R}_+^{N \times K},$$ (6.96)

$$\mathcal{B} = \{\mathbf{B}_k\} \in \mathbb{R}_+^{M \times M \times K}$$ (6.97)

where the cost function

$$\mathcal{D}_{\mathrm{LD}}(\mathcal{X}|\widehat{\mathcal{X}}) = \sum_{n=1}^{N} \mathcal{D}_{\mathrm{LD}}(\mathbf{X}_n|\widehat{\mathbf{X}}_n) \tag{6.98}$$

is minimized. We have a set of vector, matrix and tensor parameters $\Theta = \{\mathbf{w}^g, \mathbf{W}^l, \mathcal{B}\}$.

6.5.2 GAMMA PROCESS

LD-PSDTF is formulated as a Bayesian inference problem where the local parameter w_{kn}^l and basis matrix $\mathbf{B}_k$ are characterized by gamma and Wishart priors as

$$w_{kn}^l \sim \mathrm{Gam}\left(w_{kn}^l|a_0, b_0\right), \tag{6.99}$$

$$\mathbf{B}_k \sim \mathcal{W}(\mathbf{B}_k|\beta_0, \mathbf{B}_0) \tag{6.100}$$

with parameters $\{a_0, b_0\}$ and $\{\beta_0, \mathbf{B}_0\}$, respectively. The PSD matrices $\{\beta \mathbf{X}_n\}_{n=1}^{N}$ with scaling parameter β are assumed driven independently by Wishart likelihood function

$$p(\beta \mathbf{X}_n|\Theta) = \mathcal{W}\left(\beta, \sum_{k=1}^{K} w_k^g w_{kn}^l \mathbf{B}_k\right). \tag{6.101}$$

The log-likelihood function of $\mathbf{X}_n$ is then yielded by

$$\begin{aligned}
\log &p(\mathbf{X}_n|\widehat{\mathbf{X}}_n) \\
&= c(\beta) + \frac{\beta - M - 1}{2} \log |\mathbf{X}_n| - \frac{\beta}{2} \log |\widehat{\mathbf{X}}_n| - \frac{\beta}{2} \mathrm{tr}\left(\mathbf{X}_n \widehat{\mathbf{X}}_n^{-1}\right)
\end{aligned} \tag{6.102}$$

where $c(\beta)$ is a normalization depending on β and the second term of RHS is independent of Θ. It is interesting that maximizing the logarithm of Wishart distribution in Eq. (6.102) is equivalent to minimizing the log-determinant divergence in Eq. (6.94).

Importantly, the number of bases K is automatically determined by introducing the Bayesian nonparametrics via taking the infinite limit in the representation

$$\mathbf{X}_n \approx \sum_{k=1}^{\infty} w_k^g w_{kn}^l \mathbf{B}_k. \tag{6.103}$$

An effective number $\widehat{K}$ of bases should be estimated in a training procedure. Or equivalently, we should estimate the infinite-dimensional weight vector $\mathbf{w}^g = \{w_k^g\}_{k=1}^{\infty}$. To do so, a Gamma process (GaP) prior is merged to produce Bayesian nonparametrics for global parameter $\mathbf{w}^g$. Namely, the global parameter is assumed by a Gamma distribution

$$w_k^g \sim \mathrm{Gam}\left(w_k^g \left| \frac{\alpha c}{K}, \alpha \right.\right) \tag{6.104}$$

where α and c are positive numbers. Notably, this parameter is drawn from an infinite-dimensional discrete measure G, which is driven by the GaP

$$G \sim \text{GaP}(\alpha, G_0) \qquad (6.105)$$

where α is a concentration parameter and G_0 is a base measure. In Yoshii and Goto (2012), a similar work was proposed for music signal analysis using the infinite latent harmonic model where Bayesian nonparametrics were also calculated in the implementation. Nevertheless, the goal of infinite tensor factorization is to calculate the posterior distribution $p(\mathbf{w}^g, \mathbf{W}^l, \mathcal{B} | \mathcal{X})$ and estimate the effective number $\widehat{K}$ in $\mathbf{w}^g$ simultaneously.

6.5.3 VARIATIONAL INFERENCE

In implementation of Bayesian nonparametric learning with a Gamma process prior for $\mathbf{w}^g$, we use the variational Bayesian method to approximate the posterior distribution $p(\mathbf{w}^g, \mathbf{W}^l, \mathcal{B} | \mathcal{X})$ based on a factorized variational distribution, i.e.,

$$
\begin{aligned}
& q(\mathbf{w}^g, \mathbf{W}^l, \mathcal{B}) \\
& = \prod_{k=1}^{K} \left(q\left(w_k^g\right) \left(\prod_{n=1}^{N} q\left(w_{kn}^l\right) \right) q(\mathbf{B}_k) \right).
\end{aligned}
\qquad (6.106)
$$

By referring Section 3.7.3, the evidence lower bound or the variational lower bound of $\log p(\mathcal{X})$ is obtained by

$$
\begin{aligned}
\mathcal{L}(q) = & \, \mathbb{E}_q \left[\log p(\mathcal{X} | \mathbf{w}^g, \mathbf{W}^l, \mathcal{B}) \right] + \mathbb{E}_q \left[\log p(\mathbf{w}^g) \right] \\
& + \mathbb{E}_q \left[\log p(\mathbf{W}^l) \right] + \mathbb{E}_q \left[\log p(\mathcal{B}) \right] \\
& + \mathbb{H}_q \left[\mathbf{w}^g \right] + \mathbb{H}_q \left[\mathbf{W}^l \right] + \mathbb{H}_q [\mathcal{B}]
\end{aligned}
\qquad (6.107)
$$

where $\mathbb{H}_q[\cdot]$ denotes an entropy function using the factorizable variational distribution $q(\mathbf{w}^g, \mathbf{W}^l, \mathcal{B})$. The variational distribution q_j of an individual variable in vector, matrix and tensor parameters $j \in \{\mathbf{w}^g, \mathbf{W}^l, \mathcal{B}\}$ is given by

$$\log \widehat{q}_j \propto \mathbb{E}_{q_{(i \neq j)}} \left[\log p(\mathcal{X}, \mathbf{w}^g, \mathbf{W}^l, \mathcal{B} | \mathbf{\Psi}) \right] \qquad (6.108)$$

where hyperparameters consist of $\mathbf{\Psi} = \{\alpha, c, a_0, b_0, \beta_0, \mathbf{B}_0\}$. However, in variational Bayesian learning, the difficulties lie in the nonconjugacy of Bayesian model in LD-PSDTF. To pursue tractable Bayesian inference, the generalized inverse-Gaussian (GIG) distribution, as given in Eq. (5.31), is introduced to characterize the individual variational distributions $\{q(w_k^g)\}$ and $\{q(w_{kn}^l)\}$ where $1 \leq k \leq K$ and $1 \leq n \leq N$. The matrix-variate GIG (MGIG) distribution is incorporated to model the basis matrix $\mathbf{B}_k$. We have

$$q(w_k^g) = \text{GIG}\left(w_k^g | \nu_k^g, \rho_k^g, \tau_k^g\right), \qquad (6.109)$$

$$q(w_{kn}^l) = \text{GIG}\left(w_{kn}^l | v_{kn}^l, \rho_{kn}^l, \tau_{kn}^l\right), \tag{6.110}$$

$$q(\mathbf{B}_k) = \text{MGIG}\left(\mathbf{B}_k | v_k^b, \mathbf{R}_k^b, \mathbf{T}_k^b\right). \tag{6.111}$$

In Yoshii et al. (2013), closed-form solutions to three sets of variational parameters

$$\{v_k^g, \rho_k^g, \tau_k^g, v_{kn}^l, \rho_{kn}^l, \tau_{kn}^l, v_k^b, \mathbf{R}_k^b, \mathbf{T}_k^b\}$$

were derived and detailed. Attractively, the multiplicative updating rules for local weights $\{w_{kn}^l\}$ and basis matrices $\{\mathbf{B}_k\}$ are then derived by maximizing the variational lower bound $\mathcal{L}(q)$ in a form of

$$w_{kn}^l \leftarrow w_{kn}^l \sqrt{\frac{\text{tr}\left(\widehat{\mathbf{X}}_n^{-1}\mathbf{B}_k\widehat{\mathbf{X}}_n^{-1}\mathbf{X}_n\right)}{\text{tr}\left(\widehat{\mathbf{X}}_n^{-1}\mathbf{B}_k\right)}}, \tag{6.112}$$

$$\mathbf{B}_k \leftarrow \mathbf{B}_k\mathbf{L}_k\left(\mathbf{L}_k^\top\mathbf{B}_k\mathbf{P}_k\mathbf{B}_k\mathbf{L}_k\right)^{-\frac{1}{2}}\mathbf{L}_k^\top\mathbf{B}_k \tag{6.113}$$

where $\mathbf{P}_k$ and $\mathbf{Q}_k$ are two auxiliary PSD matrices given by

$$\begin{aligned}
\mathbf{P}_k &= \sum_{n=1}^{N} w_{kn}^l \widehat{\mathbf{X}}_n^{-1}, \\
\mathbf{Q}_k &= \sum_{n=1}^{N} w_{kn}^l \widehat{\mathbf{X}}_n^{-1}\mathbf{X}_n\widehat{\mathbf{X}}_n^{-1},
\end{aligned} \tag{6.114}$$

and the Cholesky decomposition $\mathbf{Q}_k = \mathbf{L}_k\mathbf{L}_k^\top$ is applied to find the lower triangular matrix $\mathbf{L}_k$. Interestingly, the multiplicative updating rules for local weight w_{kn}^l and basis vector $\mathbf{B}_k$ in Eqs. (6.112)–(6.113) using LD-PSDTF are expressed in a similar style to those for updating W_{kn} and B_{mk} using IS-NMF as shown in Eqs. (2.33)–(2.32), respectively.

6.5.4 SYSTEM EVALUATION

The Bayesian framework based on LD-PSDTF was compared with the nonparametric solution based on IS-NMF in a variety of experiments (Yoshii et al., 2013). In general, LD-PSDTF is useful for source separation of music audio signals. The traditional KL-NMF or IS-NMF was implemented in amplitude or power spectrogram domain. It is usually difficult to recover natural source signals from these spectrograms having no phase information. If the phase of observed spectrogram is merged to source spectrograms, the resulting signals have some unpleasant artifacts. In Yoshii et al. (2013), the time-domain LD-PSDTF was implemented so that real-valued source signals was directly estimated without the need of phase reconstruction. In the experiments, LD-PSDTF was carried out for the separation of piano sources or, more specifically, the separation of three source signals having different pitches. A length of 8.4 seconds was generated by concatenating seven piano sounds with an MIDI synthesizer which comprised different combination of mixed signals. The signal was sampled at

16 kHz and was split into overlapping frames by using a Gaussian window with width of 512 samples ($M = 512$) and a shifting interval of 160 samples ($N = 840$). In addition to LD-PSDTF, the comparison with KL-NMF using amplitude-spectrogram decomposition and IS-NMF using power-spectrogram decomposition was performed. The experimental results showed significant improvement of using LD-PSDTF in terms of SDR, SIR and SAR when compared with KL-NMF and IS-NMF. It was found that the initialization of w_{kn}^l and $\mathbf{B}_k$ in LD-PSDTF using the values of W_{kn} and B_{mk} obtained from IS-NMF performed very well.

6.6 SUMMARY

We have systematically addressed a variety of nonnegative tensor factorizations with applications on single-channel source separation, as well as some other task, including collaborative filtering for recommendation system. Starting from the baseline NTF method, based on CP decomposition or PARAFAC model, we have described how tensor data using the modulation spectrograms and the multiresolution spectrograms are decomposed into three-way factor matrices which convey the additional factorized information from multiple auditory channels and multiple spectral resolutions, respectively. In addition to temporal and frequency features in speech and audio signals, NTF learns additional channel information from different modulations and resolutions. For the application of recommendation system, NTF not only captures the features of users and products but also those features which vary in time. Richer information is learned from more modalities. Detailed formulation was provided for multiplicative updating of three factor matrices.

Moreover, a number of advanced tensor factorization methods were introduced. These methods provided different extensions to bring in various capabilities in multiway decomposition. First of all, the convolutive NTF was presented to learn temporal and frequency correlations in audio signals which brought helpful information for single-channel source separation. The multiplicative updating rules for the factors of channels, bases and weights were formulated according to the learning objectives of squared Euclidean distance and Kullback–Leibler divergence. The updating rules were further extended to carry out the nonnegative tensor deconvolution with sparsity constraint which mitigates the issue of overfitting in machine learning. Secondly, probabilistic tensor factorization was addressed to model uncertainty in signals and parameters. Furthermore, a hierarchical Bayesian model with rectified Gaussian priors was introduced to characterize the nonnegative factor matrices. Hierarchical Bayesian model using conjugate priors could increase the efficiency in Bayesian inference. Bayesian tensor factorization was presented and implemented according to the approximate inference based on the Gibbs sampling algorithm. At last, the Gamma process was introduced to carry out an infinite tensor factorization where the number of global weights for different bases was infinite. This tensor factorization was shown to be a generalization of Itakura–Saito nonnegative matrix factorization. Bayesian nonparameters were used for controlling model complexity in tensor factorization.

In the next section, we will present the advanced studies on deep learning approaches to single-channel source separation. The importance of deep recurrent neural network for source separation will be emphasized. State-of-the-art deep learning algorithms and their applications for source separation will be introduced. In particular, a number of newest recurrent neural networks will be presented to demonstrate the powerfulness of combining the techniques of signal processing and deep learning for source separation.

DEEP NEURAL NETWORK

Artificial intelligence and deep learning are currently riding the high wave and substantially affecting human life and industry development. Deep learning is now changing the world. Many solutions and applications have been successfully developed. Source separation based on deep learning has been attracting a large number of researchers focusing on this new trend. A variety of signal processing for deep learning algorithms have been integrated to accomplish different challenging tasks. In Chapter 2, we have introduced the fundamentals of deep neural network (DNN), recurrent neural network (RNN), deep RNN (DRNN) and long short-term memory (LSTM) in Sections 2.4, 2.5, 2.5.2 and 2.5.3, respectively. This chapter surveys state-of-the-art monaural source separation methods based on advanced deep learning models with signal processing techniques. The evolution of different methods with deep processing, modeling and learning is presented. Fig. 7.1 depicts the categorization and evolution of various deep learning approaches which have been exploited for speech separation, speech enhancement, dereverberation and segregation. First of all, Section 7.1 addresses the deep machine learning based on a regression model using DNN which is applied for speech separation and segregation. Ensemble learning combined with DNN will be mentioned. Section 7.2 highlights a number of deep signal processing methods where the advanced signal processing methods are seamlessly merged in deep models. The methods of spectral clustering, permutation invariance and matrix factorization are introduced in joint optimization. Deep clustering and discriminative embedding are presented for joint segmentation and separation. Nevertheless, learning a regression model from sequence data given by the mixed signals using DNN may be restricted. Section 7.3 accordingly focuses on the development of RNN, in particular DRNN, for source separation. The recurrent layer is implemented as an internal memory. DRNNs and mask functions are jointly optimized to improve source separation. Discriminative DRNN is realized to improve separation performance by imposing the discrimination between the separated signals in the objective function. In Section 7.4, the gradient vanishing or exploding problem in DRNN is compensated by introducing LSTM. Long and short contextual information is extracted for monaural source separation. The bidirectional LSTM is implemented by capturing contextual information from past and future events for signal separation. Section 7.4 addresses a number of modern sequential learning methods for monaural source separation, which include the variational RNN (Section 7.5), memory augmented neural network (Section 7.6) and recall neural network (Section 7.7). Stochastic modeling, external memory and attention mechanism are introduced to enhance different capabilities of RNN. Sequence-to-sequence learning is carried out for source separation.

7.1 DEEP MACHINE LEARNING

This section illustrates a number of examples showing how deep machine learning is carried out by combining machine and deep learning. For example, a deep neural network (DNN) is constructed by integrating with a regression model and combining with an ensemble learning algorithm, which will

Source Separation and Machine Learning. https://doi.org/10.1016/B978-0-12-804566-4.00019-X

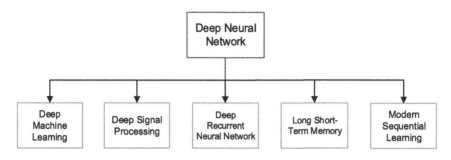

FIGURE 7.1

Categorization of different deep learning methods for source separation.

be detailed in Sections 7.1.1 and 7.1.2, respectively. DNN learning for reverberant speech segregation will be addressed in Section 7.1.3.

7.1.1 DEEP SPECTRAL MASKING

Speech separation or enhancement is seen as a regression problem which can be tackled by using the supervised learning based on a DNN model (Xu et al., 2014, Du et al., 2016, Grais et al., 2014). Deep spectral mapping or masking is performed by deep learning. Fig. 7.2 depicts how a DNN is developed for single-channel source separation or speech enhancement, which separates a noisy speech spectrum into the corresponding spectra of clean speech and additive noise. In real-world implementation, we usually calculate a 1024-point short-time Fourier transform with a window size of 64 ms and an overlap of 32 ms. A mixed spectral signal $\mathbf{x}_t^{\mathrm{mix}}$ at time t is obtained from the magnitude or log-magnitude of spectral signal. Mel-spectral data can be also adopted. Nevertheless, the input vector of DNN $\mathbf{x}_t$ at time t is composed of a window of mixed spectral signals $\mathbf{x}_t^{\mathrm{mix}}$ centered at frame t with τ neighboring frames in two sides

$$\mathbf{x}_t = \left[(\mathbf{x}_{t-\tau}^{\mathrm{mix}})^\top, \ldots, (\mathbf{x}_t^{\mathrm{mix}})^\top, \ldots, (\mathbf{x}_{t+\tau}^{\mathrm{mix}})^\top \right]^\top \in \mathbb{R}^{M(2\tau+1)} \tag{7.1}$$

where M is the number of frequency bins. Temporal dynamics of input spectra are incorporated in the input vector, which contains helpful information for source separation. DNN is applied to learn a mapping function between a mixed signal and its two source signals. Namely, our goal is to decompose the mixed signal $\mathbf{x}_t$ into those of two sources $\widetilde{\mathbf{x}}_{1,t}$ and $\widetilde{\mathbf{x}}_{2,t}$ using a fully-connected DNN with L layers of weight parameters $\mathbf{w} = \{\mathbf{w}^{(l)}\}_{l=1}^L$. The weights, connecting to output layer L, consist of those for two sources

$$\mathbf{w}^{(L)} = \{\mathbf{w}_1^{(L)}, \mathbf{w}_2^{(L)}\},$$

which are used to calculate the activations of two sources

$$\{\mathbf{a}_{1,t}^{(L)} = \{a_{1,tk}^{(L)}\}, \mathbf{a}_{2,t}^{(L)} = \{a_{2,tk}^{(L)}\}\}.$$

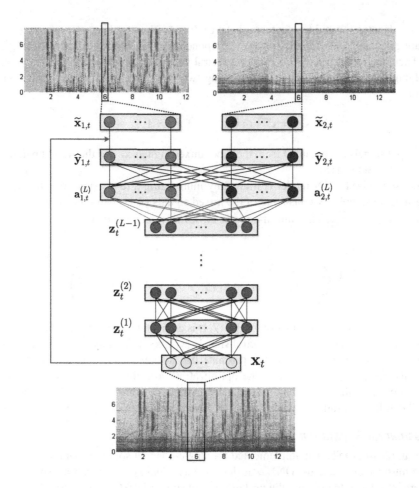

FIGURE 7.2

A deep neural network for single-channel speech separation, where $\mathbf{x}_t$ denotes the input features of the mixed signal at time step t, $\mathbf{z}_t^{(l)}$ denotes the features in the hidden layer l, $\widehat{\mathbf{y}}_{1,t}$ and $\widehat{\mathbf{y}}_{2,t}$ denote the mask function for source one and source two, and $\widetilde{\mathbf{x}}_{1,t}$ and $\widetilde{\mathbf{x}}_{2,t}$ are the estimated signals for source one and source two, respectively.

Similar to what we have addressed in Section 2.5.2, a pair of soft mask functions are introduced to improve the estimated source spectra. The ideal ratio masks (Narayanan and Wang, 2013) are calculated by

$$
\begin{aligned}
\widehat{\mathbf{y}}_{i,t} &= \frac{|\mathbf{w}_i^{(L)}\mathbf{z}_t^{(L-1)}|}{|\mathbf{w}_1^{(L)}\mathbf{z}_t^{(L-1)}| + |\mathbf{w}_2^{(L)}\mathbf{z}_t^{(L-1)}|} \\
&= \frac{|\mathbf{a}_{i,t}^{(L)}|}{|\mathbf{a}_{1,t}^{(L)}| + |\mathbf{a}_{2,t}^{(L)}|}, \qquad \text{for } i = 1, 2,
\end{aligned}
\tag{7.2}
$$

where $z_t^{(L-1)}$ denotes the hidden features in layer $L - 1$. Notably, the ratio masks $\widehat{\mathbf{y}}_{i,t} = \{\widehat{y}_{i,tk}\}$ in Eq. (7.2) are calculated individually for each component k. The Wiener gain functions $\{\widehat{\mathbf{y}}_{1,t}, \widehat{\mathbf{y}}_{2,t}\}$ are obtained. Then, the reconstructed magnitude spectral vectors of two sources $\{\widetilde{\mathbf{x}}_{1,t}, \widetilde{\mathbf{x}}_{2,t}\}$ are estimated by multiplying the mixed spectral vector $\mathbf{x}_t$ using two mask functions $\{\widehat{\mathbf{y}}_{1,t}, \widehat{\mathbf{y}}_{2,t}\}$ via element-wise computation

$$\widetilde{\mathbf{x}}_{i,t} = \mathbf{x}_t^{\text{mix}} \odot \widehat{\mathbf{y}}_{i,t}, \qquad \text{for } i = 1, 2. \tag{7.3}$$

Here, the output signals are obtained from the input mixed signals $\mathbf{x}_t$ and the weight parameters $\mathbf{w}$. The expression $\widetilde{\mathbf{x}}_{1,t}(\mathbf{x}_t, \mathbf{w})$ is used.

Next, the supervised DNN model is trained by minimizing a regression error function from a set of training spectral samples $\mathbf{X}$ consisting of the mixed signals $\{\mathbf{x}_t\}_{t=1}^T$ and the corresponding source or target signals $\{\mathbf{x}_{1,t}, \mathbf{x}_{2,t}\}_{t=1}^T$. The sum-of-squares error function is yielded by

$$
\begin{aligned}
&E(\mathbf{w}) \\
&= \frac{1}{2} \sum_{t=1}^T \left(\|\widetilde{\mathbf{x}}_{1,t}(\mathbf{x}_t, \mathbf{w}) - \mathbf{x}_{1,t}\|^2 + \|\widetilde{\mathbf{x}}_{2,t}(\mathbf{x}_t, \mathbf{w}) - \mathbf{x}_{2,t}\|^2 \right).
\end{aligned}
\tag{7.4}
$$

Minimizing the regression error function in Eq. (7.4) is equivalent to increasing the similarity between the estimated $\{\widetilde{\mathbf{x}}_{1,t}, \widetilde{\mathbf{x}}_{2,t}\}$ and the clean spectral signals $\{\mathbf{x}_{1,t}, \mathbf{x}_{2,t}\}$. This learning objective is minimized to train DNN weights $\mathbf{w}$ or, equivalently, estimate the Wiener gains $\{\widehat{\mathbf{y}}_{1,t}, \widehat{\mathbf{y}}_{2,t}\}$ in Eq. (7.2), or the source signals $\{\widetilde{\mathbf{x}}_{1,t}, \widetilde{\mathbf{x}}_{2,t}\}$ in Eq. (7.3). The optimization procedure is realized by running the error backpropagation algorithm based on the stochastic gradient descent using minibatch data as addressed in Section 2.4.1. In what follows, we address two advanced studies on DNN speech separation.

DNN Classifier for Separation

In Grais et al. (2014), DNN was implemented for single-channel source separation in the presence of speech and music source signals. DNN was designed as a *binary classifier* for two sources. A pair of outputs were introduced to measure the *probabilities* of source 1, $f_1(\cdot)$, and source 2, $f_2(\cdot)$. There were several terms in the learning objective at each time step t, that is,

$$
\begin{aligned}
&E(\mathbf{x}_{1,t}, \mathbf{x}_{2,t}, \mathbf{x}_t, u, v) \\
&= E_1(\mathbf{x}_{1,t}) + E_2(\mathbf{x}_{2,t}) + \lambda_e E_e(\mathbf{x}_{1,t}, \mathbf{x}_{2,t}, \mathbf{x}_t, u, v) + \lambda_r \sum_i R(\Theta_i)
\end{aligned}
\tag{7.5}
$$

where E_1 and E_2 denote the energy functions for source 1 and source 2 expressed by

$$E_1(\mathbf{x}_{1,t}) = \left(1 - f_1(\mathbf{x}_{1,t})\right)^2 + \left(f_2(\mathbf{x}_{1,t})\right)^2, \tag{7.6}$$

$$E_2(\mathbf{x}_{2,t}) = \left(f_1(\mathbf{x}_{2,t})\right)^2 + \left(1 - f_2(\mathbf{x}_{2,t})\right)^2. \tag{7.7}$$

Here E_e denotes the regression error function

$$E_e(\mathbf{x}_{1,t}, \mathbf{x}_{2,t}, \mathbf{x}_t, u, v) = \|u\mathbf{x}_{1,t} + v\mathbf{x}_{2,t} - \mathbf{x}_t\|^2 \tag{7.8}$$

and $R(\Theta_i)$ is an energy or ℓ_2 regularization given by

$$R(\Theta_i) = (\min(\Theta_i, 0))^2. \qquad (7.9)$$

The mixed signal is assumed to be linearly interpolated by source signals $\mathbf{x}_{1,t}$ and $\mathbf{x}_{2,t}$ using the unknown parameters u and v, respectively. The regression error term in Eq. (7.8) is minimized to find the interpolation parameters u and v for a mixed signal $\mathbf{x}_t$. The parameter set consists of all unknown parameters in source separation system

$$\Theta = \{\Theta_i\} = \{\widetilde{\mathbf{x}}_{1,t}, \widetilde{\mathbf{x}}_{2,t}, u, v\}. \qquad (7.10)$$

In Eq. (7.5), λ_e and λ_r denote the regularization parameters for regression error and regularization term, respectively.

In system evaluation, this method was investigated by a monaural source separation task for separating speech and music signals. Speech data were sampled from 20 files in TIMIT database spoken by a male speaker. Music data contained 39 pieces from pianos with total length of 185 minutes. The mixed data were obtained at different speech-to-music ratios (SMRs). A sampling rate of 16 kHz was used. Nonnegative matrix factorization (NMF) was implemented with a dictionary size of 128 for each source. DNN was constructed in a topology of 257–100–50–200–2 where three hidden layers with 100, 50 and 200 hidden nodes were used. Here, one frame ($\tau = 0$) of 257 spectral vectors from STFT was considered and two output nodes $\{f_1, f_2\}$ were arranged. Three neighbor frames ($\tau = 1$) were also used in input layer for investigation. Sigmoid nonlinear activation, as shown in Eq. (2.51), was used. DNN was realized via a deep belief network, which was initialized by using the restricted Boltzmann machine (RBM) according to the contrastive divergence algorithm as addressed in Section 2.4.2. Also $\lambda_e = 5$ and $\lambda_r = 3$ were chosen, and 150 epochs were run for training each layer using RBM. Then, 500 epochs were performed for error backpropagation training of DNN. In RBM training, the first five epochs were run to optimize the output layer while keeping the lower layer weights unchanged. The optimization method based on the limited memory Broyden–Fletcher–Goldfarb–Shanno (LBFGS) algorithm (Goodfellow et al., 2016). During optimization, the gradients of DNN outputs with respect to input vector

$$\left\{ \frac{\partial f_1(\mathbf{x})}{\partial \mathbf{x}}, \frac{\partial f_2(\mathbf{x})}{\partial \mathbf{x}} \right\}$$

were calculated to updating the separated signals $\{\widetilde{\mathbf{x}}_{1,t}, \widetilde{\mathbf{x}}_{2,t}\}$. The separated signals using nonnegative matrix factorization (NMF) $\{\widetilde{\mathbf{x}}_{1,t}, \widetilde{\mathbf{x}}_{2,t}\}$ were used as the initialization for this updating when training DNN. In the comparison, DNN performed better than NMF for monaural source separation in terms of signal-to-distortion ratio (SDR) and signal-to-interference ratio (SIR) under different speech-to-music ratios (SMRs). Considering more neighboring frames in case $\tau = 1$ obtained better results than the case of $\tau = 0$.

High-Resolution DNN

In Du et al. (2016), a high-resolution DNN was trained as a regression model for single-channel source separation where the separated target speech data were evaluated by the metrics of speech separation and recognition using the short-time objective intelligibility (STOI) and the word error rate (WER), respectively. STOI was highly correlated with the perceptual evaluation of speech quality (PESQ). The higher STOI or PESQ, the better the speech separation performance. Semisupervised learning was preformed to separate the speech of a target speaker from that of an unknown interfering speaker. This learning task was different from the supervised speech separation where target and interfering speakers were both known. There were two findings, which supported the construction of high-resolution DNN to achieve state-of-the-art performance for speech separation and recognition. First, in scenario of semisupervised learning, the DNN architecture with dual outputs for hidden features due to both target and interfering speakers performed better than that with a single set of features from the target speaker. This architecture was denoted by dual-output DNN. Second, a delicate model with multiple signal-to-noise ratio (SNR) dependent DNNs was proposed to accommodate all mixing conditions at different SNR levels so as to cope with the weakness of the conventional method based on a universal DNN. The architecture was named the SNR-dependent DNN.

In the implementation, there were 34 speakers (18 males and 16 females) both in training and test sessions. The test data consisted of two-speaker mixed speech at SNR between -9 and 6 dB. Training data covered SNR levels between -10 and 10 dB. The sampling rate was set at 16 kHz; 512-point STFT was computed for each frame with a length of 32 ms and an overlap of 16 ms. Moreover, 257-dimensional log-power spectral features were extracted to train DNN. The topology of dual-output DNN was specified as 1799–2048–2048–514. The number of input nodes was 1799 ($257 \star 7$) because three neighbor frames ($\tau = 3$) in two sides are considered. The number of output nodes was 514 for two sources. RBM pretraining was performed and 20 epochs were run for each layer of RBM. The learning rate was 0.1 for the first 10 epochs and then decreased by 10% at later epochs. Totally, 50 epochs were run. A minibatch size of 128 was used. Speech recognition system was built by using a whole-word left-to-right hidden Markov model (HMM) with 32 Gaussian mixtures per state. In the experiments, the dual-output DNN without any information about the interferer even outperformed the baseline Gaussian mixture model (GMM) (Hu and Wang, 2013) where information of both the target and interferer was known. On the other hand, SNR-dependent DNNs were implemented as a high-resolution DNN model. We should estimate SNR of a mixed utterance, which has been well studied with high accuracy in speech research areas. Furthermore, a DNN speech separation model was trained individually for different source speakers. Speaker-dependent speech separation using DNN was constructed with a speaker recognition module which was built by using a universal background model using GMM with maximum *a posteriori* adaptation. In particular, speech separation as front end and speech recognition as back end were integrated as a unified speech separation and recognition system, which conducted the following six steps to obtain the final speech recognition result:

1. Build models of all speakers for speaker recognition and speech separation;
2. First-pass recognition for finding top M source speakers;
3. First-pass separation using speaker-dependent DNN;
4. Second-pass recognition for finding top one speaker using separated speech
5. Second-pass separation using DNN associated with the most likely speaker;
6. Third-pass recognition for the dual output speech from DNN separation.

Remarkable results were achieved in terms of STOI and WER by following this procedure (Du et al., 2016). On the other hand, a DNN regression model (Xu et al., 2014) was developed for speech enhancement using a DNN with a similar experimental setup but three hidden layers. Target speech was identified from unknown noises. Such a scenario is comparable with that of semisupervised speech separation in the presence of a known target speaker and an unknown interfering speaker. Interestingly, the success of using SNR-dependent DNN for speech separation is consistent with that of using SNR-dependent probabilistic linear discriminant analysis for speaker verification as shown in Mak et al. (2016).

7.1.2 DEEP ENSEMBLE LEARNING

Source separation is known as a machine learning system where the mixed signals are provided as inputs and the demixed signals are produced as outputs. Although DNN regression model for estimating target speech and mask function, as addressed in Section 7.1.1, has achieved competing performance, a single DNN may not sufficiently leverage various acoustic information for speech separation. Combining multiple complimentary DNNs is beneficial to capture a variety of separation information. In machine learning, ensemble methods use multiple learning algorithms to obtain better predictive performance when compared with any of the constitute learning algorithms alone. Basically, ensemble learning is performed by using a set of alternative models. The goal is to utilize these models to boost learning for a rich structure and diversity among different alternatives. Source separation is seen as a task of searching a hypothesis space of demixed signals. Ensembles combine multiple hypotheses to form a better hypothesis. Multiple regression models can span hybrid hypotheses that a single model could not cover. The generalization using ensemble learning is assured for flexible representation. Ensembles tend to combine models and promote diversity. The overfitting problem of training data is reduced. There are different styles of ensembles. *Bagging* and *boosting* are two common types of ensemble learning. Very high regression or classification performance could be achieved. Basically, an ensemble model requires more computation time in prediction than a single model. A lot of extra computation is demanded to compensate the poor learning in a single model.

This section addresses an advanced study which introduces the techniques of signal processing, deep learning and ensemble learning for monaural speech separation. A so-called multicontext network was constructed in Zhang and Wang (2016) to carry out a deep ensemble learning approach to monaural speech separation. The underlying motivation of multicontext network was originated from the fact that speech signal is highly structured with rich temporal context. Concatenating neighboring frames in Eq. (7.1) instead of using a single frame is important in implementation of speech separation. However, a single DNN trained from a fixed window size $2\tau + 1$ of contextual data may be insufficient. Multicontext network was proposed tackle source separation problem by integrating different DNNs under different window lengths. Ensemble learning was performed by using the DNN which followed the *masking-based source separation* as shown in Fig. 7.2 where the soft mask function in Eq. (7.2) was estimated and the regression error in Eq. (7.4) was minimized. The deep ensemble learning was also evaluated for *mapping-based source separation* where the regression function was learned directly from mixed signal to clean speech *without* going through a soft mask function. High-resolution DNN (Du et al., 2016), as addressed in Section 7.1.1, was trained according to mapping-based method. It was found that masking-based method performed well in representation of target speaker while mapping-based method was robust to SNR variation (Zhang and Wang, 2016). Nevertheless, in what follows, we introduce two kinds of ensembles of DNNs which are feasible to both masking and mapping-based source separation.

Multicontext Averaging

The first type of multicontext network is carried out by *averaging* the outputs of multiple DNNs where the inputs under different window sizes are observed. The magnitude spectra using STFT were extracted for mixed signals $\mathbf{x}_t^{(\tau)}$ and their corresponding source signals $\{\mathbf{x}_{1,t}, \mathbf{x}_{2,t}\}$ at each time frame t. The mixed signals $\mathbf{x}_t^{(\tau)}$ are formed by

$$\mathbf{x}_t^{(\tau)} = \left[(\mathbf{x}_{t-\tau}^{\mathrm{mix}})^\top, \dots, (\mathbf{x}_t^{\mathrm{mix}})^\top, \dots, (\mathbf{x}_{t+\tau}^{\mathrm{mix}})^\top \right]^\top \qquad (7.11)$$

with different numbers of neighbor frames τ, i.e., $\{\mathbf{x}_t^{(0)}, \mathbf{x}_t^{(1)}, \mathbf{x}_t^{(2)}\}$ are included. In the training stage, the multicontext network is trained by finding different DNNs with parameters $\mathbf{w}^{(\tau)}$ estimated by using mixed signals $\mathbf{x}_t^{(\tau)}$ with varying τ or context window sizes. There are N_τ DNNs in the system. The sum-of-squares error function $E(\mathbf{w}^{(\tau)})$ is minimized for individual DNN. In the test stage, we calculate the individual soft mask functions in N_τ DNNs which are denoted by $\left\{ \mathbf{y}_{1,t}^{(\tau)}, \mathbf{y}_{2,t}^{(\tau)} \right\}$ where $\tau = 0, \dots, N_\tau - 1$. The multicontext network is run by averaging the outputs of DNNs, namely the soft mask functions under different context windows τ as

$$\bar{\mathbf{y}}_{i,t} = \frac{1}{N_\tau} \sum_{\tau=0}^{N_\tau - 1} \mathbf{y}_{i,t}^{(\tau)}, \qquad \text{for } i = 1, 2. \qquad (7.12)$$

Using this averaged soft mask function, the magnitude spectra of separated signals at time frame t are obtained by

$$\widetilde{\mathbf{x}}_{i,t}^{\mathrm{avg}} = \mathbf{x}_t^{\mathrm{mix}} \odot \bar{\mathbf{y}}_{i,t}, \qquad \text{for } i = 1, 2. \qquad (7.13)$$

The time-domain source signals are finally reconstructed via the inverse STFT using the phase information of the mixed signal.

Multicontext Stacking

In addition to multicontext averaging, the ensemble in multicontext network can be implemented by stacking DNNs as illustrated in Fig. 7.3. There are three stacks in this multicontext network. In the first stack, the mixed speech spectra $\mathbf{x}_t^{\mathrm{mix}}$ at each time frame t under different context windows τ are extracted to obtain the contextual features $\{\mathbf{x}_t^{(\tau)}\}$ and use them as inputs for multiple DNNs. The soft mask functions $\{\mathbf{y}_{1,t}^{(\tau)}(1), \mathbf{y}_{2,t}^{(\tau)}(1)\}$ are accordingly estimated under different context windows τ. The outputs of the first stack $\mathbf{z}_t(1)$ are formed by concatenating these soft mask functions with the original mixed signal $\mathbf{x}_t^{\mathrm{mix}}$. These outputs are used as the inputs of the second stack which also consists of the modules of multiple DNNs and feature concatenation. The soft mask functions of multiple DNNs in the second stack are obtained by $\{\mathbf{y}_{1,t}^{(\tau)}(2), \mathbf{y}_{2,t}^{(\tau)}(2)\}$. After concatenation, the outputs of the second stack are found as $\mathbf{z}_t(2)$. In this stack-wise multicontext network, multiple DNNs are different in different stacks. The concatenation of features in stack s is performed by

$$\mathbf{z}_t(s) = \left[(\mathbf{y}_{1,t}^{(0)}(s))^\top \; (\mathbf{y}_{2,t}^{(0)}(s))^\top (\mathbf{y}_{1,t}^{(1)}(s))^\top \; (\mathbf{y}_{2,t}^{(1)}(s))^\top \right.$$
$$\left. \cdots (\mathbf{y}_{1,t}^{(N_\tau-1)}(s))^\top \; (\mathbf{y}_{2,t}^{(N_\tau-1)}(s))^\top \; (\mathbf{x}_t^{\mathrm{mix}})^\top \right]^\top . \qquad (7.14)$$

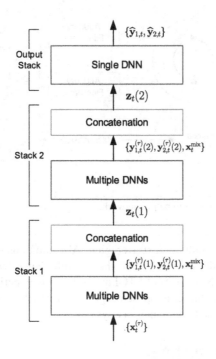

FIGURE 7.3

Illustration of stacking in deep ensemble learning for speech separation.

Such a stacking operation is run until the output stack, which employs a single DNN to regress the outputs of the previous stack $\mathbf{z}_t(s)$ to the final soft mask function $\{\widehat{\mathbf{y}}_{1,t}, \widehat{\mathbf{y}}_{2,t}\}$. The magnitude spectra of separated signals are obtained by

$$\widetilde{\mathbf{x}}_{i,t}^{\text{stack}} = \mathbf{x}_t^{\text{mix}} \odot \widehat{\mathbf{y}}_{i,t}, \qquad \text{for } i = 1, 2. \tag{7.15}$$

This work was evaluated for monaural speech separation where the observed signals were randomly mixed by the utterances from a male and a female speaker. In total, there were 34 speakers, each of which had 500 utterances. Speech signals were resampled at 8 kHz. Each frame had a length of 25 ms and shift of 10 ms. Different SNR levels were investigated. In the system configuration, each DNN had two hidden layers, each of which consisted of 2048 hidden units using ReLU as activation function. Minibatch size was set to 128. Adaptive SGD with momentum was applied and 50 learning epochs were run. Hundreds of speaker-pair dependent DNNs were trained. The metric of short-time objective intelligibility (STOI) was evaluated. In multicontext averaging, the context windows with parameters $\tau = 0, 1, 2$ were included. In multicontext stacking, the network consisted of three stacks as shown in Fig. 7.3. Three DNNs with parameters $\tau = 0, 1, 2$ were built in the first two stacks while one DNN was merged into the output stack. The number of stacks was also reduced to two for comparison. Some tuning parameters were found from a development task. The experimental results showed that the multicontext averaging outperformed baseline DNN under different SNR conditions. Multicontext

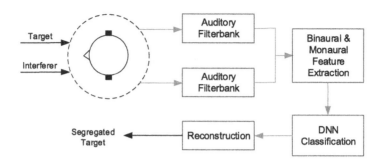

FIGURE 7.4

Speech segregation procedure by using deep neural network.

stacking performed better than multicontext averaging. Also multicontext stacking with three stacks performed better than that with two stacks. All these results have demonstrated the superiority of DNN ensemble to baseline DNN in terms of STOI under different experimental conditions. In the next section, we introduce a DNN solution to speech segregation which is a special realization of source separation in accordance with the human auditory system.

7.1.3 DEEP SPEECH SEGREGATION

One of the most challenging issue in computational auditory scene analysis is to identify source signals with emphasis on modeling an auditory scene analyzer in a human, which is basically stimulated by binaural cues. In Jiang et al. (2014), deep neural networks were developed for speech segregation in noisy and reverberant environments. Speech segregation is seen as a special type of source separation where the segregated or separated speech is identified through a binaural auditory scene. This is comparable with the way of human listener when hearing a single speaker in the presence of interferer or adverse acoustic conditions. Binaural cues are generally more useful than monaural features for speech segregation. The goal is to build a binaural hearing model to hear a target speaker or to identify his/her speech in the presence of interferences of a non-target speaker or ambient noise. Fig. 7.4 illustrates a procedure of speech segregation, which combines signal processing and deep learning in a hybrid procedure. The binaural hearing system with left and right ears (denoted by black blocks) is receiving sound sources in an auditory scene (represented by dashed circle). Target and interfering signals are present simultaneously. In what follows, we address four components in this procedure, which are implemented for deep speech segregation.

1. Auditory filterbank: First of all, the same two auditory filterbanks are used to represent the left- and right-ear inputs by time–frequency (T–F) units which are seen as a two-dimensional matrix of binary labels where one indicates that the target signal dominates the T–F unit and zero otherwise. The so-called ideal binary mask (IBM) is performed. In Jiang et al. (2014), the gammatone filterbank was used as auditory periphery where 64 channels with filter order 4 were set for each ear model. The filter's impulse response was employed in decomposing the input mixed signal into the time–frequency domain. The frame length of T–F units is 20 ms with overlap of 10 ms under

16 kHz sampling rate. This filter simulates the firing activity and saturation effect in an auditory nerve. The left- and right-ear signals in a T–F unit in channel c and at time t are denoted by

$$\{x_{c,t}^l(k), x_{c,t}^r(k)\}.$$

There are 320 samples, indexed by k, in a frame associated with a channel.

2. **Binaural and monaural feature extraction:** Next, binaural features are extracted according to the interaural time difference (ITD) and the interaural level difference (ILD) (Roman et al., 2003) by using the normalized cross-correlation function (CCF) between two ear signals. CCF is indexed by time lag τ, which is between -1 ms and 1 ms. There are 32 CCF features for each pair of T–F units, denoted by $\text{CCF}_{c,t,\tau}$ in two ears. The interaural time difference in each T–F unit (c, t) is calculated by

$$\text{ITD}_{c,t} = \arg\max_{\tau} \text{CCF}_{c,t,\tau}, \tag{7.16}$$

which captures the time lag with the largest cross-correlation function between the two ears. The interaural level difference (in dB) is defined as the energy ratio between the left and right ear for each T–F unit

$$\text{ILD}_{c,t} = 10 \log_{10} \frac{\sum_k \left(x_{c,t}^l(k)\right)^2}{\sum_k \left(x_{c,t}^r(k)\right)^2}. \tag{7.17}$$

ILD is extracted every 10 ms, i.e., two ILD features are calculated. At the same time, monaural features based on 36-dimensional gammatone frequency cepstral coefficients (GFCCs) are extracted as complementary features of the speech signal, which are helpful for speech segregation. For each T–F unit pair for two ears (c, t), the 70-dimensional feature vector consists of 32-dimensional CCF features, 2 ILD features and 36 GFCC features.

3. **DNN classification:** The success of binary masking in audio signal processing implies that the segregation problem may be treated as a binary classification problem. Speech segregation can be formulated as supervised classification by using the acoustically meaningful features. Here, 70-dimensional binaural and monaural features are employed to detect if a T–F unit (c, t) is dominated by the target signal. A binary DNN classifier is trained by supervised learning. In the training stage, the labels in DNN supervised training are provided by ideal binary mask. In the test stage, the posterior probability of a T–F unit dominating the target is calculated. A labeling criterion is used to estimate the ideal binary mask. In the experimental setup (Jiang et al., 2014), each subband or channel-dependent DNN classifier was composed of two hidden layers. The input layer had 70 units. The output layer produced the posterior probability of detecting the target signal. DNN was pretrained and initialized by the restricted Boltzmann machine (RBM). After RBM pretraining, the error backpropagation algorithm was run for supervised fine-tuning. The minibatch size was 256 and the stochastic gradient descent with momentum 0.5 was applied. The learning rate was linearly decreased from 1 to 0.001 in 50 epochs.

4. **Reconstruction:** All the T–F units with the target labels of one comprise the segregated target stream.

In the system evaluation, this approach was assessed for speech segregation under noisy and reverberant environments. The reverberant signal was generated by running binaural impulse responses.

Simulated and real-recorded BIRs were both investigated. A head related transfer function was used to simulate room acoustics for a dummy head. The speech and noise signals were convolved with binaural impulse responses to simulate individual sources in two reverberant rooms with different room sizes. The position of the listener in a room was fixed. Reflection coefficients of the wall surfaces were uniform. The reverberation time T_{60} of two rooms was 0.3 and 0.7 s, respectively. BIRs for azimuth angles between 0° and 360° were generated. Using simulated BIRs, the audio signals and BIRs were adjusted to have the same sampling rate of 44.1 kHz. Four real-recorded reverberant rooms had reverberation time of 0.32, 0.47, 0.68 and 0.89 s, respectively. Input SNR of training data was 0 dB while that of test data was varied. Babble noise was used. Number of non-target sources was increased from one to three for comparison. The performance of speech segregation was evaluated by the hit rate, which is the percentage of correctly classified target-dominated T–F units, as well as the false-alarm rate, which is the percent of wrongly classified interferer-dominated T–F units. In addition, the SNR metric, calculated by using the signals resynthesized from IBM and the estimated IBM, was examined. Experimental results show that SNRs were decreased when increasing the number of non-target sources. Merging monaural features with binaural features performed better than using binaural features alone. This method significantly outperformed the existing methods in terms of the hit and false-alarm rates and the SNRs of resynthesized speech. The desirable performance was not only limited to the target direction or azimuth but also other azimuths which were unseen but similarly trained. In the next section, a number of advanced studies and extended works are introduced. Signal processing plays a crucial role in implementation of learning machines.

7.2 DEEP PROCESSING AND LEARNING

Source separation is recognized as a challenging task which intensively involves signal processing and machine learning. This section surveys recent advances in speech separation and dereverberation. Different perspectives are incorporated in finding deep models to tackle various issues in source separation. In Section 7.2.1, a deep clustering method is presented to deal with the issue of a varying number of sources or speakers in speaker-independent monaural speech separation. Section 7.2.2 surveys a number of mask functions for learning a separation model. Section 7.2.3 introduces a permutation invariant training of deep model which handles the scenario of multitalker speech separation in an end-to-end style. Section 7.2.4 addresses the basics of matrix factorization and the deep model for speech dereverberation. Section 7.2.5 addresses a joint work of matrix factorization and deep neural network for reverberant source separation.

7.2.1 DEEP CLUSTERING

Traditionally, the source separation problem is solved by the *class-based* source-dependent methods where each source belongs to a distinct signal class and the number of sources is fixed. Sources are known beforehand. Such a scenario is not realistic in general source separation applications. It becomes challenging to develop a *speaker-independent* speech separation system where sources or speakers are arbitrary. It is also demanding to build a speech separation system which is generalizable to separate a mixed signal under different number of speakers. In Hershey et al. (2016), deep embedding and clustering was proposed to deal with speaker-independent speech separation for two- as well as three-speaker mixtures. The basic idea is to build the low-dimensional embeddings for different speakers and calculate the pairwise affinity matrix for different clusters. Decoding for segmentation of clusters is performed to carry out speaker-independent speech separation as detailed in what follows.

Learning for Embedding

Clustering method performs the unsupervised learning for data analysis where the observation data are partitioned into different clusters. The samples are represented by the corresponding clusters. Unseen but similar samples can be characterized by those seen samples and the trained clusters. For audio signal processing, *spectral clustering* is developed to partition the time-frequency samples of spectrogram of an audio signal $\mathbf{X} = \{X_{mn}\} = \{x_i\}_{i=1}^{N}$ into the corresponding regions and represent these samples separately for each region. Finding a reasonable partition enables us to build the time-frequency masks and accordingly obtain the separated sources. We seek a low-dimensional embedding $\mathbf{V} \in \mathbb{R}^{N \times D}$, where $D < N$ and perform clustering in the embedding space. Each element in the embedding matrix is based on a deep neural network (DNN)

$$
\begin{aligned}
\mathbf{V} = \{\mathbf{v}_i\} &= \{v_{id}\} \\
&= \{y_d(x_i, \mathbf{w})\} = \mathbf{y}(\mathbf{X}, \mathbf{w})
\end{aligned}
\tag{7.18}
$$

controlled by DNN parameters $\mathbf{w}$. There are D embedding outputs in DNN. The unit-norm embedding is considered as

$$
v_{id} \leftarrow \frac{v_{id}}{\|\mathbf{v}_i\|}
\tag{7.19}
$$

where $\mathbf{v}_i \in \mathbb{R}^D$ and v_{id} denotes dimension d of the embedding for time–frequency unit x_i. The embeddings in $\mathbf{V}$ are implicitly used to estimate the affinity matrix

$$
\widehat{\mathbf{A}} = \mathbf{V}\mathbf{V}^{\top} \in \mathbb{R}^{N \times N}.
\tag{7.20}
$$

This estimated affinity matrix is evaluated by the target affinity $\mathbf{A}$ which is formed by using the indicator $\mathbf{M} = \{m_{is}\} \in \mathbb{R}^{N \times S}$ mapping each time-frequency (T–F) unit x_i to each of S clusters. In practice, each cluster reflects an individual source. The target indicator matrix $\mathbf{M}$ is seen as the *ideal binary mask* (IBM). We set $m_{is} = 1$ if T–F unit i belongs to cluster, source or speaker s and $m_{is} = 0$ otherwise. Clustering assignments are performed for segmentation and separation. Furthermore,

$$
\mathbf{A} = \mathbf{M}\mathbf{M}^{\top} \in \mathbb{R}^{N \times N}
\tag{7.21}
$$

is seen as a ideal binary affinity matrix where $(\mathbf{M}\mathbf{M}^{\top})_{i,j} = 1$ if T–F units i and j belong to the same cluster or speaker, and $(\mathbf{M}\mathbf{M}^{\top})_{i,j} = 0$ otherwise. Deep learning for embedding is run to construct a DNN or find the corresponding parameters $\mathbf{w}$ which produce the embedding matrix $\mathbf{V}$ with the minimum distances between ideal and estimated affinity matrices over all training samples. Learning objective is yielded by

$$
\begin{aligned}
E(\mathbf{w}) &= \|\widehat{\mathbf{A}}(\mathbf{w}) - \mathbf{A}\|^2 \\
&= \|\mathbf{V}(\mathbf{w})\mathbf{V}^{\top}(\mathbf{w}) - \mathbf{M}\mathbf{M}^{\top}\|^2 \\
&= \|\mathbf{V}^{\top}(\mathbf{w})\mathbf{V}(\mathbf{w})\|^2 - 2\|\mathbf{V}^{\top}(\mathbf{w})\mathbf{M}\|^2 + \|\mathbf{M}^{\top}\mathbf{M}\|^2.
\end{aligned}
\tag{7.22}
$$

In the implementation, we need to calculate the derivative $\frac{\partial E(\mathbf{w})}{\mathbf{w}}$ for parameter updating, which is formulated by

$$\frac{\partial E(\mathbf{w})}{\partial \mathbf{w}}$$
$$= \sum_{i=1}^{N} \sum_{d=1}^{D} \frac{\partial E(\mathbf{w})}{\partial v_{id}(\mathbf{w})} \frac{\partial y_d(x_i, \mathbf{w})}{\partial \mathbf{w}} \tag{7.23}$$

where the derivative of DNN output d with respect to layer-wise weights

$$\frac{\partial y_d(x_i, \mathbf{w})}{\partial \mathbf{w}}$$

using input x_i is calculated according to the error backpropagation algorithm, as mentioned in Section 2.4.1, and

$$\frac{\partial E(\mathbf{w})}{\partial \mathbf{V}(\mathbf{w})} = 4\mathbf{V}(\mathbf{V}^{\top}(\mathbf{w})\mathbf{V}(\mathbf{w})) - 4\mathbf{M}(\mathbf{M}^{\top}\mathbf{V}). \tag{7.24}$$

Eq. (7.24) can be efficiently computed due to *low-rank* structure in

$$\mathbf{V}^{\top}(\mathbf{w})\mathbf{V}(\mathbf{w}) \in \mathbb{R}^{D \times D},$$
$$\mathbf{M}^{\top}\mathbf{M} \in \mathbb{R}^{D \times D} \tag{7.25}$$

where $D < N$. Deep clustering directly optimizes a low-rank embedding model.

In the test stage of deep clustering, the mask indicator of T–F units $\mathbf{M}$ or the ideal affinity matrix $\mathbf{M}\mathbf{M}^{\top}$ for pairs (i, j) of T–F units should be updated to $\widehat{\mathbf{M}}$ or $\widehat{\mathbf{M}}\widehat{\mathbf{M}}^{\top}$ by using the test signal $\mathbf{X}$. The updating is performed by minimizing the cost function of k-means algorithm and partitioning N row vectors $\mathbf{v}_i \in \mathbb{R}^D$ for different T–F units by minimizing

$$\widehat{\mathbf{M}} = \arg \min_{\mathbf{M}} \|\mathbf{V} - \mathbf{M}\mathbf{U}\|^2, \tag{7.26}$$

which produces S mean vectors for different clusters

$$\mathbf{U} = (\mathbf{M}^{\top}\mathbf{M})^{-1}\mathbf{M}^{\top}\mathbf{V} \in \mathbb{R}^{C \times D}. \tag{7.27}$$

The resulting cluster assignments $\widehat{\mathbf{M}}$ correspond to the ideal binary masks for source segmentation and separation.

System Evaluation

In Hershey et al. (2016), deep clustering was implemented for speaker-independent speech separation in the presence of two or three unknown speakers. A 30-hour training set and a 10-hour validation set consisting of two-speaker mixed signals were generated by randomly selecting utterances from different speakers and mixing them at various signal-to-noise ratios (SNR) between 0 and 10 dB. There were 16 speakers in test data. These speakers were different from those in the training and development data. The open condition evaluation was investigated. Such an evaluation was not conducted in the previous methods, which were infeasible to open speakers and required knowledge of the speakers in the evaluation. Speech data were consistently sampled or downsampled to 8 kHz. Log spectral magnitudes of mixed speech were calculated by STFT with window size of 30 ms and frame shift of 8 ms.

In the training procedure, binary masks were used and determined as the target indicator $\mathbf{M}$ where each mask m_{is} was one for the target source x_i with maximum magnitude in cluster or speaker s and zero for the others. The number of clusters S was set to be the number of speakers in the mixed signals, i.e., $S=2$ or 3. The masks were obtained by clustering $\{\mathbf{v}_i\}_{i=1}^N$, which were calculated for each segment with 100 frames. For two-source mixtures, $\mathbf{M}$ corresponded to the ideal binary mask. The network structure was constructed by two bidirectional long short-term memory (BLSTM)(with 600 hidden units in each BLSTM layers) followed by one feedforward layer (with D embedding outputs). BLSTM will be introduced in Section 7.4.3. Stochastic gradient descent algorithm with momentum 0.9 and fixed learning rate 10^{-5} was applied. A zero-mean Gaussian noise was added to the weights $\mathbf{w}$ to alleviate local optima. The weights were randomly initialized by Gaussian distribution with zero mean and 0.1 variance. Activation functions using logistic sigmoid and hyperbolic tangent functions were evaluated. Two BLSTMs with the same network architecture were built for deep clustering and source separation. The only differences between two BLSTMs were the output layers and objective functions. Different embedding dimension D was evaluated. The separation performance was evaluated in terms of averaged signal-to-distortion ratio (SDR). Experimental results showed that system performance deteriorated if the dimension of embedding D was as low as 5. This method required more dimensions. The cases of $D = 20$, 40 and 60 obtained comparable results. Deep clustering performed much better than nonnegative matrix factorization with sparseness constraint in terms of SDR. Such an improvement was observed for two- and three-speaker mixed signals. The performance of three-speaker mixed signals was good even when the system was trained by using two-speaker mixed signals.

In general, source separation suffers from the permutation problem, which is challenging as mentioned in Section 1.2.2. However, this problem is mitigated in a deep clustering algorithm. It is because the deep embedding model

$$\mathbf{y}(x_i, \mathbf{w}) = \{y_d(x_i, \mathbf{w})\}_{d=1}^D \tag{7.28}$$

for an embedding vector $\mathbf{v}_i = \{v_{id}\}_{d-1}^D$ of T–F bin x_i is trained to estimate an affinity matrix $\widehat{\mathbf{A}} = \mathbf{V}\mathbf{V}^\top$, which minimizes the distance to the ideal affinity matrix $\mathbf{A} = \mathbf{M}\mathbf{M}^\top$ where the target indicator matrix $\mathbf{M} = \{\mathbf{m}_i\}$ with masks $\mathbf{m}_i = \{m_{is}\}_{s=1}^S$ of each T–F bin x_i is insensitive to permutation due to different speakers $1 \le s \le S$. The reason is that the ideal affinity matrix using ideal binary masks satisfies the property

$$(\mathbf{M}\mathbf{P})(\mathbf{M}\mathbf{P})^\top = \mathbf{M}\mathbf{M}^\top \tag{7.29}$$

for any permutation matrix $\mathbf{P}$. Deep learning carries out a permutation independent encoding for discriminative embedding, clustering and separation. The outputs of BLSTMs for embeddings and sources are seen as the realization of class-based models, which assure discriminative performance in clustering and separation, respectively. In what follows, we continue the issue of permutation and present a new permutation-invariant training for speaker-independent speech separation based on an end-to-end and deep-learning based solution.

7.2.2 MASKING AND LEARNING

Masks are known as a crucial intermediate step towards estimating the magnitude spectra of source signals. In Section 7.2.1, deep embedding and clustering was developed by minimizing the error function of affinities of time-frequency bins $\mathbf{X} = \{x_i\}$ in mixed signals. The ideal affinity matrix $\mathbf{A}$ was estimated by ideal binary masks $\mathbf{M} = \{m_{si}\}$, which contained binary values to indicate the activities of N individual T–F units x_i in S different clusters or sources s. Generally, we may develop a new training algorithm by incorporating different mask functions into the learning objective for high-performance speech separation. Three mask functions are introduced in what follows:

1. Ideal ratio mask (IRM): Consider an S-speaker mixed signal represented by magnitude spectra $\mathbf{X}^{\text{mix}} = \{x_i^{\text{mix}}\}$. The magnitude spectra of the corresponding source signals are denoted by $\mathbf{X}_s = \{x_{si}\}$ for different speakers $1 \leq s \leq S$. Each signal is indexed by i, which corresponds to a time-frequency unit (t, f). There are N time-frequency units. Ideal ratio masks $\mathbf{M}^{\text{IRM}} = \{m_{si}^{\text{IRM}}\}$ are defined as

$$m_{si}^{\text{IRM}} = \frac{|x_{si}|}{\sum_{j=1}^{S} |x_{ji}|} \tag{7.30}$$

for S speakers and N time-frequency units. IRM defined in Eq. (7.2) is only for two-speaker mixtures. Eq. (7.30) is a general IRM for S speakers. IRMs are constrained by

$$0 \leq m_{si}^{\text{IRM}} \leq 1 \tag{7.31}$$

and

$$\sum_{s=1}^{S} m_{si}^{\text{IRM}} = 1 \tag{7.32}$$

for all T–F units $\{x_{si}\}_{i=1}^{N}$ in different sources s. This IRM achieves the highest signal-to-distortion ratio (SDR) when all sources s have the same phase. However, different sources vary in phase information.

2. Ideal amplitude mask (IAM): The IAM matrix is denoted by $\mathbf{M}^{\text{IAM}} = \{m_{si}^{\text{IAM}}\}$ where each mask is defined as

$$m_{si}^{\text{IAM}} = \frac{|x_{si}|}{|x_i^{\text{mix}}|}. \tag{7.33}$$

If the phase of each source equals that of the mixed signal, this IAM obtains the highest SDR, although this assumption is not met in practice. Empirically, IAM is set in the range $0 \leq m_{si}^{\text{IAM}} \leq 1$.

3. Ideal phase sensitive mask (IPSM): In general, IRM and IAM do not work perfectly for reconstruction because the phase difference between source signals and the phase difference between source and mixed signals are not considered, respectively. The ideal phase-sensitive mask was proposed to deal with this consideration. The IPSM matrix $\mathbf{M}^{\text{IPSM}} = \{m_{si}^{\text{IPSM}}\}$ is yielded by calculating the mask (Erdogan et al., 2015)

$$m_{si}^{\text{IPSM}} = \frac{|x_{si}| \cos\left(\theta_i^{\text{mix}} - \theta_{si}\right)}{|x_i^{\text{mix}}|} \tag{7.34}$$

where θ_i^{mix} and θ_{si} denote the phases of mixed signal x_i^{mix} and source signal x_{si}, respectively. The phase difference $\theta_i^{\mathrm{mix}} - \theta_{si}$ is taken into account for phase correction. To avoid a negative mask, it is reasonable to use the nonnegative IPSM (NIPSM) which is expressed by

$$m_{si}^{\mathrm{NIPSM}} = \max\left(0, m_{si}^{\mathrm{IPSM}}\right). \tag{7.35}$$

In the implementation, we need to estimate the mask functions for source separation. Basically, the softmax function in Eq. (2.82), sigmoid function in Eq. (2.51), hyperbolic function in Eq. (2.11) and rectified linear unit in Eq. (2.50) are suitable as the activation functions for estimating these mask functions with a reasonable range of mask values.

7.2.3 PERMUTATION-INVARIANT TRAINING

This section introduces a permutation-invariant training algorithm (Kolbæk et al., 2017) for monaural speech separation where speakers are unknown beforehand similar to the task in Section 7.2.1. This algorithm is developed for multitalker speech separation, which is a challenging task in recent years. In Hershey et al. (2016), deep clustering was proposed for multitalker speaker-independent speech separation which was implemented by estimating the discriminative embeddings or affinities in different time–frequency units of mixed signals. Differently, this method emphasizes an utterance-level permutation-invariant training (PIT), which minimizes the utterance-level separation error to force the separated frames belonging to the same speaker to be aligned to the same output stream. This PIT algorithm is incorporated in training the recurrent neural network (RNN) for speech separation in the presence of unknown speakers and an unknown speaker number. This section is organized as two parts. In addition to system evaluation, we address how various mask functions are merged when constructing the training criterion for deep processing and learning in a multitalker speech separation system.

Training Criterion

Mask functions of different sources s in T–F units i are estimated by optimizing a learning objective. It is straightforward to build a training criterion based on a sum-of-squares error function

$$\begin{aligned} E_m(\mathbf{w}) &= \|\widehat{\mathbf{M}}(\mathbf{w}) - \mathbf{M}\|^2 \\ &= \frac{1}{SN} \sum_{s=1}^{S} \sum_{i=1}^{N} (\widehat{m}_{si}(\mathbf{w}) - m_{si})^2 \end{aligned} \tag{7.36}$$

where $\mathbf{M} = \{m_{si}\}$ is an ideal mask function and $\widehat{\mathbf{M}}(\mathbf{w}) = \{\widehat{m}_{si}(\mathbf{w})\}$ denotes the estimated mask function, which is driven by a deep model with parameters $\mathbf{w}$. This error function is accumulated over all sources s in different T–F units i.

However, in the task of speech separation, the ideal masks m_{si} using silence segments of source signals x_{si} and mixed signal x_i^{mix} are not well defined. A meaningful training criterion can be constructed according to a sum-of-squares error function between the reconstructed source signal $\widehat{\mathbf{X}}(\mathbf{w})$ and the true source signal $\mathbf{X}$. The reconstruction is driven by a mask function which is obtained by a deep model with parameters $\mathbf{w}$. This learning objective is expressed by

$$E_x(\mathbf{w}) = \|\widehat{\mathbf{X}}(\mathbf{w}) - \mathbf{X}\|^2$$

$$= \frac{1}{SN} \sum_{s=1}^{S} \sum_{i=1}^{N} (\widehat{x}_{si}(\mathbf{w}) - x_{si})^2$$

$$= \frac{1}{SN} \sum_{s=1}^{S} \sum_{i=1}^{N} (\widehat{m}_{si}(\mathbf{w})x_i^{\text{mix}} - x_{si})^2.$$

(7.37)

More attractively, the phase sensitive mask can be used as a training objective to compensate the phase difference between the reconstructed source and the target source. This objective is formulated by

$$E_p(\mathbf{w}) = \frac{1}{SN} \sum_{s=1}^{S} \sum_{i=1}^{N} \left(\widehat{m}_{si}(\mathbf{w})x_i^{\text{mix}} - x_{si} \cos \left(\theta_i^{\text{mix}} - \theta_{si} \right) \right)^2.$$

(7.38)

Compared with Eq. (7.37), this phase-sensitive mask objective function considers the phase discounted target source in the training procedure.

Traditionally, we adopt a deep model $\mathbf{y}(\cdot)$ to estimate the mask functions

$$\mathbf{y}(\mathbf{X}^{\text{mix}}, \mathbf{w}) = \widehat{\mathbf{M}}(\mathbf{w}) = \{\widehat{m}_{si}(\mathbf{w})\}$$

(7.39)

for all T–F units i in different sources s. These masks are then used to estimate the source signal $\mathbf{X} = \{x_{si}\}$ where each T–F unit i is separated by

$$\widehat{x}_{si} = \widehat{m}_{si}(\mathbf{w})x_i^{\text{mix}}$$

(7.40)

for an individual source i. In real-world implementation, as addressed in Section 1.2.2, a source separation system could not tell the order or the arrangement of separated sources in advance. Permutation problem happens and affects system performance especially under a challenging condition where the mixed signals are observed in the presence of multiple speakers or sources and the source identities are unknown in the test stage.

To deal with the permutation problem, an additional learning objective is merged to comply with the so-called permutation-invariant training for source separation. Namely, in addition to mask-based training objective $E_m(\mathbf{w})$ in Eq. (7.36), $E_x(\mathbf{w})$ in Eq. (7.37) or $E_p(\mathbf{w})$ in Eq. (7.38), the sum-of-squares error function due to permutation ambiguity is introduced for joint minimization. This permutation loss function is measured and minimized when training a deep neural network. The measurement is based on the association between the target source signals $\{x_{si}\}$ and the estimated masks $\{\widehat{m}_{si}(\mathbf{w})\}$, which is computed as the pairwise sum-of-squares error function over S^2 pairs of target and estimated sources $\{x_{si}, \widehat{x}_{si}\}$ for S sources. There are $S!$ possible permutations considered in the PIT method. Furthermore, since the permutation of a frame using PIT is affected by that of previous frames, an incorrect permutation may continuously degrade the assignment decisions in later frames. An efficient solution is to conduct PIT at the utterance level to avoid the continuously wrong assignment decision. In general, the trained deep model will guide the permutation, staying in the same frames inside the same output for each test utterance. Notably, this model is feasible to apply for speaker-independent source separation while the previous solution (Shao and Wang, 2006) to permutation ambiguity needs prior knowledge about source signals.

System Evaluation

Permutation-invariant training was evaluated for monaural source separation in the presence of two source speakers where the identities of source speakers were unknown beforehand. An unknown speaker may be seen or unseen in a test session to setup the so-called closed or open condition, respectively. Similar to the evaluation in deep clustering in Section 7.2.1, there were 30 hours of training data and 10 hours of validation data. Then a 129-dimensional STFT magnitude spectral vector was extracted with the sampling rate of 8 kHz, frame size of 32 ms and frame shift of 16 ms. Two-speaker mixed signals were randomly generated from 49 males and 51 females at various SNRs ranging from 0 to 5 dB. Test data in a form of two- and three-speaker mixtures were similarly obtained from 16 speakers. Bidirectional long short-term memory (BLSTM), as mentioned in Section 7.4.3, was employed in training a deep separation model based on permutation-invariant training. Different activation functions were examined. The DNN with 3 hidden layers and each with 1024 ReLU units was implemented. The results of using convolutional neural network (CNN) (LeCun et al., 1998) were included for comparison.

CNN is known as a variant of feedforward deep neural network (DNN) where the affine transformation in layer-wise DNN is replaced by a number of convolutions to extract a set of feature maps. In addition to this *convolution* stage, a pooling stage is performed in each layer so as to avoid the curse of dimensionality. A deep CNN model can be realized with several layers of convolution and pooling. To calculate the class posteriors based on CNN, a full-connected layer is arranged before output layer. CNN has achieved significant performance in computer vision and many other tasks. Key benefits of using a CNN include the *sparse representation* and the *shared parameterization*, which assure a well-regularized model for prediction.

In the evaluation, the signal-to-distortion (SDR) and the perceptual evaluation of speech quality (PESQ) were measured. The learning rate in optimization was initialized by 2×10^{-5} and scaled down by 0.7 per sample until the learning rate got below 10^{-10}. A minibatch of 8 sentences was used. The results of estimating different mask functions based on IRM, IAM, IPSM and NIPSM were evaluated. Experimental results on PIT showed that SDRs of open condition were close to those of closed condition. The robustness to unknown and unseen speaker sources was assured. Using IPSM was better than using IAM and IRM when estimating mask functions based on PIT. The best solution was obtained by using NIPSM as the mask function. The results when using a logistic sigmoid function, hyperbolic tangent function or rectified linear unit did not change too much. In this task, CNN obtained higher SDR than DNN. The highest SDR was obtained by using BLSTM. BLSTM performed better than unidirectional LSTM. In monaural source separation, SDRs with three-speaker mixed signals were lower than those with two-speaker mixed signals. Attractively, such a single model can handle both two- and three-speaker mixed signals. In what follows, we switch to an advanced model which incorporates matrix factorization in DNN for reverberant source separation based on the spectro-temporal neural factorization.

7.2.4 MATRIX FACTORIZATION AND NEURAL NETWORK

Matrix factorization and DNN have been extensively developing in the areas of signal processing and machine learning with numerous applications. Matrix factorization performs a two-way decomposition, which is generalizable to tensor factorization for multiway observations (De Lathauwer et al., 2000a) as addressed in Sections 2.2 and 2.3. The *front-end* processing based on signal decomposition helps extract meaningful features in *back-end* modeling for source separation. This section introduces the basics of matrix factorization and the construction of deep recurrent neural network for reverberant source separation.

Matrix Factorization

A time-series audio signal $\mathbf{X} = \{X_{mn}\}$ or $\mathbf{X} = \{X_{ft}\} \in \mathbb{R}^{F \times T}$ contains the log-magnitude spectra with F frequency bins and T time frames, which are calculated by STFT. Without loss of generality, we use the notation X_{ft} for an observed T–F unit by following that used for speech dereverberation as seen in Sections 1.1.3 and 5.1.3. This matrix $\mathbf{X}$ can be factorized by Tucker decomposition to extract a core matrix $\mathbf{A} = \{A_{ij}\} \in \mathbb{R}^{I \times J}$ via

$$\mathbf{X} = \mathbf{A} \times_1 \mathbf{U} \times_2 \mathbf{V} \tag{7.41}$$

where

$$\mathbf{U} = \{U_{fi}\} \in \mathbb{R}^{F \times I}, \tag{7.42}$$

$$\mathbf{V} = \{V_{tj}\} \in \mathbb{R}^{T \times J} \tag{7.43}$$

denote the factor matrices; I and J indicate the reduced dimensions corresponding to F and T, respectively. Each entry in the core matrix is expressed by

$$X_{ft} = \sum_i \sum_j A_{ij} U_{fi} V_{tj}. \tag{7.44}$$

The *inverse* of Tucker decomposition has the property

$$\mathbf{A} = \mathbf{X} \times_1 \mathbf{U}^\dagger \times_2 \mathbf{V}^\dagger \tag{7.45}$$

where

$$\mathbf{U}^\dagger = (\mathbf{U}^\top \mathbf{U})^{-1} \mathbf{U}^\top \tag{7.46}$$

is the pseudo-inverse of $\mathbf{U}$. Importantly, the core matrix $\mathbf{A}$ is viewed as a *factorized* feature matrix of the spectro-temporal data $\mathbf{X}$.

Deep Learning and Dereverberation

Fig. 7.5 depicts a deep recurrent neural network (DRNN) (Wang et al., 2016) for speech dereverberation, which is developed for spectral mapping from a reverberant speech $\mathbf{x}_t^{\text{rev}} = \{X_{ft}^{\text{rev}}\}$ to a clean speech $\mathbf{y}_t = \{Y_{ft}\}$. The input vector

$$\mathbf{x}_t = [(\mathbf{x}_{t-\tau}^{\text{rev}})^\top, \dots, (\mathbf{x}_t^{\text{rev}})^\top, \dots, (\mathbf{x}_{t+\tau}^{\text{rev}})^\top]^\top \in \mathbb{R}^{F(2\tau+1)} \tag{7.47}$$

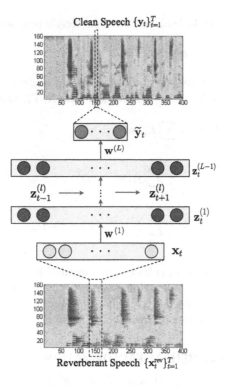

FIGURE 7.5

A deep recurrent neural network for speech dereverberation.

consists of a window of $2\tau + 1$ frames of reverberant speech centered at time t. Each frame $\mathbf{x}_t$ has values in F frequency bins. The forward and backward calculations are required to train this vector-based neural network. In the forward pass, the affine transformation and nonlinear activation are calculated by

$$\mathbf{z}_t^{(l)} = h(\mathbf{a}_t^{(l)}) = h(\mathbf{w}^{(l)}\mathbf{z}_t^{(l-1)}) \tag{7.48}$$

in each fully-connected (FC) layer l. The activation function $h(\cdot)$ can be a logistic sigmoid function or a rectified linear unit. The recurrent layer m is calculated by

$$\begin{aligned} \mathbf{z}_t^{(m)} &= h(\mathbf{a}_t^{(m)}) \\ &= h(\mathbf{w}^{(m)}\mathbf{z}_t^{(m-1)} + \mathbf{w}^{(mm)}\mathbf{z}_{t-1}^{(m)}) \end{aligned} \tag{7.49}$$

where the forward weights $\mathbf{w}^{(m)}$ and recurrent weights $\mathbf{w}^{(mm)}$ are individually used to connect with the previous layer $m-1$ and the previous time $t-1$, respectively. Temporal information is stored for speech dereverberation. Notably, the input signal $\mathbf{x}_t$, hidden features $\{\mathbf{z}_t^{(l)}, \mathbf{z}_t^{(m)}\}$ and output signal $\widetilde{\mathbf{y}}_t$

are all time-dependent vectors. In the backward pass, the weight parameters

$$\mathbf{w} = \{\mathbf{w}^{(l)}, \mathbf{w}^{(m)}, \mathbf{w}^{(mm)}\} \tag{7.50}$$

in different layers are estimated according to the error backpropagation algorithm where the sum-of-squares error function using minibatches of training samples $\{\mathbf{X}, \mathbf{Y}\}$, namely

$$
\begin{aligned}
E(\mathbf{w}) &= \sum_n E_n(\mathbf{w}) \\
&= \frac{1}{2} \sum_{t=1}^{T} (\widetilde{\mathbf{y}}_t(\mathbf{w}) - \mathbf{y}_t)^2 \\
&= \frac{1}{2} \sum_{t=1}^{T} \sum_{f=1}^{F} (\widetilde{Y}_{ft}(\mathbf{w}) - Y_{ft})^2,
\end{aligned}
\tag{7.51}
$$

is minimized. In Eq. (7.51), $E_n(\mathbf{w})$ is calculated over a minibatch data $t \in \{\mathbf{X}_n, \mathbf{Y}_n\}$; $\widetilde{\mathbf{y}}_t = \{\widetilde{Y}_{ft}\}$ is the output of dereverberation neural network at time t while $\mathbf{y}_t = \{Y_{ft}\}$ is the associated clean output. In LeCun et al. (1998), CNN was developed to catch spatial features by means of convolution and pooling layers. There were no factorized features extracted in different ways for classification or regression.

7.2.5 SPECTRO-TEMPORAL NEURAL FACTORIZATION

This section presents the spectro-temporal neural factorization in a layer-wise regression model for speech dereverberation (Chien and Kuo, 2018, Chien and Bao, 2018).

Factorized Neural Network

First of all, we develop the factorized neural network where the hybrid spectro-temporal features are extracted for speech dereverberation as detailed in what follows. Given a spectro-temporal input matrix $\mathbf{X} = \{X_{ft}\}$, the feature matrix $\mathbf{A}^{(1)} = \{A_{ji}^{(1)}\}$ in the first hidden layer of the factorized neural network is obtained by a bilinear transformation through two factor matrices $\mathbf{U}^{(1)} = \{U_{if}^{(1)}\} \in \mathbb{R}^{I \times F}$ and $\mathbf{V}^{(1)} = \{V_{jt}^{(1)}\} \in \mathbb{R}^{J \times T}$ as

$$
\begin{aligned}
\mathbf{A}^{(1)} &= \mathbf{X} \times_1 \mathbf{U}^{(1)} \times_2 \mathbf{V}^{(1)} \\
&= \sum_{f=1}^{F} \sum_{t=1}^{T} X_{ft} \left(\mathbf{u}_f^{(1)} \circ \mathbf{v}_t^{(1)} \right).
\end{aligned}
\tag{7.52}
$$

Eq. (7.52) is seen as a sum of outer products of all individual columns of factor matrices $\mathbf{u}_f^{(1)}$ and $\mathbf{v}_t^{(1)}$ along frequency and time axes, respectively. A spectro-temporal feature matrix $\mathbf{Z}^{(1)} = \{Z_{ij}^{(1)}\} \in \mathbb{R}^{I \times J}$ is obtained through an activation function $\mathbf{Z}^{(1)} = h(\mathbf{A}^{(1)})$. This feature matrix is then factorized and activated again as $\mathbf{A}^{(2)}$ and $\mathbf{Z}^{(2)}$ in the second hidden layer, respectively. The feedforward computation is run layer by layer until layer L to estimate the dereverberant speech $\widetilde{\mathbf{Y}} \in \mathbb{R}^{F \times T}$ corresponding to reverberant speech $\mathbf{X}$. In the last layer L, the dereverberant speech matrix is calculated by

$$
\begin{aligned}
\widetilde{\mathbf{Y}} &= h(\mathbf{A}^{(L)}) \\
&= h(\mathbf{Z}^{(L-1)} \times_1 \mathbf{U}^{(L)} \times_2 \mathbf{V}^{(L)})
\end{aligned}
\tag{7.53}
$$

where $\mathbf{A}^{(L)} = \{A_{ft}^{(L)}\}$ denotes the activation matrix in layer L and $\mathbf{Z}^{(L-1)} = \{Z_{ij}^{(L-1)}\}$ denotes the output matrix of hidden units in layer $L - 1$. Using the dereverberant output matrix $\widetilde{\mathbf{Y}}$ and the associated clean speech matrix $\mathbf{Y}$, the regression error function $E(\mathbf{w})$ in Eq. (7.51) is minimized to estimate model parameters $\mathbf{w}$, which contain the spectral parameters $\mathbf{U}^{(l)}$ and the temporal parameters $\mathbf{V}^{(l)}$ in a different layer l. Different from the traditional neural network calculating the propagation of input vector $\mathbf{x}_t$ independently at each time t, the factorized neural network conducts the layer-wise calculation over a window or a matrix of reverberant speech frames $\mathbf{X}$. The hybrid spectro-temporal features are characterized.

Factorized Error Backpropagation

To estimate model parameters $\mathbf{w} = \{\mathbf{U}^{(l)}, \mathbf{V}^{(l)}\}_{l=1}^{L}$, we need to calculate the gradients of E_n over a minibatch data $t \in \{\mathbf{X}_n, \mathbf{Y}_n\}$ with respect to individual parameters in $\mathbf{w}$. Starting from the regression layer L, we calculate the local gradient $\mathcal{D}_{ft}^{(L)}$ of an output neuron at a T–F unit (t, f) by

$$
\begin{aligned}
\frac{\partial E_n}{\partial A_{ft}^{(L)}} &= \frac{\partial E_n}{\partial \widetilde{Y}_{ft}} \frac{\partial \widetilde{Y}_{ft}}{\partial A_{ft}^{(L)}} \\
&= \left(\widetilde{Y}_{ft} - Y_{ft}\right) h'\left(A_{ft}^{(L)}\right) \triangleq \mathcal{D}_{ft}^{(L)}.
\end{aligned}
\tag{7.54}
$$

Then, we can derive the gradients

$$
\begin{aligned}
\frac{\partial E_n}{\partial U_{if}^{(L)}} &= \sum_{t \in \{\mathbf{X}_n, \mathbf{Y}_n\}} \frac{\partial E_n}{\partial A_{ft}^{(L)}} \frac{\partial A_{ft}^{(L)}}{\partial U_{if}^{(L)}} \\
&= \sum_{t \in \{\mathbf{X}_n, \mathbf{Y}_n\}} \mathcal{D}_{ft}^{(L)} \left(\sum_k Z_{ki}^{(L-1)} V_{kt}^{(L)}\right) \\
&= \langle \mathcal{D}_{f:}^{(L)}, \mathbf{Z}_{:i}^{(L-1)} \times_2 \mathbf{V}^{(L)}\rangle
\end{aligned}
\tag{7.55}
$$

$$
\frac{\partial E_n}{\partial V_{jt}^{(L)}} = \langle \mathcal{D}_{:t}^{(L)}, \mathbf{Z}_{j:}^{(L-1)} \times_1 \mathbf{U}^{(L)}\rangle
\tag{7.56}
$$

for updating parameters $\mathbf{U}^{(L)}$ and $\mathbf{V}^{(L)}$, respectively. We assume $\mathbf{Z}^{(L-1)} = \{Z_{ji}^{(L-1)}\} \in \mathbb{R}^{J \times I}$. After updating $\{\mathbf{U}^{(L)}, \mathbf{V}^{(L)}\}$, we propagate local gradients from $\mathcal{D}^{(L)} = \{\mathcal{D}_{ft}^{(L)}\}$ in layer L back to $\mathcal{D}^{(L-1)} =$

$\{\mathcal{D}_{ji}^{(L-1)}\}$ in layer $L-1$ through

$$\frac{\partial E_n}{\partial A_{ji}^{(L-1)}}$$

$$= \sum_{t \in \{\mathbf{X}_n, \mathbf{Y}_n\}} \sum_{f=1}^{F} \frac{\partial E_n}{\partial A_{ft}^{(L)}} \frac{\partial A_{ft}^{(L)}}{\partial Z_{ji}^{(L-1)}} \frac{\partial Z_{ji}^{(L-1)}}{\partial A_{ji}^{(L-1)}} \triangleq \mathcal{D}_{ji}^{(L-1)}, \tag{7.57}$$

which can be written in a matrix form as

$$\mathcal{D}^{(L-1)} = h'(\mathbf{A}^{(L-1)}) \odot \left(\mathcal{D}^{(L)} \times_1 (\mathbf{U}^{(L)})^{\top} \times_2 (\mathbf{V}^{(L)})^{\top}\right) \tag{7.58}$$

where $\odot$ denotes the element-wise multiplication. These local gradients $\mathcal{D}^{(L-1)}$ will be used to calculate gradients for updating the spectral and temporal factor matrices $\{\mathbf{U}^{(L-1)}, \mathbf{V}^{(L-1)}\}$ in layer $L-1$. Interestingly, the local gradient $\mathcal{D}^{(L-1)}$ in layer $L-1$ is calculated in a form of *transpose factorization* by using $\mathcal{D}^{(L)}$, $(\mathbf{U}^{(L)})^{\top}$ and $(\mathbf{V}^{(L)})^{\top}$ in layer L. It is meaningful that the gradients with respect to spectral factor $U_{if}^{(L)}$ at frequency f and temporal factor $V_{jt}^{(L)}$ at time t are calculated by summing up all information over time frames t and frequency bins f, respectively.

System Implementation

The factorized neural network was evaluated for speech denoising and dereverberation. The reverberant speech (SimData) was simulated by convolving a speech corpus with three room impulse responses and adding with the stationary noise recordings in different SNR levels. The experimental setup was the same as that in Section 5.1.4, which was developed for speech dereverberation using the convolutive nonnegative matrix factorization. Training data had 7862 utterances from 92 speakers while development and evaluation data contained 1484 and 2176 utterances from 20 and 28 speakers, respectively. Then 320-point STFT was calculated, i.e., $F = 160$. At each time frame, an input matrix $\mathbf{X} = \{X_{ft}\}$ consisting of five neighboring frames ($\tau = 5$) with dimension 160×11 was formed and regressed into a target frame 160×1 in dereverberant speech. STNF layers were empirically configured with a fixed size of 90×8. Hidden and output layers were implemented by using ReLU and sigmoid activations, respectively. The SGD algorithm was run using a minibatch size of 128 frames with ℓ_2-regularization where the regularization parameter was selected from validation data. The Adam optimization algorithm (Kingma and Ba, 2014) was applied. Weights were randomly initialized by a uniform distribution. Dereverberation performance was evaluated by the speech-to-reverberation modulation energy ratio (SRMR) (Falk et al., 2010) (higher is better) and the perceptual evaluation of speech quality (PESQ) (higher is better).

Fig. 7.6 illustrates how an input matrix $\mathbf{X}$ of a short segment of reverberant speech is transformed to estimate the activation matrices $\mathbf{A}^{(l)}$ in the first hidden layer $l = 1$ and the output matrix of dereverberant speech $\widetilde{\mathbf{Y}}$. The second and third rows show the visualization in spectral and temporal factorizations:

$$\mathbf{A}_f^{(l)} = \mathbf{Z}^{(l-1)} \times_1 \mathbf{U}^{(l)}, \tag{7.59}$$

$$\mathbf{A}_t^{(l)} = \mathbf{Z}^{(l-1)} \times_2 \mathbf{V}^{(l)}, \tag{7.60}$$

respectively. The factorized features in the spectral and temporal domains are obvious.

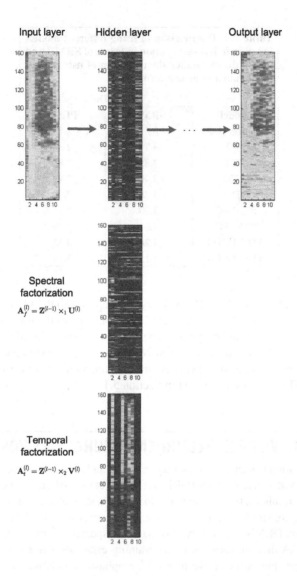

FIGURE 7.6

Factorized features in spectral and temporal domains for speech dereverberation.

A neural network topology can be realized by combining the spectro-temporal factorization (STF) layer, long short-term memory (LSTM) layer (addressed in Section 2.5.3) and fully-connected (FC) layer. The baseline DNN was built by using two or three FC layers (FC2 and FC3). A DRNN was realized by three LSTM layers (LSTM3) for comparison. There were two or three STF layers followed by one FC layer to carry out STF2-FC or STF3-FC, or followed by one LSTM layer to execute STF2-LSTM or STF3-LSTM. The results in Table 7.1 show that LSTM increases the values of SRMR

Table 7.1 Comparison of using different models for speech dereverberation in terms of SRMR and PESQ (in dB) under the condition of using simulated data and near microphone

Model	SimData	
	SRMR	PESQ
Unprocessed	3.57	2.18
FC2	3.79	2.32
FC3	3.85	2.83
LSTM3	3.90	2.92
STF3	3.78	2.29
STF2-FC	3.89	2.84
STF3-FC	4.09	3.19
STF2-LSTM	**4.20**	**3.33**
STF3-LSTM	4.13	3.29

and PESQ when compared with DNN. The STF layer is superior to the FC layer with higher SRMR and PESQ. The stand-alone three STF layers (STF3) do not work well. However, the improvement of combining STF layers with FC layer is obvious. The best results are achieved by connecting the LSTM layer with STF layers. In addition, the result of SRMR using ST2-LSTM is even better than that using full Bayesian approach to nonnegative convolutive transfer function combined with nonnegative matrix factorization (FB-NCTF-NMF) as mentioned in Section 5.1.2.

7.3 DISCRIMINATIVE DEEP RECURRENT NEURAL NETWORK

Monaural source separation is seen as a sequential learning and regression problem. It is meaningful to introduce a deep recurrent neural network (DRNN) (Huang et al., 2014a) to deal with monaural speech separation. In particular, this section presents the extended and related studies on hybrid learning and discriminative learning for single-channel speech separation based on DRNN. Section 7.3.1 describes how mask functions and DRNNs are jointly optimized for speech separation, singing voice separation and speech denoising. A discriminative network training criterion is introduced to reduce the interference between sources. Furthermore, Section 7.3.2 emphasizes a number of discriminative learning objectives for speech separation where the mask function is also jointly estimated with DRNN.

7.3.1 JOINT OPTIMIZATION AND MASK FUNCTION

As mentioned in Sections 7.2.1 and 7.2.3, it is crucial to exploit solutions to monaural source separation by identifying source signals based on an affinity matrix and a mask function. The time–frequency (T–F) mask function using deep learning is applied. These T–F masks should be internally learned for different sources by directly optimizing the separated results with respect to ground truth signals in an *end-to-end* fashion. A joint optimization of the mask function and deep model is required to achieve desirable performance (Huang et al., 2014a, 2015, Weninger et al., 2014a). A topology of deep recurrent neural network for single-channel source separation has been shown in Fig. 2.19. A DRNN is constructed to estimate the soft mask functions $\{\widehat{\mathbf{y}}_{i,t}(\mathbf{x}_t, \mathbf{w})\}$ for two sources $i = 1, 2$ or the DRNN parameters $\mathbf{w}$ from a set of mixed signals $\mathbf{x}_t^{\text{mix}}$ (or $\mathbf{x}_t$) and the corresponding source signals $\{\mathbf{x}_{1,t}, \mathbf{x}_{2,t}\}$. The separated signals are then obtained by using the estimated masks

$$\widetilde{\mathbf{x}}_{i,t} = \mathbf{x}_t^{\text{mix}} \odot \widehat{\mathbf{y}}_{i,t}.$$

DRNN parameters

$$\mathbf{w} = \{\mathbf{w}^{(l)}, \mathbf{w}^{(m)}, \mathbf{w}^{(mm)}\}$$

consist of the weights $\mathbf{w}^{(l)}$ in forward layers l and the weights $\{\mathbf{w}^{(m)}, \mathbf{w}^{(mm)}\}$ in recurrent layer m as expressed in Eqs. (7.48) and (7.49), respectively. These mask functions are calculated in the output layer of an L-layer DRNN as

$$\widehat{\mathbf{y}}_{i,t} = \frac{|\mathbf{w}_i^{(L)} \mathbf{z}_t^{(L-1)}|}{|\mathbf{w}_1^{(L)} \mathbf{z}_t^{(L-1)}| + |\mathbf{w}_2^{(L)} \mathbf{z}_t^{(L-1)}|}, \qquad \text{for } i = 1, 2, \tag{7.61}$$

where $\mathbf{z}_t^{(L-1)}$ denotes the neuron outputs in layer $L - 1$ and $\mathbf{w}^{(L)} = \{\mathbf{w}_i^{(L)}\}_{i=1,2}$ denotes the weight connections to two sources in output layer L. In Eq. (7.61), the vector is calculated in a component-wise manner.

Discriminative Training Criterion

Monaural source separation can be implemented according to different training objectives. Section 7.2.3 addresses a number of training objectives based on a sum-of-squares error function between the estimated masks (or the estimated spectra) and the ideal masks (or the target source spectra). Considering the mask functions, we may use $\widehat{\mathbf{M}}$ in Eq. (7.36) or $\widehat{\mathbf{y}}_{i,t}$ in Eq. (7.61) to construct a learning objective $E(\mathbf{w})$ for monaural source separation. However, an end-to-end learning procedure can be implemented by directly minimizing the sum-of-squares error function between the predicted source spectra $\{\widetilde{\mathbf{x}}_{1,t}, \widetilde{\mathbf{x}}_{2,t}\}$ and the target source spectra $\{\mathbf{x}_{1,t}, \mathbf{x}_{2,t}\}$ as given in Eq. (7.51). Soft mask functions $\{\widehat{\mathbf{y}}_{1,t}, \widehat{\mathbf{y}}_{2,t}\}$ have been internally embedded in the estimated source signals $\{\widetilde{\mathbf{x}}_{1,t}, \widetilde{\mathbf{x}}_{2,t}\}$. A joint optimization of mask functions and DRNNs is performed.

Basically, the regression error in Eq (7.51) is composed of the terms from two individual sources. When two targets have similar spectra, e.g., spectra of speakers from the same gender, minimizing Eq (7.51) could not obtain salient and separate features for individual speakers or sources $i = 1, 2$. This leads to the reduced signal-to-interference ratio (SIR). To improve system performance for such a class-based monaural source separation, it is important to reduce the interference between two sources or two classes by incorporating the discriminative capability in training objective. A discriminative training criterion is constructed with two terms. One is for generation of individual sources while the other is for the reduction of mutual interference. In data generation, we assume that the residual noise $\mathbf{x}_{i,t} - \widetilde{\mathbf{x}}_{i,t}$ in each source i is represented by a zero-mean Gaussian with unit variance. The log-likelihood of generating two source signals $\{\mathbf{x}_{1,t}, \mathbf{x}_{2,t}\}$ given by mixed signal $\mathbf{x}_t$ or $\mathbf{x}_t^{\text{mix}}$, or representing the training data

$$\mathbf{X} = \{\mathbf{x}_t, \mathbf{x}_{1,t}, \mathbf{x}_{2,t}\} \tag{7.62}$$

is measured by

$$\log p_w(\mathbf{X})$$

$$= -\frac{1}{2} \sum_{t=1}^{T} \left(\|\mathbf{x}_{1,t} - \tilde{\mathbf{x}}_{1,t}(\mathbf{x}_t, \mathbf{w})\|^2 + \|\mathbf{x}_{2,t} - \tilde{\mathbf{x}}_{2,t}(\mathbf{x}_t, \mathbf{w})\|^2 \right). \tag{7.63}$$

This measure is seen as a within-source likelihood which is calculated within individual sources. On the other hand, the discriminative term is defined as Kullback–Leibler (KL) divergence between within-source likelihood $p_w(\mathbf{X})$ and between-source likelihood $p_b(\mathbf{X})$

$$\mathcal{D}_{\text{KL}}(p_w \| p_b) = \frac{1}{2} \sum_{t=1}^{T} \left(\|\mathbf{x}_{1,t} - \tilde{\mathbf{x}}_{2,t}\|^2 + \|\mathbf{x}_{2,t} - \tilde{\mathbf{x}}_{1,t}\|^2 \right.$$

$$\left. - \|\mathbf{x}_{1,t} - \tilde{\mathbf{x}}_{1,t}\|^2 - \|\mathbf{x}_{2,t} - \tilde{\mathbf{x}}_{2,t}\|^2 \right). \tag{7.64}$$

The between-source likelihood is calculated by swapping between two sources in generative likelihood of Eq. (7.63), i.e., the swapped terms

$$\mathbf{x}_{1,t} - \tilde{\mathbf{x}}_{2,t}(\mathbf{x}_t, \mathbf{w})$$

and

$$\mathbf{x}_{2,t} - \tilde{\mathbf{x}}_{1,t}(\mathbf{x}_t, \mathbf{w})$$

are used. As a result, a discriminative training criterion is established for minimization based on (Huang et al., 2015)

$$E(\mathbf{w}) = -(1 - \lambda) \log p_w(\mathbf{X}) - \lambda \mathcal{D}_{\text{KL}}(p_w \| p_b)$$

$$= \frac{1}{2} \sum_{t=1}^{T} \left(\|\mathbf{x}_{1,t} - \tilde{\mathbf{x}}_{1,t}\|^2 + \|\mathbf{x}_{2,t} - \tilde{\mathbf{x}}_{2,t}\|^2 \right.$$

$$\left. - \lambda \|\mathbf{x}_{1,t} - \tilde{\mathbf{x}}_{2,t}\|^2 - \lambda \|\mathbf{x}_{2,t} - \tilde{\mathbf{x}}_{1,t}\|^2 \right) \tag{7.65}$$

where λ is a tradeoff parameter between generative and discriminative terms.

Physical Interpretation

It is crucial to highlight the physical meaning of discriminative training criterion in Eq. (7.65). Basically, this learning objective is optimized to achieve two goals. One is to minimize the within-source regression error while the other is to maximize the between-source regression error. These two criteria are jointly optimized to fulfill the optimal discrimination for monaural source separation. Interestingly, this discriminative training in source separation is coincident with the *linear discriminant analysis* (Fukunaga, 1990, Chien and Wu, 2002) in general pattern classification where a discriminant transformation function is estimated by maximizing a Fisher criterion which promotes maximum between-class separation as well as minimum within-class separation for the transformed data. The regression problem in monaural source separation with two sources in mixed signals is comparable with the classification problem in object recognition with two classes in observed data. In Huang et al. (2014a, 2014b, 2015), joint estimation of mask functions and DRNN parameters was evaluated for monaural source separation and applied for tasks including speech separation, singing voice separation and speech denoising.

7.3.2 DISCRIMINATIVE SOURCE SEPARATION

Discriminative learning and deep learning provides powerful machine learning approaches to monaural source separation. In addition to the discriminative source separation using learning objective of Eq. (7.65) as mentioned in Section 7.3.1, this section presents an extended study on an advanced optimization procedure based on a new discriminative training criterion (Section 7.3.2) (Wang et al., 2016).

New Training Criterion

Similar to Section 7.3.1, a joint optimization of mask functions

$$\{\widehat{\mathbf{y}}_{i,t}(\mathbf{x}_t, \mathbf{w})\}$$

and DRNN parameters

$$\mathbf{w} = \{\mathbf{w}^{(l)}, \mathbf{w}^{(m)}, \mathbf{w}^{(mm)}\}$$

is performed here by following the mask functions as provided in Eq. (7.61) and the DRNN parameters $\mathbf{w}^{(l)}$ in forward layer l and $\{\mathbf{w}^{(m)}, \mathbf{w}^{(mm)}\}$ in recurrent layer as given in Eqs. (7.48) and (7.49), respectively. This discriminative source separation focuses on a new learning objective for monaural source separation, which preserves mutual difference between two source spectra during the separation procedure. The training criterion is constructed by

$$
\begin{aligned}
E(\mathbf{w}) &= \sum_n E_n(\mathbf{w}) \\
&= \frac{1}{2} \sum_{t=1}^{T} \bigg(\|\mathbf{x}_{1,t} - \widetilde{\mathbf{x}}_{1,t}(\mathbf{x}_t, \mathbf{w})\|^2 + \|\mathbf{x}_{2,t} - \widetilde{\mathbf{x}}_{2,t}(\mathbf{x}_t, \mathbf{w})\|^2 \\
&\quad + \lambda \|\mathbf{d}_t - \widetilde{\mathbf{d}}_t(\mathbf{x}_t, \mathbf{w})\|^2 \bigg).
\end{aligned}
\tag{7.66}
$$

Instead of a discriminative measure based on between-source regression error in Eq. (7.65), we present a new discriminative measure using the *difference vectors* based on the reconstructed spectra $\widetilde{\mathbf{d}}_t(\mathbf{x}_t, \mathbf{w})$ and the true spectra $\mathbf{d}_t$, which are expressed by

$$\mathbf{d}_t = \mathbf{x}_{1,t} - \mathbf{x}_{2,t}, \tag{7.67}$$

$$\widetilde{\mathbf{d}}_t(\mathbf{x}_t, \mathbf{w}) = \widetilde{\mathbf{x}}_{1,t}(\mathbf{x}_t, \mathbf{w}) - \widetilde{\mathbf{x}}_{2,t}(\mathbf{x}_t, \mathbf{w}). \tag{7.68}$$

Basically, the first two terms represent the within-source reconstruction errors due to individual source spectra $\{\mathbf{x}_{1,t}, \mathbf{x}_{2,t}\}$. The third term conveys the discriminative information, which measures mutual difference between two source spectra. The discriminative term acts as regularization information to impose between-source separation for monaural source separation.

Optimization Procedure

This section addresses an error backpropagation algorithm to implement a discriminative source separation procedure based on DRNN in accordance with the training criterion in Eq. (7.66). The learning algorithm using stochastic gradient descent (SGD) is applied. To do so, we first calculate the learning objective using a minibatch $t \in \{\mathbf{X}_n\}$ from the training set

$$\mathbf{X} = \{\mathbf{X}_n\} = \{\mathbf{x}_t, \mathbf{x}_{1,t}, \mathbf{x}_{2,t}\}$$

in a form of

$$E_n(\mathbf{w}) = \frac{1}{2} \sum_{t \in \mathbf{X}_n} \sum_{k=1}^{K} \Big((x_{1,tk} - \tilde{x}_{1,tk}(\mathbf{x}_t, \mathbf{w}))^2 + (x_{2,tk} - \tilde{x}_{2,tk}(\mathbf{x}_t, \mathbf{w}))^2$$
$$+ \lambda(d_{tk} - \tilde{d}_{tk}(\mathbf{x}_t, \mathbf{w}))^2 \Big) \tag{7.69}$$

where K denotes the number of output units for each source spectra. In the forward pass, each minibatch is used to calculate the activations and outputs $\{a_{tj}^{(l)}, z_{tj}^{(l)}\}$ in different layer l where the initialization $\mathbf{z}_t^{(0)} = \mathbf{x}_t$ is specified. This calculation is run towards output layer L to find activations for two sources $\{\mathbf{a}_{1,t}^{(L)}, \mathbf{a}_{2,t}^{(L)}\}$ where

$$\mathbf{a}_{i,t}^{(L)} = \mathbf{w}_i^{(L)} \mathbf{z}_t^{(L-1)}, \quad \text{for } i = 1, 2. \tag{7.70}$$

After the feedforward computation, the error backpropagation algorithm with stochastic gradient learning is developed so as to estimate the DRNN parameters $\mathbf{w} = \{\mathbf{w}^{(l)}, \mathbf{w}^{(m)}, \mathbf{w}^{(mm)}\}$.

To estimate the parameters corresponding to the *first* source signal, in backward computation, we calculate the derivative of $E_n(\mathbf{w})$ with respect to the weights $w_{1,kj}^{(L)}$ for connecting neuron j in hidden layer $L - 1$ to neuron k in output layer L as

$$\frac{\partial E_n(\mathbf{w})}{\partial w_{1,kj}^{(L)}} = \sum_{t \in \mathbf{X}_n} \sum_{i=1}^{2} \left(\frac{\partial E_n(\mathbf{w})}{\partial \tilde{x}_{i,tk}(\mathbf{w})} \frac{\partial \tilde{x}_{i,tk}(\mathbf{w})}{\partial \hat{y}_{i,tk}(\mathbf{w})} \frac{\partial \hat{y}_{i,tk}(\mathbf{w})}{\partial a_{1,tk}^{(L)}(\mathbf{w})} \right) \frac{\partial a_{1,tk}^{(L)}(\mathbf{w})}{\partial w_{1,kj}^{(L)}}$$
$$= \sum_{t \in \mathbf{X}_n} \delta_{1,tk}^{(L)} z_{tj}^{(L-1)}. \tag{7.71}$$

The local gradient $\delta_{1,tk}^{(L)}$ and the output of neuron at layer $L - 1$, $z_{tj}^{(L-1)}$, are defined for this derivative

$$\delta_{1,tk}^{(L)} \triangleq \frac{\partial E_n(\mathbf{w})}{\partial a_{1,tk}^{(L)}(\mathbf{w})} = \sum_{i=1}^{2} \frac{\partial E_n(\mathbf{w})}{\partial \tilde{x}_{i,tk}(\mathbf{w})} \frac{\partial \tilde{x}_{i,tk}(\mathbf{w})}{\partial \hat{y}_{i,tk}(\mathbf{w})} \frac{\partial \hat{y}_{i,tk}(\mathbf{w})}{\partial a_{1,tk}^{(L)}(\mathbf{w})}, \tag{7.72}$$

$$z_{tj}^{(L-1)} = \frac{\partial a_{1,tk}^{(L)}(\mathbf{w})}{\partial w_{1,kj}^{(L)}} \tag{7.73}$$

where the terms in Eq. (7.72) are yielded as

$$\frac{\partial E_n(\mathbf{w})}{\partial \tilde{x}_{1,tk}(\mathbf{w})} = \sum_{t \in \mathbf{X}_n} (1+\lambda)(x_{1,tk} - \tilde{x}_{1,tk}(\mathbf{w}))$$
$$- \lambda(x_{2,tk} - \tilde{x}_{2,tk}(\mathbf{w})), \tag{7.74}$$

$$\frac{\partial E_n(\mathbf{w})}{\partial \tilde{x}_{2,tk}(\mathbf{w})} = \sum_{t \in \mathbf{X}_n} (1+\lambda)(x_{2,tk} - \tilde{x}_{2,tk}(\mathbf{w}))$$
$$- \lambda(x_{1,tk} - \tilde{x}_{1,tk}(\mathbf{w})), \tag{7.75}$$

$$\frac{\partial \tilde{x}_{1,tk}(\mathbf{w})}{\partial \hat{y}_{1,tk}(\mathbf{w})} = x_{tk}, \tag{7.76}$$

$$\frac{\partial \tilde{x}_{2,tk}(\mathbf{w})}{\partial \hat{y}_{2,tk}(\mathbf{w})} = x_{tk}, \tag{7.77}$$

$$\frac{\partial \hat{y}_{1,tk}(\mathbf{w})}{\partial a_{1,tk}^{(L)}(\mathbf{w})} = \mathrm{sgn}\left(a_{1,tk}^{(L)}(\mathbf{w})\right) \frac{\hat{y}_{2,tk}(\mathbf{w})}{|a_{1,tk}^{(L)}(\mathbf{w})| + |a_{2,tk}^{(L)}(\mathbf{w})|}, \tag{7.78}$$

$$\frac{\partial \hat{y}_{2,tk}(\mathbf{w})}{\partial a_{1,tk}^{(L)}(\mathbf{w})} = -\mathrm{sgn}\left(a_{1,tk}^{(L)}(\mathbf{w})\right) \frac{\hat{y}_{2,tk}(\mathbf{w})}{|a_{1,tk}^{(L)}(\mathbf{w})| + |a_{2,tk}^{(L)}(\mathbf{w})|} \tag{7.79}$$

where the function $\mathrm{sgn}(\cdot)$ extracts the sign of a real number. In Eqs. (7.78) and (7.79), the derivatives

$$\left\{ \frac{\partial \hat{y}_{1,tk}(\mathbf{w})}{\partial a_{1,tk}^{(L)}(\mathbf{w})}, \frac{\partial \hat{y}_{2,tk}(\mathbf{w})}{\partial a_{1,tk}^{(L)}(\mathbf{w})} \right\}$$

have the same value but different sign. By combining Eqs. (7.74)–(7.79), we can derive the local gradient for the output-layer weight as

$$\delta_{1,tk}^{(L)} = \left\{ (1+2\lambda)\left[(x_{1,tk} - \tilde{x}_{1,tk}(\mathbf{w})) \right. \right.$$
$$\left. - (x_{2,tk} - \tilde{x}_{2,tk}(\mathbf{w})) \right] \right\}$$
$$\times x_{tk}\hat{y}_{2,tk}(\mathbf{w}) \frac{\mathrm{sgn}\left(a_{1,tk}^{(L)}(\mathbf{w})\right)}{|a_{1,tk}^{(L)}(\mathbf{w})| + |a_{2,tk}^{(L)}(\mathbf{w})|}. \tag{7.80}$$

Similarly, we can derive the derivative of $E_n(\mathbf{w})$ with respect to the output-layer weight for the *second* source $\partial E_n(\mathbf{w})/\partial w_{2,kj}^{(L)}$.

On the other hand, we need to calculate the local gradient $\delta_{tj}^{(m)}$ for a recurrent layer m for updating weights $w_{kj}^{(m+1)}$ and $w_{js}^{(mm)}$, which is computed along two consecutive time steps t and $t+1$ with two different neighboring layers $m + 1$ and m where the corresponding neurons are indexed by k and s, respectively. The calculation of the local gradient over two time steps is because of the recurrence in DRNN with forward weights $\mathbf{w}^{(m+1)}$ and recurrent weights $\mathbf{w}^{(mm)}$. This local gradient is yielded by

$$
\begin{aligned}
\delta_{tj}^{(m)} &\triangleq \frac{\partial E_n(\mathbf{w})}{\partial a_{tj}^{(m)}} \\
&= \sum_{t \in \mathbf{X}_n} \left(\sum_k \frac{\partial E_n(\mathbf{w})}{\partial a_{tk}^{(m+1)}} \frac{\partial a_{tk}^{(m+1)}}{\partial a_{tk}^{(m)}} + \sum_s \frac{\partial E_n(\mathbf{w})}{\partial a_{t+1,s}^{(m)}} \frac{\partial a_{t+1,s}^{(m)}}{\partial a_{tk}^{(m)}} \right) \\
&= h'(a_{tj}^{(m)}) \sum_{t \in \mathbf{X}_n} \left(\sum_k w_{kj}^{(m+1)} \delta_{tk}^{(m+1)} + \sum_s w_{sj}^{(mm)} \delta_{t+1,s}^{(m)} \right)
\end{aligned}
\tag{7.81}
$$

where $h(\cdot)$ is the activation function of a neuron in the recurrent layer m as used in Eq. (7.49). Such gradients are then propagated backwards through all layers to adjust the weights $\mathbf{w} = \{\mathbf{w}^{(l)}, \mathbf{w}^{(m)}, \mathbf{w}^{(mm)}\}$ in the forward layer l as well as in recurrent layer m.

7.3.3 SYSTEM EVALUATION

Deep recurrent neural networks (DRNN) based on different discriminative training criteria were evaluated for single-channel speech separation under the same system configuration. Discriminative DRNN (DDRNN) was implemented in two variants. The DDRNN based on a between-source vector (denoted by DDRNN-bw), as mentioned in Eq. (7.65) and Section 7.3.1, was compared with DDRNN based on a difference vector (denoted by DRNN-diff), as mentioned in Eq. (7.66) and Section 7.3.2. The regularization parameter λ in DDRNN-bw and DDRNN-diff were determined from validation data. NMF was carried out for comparison. The number of bases was determined by using validation data. DRNN based on the training criterion in Eq. (7.63) was also implemented in the experiments. Relative to DDRNN-bw and DDRNN-diff, without loss of generality, baseline DRNN is seen as a generative DRNN.

In system evaluation, the single-channel mixed speech signals from two TIMIT speakers were collected to evaluate the effectiveness of using NMF, DRNN, DDRNN-bw and DDRNN-diff (Wang et al., 2016). Each TIMIT speaker provided ten sentences. In the training phase, eight sentences were randomly chosen from one male and one female speaker for signal mixing. Another sentence was used for cross validation, and the remaining sentence was used for testing. All sentences were normalized to be of equal power. In order to enrich the variety of the training samples, the sentences from one speaker were circularly shifted and added to the sentences from the other speaker as the training data. The 1024-point STFT was calculated to extract the spectral features with frame length of 64 ms and frame shift of 32 ms. The features were used as inputs to different models. The DRNN architecture used in the experiments was fixed as 513–150–150–1026 where two hidden layers with 150 units in each layer were specified and two sources with 1026 (513 $\star$ 2) units were arranged. The contextual features, as addressed in Eq. (7.1) were ignored. The activation function using ReLU was adopted. The limited-memory Broyden–Fletcher–Goldfarb–Shanno (L-BFGS) method (Ngiam et al., 2011) was employed in the optimization procedure for DRNN, DDRNN-bw and DDRNN-diff. The separation performance was examined by using the signal-to-distortion ratio (SDR), the signal-to-interferences

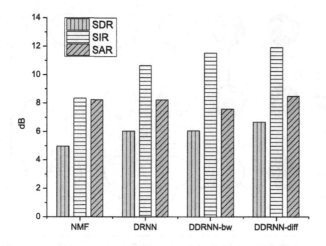

FIGURE 7.7

Comparison of SDR, SIR and SAR of the separated signals by using NMF, DRNN, DDRNN-bw and DDRNN-diff.

ratio (SIR), and the source-to-artifacts ratio (SAR) (Vincent et al., 2006). Fig. 7.7 compares SDR, SIR and SAR by using NMF and three DRNNs based on different training criteria Eqs. (7.63), (7.65) and (7.66). We can see that discriminative DRNNs (DDRNN-bw and DDRNN-diff) perform better than generative DRNN in terms of SIR mainly because of discriminative terms from between-source and difference vectors, which can reduce the interference between two source signals. However, the artifacts are *not* well treated by using DDRNN-bw, yet could be properly tackled by using DRNN-diff when compared with DRNN.

7.4 LONG SHORT-TERM MEMORY

This section focuses on the extension of monaural source separation from standard recurrent neural network (RNN) and deep RNN to long short-term memory (LSTM) (Hochreiter and Schmidhuber, 1997) and bidirectional LSTM. We first address the issues of RNN and the motivation of LSTM in Section 7.4.1. Section 7.4.2 mentions how LSTM is superior to a feedforward neural network to tackle the issue of generalization to unseen speakers for speech separation. Section 7.4.3 introduces the extension to bidirectional LSTM. A couple of studies on comparison of LSTM and bidirectional LSTM are also surveyed.

7.4.1 PHYSICAL INTERPRETATION

As mentioned in Section 2.5.3, the standard recurrent neural network (RNN) suffers from the problem of gradient vanishing or gradient exploding. This is because of the repeated multiplication of the same weights along a number of time steps in a long unfolded network as shown in Eq. (2.72). Fig. 7.8 illustrates the phenomenon of vanishing gradients in RNN where the gradient at time step 1 is fully

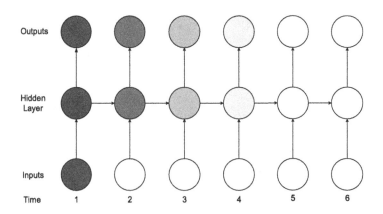

FIGURE 7.8

Vanishing gradients in a recurrent neural network. Degree of lightness means the level of vanishing in the gradient. This figure is a counterpart to Fig. 2.21 where vanishing gradients are mitigated by a gating mechanism.

propagated to the output layer through hidden state z_t but is gradually vanishing at time steps 2, 3 and 4. Here, the magnitude of the gradient is revealed by darkness (the largest) and lightness (the smallest) of blue in different levels (gray in print version). Such a gradient is totally vanished at time step 5. The information extracted at time step 1 is decaying rapidly in time and only holds for the first 4 time steps.

LSTM is accordingly introduced to handle the difficulty of gradient propagation along multiple time steps. The goal of LSTM is to preserve the activations of hidden nodes at an earlier time for prediction at the current time t and extract the short-term features existing in long history for application in monaural source separation. As illustrated in Fig. 2.22 and Eqs. (2.75)–(2.81), an LSTM block, consisting of a memory cell c_t and three sigmoid gates, including *input* gate i_t, *output* gate o_t and *forget* gate f_t, is implemented to preserve the gradient propagation at different time steps t using x_t. A simplified view is also provided in Fig. 7.9. Then z_t is calculated as a hidden state, which is used to run for recurrent updating of three gates $\{i_t, o_t, f_t\}$ at the next time step $t+1$. LSTM is equivalently implemented as a *composite* function for estimating mask functions $\{\widehat{y}_{1,t}, \widehat{y}_{2,t}\}$ for two sources by using the hidden state z_t and the corresponding connection weights. Definitely, LSTM can be incorporated into a deep model to improve gradient descent learning for deep recurrent neural network (DRNN). In Boulanger-Lewandowski et al. (2014), a recurrent neural network (RNN) was used to extract long-term temporal dependencies in audio signals for temporally constrained nonnegative matrix factorization (NMF) which was viewed as a hybrid NMF and RNN model.

In Fig. 7.10, the *input, output* and *forget* gates of an LSTM memory block are demonstrated as the switches for *receiving, producing* and *propagating* gradients at each time t, respectively. Here, o and − represent the gate opening and closing, respectively. The gating mechanism automatically drives the propagation of gradients and controls the receiving from inputs and producing for outputs. In this example, only time step 1 receives gradient from input x_t. The gradient in the hidden state z_t is propagated and used to produce outputs $\{\widehat{y}_{1,t}, \widehat{y}_{2,t}\}$ at time steps 3 and 5. This gradient is terminated at time step 6. Before the termination, the amount of gradient is well preserved in the hidden states for affecting and producing the outputs at time steps 3 and 5 where the output gate is opened.

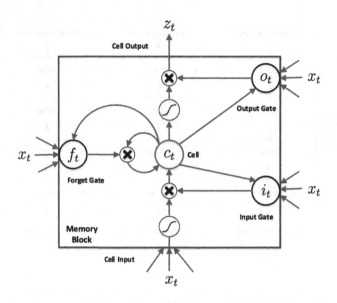

FIGURE 7.9

A simplified view of long short-term memory. The red (light gray in print version) line shows recurrent state z_t. The detailed view was provided in Fig. 2.22.

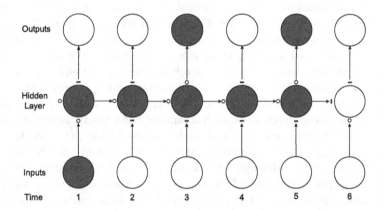

FIGURE 7.10

Illustration of preserving gradients in long short-term memory. There are three gates in a memory block. ○ means gate opening while − denotes gate closing.

7.4.2 SPEAKER GENERALIZATION

Speech separation is seen as a learning task where the time–frequency mask functions of different sources are estimated according to a predefined model, e.g., NMF, DNN, RNN or DRNN. Assuming

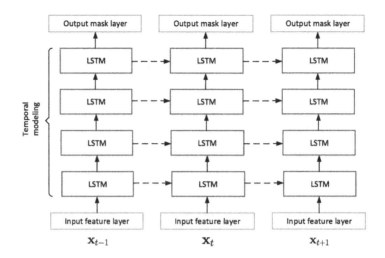

FIGURE 7.11

A configuration of four stacked long short-term memory layers along three time steps for monaural speech separation. Dashed arrows indicate the modeling of the same LSTM across time steps while solid arrows mean the modeling of different LSTMs in deep structure.

test conditions are known in advance is basically unrealistic for source separation. In Sections 7.2.1 and 7.2.3, deep clustering (Hershey et al., 2016) and permutation-invariant training (Kolbæk et al., 2017) have been presented to implement a speaker-independent speech separation system where the number and identity of test speakers are unknown in the test session. This section introduces an alternative solution to speaker generalization for monaural speech separation based on LSTM (Chen and Wang, 2017) where the noise condition and test speaker are unseen in training environments.

Traditionally, the deep neural network (DNN) (Xu et al., 2014, Du et al., 2016, Grais et al., 2014) performed very well in speaker-dependent speech separation where test speakers were seen in training data. However, the separation performance was degraded rapidly for unseen test speakers. Such performance was even worse than that of unprocessed system in terms of short-time objective intelligibility (STOI) (Chen and Wang, 2017). Although DNN was implemented by considering the input features within a context window, those temporal information outside window was ignored. The generalization using DNN was accordingly constrained. Recurrent neural network is feasible to capture temporal dynamics in audio signals. As a result, RNN based on long short-term memory is strongly recommended to overcome the weakness of DNN.

Fig. 7.11 depicts a configuration of a deep recurrent neural network (DRNN) consisting of one input feature layer $\mathbf{x}_t$, four stacked LSTM layers $\{\mathbf{z}_t^{(l)}\}_{l=1}^{4}$, and one output mask layer $\{\widehat{\mathbf{y}}_{1,t}, \widehat{\mathbf{y}}_{2,t}\}$. The stacking of LSTMs is designed to capture deep temporal dependencies in audio signals. The top LSTM layer $\mathbf{z}_t^{(L)}$ is used to estimate soft mask functions $\{\widehat{\mathbf{y}}_{1,t}, \widehat{\mathbf{y}}_{2,t}\}$ similar to the mask functions in Eq. (7.2). Backpropagation through time (BPTT) with three time steps is arranged. In system evaluation, this method was evaluated by using WSJ utterances from 83 speakers. Utterances from 77 speakers were used for training while those from the remaining 6 speakers were used for test. The mixed signals were seen as the noisy speech signals which were created by mixing WSJ utterances with a large set of training noises at a signal-to-noise ratio (SNR) between −5 and 0 dB. A large training set of around 5000 hours was used for training the speaker-independent model.

In the implementation, a context window of 23 frames ($\tau = 11$) was utilized to build an input vector of a mixed signal frame $\mathbf{x}_t$, which was used to estimate soft mask functions $\{\widehat{\mathbf{y}}_{1,t}, \widehat{\mathbf{y}}_{2,t}\}$. Each frame had a length of 20 ms with 10 ms of frame shift. The 64-dimensional gammatone filterbank energies were used as the input features, which met the human auditory system. The DRNN topology with four 1024-unit LSTM layers, denoted by 23⋆64–1024–1024–1024–1024–64, was implemented. This LSTM system was compared with a DNN system consisting of five 2048-unit fully-collected (FC) layers. The topology 23⋆64–2048–2048–2048–2048–2048–64 was implemented. Activations in FC-layers were run using ReLUs. In the optimization procedure, Adam optimization (Kingma and Ba, 2014) was run with a minibatch size of 256 and truncated BPTT of 250 time steps. The experimental results showed that training a DNN with many speakers did not perform well on both seen and unseen speakers. A separation model based on LSTM did improve speaker generalization because of the capability of modeling temporal dynamics and memorizing a target speaker. The visualization of LSTM cell reflected the property of speech patterns (Chen and Wang, 2017), which captured different contexts to improve mask estimation.

In addition, deep recurrent neural networks (DRNNs) using LSTM were successfully developed for single-channel speech separation under the task of CHiME Speech Separation and Recognition Challenge (Weninger et al., 2014a). There are two findings, which are consistent with what we have addressed in this section and Section 7.3. First, similar to the idea in Section 7.3, the incorporation of discriminative training criterion into recurrent neural networks is beneficial to single-channel speech enhancement which restores clean speech from noise interference. Second, DRNN using two-layer LSTM network with mask approximation and signal approximation significantly outperformed the NMF with 100 bases and the DNN with three 1024-unit FC layers. The highest signal-to-distortion ratio (SDR) was achieved by using a DRNN with two 256-unit LSTM layers. More attractively, DRNN with LSTM layers had a smaller number of parameters than DNN with FC layers. The significant improvement using DRNN was assured when using LSTM layers (Weninger et al., 2014a) rather than conventional recurrent layers (Huang et al., 2014a). The gradient vanishing problem was substantially tackled by using LSTM layers.

7.4.3 BIDIRECTIONAL LONG SHORT-TERM MEMORY

Recurrent neural networks and long short-term memories, as addressed in Section 2.5, and different variants of deep recurrent neural networks, as surveyed in Sections 7.3 and 7.4, only make use of *previous* context of the mixed signal to predict future events in source separation. The modeling of sequential data is unidirectional. However, in real world, the sequential samples in audio signals are correlated in a bidirectional fashion. Traditional RNNs make prediction by using the state information $\mathbf{z}_t$ from past input data $\mathbf{x}_{<t}$. System performance is restricted because future input information $\mathbf{x}_{>t}$ cannot be reached from the current state $\mathbf{z}_t$. In fact, future mixed signals $\mathbf{x}_{>t}$ are available for estimating the masks or demixed signals $\{\widehat{\mathbf{y}}_{1,t}, \widehat{\mathbf{y}}_{2,t}\}$ at current frame t for single-channel source separation in the presence of two sources. Future input information is reachable from the current state $\mathbf{z}_t$.

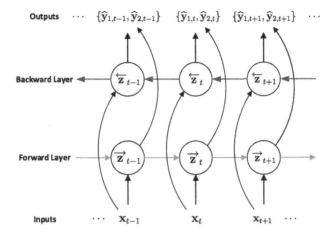

Outputs $\cdots$ $\{\widehat{y}_{1,t-1}, \widehat{y}_{2,t-1}\}$ $\{\widehat{y}_{1,t}, \widehat{y}_{2,t}\}$ $\{\widehat{y}_{1,t+1}, \widehat{y}_{2,t+1}\}$ $\cdots$

Backward Layer

Forward Layer

Inputs $\cdots$ $\mathbf{x}_{t-1}$ $\mathbf{x}_t$ $\mathbf{x}_{t+1}$ $\cdots$

FIGURE 7.12

Illustration of bidirectional recurrent neural network for monaural source separation. Two hidden layers are configured to learn bidirectional features for finding soft mask functions for two sources. Forward and backward directions are shown in different colors.

Model Construction

To deal with the restriction in RNN, the bidirectional recurrent neural network (BRNN), proposed in Schuster and Paliwal (1997), not only uses contextual information from *past* events but also from *future* events. BRNNs are flexible and do not require their input data to be fixed. The amount of training data can be increased to improve model estimation. In the implementation, two separate hidden layers are constructed in BRNN to merge the hidden information from two opposite directions. This information is fed forward towards the same output layer. As illustrated in Fig. 7.12, the forward and backward layers are configured to learn forward features $\overrightarrow{\mathbf{z}}_t$ and backward features $\overleftarrow{\mathbf{z}}_t$, respectively, from input signals $\{\mathbf{x}_t\}_{t=1}^T$. Forward features $\overrightarrow{\mathbf{z}}_t$ are calculated iteratively from $t=1$ towards T while backward features $\overleftarrow{\mathbf{z}}_t$ are extracted recursively from $t=T$ back to 1. These features are then forwarded to calculate the soft mask functions $\{\widehat{y}_{1,t}, \widehat{y}_{2,t}\}$. Such an operation is performed recursively at different time steps $t = 1, \ldots, T$. Different from standard RNN using a single set of forward features $\overrightarrow{\mathbf{z}}_t$, BRNN adopts both sets of forward and backward features

$$\{\overrightarrow{\mathbf{z}}_t, \overleftarrow{\mathbf{z}}_t\}$$

for predicting the masks

$$\{\widehat{y}_{1,t}, \widehat{y}_{2,t}\}$$

at each time step t.

Again, the difficulty of gradient vanishing or exploding does not only happen in RNN but also in BRNN. Consequently, the bidirectional long short-term memory (BLSTM) (Graves et al., 2013) was proposed to mitigate this difficulty and applied for speech recognition. For the task of monaural source separation, bidirectional LSTM was developed to learn relevant past and future events for prediction of mask functions of individual source signals at different time steps (Erdogan et al., 2015, Hershey et al., 2016, Kolbæk et al., 2017). Bidirectional LSTM is implemented by

$$\overrightarrow{\mathbf{z}}_t = \mathcal{Z}\left(W_{x\overrightarrow{z}}\mathbf{x}_t + W_{\overrightarrow{z}\overrightarrow{z}}\overrightarrow{\mathbf{z}}_{t-1} + \mathbf{b}_{\overrightarrow{z}}\right), \tag{7.82}$$

$$\overleftarrow{\mathbf{z}}_t = \mathcal{Z}\left(W_{x\overleftarrow{z}}\mathbf{x}_t + W_{\overleftarrow{z}\overleftarrow{z}}\overleftarrow{\mathbf{z}}_{t+1} + \mathbf{b}_{\overleftarrow{z}}\right), \tag{7.83}$$

$$\widehat{\mathbf{y}}_{1,t} = W_{\overrightarrow{z}y1}\overrightarrow{\mathbf{z}}_t + W_{\overleftarrow{z}y1}\overleftarrow{\mathbf{z}}_t + \mathbf{b}_{y1}, \tag{7.84}$$

$$\widehat{\mathbf{y}}_{2,t} = W_{\overrightarrow{z}y2}\overrightarrow{\mathbf{z}}_t + W_{\overleftarrow{z}y2}\overleftarrow{\mathbf{z}}_t + \mathbf{b}_{y2}. \tag{7.85}$$

In Eqs. (7.82) and (7.83), $\mathcal{Z}$ denotes the *composite* function in a hidden layer as shown in Fig. 7.12, which integrates all calculations in a standard LSTM memory block as given in Eqs. (2.75)–(2.81). The parameters $W_{x\overrightarrow{z}}$ and $W_{x\overleftarrow{z}}$ are used in the connections between input vector $\mathbf{x}_t$ and forward feature vector $\overrightarrow{\mathbf{z}}_t$ and backward feature vector $\overleftarrow{\mathbf{z}}_t$, respectively. The parameters $W_{\overrightarrow{z}\overrightarrow{z}}$ and $W_{\overleftarrow{z}\overleftarrow{z}}$ are arranged to model the connections between hidden features $\overrightarrow{\mathbf{z}}_t$ and $\overleftarrow{\mathbf{z}}_t$ at consecutive frames $t-1$ and t in forward and backward directions, respectively. Parameters $\mathbf{b}_{\overrightarrow{z}}$ and $\mathbf{b}_{\overleftarrow{z}}$ are seen as the bias parameters in forward and backward directions, respectively. Output masks $\widehat{\mathbf{y}}_{i,t}$ are estimated for the two sources $i = 1, 2$ by combining those statistics from two opposite directions by using the corresponding bias parameter $\mathbf{b}_{yi}$ and weight parameters $W_{\overrightarrow{z}yi}$ and $W_{\overleftarrow{z}yi}$, which connect to forward and backward states, respectively.

In addition, deep bidirectional LSTM can be constructed by staking L BLSTM layers and calculating forward and backward features

$$\{\overrightarrow{\mathbf{z}}_t^{(l)}, \overrightarrow{\mathbf{z}}_t^{(l)}\}$$

for each individual BLSTM layer l ($l = 1, \dots, L$) at each time step t ($t = 1, \dots, T$) as (Graves et al., 2013)

$$\overrightarrow{\mathbf{z}}_t^{(l)} = \mathcal{Z}\left(W_{\overrightarrow{z}^{(l-1)}\overrightarrow{z}^{(l)}}\overrightarrow{\mathbf{z}}_t^{(l-1)} + W_{\overrightarrow{z}^{(l)}\overrightarrow{z}^{(l)}}\overrightarrow{\mathbf{z}}_{t-1}^{(l)} + \overrightarrow{\mathbf{b}}_z^{(l)}\right), \tag{7.86}$$

$$\overleftarrow{\mathbf{z}}_t^{(l)} = \mathcal{Z}\left(W_{\overleftarrow{z}^{(l-1)}\overleftarrow{z}^{(l)}}\overleftarrow{\mathbf{z}}_t^{(l-1)} + W_{\overleftarrow{z}^{(l)}\overleftarrow{z}^{(l)}}\overleftarrow{\mathbf{z}}_{t-1}^{(l)} + \overleftarrow{\mathbf{b}}_z^{(l)}\right). \tag{7.87}$$

Here, $\mathcal{Z}$ denotes the composite function, $\mathbf{b}_z^{(l)}$ denotes the bias parameters and

$$\{W_{z^{(l-1)}z^{(l)}}, W_{z^{(l)}z^{(l)}}\}$$

denote the weight parameters for connecting to previous layer $l-1$ and previous time step $t-1$, respectively. These notations are used for both forward and backward layers. The stacked BLSTMs are similar to the stacked LSTMs, which have been displayed in Fig. 7.11. Finally, the output masks are calculated by using the information collected in BLSTM in layer L as

$$\widehat{\mathbf{y}}_{1,t} = W_{\overrightarrow{z}^{(L)}y1}\overrightarrow{\mathbf{z}}_t^{(L)} + W_{\overleftarrow{z}^{(L)}y1}\overleftarrow{\mathbf{z}}_t^{(L)} + \mathbf{b}_{y1}, \tag{7.88}$$

$$\widehat{\mathbf{y}}_{2,t} = W_{\overrightarrow{z}^{(L)}y2}\overrightarrow{\mathbf{z}}_t^{(L)} + W_{\overleftarrow{z}^{(L)}y2}\overleftarrow{\mathbf{z}}_t^{(L)} + \mathbf{b}_{y2}. \tag{7.89}$$

The formulas for extending from bidirectional LSTM to deep bidirectional LSTM are straightforward.

Source Separation Application

Bidirectional long short-term memory (BLSTM) has been successfully explored and applied for monaural speech separation (Hershey et al., 2016, Kolbæk et al., 2017, Erdogan et al., 2015). In Erdogan et al. (2015), tightly-integrated speech separation and recognition was developed through deep recurrent networks. The idea of joint speech separation and recognition could be also extended for joint music separation and transcription. Basically, the hybrid system was constructed by adopting the alignment information from speech recognition and then utilizing this information to conduct speech separation where different segments and phones were treated differently. Speech recognition and speech separation were jointly performed. Attractively, the EM algorithm, as addressed in Section 3.7.1, was introduced here to iteratively and alternatively estimate two sets of parameters for speech recognition and separation. Accordingly, the hybrid model in separation and recognition could be consistently specified, mutually combined, and jointly optimized. Such a speech separation and recognition system was similarly created in Chien and Chen (2006), as addressed in Section 4.2.4, where speech separation and recognition were separately performed according to the same independent component analysis (ICA) algorithm. In Erdogan et al. (2015), BLSTM was used to estimate the phase-sensitive soft mask function for speech separation, which has been addressed in Section 7.2.2. Two hidden layers of BLSTM were implemented for comparison in the experiments. In the comparison with the standard BLSM for estimating different mask functions, BLSTM consistently achieved the best speech separation performance in this hybrid speech separation and recognition system in terms of signal-to-distortion ratio (SDR). The alignment information from speech recognition did help the estimation of mask functions for speech separation. In what follows, we will be further introducing a variety of advanced deep machine learning algorithms for monaural speech separation.

Modern Sequential Learning

In addition to RNN, LSTM, BLSTM and their deep variants including DRNN, deep LSTM and deep BLSTM, there are a number of advanced approaches to constructing other variants of deep recurrent neural networks, which are feasible to deal with modern sequential learning problems. The advances in recurrent neural networks based on variational learning, external memorization, addressing mechanism, sequence-to-sequence learning are described. Starting from the next section, we will address a number of modern deep sequential learning theories and algorithms, which were recently proposed for speech separation. First, in Section 7.5, we present a new variational recurrent neural network for speech separation where variational inference is introduced to carry out stochastic learning for a latent variable model. The randomness of latent features in deep model is characterized. In Section 7.6, an external memory is incorporated into construction of a recurrent neural network (RNN) for monaural speech separation. This memory is extended to compensate the limitation in RNN or LSTM, which only holds the internal memory and disregards the vanishing problem of memory information in signal prediction. Furthermore, in Section 7.7, a recall neural network is proposed to carry out a sequence-to-sequence learning, which mimics the human listening mechanism and performs memory recall for speech listening and then separation.

7.5 **VARIATIONAL RECURRENT NEURAL NETWORK**

This section presents a new stochastic learning machine for speech separation based on the variational recurrent neural network (VRNN) (Chung et al., 2015, Chien and Kuo, 2017). This VRNN is constructed from the perspectives of *generative stochastic network* and *variational autoencoder*. The basic idea is to faithfully characterize the randomness of the hidden state of a recurrent neural network through variational learning. The neural parameters under this latent variable model are estimated by maximizing the variational lower bound of the log marginal likelihood. A recurrent latent variable model driven by the variational distribution is trained from a set of mixed signals and the associated source targets. This section will first survey the variational autoencoder in Section 7.5.1 and then address how to build a supervised variational recurrent neural network and make inference for monaural source separation in Section 7.5.2.

7.5.1 **VARIATIONAL AUTOENCODER**

Variational autoencoder (VAE) (Kingma and Welling, 2013) was proposed to estimate the distribution of hidden variables $\mathbf{z}$ and use this information to reconstruct the original signal $\mathbf{x}$ where the temporal information in different $\mathbf{x}$ is disregarded. This distribution characterizes the randomness of hidden units, which provides a means to reconstruct different realizations of output signals rather than a point estimate of outputs in a traditional autoencoder. Accordingly, it makes possible synthesizing the generative samples and analyzing the statistics of hidden information of a neural network. Fig. 7.13(A) shows how the output $\hat{\mathbf{x}}$ is reconstructed from the original input $\mathbf{x}$. The graphical model of VAE is depicted in Fig. 7.13(B) which consists of an encoder and a decoder. The encoder is seen as a recognition model, which identifies the stochastic latent variables $\mathbf{z}$ using a variational posterior $q_{\boldsymbol{\phi}}(\mathbf{z}|\mathbf{x})$ with parameters $\boldsymbol{\phi}$. Latent variables $\mathbf{z}$ are sampled by using a variational posterior. These samples $\mathbf{z}$ are then used to generate or reconstruct the original signal $\hat{\mathbf{x}}$ based on the decoder or generative model using the likelihood function $p_{\boldsymbol{\theta}}(\mathbf{x}|\mathbf{z})$ with parameters $\boldsymbol{\theta}$. The whole model is formulated by using the variational Bayesian expectation maximization algorithm. Variational parameters $\boldsymbol{\phi}$ and model parameters $\boldsymbol{\theta}$ are estimated by maximizing the *variational lower bound* of log-likelihood $\log p(\mathbf{x}_{\leq T})$ from a collection of samples

$$\mathbf{x}_{\leq T} = \{\mathbf{x}_t\}_{t=1}^{T}.$$

Stochastic error backpropagation is implemented for variational learning. This VAE was extended to other unsupervised learning tasks (Miao et al., 2016) for finding the synthesized images. This study develops a variational recurrent neural network (VRNN) to tackle the supervised regression learning for speech separation.

7.5.2 **MODEL CONSTRUCTION AND INFERENCE**

This work is motivated by introducing VAE into construction of RNN to implement a stochastic realization of RNN. Figs. 7.14(A) and 7.14(B) compare graphical representations of RNN and VRNN (Chung et al., 2015), respectively. In VRNN, we estimate a set of T time-dependent hidden units $\mathbf{h}_{\leq T}$ corresponding to the observed mixture signals $\mathbf{x}_{\leq T}$, which are used to produce the RNN outputs $\mathbf{y}_{\leq T}$ as the demixed signals for regression task. The hidden units $\mathbf{h}_{\leq T}$ are characterized and generated by

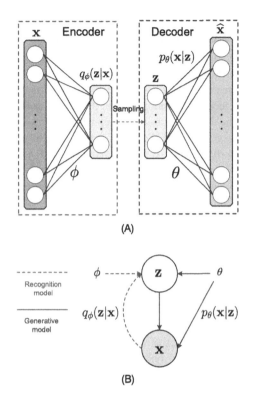

(A)

(B)

FIGURE 7.13

(A) Encoder and decoder in a variational autoencoder. (B) Graphical representation for a variational autoencoder.

hidden variables $\mathbf{z}_{\leq T}$. Similar to VAE, VRNN is equipped with an encoder and a decoder. This VRNN aims to capture the temporal and stochastic information in time-series observations and hidden features. The encoder in VRNN is designed to encode or identify the distribution

$$q_{\boldsymbol{\phi}}(\mathbf{z}_t \,|\, \mathbf{x}_t, \mathbf{y}_t, \mathbf{h}_{t-1})$$

of latent variable $\mathbf{z}_t$ from input–output pair $\{\mathbf{x}_t, \mathbf{y}_t\}$ at each time t and hidden feature $\mathbf{h}_{t-1}$ at the previous time $t-1$ as shown by dashed red lines (light gray in print version). Given the random samples $\mathbf{z}_t$ from variational distribution $q_{\boldsymbol{\phi}}(\cdot)$, the decoder in VRNN is introduced to realize the hidden units

$$\mathbf{h}_t = \mathcal{F}(\mathbf{x}_t, \mathbf{z}_t, \mathbf{h}_{t-1}) \tag{7.90}$$

at the current time t as shown in solid black lines (dark gray in print version). Hidden unit $\mathbf{h}_t$ acts as the realization or surrogate of hidden variable $\mathbf{z}_t$. The generative likelihood $p_{\boldsymbol{\theta}}(\cdot)$ is estimated to obtain the random output $\mathbf{y}_t$. Comparable with standard RNN, the hidden units $\mathbf{h}_t$ in VRNN are used to generate the outputs $\hat{\mathbf{y}}_t$ by

$$\hat{\mathbf{y}}_t \sim p_{\boldsymbol{\theta}}(\mathbf{y} \,|\, \mathbf{h}_t) \tag{7.91}$$

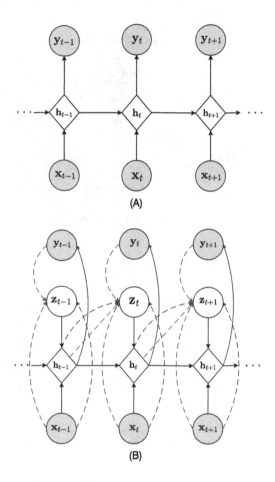

FIGURE 7.14

Graphical representation for (A) recurrent neural network and (B) variational recurrent neural network.

as shown by solid blue lines (middle gray in print version). VRNN pursues the *random* generation of regression outputs guided by the variational learning of hidden features. A stochastic learning of RNN is fulfilled by the following inference procedure.

Fig. 7.15 shows the inference procedure of VRNN. In model inference of supervised VRNN, we maximize the variational lower bound $\mathcal{L}$ of the logarithm of conditional likelihood

$$
p(\mathbf{y}_{\leq T} | \mathbf{x}_{\leq T}) = \prod_{t=1}^{T} \sum_{\mathbf{z}_t} p_{\boldsymbol{\theta}}(\mathbf{y}_t | \mathbf{x}_{\leq t}, \mathbf{z}_{\leq t})
$$
$$
\times \, p_{\boldsymbol{\omega}}(\mathbf{z}_t | \mathbf{x}_{\leq t}, \mathbf{z}_{<t})
$$

(7.92)

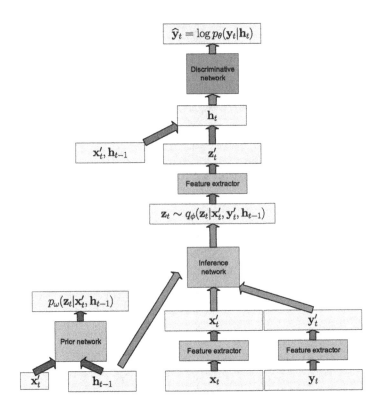

FIGURE 7.15

Inference procedure for variational recurrent neural network.

which is decomposed into two terms containing parameters θ and ω. The first term is a negative sum-of-squares error function which is obtained due to the Gaussian assumption for regression errors. The second term is an expected Kullback–Leibler (KL) divergence between distributions $q_\phi(\cdot)$ and $p_\omega(\cdot)$. The lower bound is expressed by

$$
\mathcal{L} \triangleq \mathbb{E}_{q_\phi(\mathbf{z}_{\leq T}|\mathbf{x}_{\leq T}, \mathbf{y}_{\leq T})} \left[\sum_{t=1}^{T} \left(\log p_\theta(\mathbf{y}_t|\mathbf{x}_{\leq t}, \mathbf{z}_{\leq t}) \right. \right.
$$

$$
\left. \left. - \mathcal{D}_{\mathrm{KL}}(q_\phi(\mathbf{z}_t|\mathbf{x}_{\leq t}, \mathbf{y}_{\leq t}, \mathbf{z}_{<t}) \| p_\omega(\mathbf{z}_t|\mathbf{x}_{\leq t}, \mathbf{z}_{<t})) \right) \right]. \tag{7.93}
$$

In maximization of Eq. (7.93), we first calculate the prior distribution of latent variable $\mathbf{z}_t$, which is a Gaussian

$$
p_\omega(\mathbf{z}_t|\mathbf{x}_t', \mathbf{h}_{t-1}) = \mathcal{N}(\boldsymbol{\mu}_{0,t}, \mathrm{diag}(\sigma_{0,t}^2)) \tag{7.94}
$$

where the mean and variance are calculated by a prior network

$$[\mu_{0,t}, \sigma_{0,t}^2] = \psi_\omega^{\text{prior}}(\mathbf{x}_t', \mathbf{h}_{t-1}) \tag{7.95}$$

using encoding weights ω. Then, the variational distribution is calculated at each time frame by using a Gaussian

$$q_\phi(\mathbf{z}_t | \mathbf{x}_t', \mathbf{y}_t', \mathbf{h}_{t-1}) = \mathcal{N}(\mu_{z,t}, \text{diag}(\sigma_{z,t}^2)) \tag{7.96}$$

with mean and variance calculated by an inference network using the encoder weights, i.e.,

$$[\mu_{z,t}, \sigma_{z,t}^2] = \psi_\phi^{\text{enc}}(\mathbf{x}_t', \mathbf{y}_t', \mathbf{h}_{t-1}). \tag{7.97}$$

Here, $\mathbf{x}_t'$ and $\mathbf{y}_t'$ denote the encoded features of $\mathbf{x}_t$ and $\mathbf{y}_t$ with reduced dimensions by using feature extractors $\psi^x(\mathbf{x}_t)$ and $\psi^y(\mathbf{y}_t)$ with parameters ϕ^x and ϕ^y, respectively. This is called the recognition or encoding phase with four sets of encoding weights

$$\{\phi^x, \phi^y, \phi^{\text{enc}}, \omega\}.$$

In the generation or decoding phase, we first apply the feature extractor $\psi^z(\mathbf{z}_t)$ with parameters θ^z to estimate the features $\mathbf{z}_t'$ corresponding to the latent variables $\mathbf{z}_t$. The variables $\mathbf{z}_t$ are sampled from the Gaussian distribution

$$q_\phi(\mathbf{z}_t | \mathbf{x}_t', \mathbf{y}_t', \mathbf{h}_{t-1}),$$

which was obtained in encoding phase. Then, we calculate the conditional likelihood $p_\theta(\mathbf{y}_t | \mathbf{h}_t)$ at each time t. This likelihood is estimated from the outputs of a decoder or discriminative network $\psi_\theta^{\text{dec}}(\mathbf{h}_t)$ with parameters θ^{dec}. This likelihood is used to calculate the regression output $\hat{\mathbf{y}}_t$ by using the inputs from

$$\mathbf{h}_t = \mathcal{F}(\mathbf{x}_t', \mathbf{z}_t', \mathbf{h}_{t-1}), \tag{7.98}$$

which is function of $\mathbf{x}_t'$, $\mathbf{z}_t'$ and $\mathbf{h}_{t-1}$ with parameters θ^h. There are three sets of parameters $\{\theta^z, \theta^{\text{dec}}, \theta^h\}$ in the VRNN decoder. Importantly, the expectation in the variational lower bound Eq. (7.93) is calculated by using L samples $\mathbf{z}_t$ obtained via variational distribution

$$q_\phi(\mathbf{z}_t | \mathbf{x}_t', \mathbf{y}_t', \mathbf{h}_{t-1}).$$

However, directly sampling $\mathbf{z}_t$ using the Gaussian distribution with the mean $\mu_{z,t}$ and variance $\sigma_{z,t}^2$ obtained by encoding network

$$\psi_\phi^{\text{enc}}(\mathbf{x}_t', \mathbf{y}_t', \mathbf{h}_{t-1})$$

is unstable with high variance. We follow Rezende et al. (2014) and use the reparameterization trick to resolve this problem. Using this trick, we sample $\epsilon \sim \mathcal{N}(\mathbf{0}, \mathbf{I})$ and use this sample to determine the sample for latent variable

$$\mathbf{z}_t \leftarrow \mu_{z,t} + \sigma_{z,t} \odot \epsilon \tag{7.99}$$

where $\odot$ denotes the element-wise multiplication. The stochastic training procedure for encoding weights

$$\{\phi^x, \phi^y, \phi^{enc}, \omega\}$$

and decoding weights

$$\{\theta^z, \theta^{dec}, \theta^h\}$$

is illustrated in Algorithm 7.1 which is seen as a stochastic gradient variational Bayes estimator.

Algorithm 7.1 Variational Recurrent Neural Network.

Initialize with hidden state $\mathbf{h}_0$, parameters ω, ϕ, θ
For $t = 1, 2, \ldots, T$
    Feedforward computation
    $\mathbf{x}_t' \leftarrow \psi^x(\mathbf{x}_t)$ with parameter ϕ^x
    $\mathbf{y}_t' \leftarrow \psi^y(\mathbf{y}_t)$ with parameter ϕ^y
    $\{\mu_{z,t}, \sigma_{z,t}^2\} \leftarrow \psi_\phi^{enc}(\mathbf{x}_t', \mathbf{y}_t', \mathbf{h}_{t-1})$ with parameter ϕ^{enc}
    $\{\mu_{0,t}, \sigma_{0,t}\} \leftarrow \psi_\omega^{prior}(\mathbf{x}_t', \mathbf{h}_{t-1})$ with parameter ω
    ϵ sampled from $\mathcal{N}(\mathbf{0}, \mathbf{I})$
    $\mathbf{z}_t \leftarrow \mu_{z,t} + \sigma_{z,t} \odot \epsilon$
    $\mathbf{z}_t' \leftarrow \psi^z(\mathbf{z}_t)$ with parameter θ^z
    $\mathbf{h}_t \leftarrow \mathcal{F}(\mathbf{x}_t', \mathbf{z}_t', \mathbf{h}_{t-1})$ with parameter θ^h
    $\hat{\mathbf{y}}_t \leftarrow \log p_\theta(\mathbf{y}_t|\mathbf{h}_t)$ via $\psi_\theta^{dec}(\mathbf{h}_t)$ with parameter θ^{dec}
    Backward computation
    Calculate the variational lower bound $\mathcal{L}$
    Compute the gradients $\frac{\partial \mathcal{L}}{\partial \phi}, \frac{\partial \mathcal{L}}{\partial \omega}, \frac{\partial \mathcal{L}}{\partial \theta}$
    Update the parameters
      $\phi \leftarrow \phi + \eta_\phi \odot \frac{\partial \mathcal{L}}{\partial \phi}$
      $\omega \leftarrow \omega + \eta_\omega \odot \frac{\partial \mathcal{L}}{\partial \omega}$
      $\theta \leftarrow \theta + \eta_\theta \odot \frac{\partial \mathcal{L}}{\partial \theta}$
    Return $\{\phi^x, \phi^y, \phi^{enc}, \omega\}$ and $\{\theta^z, \theta^{dec}, \theta^h\}$

A new type of supervised VRNN is developed for speech separation. This VRNN provides a stochastic point of view which accommodates the uncertainty in hidden states and facilitates the analysis of model construction. The masking function is further employed in network outputs for speech separation. The benefit of using VRNN is demonstrated by the experiments on monaural speech separation.

7.5.3 SYSTEM EVALUATION

The experiments on single-channel speech separation was evaluated by using the mixed speech signals from TIMIT corpus (Wang et al., 2016) as addressed in Section 7.3.3. The separation performance was examined by using the signal-to-distortion ratio (SDR), signal-to-interferences ratio

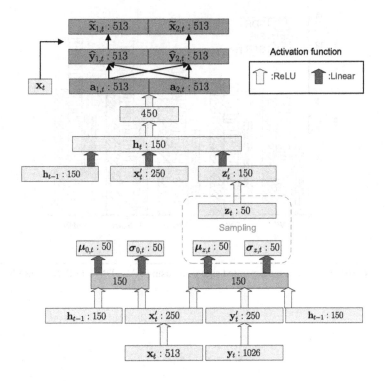

FIGURE 7.16

Implementation topology for variational recurrent neural network.

(SIR), and source-to-artifacts ratio (SAR) (Vincent et al., 2006). The adaptive SGD algorithm using Adam (Kingma and Ba, 2014) was performed. For each frame t, the spectral signal of the mixed speech $\mathbf{x}_t$ with dimension 513 was fed into VRNN. The outputs of VRNN corresponded to the spectra of demixed signals $\{\widetilde{\mathbf{x}}_{1,t}, \widetilde{\mathbf{x}}_{2,t}\}$ with dimension 1026. Fig. 7.16 depicts the network topology for implementation of VRNN. The feature extractors calculated $\mathbf{x}'_t$ and $\mathbf{y}'_t$ with the same dimension 250. Single layer with ReLU activation function was specified. Using $\mathbf{x}'_t$ and $\mathbf{y}'_t$ at time t and hidden state $\mathbf{h}_{t-1}$ at time $t-1$, the prior network and the inference network were constructed using ReLU activation with 150 dimensional hidden units. The outputs of two networks were calculated by linear activation and viewed as 50 dimensional mean and variance for Gaussians in prior distribution $p_\omega(\mathbf{z}_t | \mathbf{x}'_t)$ and variational distribution $q_\phi(\mathbf{z}_t | \mathbf{x}'_t, \mathbf{y}'_t, \mathbf{h}_{t-1})$. The latent variable $\mathbf{z}_t$ was then sampled and transformed to 150-dimensional $\mathbf{z}'_t$. The 150-dimensional hidden state $\mathbf{h}_t$ was accordingly obtained by RNN with linear activation. This hidden state was then forwarded through a hidden layer with 450 units for calculating the 513-dimensional activations $\{\mathbf{a}_{1,t}, \mathbf{a}_{2,t}\}$ for two sources. ReLU activation function was applied. An ideal ratio mask (Narayanan and Wang, 2013, Wang et al., 2016) was implemented to find 513-dimensional masking functions $\{\widehat{\mathbf{y}}_{1,t}, \widehat{\mathbf{y}}_{2,t}\}$ for two sources as given in Eq. (7.2). Finally, the demixed spectral signals of two sources $\{\widetilde{\mathbf{x}}_{1,t}, \widetilde{\mathbf{x}}_{2,t}\}$ were

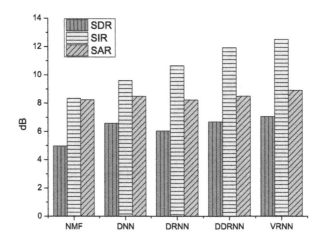

FIGURE 7.17

Comparison of SDR, SIR and SAR of the separated signals by using NMF, DNN, DRNN, DDRNN and VRNN.

calculated by

$$\widetilde{\mathbf{x}}_{i,t} = \mathbf{x}_t \odot \widehat{\mathbf{y}}_{i,t}$$

where $i \in \{1, 2\}$. In the implementation, we first trained the inference network and discriminative network by maximizing the first term in Eq. (7.93). After convergence, we used the trained parameters as the initialization to optimize the first and second terms of Eq. (7.93) to find prior network and fine-tune inference and discriminative networks.

The results of nonnegative matrix factorization (NMF) (Schmidt and Olsson, 2007), DNN (Grais et al., 2014), DRNN (Huang et al., 2014a), deep discriminative RNN (DDRNN) (Wang et al., 2016) (addressed in Section 7.3.2) and VRNN were carried out for comparison. The same spectral input and masking function were employed in different methods. NMF was implemented with the number of bases selected from validation data. Using DNN, the topology of 513–150–150–150–1026 with three 150-unit hidden layers was employed. In the implementation of DRNN, two recurrent hidden layers were constructed in a topology of 513–150–150–1026 (Wang et al., 2016). The DDRNN was implemented in a topology of 513–150–150–1026 according to a discriminative objective. ReLU activation was used in DNN, DRNN and DDRNN. Fig. 7.17 shows a comparison of SDR, SIR and SAR of using NMF, DNN, DRNN, DDRNN and VRNN. Four neural network methods consistently perform better than NMF in terms of SDR and SIR. The improvement in terms of SAR is limited. The recurrent neural networks using DRNN, DDRNN and VRNN achieve higher SIR than DNN. Also DNN slightly performs better than DRNN but worse than DDRNN and VRNN in terms of SDR and SAR. Nevertheless, VRNN outperforms the other methods in terms of SDR, SIR and SAR. These results show that the stochastic modeling in DRNN using variational learning can partially compensate the deterministic assumption in conventional DRNN for monaural source separation.

7.6 NEURAL TURING MACHINE

Traditionally, the recurrent neural network based on long short-term memory (LSTM) explores the temporal information by learning dynamic states which are evolved through time and stored as an *internal* memory. The performance of source separation is constrained due to the limitation of internal memory which could not sufficiently preserve long-term characteristics from different sources. This section presents a new paradigm of recurrent neural network, called the memory augmented neural network (MANN), and illustrates how this paradigm is capable of enhancing the memorability of a regression system for monaural source separation (Tsou and Chien, 2017). In the literature, there are two streams of memory-augmented neural networks. One is the neural Turing machine (Graves et al., 2014) and the other is the memory network or end-to-end memory network (Sukhbaatar et al., 2015, Weston et al., 2014). This section focuses on implementation of neural Turing machine (NTM) to learn a separation model for sequential signals of speech and noise in presence of different speakers and noise types. The memory augmented neural network based on end-to-end memory network will be addressed and used in Section 7.7.

7.6.1 MEMORY AUGMENTED SOURCE SEPARATION

Fig. 7.18(A) illustrates a recurrent neural network for single-channel source separation. In this neural network, the input features of a mixed signal $\mathbf{x}_t^{\text{mix}}$ or simply $\mathbf{x}_t$ at time t consist of a 513-dimensional spectral vector calculated by using the short-time Fourier transform with a window size of 64 ms and overlap of 32 ms. We use a fully-connected layer for feature extraction followed by two recurrent layers for recurrent memorization, one layer for estimation of ideal ratio mask, and one output layer for construction of the estimated source signals. Rectified linear units are used as the activation function in a fully-connected layer. Recurrent memorization is governed by the dynamic state

$$\mathcal{Z}_t = \{\mathbf{z}_t, \mathbf{r}_t, \mathbf{w}_{r,t}, \mathbf{w}_{w,t}\} \tag{7.100}$$

containing the hidden feature vector, read vector, read head and write head at layer $L-1$, respectively. Reading and writing over memory matrix $\mathbf{M}_t$ are performed. NTM layer is capable of learning the long-term dependency based on external memory. Different from LSTM, NTM is geared with the memorization of $\mathbf{M}_t$. The memory cell in LSTM is prone to be overwritten in long-term time steps. Some important information will be lost. But NTM allows a large amount of information stored in memory, which is read when retrieving the relevant information and written when disregarding the irrelevant information according to the attention mechanism. For example, if a mixed signal or a noisy speech signal is *nonstationary* with different source or noise signals in different time periods, the memory cell of LSTM could not handle the changing source signals or noise types. But NTM is able to preserve some features in external memory which characterizes both speakers and noises. The recurrent layer $L-1$ in NTM is used to find the activation vectors $\{\mathbf{a}_{1,t}^{(L)}, \mathbf{a}_{2,t}^{(L)}\}$ corresponding to two sources. The output layer L consists of $\{\widehat{\mathbf{y}}_{1,t}, \widehat{\mathbf{y}}_{2,t}\}$, which are obtained via predictive masking function by using the activation vectors as provided in Eq. (7.2). The output signals are then obtained by

$$\widetilde{\mathbf{x}}_{i,t} = \widehat{\mathbf{y}}_{i,t} \odot \mathbf{x}_t^{\text{mix}}$$

where $i = 1, 2$. Given the estimated signals $\widetilde{\mathbf{x}}_{1,t}$ and $\widetilde{\mathbf{x}}_{2,t}$ of original sources $\mathbf{x}_{1,t}$ and $\mathbf{x}_{2,t}$ over T frames, the regression objective is constructed as a sum-of-squares error function as given in Eq. (7.51). Min-

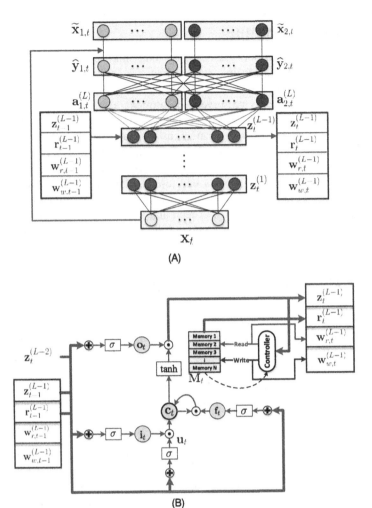

FIGURE 7.18

(A) Single-channel source separation with dynamic state $\mathcal{Z}_t = \{\mathbf{z}_t, \mathbf{r}_t, \mathbf{w}_{r,t}, \mathbf{w}_{w,t}\}$ in recurrent layer $L - 1$. (B) Recurrent layers are driven by a cell $\mathbf{c}_t$ and a controller for memory $\mathbf{M}_t$ where dashed line denotes the connection between cell and memory at previous time step and bold lines denote the connections with weights.

imizing Eq. (7.51) is equivalent to increasing the similarity between target outputs $\{\mathbf{x}_{i,t}\}$ and actual outputs $\{\widetilde{\mathbf{x}}_{i,t}\}$. More generally, model parameters include the *feedforward* weight parameters

$$\Theta_f = \{\mathbf{w}^{(1)}, \ldots, \mathbf{w}^{(L-3)}, \mathbf{w}^{(L)}\} \tag{7.101}$$

in fully-connected layers and the NTM *recurrent* weight parameters Θ_r in layers $L - 1$ and $L - 2$.

We are presenting NTM for single-channel source separation which is particularly realized for the scenario of *speaker-* and *noise-independent* speech enhancement in the presence of a variety of speakers and noise conditions. The recurrent layers in NTM are extended from those in LSTM, which are augmented with an external memory $\mathbf{M}_t$ and functioned by a controller with read and write heads.

7.6.2 LSTM CONTROLLER AND ADDRESSING MECHANISM

Fig. 7.18(B) illustrates the controller network, which is developed for updating the *dynamic state* of recurrent layers $\mathcal{Z}_t = \{\mathbf{z}_t, \mathbf{r}_t, \mathbf{w}_{r,t}, \mathbf{w}_{w,t}\}$ from time step $t - 1$ to t for memory augmented source separation. This is different from LSTM where the dynamic state $\mathcal{Z}_t = \mathbf{z}_t$ disregards the access to external memory $\mathbf{M}_t$. Such a network learns for long-term dependencies in a demixing system and stores them in external memory $\mathbf{M}_t$. Compared with LSTM, NTM can preserve temporal information from much longer sequential data (Graves et al., 2014). There are two basic components in NTM. One is the controller network, which is run as a variant of LSTM with the specialized configuration for interactions with memory. The other is the addressing mechanism where the real-valued demixing measure is sequentially read and written based on content-based soft attention. These two components are addressed hereafter.

Controller Network

The controller network is built by using a new LSTM variant, which is computed by

$$\mathbf{i}_t = \sigma(W_{ix}\mathbf{z}_t^{(L-2)} + W_{iz}\mathbf{z}_{t-1}^{(L-1)} + W_{ir}\mathbf{r}_{t-1} + \mathbf{b}_i), \tag{7.102}$$

$$\mathbf{f}_t = \sigma(W_{fx}\mathbf{z}_t^{(L-2)} + W_{fz}\mathbf{z}_{t-1}^{(L-1)} + W_{fr}\mathbf{r}_{t-1} + \mathbf{b}_f), \tag{7.103}$$

$$\mathbf{u}_t = \tanh(W_{ux}\mathbf{z}_t^{(L-2)} + W_{uz}\mathbf{z}_{t-1}^{(L-1)} + W_{ur}\mathbf{r}_{t-1} + \mathbf{b}_u), \tag{7.104}$$

$$\mathbf{c}_t = \mathbf{f}_t \odot \mathbf{c}_{t-1} + \mathbf{i}_t \odot \mathbf{u}_t, \tag{7.105}$$

$$\mathbf{o}_t = \sigma(W_{ox}\mathbf{z}_t^{(L-2)} + W_{oz}\mathbf{z}_{t-1}^{(L-1)} + W_{or}\mathbf{r}_{t-1} + \mathbf{b}_o), \tag{7.106}$$

$$\mathbf{z}_t^{(L-1)} = \mathbf{o}_t \odot \tanh(\mathbf{c}_t) \tag{7.107}$$

where W_{ix}, W_{fx}, W_{ox} and W_{ux} denote the weight parameters from the input $\mathbf{z}_t^{(L-2)}$ at current time t to input gate, forget gate, output gate and cell, respectively; W_{iz}, W_{fz}, W_{oz} and W_{uz} denote the parameters for output of memory block $\mathbf{z}_{t-1}^{(L-1)}$ at a previous time $t - 1$, and W_{ir}, W_{fr}, W_{or} and W_{ur} denote the parameters for peephole connections from read vector $\mathbf{r}_{t-1}$ at the same previous time; $\{\mathbf{b}_i, \mathbf{b}_f, \mathbf{b}_o, \mathbf{b}_u\}$ denote the corresponding bias parameters. Different from standard LSTM in Eqs. (2.75)–(2.81), we arrange the input and output of recurrent layer $L - 1$ for source separation by using hidden units $\mathbf{z}_t^{(L-2)}$ and $\mathbf{z}_{t-1}^{(L-1)}$ in consecutive two layers, respectively, as shown in Eqs. (7.102)–(7.106). Instead of using memory cell $\mathbf{c}_t$, the read vector $\mathbf{r}_{t-1}$ is adopted to update four gates $\{\mathbf{i}_t, \mathbf{f}_t, \mathbf{o}_t, \mathbf{u}_t\}$. Memory cell $\mathbf{c}_t$ is applied to update the output of recurrent layer from $\mathbf{z}_{t-1}^{(L-1)}$ to $\mathbf{z}_t^{(L-1)}$ by setting

$$\mathbf{z}_t^{(L-1)} \leftarrow \mathbf{c}_t. \tag{7.108}$$

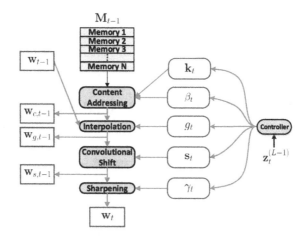

FIGURE 7.19

Four steps of addressing procedure which is driven by parameters $\{\mathbf{k}_t, \beta_t, g_t, \mathbf{s}_t, \gamma_t\}$.

This process is therefore denoted by

$$\mathbf{z}_t^{(L-1)} = \text{LSTM}(\mathbf{z}_t^{(L-2)}, \mathbf{z}_{t-1}^{(L-1)}, \mathbf{r}_{t-1}). \tag{7.109}$$

Importantly, a controller is formed by a fully-connected layer for mapping from $\mathbf{z}_t^{(L-1)}$ to addressing parameters

$$\{\mathbf{k}_t, \beta_t, g_t, \mathbf{s}_t, \gamma_t, \mathbf{e}_t, \mathbf{a}_t\} = \text{Controller}(\mathbf{z}_t^{(L-1)}), \tag{7.110}$$

which are known as the memory key, key strength, interpolation gate, shift weights, sharpening factor, erase vector, and add vector, respectively. Overall, the recurrent parameters Θ_r in NTM recurrent layers include the LSTM parameters Θ_l in Eqs. (7.102)–(7.106) and the fully-connected weights $\mathbf{w}_c$ in the controller.

Addressing Mechanism

Fig. 7.19 illustrates the four steps of the addressing process, which are controlled by addressing parameters $\{\mathbf{k}_t, \beta_t, g_t, \mathbf{s}_t, \gamma_t\}$ in recurrent layers. The external memory at time t is expressed as a matrix $\mathbf{M}_t \in \mathbb{R}^{N \times M}$ consisting of N memory slots with an M-dimensional vector in each slot. The ith memory slot is denoted by $\mathbf{M}_t(i)$. Content-based soft attention is maintained. In the first step, the *content*-based addressing is performed by calculating an initial attention

$$w_{c,t}(i) = \frac{\exp\left(\beta_t \cos\left(\mathbf{M}_t(i), \mathbf{k}_t\right)\right)}{\sum_j \exp\left(\beta_t \cos\left(\mathbf{M}_t(j), \mathbf{k}_t\right)\right)} \tag{7.111}$$

where the cosine similarity

$$\cos(\mathbf{M}_t(i), \mathbf{k}_t)$$

measures the strength of a memory slot $\mathbf{M}_t(i)$ accessed by a key $\mathbf{k}_t \in \mathbb{R}^N$. This similarity is weighted by β_t to find the normalized attention $0 \le w_{c,t}(i) \le 1$.

Secondly, an interpolation step is performed to update the attention as $\mathbf{w}_{g,t} = \{w_{g,t}(i)\}$ by using a scalar interpolation gate $0 \le g_t \le 1$, which balances between the content-based attention $\mathbf{w}_{c,t} = \{w_{c,t}(i)\}$ and the read or write head

$$\mathbf{w}_{t-1} = \{w_{t-1}(i)\} \in \{\mathbf{w}_{r,t-1}, \mathbf{w}_{w,t-1}\} \tag{7.112}$$

at consecutive time steps $t - 1$ and t by

$$w_{g,t}(i) = g_t w_{c,t}(i) + (1 - g_t)w_{t-1}(i). \tag{7.113}$$

In the third step, a convolutional shift is performed for *location*-based attention by blurring the attention over S points instead of spotting at a single point or location. The attention is then updated to $\mathbf{w}_{s,t} = \{w_{s,t}(i)\}$ where

$$w_{s,t}(i) = \sum_{j=0}^{S-1} w_{g,t}(j)s_t(i - j). \tag{7.114}$$

Here, $\mathbf{s}_t = \{s_t(i)\}$ denotes a set of S shift weights. In the fourth step, the read or write head is updated by

$$w_t(i) = \frac{w_{s,t}(i)^{\gamma_t}}{\sum_j w_{s,t}(j)^{\gamma_t}} \tag{7.115}$$

where the dispersed vector $\mathbf{w}_{s,t}$ due to convolutional shift is *sharpened* by using $\gamma_t \ge 1$. The property of soft attention $0 \le w_t(i) \le 1$ with $\sum_i w_t(i) = 1$ is met. We therefore obtain the read head $\mathbf{w}_{r,t}$ and write head $\mathbf{w}_{w,t}$ at each time t, which are employed to *travel* over N locations of an external memory $\mathbf{M}_t$ for reading and writing, respectively.

Reading and Writing Mechanism

Using this NTM, we carry out the reading and writing at each time t to retrieve and store the demixing information in recurrent layers for monaural source separation, respectively. In reading phase, a read vector of length M is calculated by

$$\mathbf{r}_t = \sum_{i=1}^{N} w_{r,t}(i)\mathbf{M}_t(i), \tag{7.116}$$

which is obtained by attending N memory slots using the weights of the read head $\mathbf{w}_{r,t} = \{w_{r,t}(i)\}$. This read vector $\mathbf{r}_t$ is incorporated into the controller network in Fig. 7.18(B) for updating the recurrent variables $\mathbf{z}_{t+1}^{(L-1)}$ at the next time step $t + 1$. In addition, the writing process consists of two operations, *erase* and *add*, which are controlled by the write head $\mathbf{w}_{w,t}$ using the erase vector $\mathbf{e}_t$ and add vector $\mathbf{a}_t$, respectively. The ith memory slot is updated by erasing and adding

$$\mathbf{M}_t(i) = \underbrace{\mathbf{M}_{t-1}(i) \odot [1 - w_{w,t}(i)\mathbf{e}_t]}_{\text{erasing}}$$

$$+ \underbrace{w_{w,t}(i)\mathbf{a}_t}_{\text{adding}} \tag{7.117}$$

where **1** is a vector of 1's. Notably, the addressing parameters $\{\mathbf{e}_t, \mathbf{a}_t\}$ are obtained from the outputs of the controller given by the input $\mathbf{z}_{t-1}^{L-1}$. The entire model parameters

$$\{\boldsymbol{\Theta}_f, \boldsymbol{\Theta}_r = \{\boldsymbol{\Theta}_l, \mathbf{w}_c\}\} \tag{7.118}$$

of NTM are estimated by minimizing the regression loss in Eq. (7.51) from the minibatches of training data by using the SGD algorithm. Algorithms 7.2 and 7.3 show the procedures for recurrent layers in NTM and memory augmented source separation, respectively.

Algorithm 7.2 Recurrent layers in neural Turing machine.

Require $\mathbf{z}_t^{(L-2)}, \mathbf{z}_{t-1}^{(L-1)}, \mathbf{r}_{t-1}, \mathbf{w}_{t-1} = \{\mathbf{w}_{t-1,r}, \mathbf{w}_{t-1,w}\}$
$\quad \mathbf{z}_t^{(L-1)} = \text{LSTM}(\mathbf{z}_t^{(L-2)}, \mathbf{z}_{t-1}^{(L-1)}, \mathbf{r}_{t-1})$
$\quad \mathbf{w}_t = \text{Address}(\text{Controller}(\mathbf{z}_t^{(L-1)}), \mathbf{w}_{t-1}, \mathbf{M}_{t-1})$
$\quad \{\mathbf{e}_t, \mathbf{a}_t\} = \text{Controller}(\mathbf{z}_t^{(L-1)})$
$\quad \mathbf{r}_t = \text{Read}(\mathbf{w}_{r,t}, \mathbf{M}_{t-1})$
$\quad \mathbf{M}_t = \text{Write}(\mathbf{w}_{w,t}, \mathbf{e}_t, \mathbf{a}_t, \mathbf{M}_{t-1})$
Return $\mathbf{z}_t^{(L-1)}, \mathbf{r}_t, \mathbf{w}_t = \{\mathbf{w}_{r,t}, \mathbf{w}_{w,t}\}$

Algorithm 7.3 Monaural source separation using neural Turing machine.

Require $\mathbf{x}_t^{\text{mix}}$: mixed signal
$\quad$ Feature extraction: $\mathbf{z}_t^{(L-1)} = \text{NTM}(\mathbf{x}_t^{\text{mix}})$
$\quad$ Mask prediction: $\widehat{\mathbf{y}}_{1,t}, \widehat{\mathbf{y}}_{2,t}$
$\quad$ Source signal estimation: $\widetilde{\mathbf{x}}_{i,t} = \widehat{\mathbf{y}}_{i,t} \odot \mathbf{x}_t^{\text{mix}}$ for $i = 1, 2$
Return $\widetilde{\mathbf{x}}_{1,t}, \widetilde{\mathbf{x}}_{2,t}$: estimated source signals

---

7.6.3 SYSTEM EVALUATION

In system evaluation, speech signals were sampled from 83 speakers and noise signals were collected with 88 noise types (Tsou and Chien, 2017). Among these speakers, 77 speakers were chosen as training speakers and the remaining 6 speakers were treated as unseen test speakers. The utterances from different speakers were randomly mixed with various nonstationary noises. There were 7768 training utterances which were mixed by using 86 noise types. Noisy training data were generated by SNR of

Table 7.2 Comparison of STOIs using DNN, LSTM and NTM under different SNRs with seen speakers

Model	−5 dB	0 dB	5 dB
Unprocessed	0.614	0.717	0.808
DNN	0.662	0.763	0.820
LSTM	0.688	0.788	0.847
NTM	**0.702**	**0.797**	**0.851**

Table 7.3 Comparison of STOIs using DNN, LSTM and NTM under different SNRs with unseen speakers

Model	−5 dB	0 dB	5 dB
Unprocessed	0.647	0.746	0.829
DNN	0.678	0.785	0.844
LSTM	0.718	0.815	0.872
NTM	**0.731**	**0.825**	**0.877**

−5 dB while the noisy test data were generated by SNR of −5, 0 and 5 dB with two unseen noise types (cafeteria and bus). The conditions of seen and unseen speakers were also investigated. There were 300 noisy test utterances collected in these two conditions. The task of single-channel speech enhancement was evaluated. In the implementation, 1024-point STFT was calculated for mixed signals $\mathbf{x}_t^{mix} \in \mathbb{R}^{513}$. A DNN with the topology 513–1000–1000–1000–700–{513–513} was realized. For consistency in comparison with DNN, LSTM and NTM followed the same topology with the same number of neurons but differed in the style of the 3rd and 4th hidden layers. LSTM implemented two recurrent layers of LSTM while NTM implemented the 3rd hidden layer as an LSTM layer and the 4th layer as an NTM layer. The memory size of $\mathbf{M}_t$ was fixed as 32. All models were trained by using the SGD algorithm with mini-batch size of 50 frames. The step size of 20 frames was used in backpropagation through time. Adam optimizer was applied. Speech enhancement was evaluated by using short-time objective intelligibility (STOI). The higher the STOI, the better the source separation.

We compare the performance of monaural source separation by using DNN, LSTM and NTM in terms of STOI in the presence of seen and unseen test speakers as shown in Tables 7.2 and 7.3, respectively. There are 77 training speakers. The demixing results under different SNRs and test noises are averaged. The improvement of NTM over DNN and LSTM is consistently obtained for seen and unseen speakers under different noises and SNRs. The improvement is significant for −5 dB while slight improvement is observed for 0 and 5 dB.

7.7 END-TO-END MEMORY NETWORK

There are two issues raised in the implementation of a neural Turing machine, which considerably affect the practical use of NTM in reality. The first one is the high computation cost due to the intensive calculation for read and write with external memory. The second one is the limited size of memory, which causes fast expiration of useful information due to the erase operation at each time. To deal with these issues, this section presents a variant of memory-augmented neural network, called the end-to-end memory network, where the external memory is constructed according to a *sequence-to-sequence learning* based on an encoder–decoder network. There is no writing of the memory matrix. The frequent interaction with external memory is mitigated. In particular, a so-called recall neural network (Chien and Tsou, 2018) is addressed to effectively preserve and utilize useful information for a period of time without fast expiration due to a limited size of memory.

7.7.1 RECALL NEURAL NETWORK

The end-to-end memory network based on recall neural network (RCNN) is presented to mimic the human listening for signal-channel speech enhancement. In the implementation, a couple of external memories are implemented to build an end-to-end memory network (Sukhbaatar et al., 2015) for sequence-to-sequence learning based on an encoder and a decoder. These memories are learned in a two-pass listening procedure where the mixed signal is encoded and then decoded (or recalled) as context vectors by using a bidirectional long short-term memory (bidirectional LSTM or BLSTM) (as seen in Section 7.4.3) and an LSTM (as given in Section 2.5.3), respectively. These context vectors are integrated in a gating layer. A set of attention weights are calculated to attend the hidden state of decoder to implement a recurrent neural network for source separation. A gated attention mechanism is carried out to fulfill a specialized memory network. The regression errors due to two passes of sensing procedure and one pass of gated attention are jointly minimized in an end-to-end manner so as to estimate the weight parameters of different components in different layers.

Fig. 7.20 depicts the overall architecture of a deep sequence-to-sequence model for monaural source separation from a mixed signal $\{\mathbf{x}_t^{\text{mix}}\}_{t=1}^T$ or simply $\{\mathbf{x}_t\}_{t=1}^T$ to the separated signals $\{\widetilde{\mathbf{x}}_{1,t}\}_{t=1}^T$ and $\{\widetilde{\mathbf{x}}_{2,t}\}_{t=1}^T$. This end-to-end memory network, consisting of three LSTMs, is established to conduct a two-pass sensing and one-pass separating. The first layer is a fully-connected (FC) layer which transforms an input $\mathbf{x}_t^{\text{mix}}$ into two hidden states $\{\mathbf{h}_t^{(1)}, \mathbf{s}_t^{(1)}\}$ at each frame t. One is for the encoder on the left and the other is for the decoder on the right. The second hidden layer of the encoder is run to obtain hidden units $\mathbf{h}_t^{(2)}$ by using a bidirectional LSTM, which is composed of a memory matrix for forward and backward directions

$$\mathbf{M}_e = \{\mathbf{h}_{f,t}^{(2)}, \mathbf{h}_{b,t}^{(2)}\}, \tag{7.119}$$

which is shown by the left panel in green with orange nodes (the left panel in gray with light gray nodes in print version). At the same time, the second hidden layer of the decoder is formed by an LSTM layer, which produces the hidden units to form the memory matrix

$$\mathbf{M}_d = \{\mathbf{s}_t^{(2)}\} \tag{7.120}$$

displayed by right panel with light blue nodes (the right panel in gray with mid gray nodes in print version). Two memories $\{\mathbf{M}_e, \mathbf{M}_d\}$ are updated in each time frame t. The context vectors of encoder $\mathbf{c}_t^e$ and decoder $\mathbf{c}_t^d$ are then obtained and used to carry out the gated attention mechanism to calculate the hidden unit in the third hidden layer $\mathbf{s}_t^{(3)}$ which is built as an LSTM layer and run for source separation. Basically, the mixed signal frames $\{\mathbf{x}_t^{\text{mix}}\}_{t=1}^T$ are perceived or listened two times before going to a

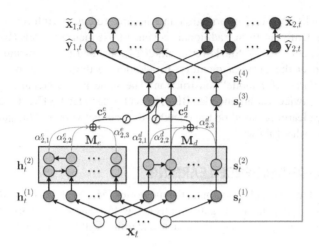

FIGURE 7.20

An end-to-end memory network for monaural source separation containing a bidirectional LSTM on the left as an encoder, an LSTM on the right as a decoder and an LSTM on the top as a separator.

separation task. The second-time sensing, run by a decoder, is seen as a listening *recall* over the information learned in the first-time sensing or listening, run by an encoder. In case of speech separation, these three LSTMs are learned to perform listening, listening and separating. This procedure is fitted to how human enhances a noisy speech. Human usually asks to listen the noisy speech again to make sure the result of the separated speech. We therefore call this model the *recall neural network*. Interestingly, the style of *listen*, *listen* and *separate* in end-to-end memory network for speech separation is similar to the style of *show*, *attend* and *tell*, which was proposed for neural image caption in Xu et al. (2015).

In the implementation, the fourth hidden layer is formed by an FC layer with parameters $\mathbf{w}^{(4)}$ to find hidden unit $\mathbf{s}_t^{(4)}$ before stacking with the reconstruction layer to estimate the soft mask function for each source. The soft mask function in RCNN is yielded by

$$\widehat{\mathbf{y}}_{i,t} = \frac{|\mathbf{w}_i^{(5)}\mathbf{s}_t^{(4)}|}{|\mathbf{w}_1^{(5)}\mathbf{s}_t^{(4)}| + |\mathbf{w}_2^{(5)}\mathbf{s}_t^{(4)}|}, \qquad \text{for } i \in \{1, 2\}, \tag{7.121}$$

where $\mathbf{w}_i^{(5)}$ denotes the weight parameters at the fifth hidden layer corresponding to two sources, and $\widehat{\mathbf{y}}_{i,t}$ is the estimated soft mask function corresponding to source i. The output signals are then obtained by

$$\widetilde{\mathbf{x}}_t^i = \widehat{\mathbf{y}}^{i,t} \odot \mathbf{x}_t^{\text{mix}}.$$

A sequence-to-sequence model is constructed and will be driven by an attention mechanism which allows the model to spotlight on complementary features from an input mixed signal based on a two-pass sensing or listening procedure and use these latent features for a demixing process. Such an attention using weights $\{\alpha_{t,i}^e, \alpha_{t,i}^d\}$ for encoder and decoder is similar to a *random dropout* in standard deep neural network which picks up useful information and discards redundant information over two external memories $\{\mathbf{M}_e, \mathbf{M}_d\}$. Different from traditional sequence-to-sequence models (Bahdanau et al., 2014, Chan et al., 2016, Xu et al., 2015, Sutskever et al., 2014), this end-to-end memory network performs the gated attention over the *external* memory. In addition to the decoder loss, the encoder loss is considered as a regularizer for model construction. The same mixed signal is fed into this end-to-end memory network twice via an encoder BLSTM and a decoder LSTM. Detailed descriptions of sequence-to-sequence learning based on the encoder–decoder network and the gated attention mechanism are addressed in what follows.

7.7.2 SEQUENCE-TO-SEQUENCE LEARNING

This section presents the approach to estimate the memories $\{\mathbf{M}_e, \mathbf{M}_d\}$, the context vectors $\{\mathbf{c}_t^e, \mathbf{c}_t^d\}$ and the attention weights $\{\alpha_{t,i}^e, \alpha_{t,i}^d\}$, and the gating units $\{\mathbf{g}_t^e, \mathbf{g}_t^d\}$ for encoder and decoder, respectively. The objective function to carry out sequence-to-sequence learning will be addressed.

Encoder–Decoder Network

An encoder–decoder network is introduced to carry out a pair of external memories in RCNN in accordance with a sequence-to-sequence learning. These memories provide the external latent codes and context vectors to demix an observed signal into two source signals. First, the encoder is formed by an FC layer followed by a BLSTM layer. The hidden units in the FC layer and in the forward and backward directions of BLSTM layer are expressed by

$$\mathbf{h}_t^{(1)} = \text{FC}(\mathbf{x}_t^{\text{mix}}, \mathbf{w}_e^{(1)}), \tag{7.122}$$

$$\begin{aligned}\mathbf{h}_t^{(2)} &= \{\mathbf{h}_{f,t}^{(2)}, \mathbf{h}_{b,t}^{(2)}\} \\ &= \text{BLSTM}(\mathbf{h}_t^{(1)}, \mathbf{h}_{t-1}^{(2)}, \mathbf{h}_{t+1}^{(2)}, \mathbf{w}_e^{(2)}) \\ &\triangleq \mathbf{M}_e\end{aligned} \tag{7.123}$$

where $\mathbf{w}_e^{(1)}$ and $\mathbf{w}_e^{(2)}$ denote the parameters in FC and BLSTM layers, respectively. The encoder memory $\mathbf{M}_e = \{\mathbf{h}_{f,t}^{(2)}, \mathbf{h}_{b,t}^{(2)}\}$ is obtained. Owing to the bidirectional sensing, the *whole* mixed signal should be encoded before moving to the next layer to combine with the information learned from decoder. Then, the decoder is built by an FC layer followed by an LSTM layer to obtain hidden codes

$$\mathbf{s}_t^{(1)} = \text{FC}(\mathbf{x}_t^{\text{mix}}, \mathbf{w}_d^{(1)}), \tag{7.124}$$

$$\mathbf{s}_t^{(2)} = \text{LSTM}(\mathbf{s}_t^{(1)}, \mathbf{s}_{t-1}^{(2)}, \mathbf{w}_d^{(2)}), \tag{7.125}$$

using decoder parameters $\mathbf{w}_d^{(1)}$ and $\mathbf{w}_d^{(2)}$, respectively. The decoder memory $\mathbf{M}_d = \{\mathbf{s}_t^{(2)}\}$ is stored. Importantly, a gated attention method is implemented to calculate the context vectors $\{\mathbf{c}_t^e, \mathbf{c}_t^d\}$ and the gating weights $\{\mathbf{g}_t^e, \mathbf{g}_t^d\}$ to carry out the attention-based LSTM layer

$$\mathbf{s}_t^{(3)} = \text{LSTM}(\mathbf{s}_t^{(2)}, \mathbf{s}_{t-1}^{(3)}, \mathbf{g}_t^e \odot \mathbf{c}_t^e, \mathbf{g}_t^d \odot \mathbf{c}_t^d, \mathbf{w}^{(3)}) \tag{7.126}$$

using the parameters $\mathbf{w}^{(3)}$. We build one BLSTM for encoder, one LSTM for decoder and one LSTM for separator. In LSTMs, we use the hidden states from previous time $\{\mathbf{s}_{t-1}^{(2)}, \mathbf{s}_{t-1}^{(3)}\}$ as inputs. In BLSTM, the hidden states in two directions $\{\mathbf{h}_{t-1}^{(2)}, \mathbf{h}_{t+1}^{(2)}\}$ are used as inputs. The gating weights $\{\mathbf{g}_t^e, \mathbf{g}_t^d\}$ reflect how much information in the context vectors $\{\mathbf{c}_t^e, \mathbf{c}_t^d\}$ is used or ignored. The context vectors and the gating weights are estimated as mentioned hereafter.

Gated Attention Mechanism

We present a gated attention mechanism which allows end-to-end memory network to select useful information for source separation. Using this mechanism, the context vectors of the encoder and decoder at each time t are calculated by linearly combining all entries in external memories of the encoder and decoder

$$\mathbf{c}_t^e = \sum_{i=1}^{N_e} \alpha_{t,i}^e \mathbf{M}_e(i), \tag{7.127}$$

$$\mathbf{c}_t^d = \sum_{i=1}^{N_d} \alpha_{t,i}^d \mathbf{M}_d(i). \tag{7.128}$$

In Eqs. (7.127)–(7.128), memories $\mathbf{M}_e = \{\mathbf{M}_e(i)\}$ and $\mathbf{M}_d = \{\mathbf{M}_d(i)\}$ have N_e and N_d vectors, respectively, and $\alpha_{t,i}^e$ and $\alpha_{t,i}^d$ denotes the attention weights given by

$$\alpha_{t,i}^e = \frac{\exp(a_{t,i}^e)}{\sum_{j=1}^{T} \exp(a_{t,j}^e)}, \tag{7.129}$$

$$\alpha_{t,i}^d = \frac{\exp(a_{t,i}^d)}{\sum_{j=1}^{T} \exp(a_{t,j}^d)} \tag{7.130}$$

where

$$a_{t,i}^e = (\mathbf{v}_e)^\top \tanh \left(W_{ea} \mathbf{s}_{t-1}^{(3)} + U_{ea} \mathbf{M}_e(i) \right), \tag{7.131}$$

$$a_{t,i}^d = (\mathbf{v}_d)^\top \tanh \left(W_{da} \mathbf{s}_{t-1}^{(3)} + U_{da} \mathbf{M}_d(i) \right) \tag{7.132}$$

where $\{\mathbf{v}_e, W_{ea}, U_{ea}\}$ and $\{\mathbf{v}_d, W_{da}, U_{da}\}$ denote the connecting weights between recurrent hidden codes $\mathbf{s}_{t-1}^{(3)}$ at the previous time $t-1$ and memories $\mathbf{M}_e$ and $\mathbf{M}_d$ at the current time t for the encoder and decoder, respectively. Given context vectors $\{\mathbf{c}_t^e, \mathbf{c}_t^d\}$ and previous hidden state $\mathbf{s}_{t-1}^{(3)}$, the gating units are computed by

$$\mathbf{g}_t^e = \sigma (W_{eg} \mathbf{s}_{t-1}^{(3)} + U_{eg} \mathbf{c}_t^e + \mathbf{b}_{eg}), \tag{7.133}$$

$$\mathbf{g}_t^d = \sigma (W_{dg} \mathbf{s}_{t-1}^{(3)} + U_{dg} \mathbf{c}_t^d + \mathbf{b}_{dg}) \tag{7.134}$$

where $\sigma(\cdot)$ denotes the sigmoid function and $\{W_{eg}, U_{eg}, \mathbf{b}_{eg}\}$ and $\{W_{dg}, U_{dg}, \mathbf{b}_{dg}\}$ denote the connecting weights between recurrent hidden nodes $\mathbf{s}_{t-1}^{(3)}$ at the previous time $t-1$ and context vectors $\{\mathbf{c}_t^e, \mathbf{c}_t^d\}$

at the current time t for the encoder and decoder, respectively. The parameters of end-to-end memory network for monaural source separation

$$
\begin{aligned}
\boldsymbol{\Theta} = \{ & \mathbf{w}_e^{(1)}, \mathbf{w}_d^{(1)}, \mathbf{w}_e^{(2)}, \mathbf{w}_d^{(2)}, \mathbf{w}^{(3)}, \mathbf{v}_e, \mathbf{v}_d, W_{ea}, \\
& W_{da}, U_{ea}, U_{da}, W_{eg}, W_{dg}, U_{eg}, U_{dg}, \mathbf{b}_{eg}, \mathbf{b}_{dg}, \\
& \mathbf{w}^{(4)}, \mathbf{w}^{(5)} \}
\end{aligned}
\tag{7.135}
$$

are estimated by minimizing the sum-of-squares error function between clean and estimated spectra

$$
\mathcal{L}(\boldsymbol{\Theta}) = \frac{1}{2} \sum_{t=1}^{T} \bigg(\underbrace{||\mathbf{x}_{1,t} - \widehat{\mathbf{x}}_{1,t}|| + ||\mathbf{x}_{2,t} - \widehat{\mathbf{x}}_{2,t}||}_{\mathcal{L}_{\mathrm{enc}}(\boldsymbol{\Theta})}
$$

$$
+ \underbrace{||\mathbf{x}_{1,t} - \widetilde{\mathbf{x}}_{1,t}|| + ||\mathbf{x}_{2,t} - \widetilde{\mathbf{x}}_{2,t}||}_{\mathcal{L}_{\mathrm{sep}}(\boldsymbol{\Theta})} \bigg).
\tag{7.136}
$$

This loss function is not only calculated from separation outputs $\{\widehat{\mathbf{x}}_{1,t}(\boldsymbol{\Theta}), \widehat{\mathbf{x}}_{2,t}(\boldsymbol{\Theta})\}$ but also from the encoder outputs $\{\widetilde{\mathbf{x}}_{1,t}(\boldsymbol{\Theta}), \widetilde{\mathbf{x}}_{2,t}(\boldsymbol{\Theta})\}$.

7.7.3 SYSTEM EVALUATION

The task of single-channel speech enhancement was performed for a comparative study. Experimental setup is the same as that used in the neural Turing machine (NTM) for source separation in Section 7.6.3. To evaluate the robustness of different methods, the noisy training data were generated by SNR of -5 dB while the noisy test data were generated by SNR of -5, 0 and 5 dB with two unseen noise types (cafeteria and bus). The mixed signal was expressed by $\mathbf{x}_t^{\mathrm{mix}} \in \mathbb{R}^{513}$. DNN, LSTM and NTM were carried out by referring (Tsou and Chien, 2017). The NTM with memory size $N = 32$ was examined. Using RCNN, the encoder and decoder with topologies 513–1000 (FC)–800 (BLSTM)–600 (FC)–{513–513} and 513–1000 (FC)–800 (LSTM)–700 (LSTM)–600 (FC)–{513–513} were implemented, respectively. Speech enhancement was evaluated by using STOI.

Table 7.4 reports the STOIs of using DNN, LSTM, NTM and RCNN under different SNRs of test samples. STOI is increased by applying DNN, LSTM, NTM and RCNN. Different RNN variants using LSTM, NTM and RCNN outperform DNN. NTM consistently performs better than LSTM under different SNRs. The improvement of RCNN over NTM is obvious in different SNRs. This is partially because RCNN is *not* constrained by memory size. RCNN stores the information without erasing and adding. To evaluate different functions in RCNN, three simplified variants are realized for comparison. RCNN without attending decoder memory $\mathbf{M}_d$ corresponds to the RCNN without using context vector $\mathbf{c}_t^d$. RCNN without $\mathcal{L}_{\mathrm{enc}}$ expresses the RCNN without including loss function from encoder. RCNN can be also reduced by disregarding both $\mathbf{c}_t^d$ and $\mathcal{L}_{\mathrm{enc}}$. Attending the decoder memory and including the encoder loss do improve the separation performance. Especially, including encoder loss enables RCNN to speed up convergence while attending the decoder memory helps little in terms of STOI. The information in context vector $\mathbf{c}_t^d$ contains redundancy. RCNN without considering $\mathbf{c}_t^d$ and $\mathcal{L}_{\mathrm{enc}}$ consistently worse than the other RCNN variants.

Table 7.4 Comparison of STOIs under different SNRs by using DNN, LSTM and different variants of NTM and RCNN

Model	−5 dB	0 dB	5 dB
Unprocessed	0.647	0.746	0.829
DNN	0.678	0.785	0.844
LSTM	0.718	0.815	0.872
NTM	0.731	0.825	0.877
RCNN w/o $\mathbf{c}_t^d$ & $\mathcal{L}_{enc}$	0.738	0.833	0.883
RCNN w/o $\mathcal{L}_{enc}$	0.742	0.835	0.887
RCNN w/o $\mathbf{c}_t^d$	**0.747**	**0.838**	0.886
RCNN	0.744	0.837	**0.888**

7.8 SUMMARY

This chapter has presented a variety of deep learning methods, expanding from a deep neural network to recurrent neural network, long short-term memory, deep recurrent neural network, deep long short-term memory, bidirectional long short-term memory, neural Turing machine and end-to-end memory network. Signal processing methods including Tucker decomposition and spectral masking were incorporated in deep learning for preserving the structural information and estimating the separated spectra, respectively. Spectro-temporal factorization preserves two-way features for estimation of mask function. The learning solutions to different kinds of mask function were developed for monaural source separation. The mask function was merged in training for speech separation. Speech separation was developed for generalization over different speakers and noise types. Speaker-independent source separation was performed. Deep clustering was introduced to learn the embedding and masking for multiple speakers. Speech separation was improved by using the alignment information from speech recognizer. Joint training of speech separation and recognition was recommended. Permutation-invariant training was further presented to improve system performance by minimizing the regression and permutation error functions due to the mask function and permutation ambiguity. Ensemble learning and discriminative learning were merged in a deep model to acquire generalization information and discriminative evidence to carry out deep machine learning for source separation. The ensemble learning using multicontext averaging and multicontext stacking were presented. The discriminative learning objectives based on the between-source likelihood and the mutual information between the reconstructed and true spectra were described. In addition to monaural speech separation and speech enhancement, deep learning methods were also developed for reverberant source separation and speech segregation. Soft and binary mask functions were introduced in speech separation and segregation, respectively. Long short-term memory was built in a stacked deep model to sufficiently capture temporal information for speaker generalization in monaural source separation. Bidirectional long short-term memory was successfully explored for a joint framework of speech separation and recognition. Modern sequential learning methods based on variational autoencoder and memory augmented neural network. The implementations of variational recurrent neural network, neural Turing machine and end-to-end memory network were developed for monaural source separation. The randomness in latent deep model was reflected and implemented via variational inference and a sampling scheme. The external memorization was incorporated in a recurrent neural network based on a recurrent controller and an addressing mechanism. Such a memorability could be also realized via embedding in an end-to-end manner through a two-pass sensing or listening procedure, which mimics the human auditory system to substantially improve short-time objective intelligence of speech enhancement. Sequence-to-sequence learning was implemented based on an encoder–decoder-separator. An attention mechanism was carried out to select the desirable memory slots for demixing a speech frame. The gating mechanism from two complimentary memories was built to fuse various information sources to benefit the performance for monaural source separation. Different types of error backpropagation algorithm were developed to implement the above-mentioned deep learning models from different perspectives.

In the next chapter, we will summarize different machine learning paradigms for source separation and point out the capabilities and limitations when developing and applying various approaches to different scenarios and issues in source separation. We will also address future remarks or research trends in signal processing and machine learning to deal with different settings of source separation and achieve desirable performance in terms of different metrics which measure intelligence objectives from a variety of perspectives.

SUMMARY AND FUTURE TRENDS

8

At last, we summarize various learning-based source separation methods and point out possible directions towards future studies and developments for source separation. The newest progresses on deep sequential learning are tremendous. This chapter emphasizes these progresses and summarizes a variety of powerful learning machines, which meet various conditions in source separation to achieve precise awareness of mixing conditions and strong capabilities of representation, memorization, reconstruction, masking, decoding and learning. The following sections summarize different learning theories and models for source separation and presents a number of potential directions to pursue state-of-the-art performance in future practical systems.

8.1 MACHINE LEARNING FOR SOURCE SEPARATION

Source separation is seen as a joint framework of signal processing and machine learning, which ranges from front-end processing of signal inputs with the mixed signals to back-end learning of regression outputs for the demixed signals. The importance and challenges in speech and audio source separation have been attracting numerous researchers exerting tremendous efforts towards this research trend. Such an emerging trend has been sufficiently reflected in recent years due to the rapidly increasing population in the areas of source separation in two major conferences, including the *IEEE International Conference on Acoustics, Speech and Signal Processing* (ICASSP) and the *Annual Conference of International Speech Communication Association* (INTERSPEECH). The number of accepted papers and attendees in the areas of speech enhancement, audio and source separation are comparable with those of speech recognition. Source separation and speech recognition have been recognized as two of the most popular research directions in ICASSP and INTERSPEECH. In the activities of the IEEE Signal Processing Society, source separation papers have been intensively proposed in recent years in three technical directions, including *Speech and Language Processing* (SLP), *Machine Learning for Signal Processing* (MLSP) and *Audio and Acoustic Signal Processing* (AASP). Many references of this book are authored by various past and current members of technical committees of SLP, MLSP and AASP. The impacts of machine learning approaches to source separation problems have been influential in signal processing community for many years. Ubiquitous applications and systems developed by Google, Amazon, Apple and many others have been widely deployed and currently affect our daily life.

This book has surveyed a number of fundamental theories and advanced practices, which pave an avenue to deal with different challenges and build various capacities for heterogeneous source separation. We have introduced source separation applications and focused on the separation tasks for instrument, singing-voice and speech/music separation. Adaptive signal processing was first addressed for front-end processing where we considered the analysis of information on each source, described the time–frequency modeling and masking, and identified the types of a mixing system. More importantly, considering the model-based source separation, we emphasized the importance of the machine learning perspective and illustrated how source separation was tackled through adaptive learning algorithms and model-based approaches. In back-end learning, we only used the information about mixture signals to build a source separation model, which was seen as a statistical model for the whole system. The inference procedure was implemented to construct such a model from a set of observed audio samples. This book highlighted the fundamentals and advances in back-end learning for source separation which are summarized as follows.

- Fundamental theories – We systematically organized and surveyed various machine learning approaches to source separation according to two categories. One is the separation models and the other is the learning algorithms. In separation models, we presented modern source separation models consisting of the independent component analysis (ICA), nonnegative matrix factorization (NMF) or nonnegative tensor factorization (NTF), and deep neural network (DNN) or recurrent neural network (RNN). We addressed how these models evolved to deal with multichannel and single-channel source separation, and explained the weakness and the strength of these models in different mixing conditions. A number of extensions of different models with convolutive processing, recurrent modeling, long short-term memory and probabilistic interpretation have been introduced. In learning algorithms, we addressed a series of adaptive learning algorithms for source separation including the information-theoretic, Bayesian, sparse, online, discriminative and deep ensemble learning. We illustrated how the information theory is developed for optimization of ICA and NMF objective functions and why the uncertainty and the sparsity of system parameters should be faithfully characterized in different separation models to assure the robustness of source separation against adverse environments. The approximate Bayesian inference procedures using variational inference and Markov chain Monte Carlo were introduced. The dynamics in source signals and sensor networks were captured via the online learning and tracking. The merits of discriminative learning for individual sources and deep learning over the complicated mixing system were demonstrated.
- Advanced practices – This book has presented a number of advanced studies, which were recently published for learning-based or model-based source separation. These studies are categorized based on three modern models: ICA, NMF or NTF, and DNN or RNN. A variety of learning algorithms are employed in these models so as to carry out different solutions to tackle different issues in the construction of source separation systems. Various models and algorithms are connected with potential extensions. In the ICA model, we introduced the independent component space for maximizing the information redundancy reduction, the nonparametric likelihood ratio ICA and the convex divergence ICA for multichannel source separation, the nonstationary Bayesian ICA and the online Gaussian process ICA for nonstationary speech and music separation. In the NMF or NTF model, we surveyed the convolutive NMF, probabilistic nonnegative factorization, Bayesian NMF, group sparse NMF, discriminative NMF, deep NMF, multichannel NMF, deep multiway factorization, convolutive NTF, infinite NTF, and stereo channel NTF, which were implemented in different scenarios for source separation. In the DNN or RNN model for single-channel source separation, we started from the baseline DNN for estimation of a soft mask function and then extended it to use it for the deep clustering and embedding, deep RNN (DRNN), DRNN with long short-term memory (LSTM), bidirectional LSTM, permutation invariant training, variational RNN, memory-augmented RNN, and end-to-end RNN.

8.2 POTENTIAL TOPICS AND DIRECTIONS

A number of directions and outlooks are pointed out for future studies. The potential topics for future studies are summarized in the following categories:

- Integrated solutions – Pure machine learning algorithms could not entirely stand alone to deal with various problems in a source separation system. Back-end machine learning needs to be seamlessly combined with front-end signal processing. The in-domain knowledge is introduced to carry out a specialized solution to high-performance source separation. The multidisciplinary approaches based on signal processing and machine learning are required. For instance, the regression error function can be adjusted to develop a perceptually meaningful objective function so as to learn a separation model to achieve the goal in terms of the interest of measures. It is crucial to optimize the learning objective, which directly reflects the performance for source separation. In the speech enhancement task, the learning objective based on a short-time objective intelligibility measure can be applied (Kolbæk et al., 2018). From the perspective of machine learning, source separation models based on the combined separation and classification and augmented with discriminative training can be established.

- Challenging tasks – Creating a source separation system under heterogeneous conditions is challenging in real-world applications. This direction has been attracting increased attention by scientists and engineers. The challenging problems in source separation range from the temporally-correlated sources to the nonstationary mixing conditions, the adaptive model complexity, as well as the unknown sources or confusing speakers. The speaker-independent source separation is helpful. To tackle these challenges, source separation can be guided with user interaction and side information (Vincent et al., 2014). Ubiquitous extensions and applications could be developed on the basis of what we have presented in different chapters. The source separation tasks are even challenging when extended to cope with the separation of observation data in the presence of multimodalities, multimodels and multiways. Universal models using various speakers, languages and noise conditions have impact on the development of future speech separation systems where the test conditions are totally unknown. A generalized model, which covers a variety of mixing conditions, acoustic signals and ambient noises, may be trained from abundant speech and noise signals under various mixing scenarios. Domain adaptation methods can be incorporated to adapt a source separation model to a specific environment.

- Deep sequential learning – Deep learning has been successfully developed and intensively investigated for source separation in recent years. Again, deep learning cannot work well if the front-end signal processing and in-domain information representation are not clearly taken into account. Sequence-to-sequence learning is performed to implement a state-of-the-art learning machine for source separation. The attention mechanism and the gating calculation should be naturally embedded in the implementation procedure. To reveal a complicated mapping between mixed and source signals, the deep structure of a separation model can be sufficiently deep while the tricks of residual computation and skip connection are required to alleviate the gradient vanishing problem in a very deep model. In addition, modern deep learning methods have potential in strengthening the generation of demixed signals and improving the perceptual precision of audio signals in source separation. For example, a generative adversarial network (GAN) (Goodfellow et al., 2014) has been extensively studied in the machine learning community. GAN is composed of a generator and a discriminator. The power of GAN is in estimating a competitive generative model where it is hard to judge if the generated sample is fake or true. Adversarial training is implemented applying minimax optimization for finding parameters in the generator and discriminator. The source separation model can be optimized via GAN (Sübakan and Smaragdis, 2018) to assure high quality of the demixed signals. However, the incorporation of signal processing and masking methods is required in system implementation. On the other hand, the study of end-to-end learning is nowadays a trend in different research areas. Developing an end-to-end learning procedure based on various deep learning modules is an attractive direction for future studies. For example, speech separation and recognition can be jointly optimized under consistent learning modules and optimization objectives (Settle et al., 2018). The perspective of end-to-end learning is also useful in construction of source separation combined with other applications.

BASIC FORMULAS

This appendix lists some basic formulas and useful vector, matrix and tensor formulas. Note that these formulas are selected for the purpose of deriving equations in this book, and do not cover the whole vector, matrix and tensor formulas.

A.1 EXPECTATION

$$\mathbb{E}_{(a)}\left[f(a) + g(a)\right] = \mathbb{E}_{(a)}\left[f(a)\right] + \mathbb{E}_{(a)}\left[g(a)\right] \tag{A.1}$$

$$\mathbb{E}_{(a)}\left[bf(a)\right] = b\mathbb{E}_{(a)}\left[f(a)\right] \tag{A.2}$$

A.2 JENSEN'S INEQUALITY

For a concave function f, X is a probabilistic random variable sampled from a distribution function, and an arbitrary function $g(X)$, we have the following inequality:

$$f\left(\mathbb{E}_{(X)}[g(X)]\right) \geq \mathbb{E}_{(X)}[f(g(X))]. \tag{A.3}$$

In the special case of a concave function $f(\cdot) = \log(\cdot)$, (A.3) is rewritten as follows:

$$\log\left(\mathbb{E}_{(X)}[g(X)]\right) \geq \mathbb{E}_{(X)}[\log(g(X))]. \tag{A.4}$$

A.3 GAMMA FUNCTION

$$\Gamma(x) \triangleq \int_0^\infty t^{x-1}e^{-t}dt \tag{A.5}$$

and

$$\Gamma(x+1) = x\Gamma(x), \tag{A.6}$$

$$\Gamma\left(\frac{1}{2}\right) = \pi^{\frac{1}{2}}. \tag{A.7}$$

A.4 TRACE

$$\text{tr}[a] = a \tag{A.8}$$

$$\text{tr}[\mathbf{ABC}] = \text{tr}[\mathbf{BCA}] = \text{tr}[\mathbf{CAB}] \tag{A.9}$$

$$\text{tr}[\mathbf{A} + \mathbf{B}] = \text{tr}[\mathbf{A}] + \text{tr}[\mathbf{B}] \tag{A.10}$$

$$\text{tr}[\mathbf{A}^\mathsf{T}] = \text{tr}[\mathbf{A}] \tag{A.11}$$

$$\text{tr}[\mathbf{A}(\mathbf{B} + \mathbf{C})] = \text{tr}[\mathbf{AB} + \mathbf{AC}] \tag{A.12}$$

A.5 TRANSPOSE

$$(\mathbf{ABC})^\mathsf{T} = \mathbf{C}^\mathsf{T}\mathbf{B}^\mathsf{T}\mathbf{A}^\mathsf{T} \tag{A.13}$$

$$(\mathbf{A} + \mathbf{B})^\mathsf{T} = \mathbf{A}^\mathsf{T} + \mathbf{B}^\mathsf{T} \tag{A.14}$$

A.6 DERIVATIVE

$$\frac{\partial \log |\mathbf{A}|}{\partial \mathbf{A}} = (\mathbf{A}^\mathsf{T})^{-1} \tag{A.15}$$

$$\frac{\partial \mathbf{a}^\mathsf{T}\mathbf{b}}{\partial \mathbf{a}} = \frac{\partial \mathbf{b}^\mathsf{T}\mathbf{a}}{\partial \mathbf{a}} = \mathbf{b} \tag{A.16}$$

$$\frac{\partial \mathbf{a}^\mathsf{T}\mathbf{Cb}}{\partial \mathbf{C}} = \mathbf{ab}^\mathsf{T} \tag{A.17}$$

$$\frac{\partial \text{tr}(\mathbf{AB})}{\partial \mathbf{A}} = \mathbf{B}^\mathsf{T} \tag{A.18}$$

$$\frac{\partial \text{tr}(\mathbf{ACB})}{\partial \mathbf{C}} = \mathbf{A}^\mathsf{T}\mathbf{B}^\mathsf{T} \tag{A.19}$$

$$\frac{\partial \text{tr}(\mathbf{AC}^{-1}\mathbf{B})}{\partial \mathbf{C}} = -(\mathbf{C}^{-1}\mathbf{BAC}^{-1})^\mathsf{T} \tag{A.20}$$

$$\frac{\partial \mathbf{a}^\mathsf{T}\mathbf{Ca}}{\partial \mathbf{a}} = (\mathbf{C} + \mathbf{C}^\mathsf{T})\mathbf{a} \tag{A.21}$$

$$\frac{\partial \mathbf{b}^\mathsf{T}\mathbf{A}^\mathsf{T}\mathbf{DAc}}{\partial \mathbf{A}} = \mathbf{D}^\mathsf{T}\mathbf{Abc}^\mathsf{T} + \mathbf{DAcb}^\mathsf{T} \tag{A.22}$$

A.7 COMPLETE SQUARE

When $\mathbf{A}$ is a symmetric matrix, we have

$$\mathbf{x}^\mathsf{T}\mathbf{A}\mathbf{x} - 2\mathbf{x}^\mathsf{T}\mathbf{b} + c = (\mathbf{x} - \mathbf{u})^\mathsf{T} \mathbf{A} (\mathbf{x} - \mathbf{u}) + v \tag{A.23}$$

where

$$\begin{aligned} \mathbf{u} &\triangleq \mathbf{A}^{-1}\mathbf{b}, \\ v &\triangleq c - \mathbf{b}^\mathsf{T}\mathbf{A}^{-1}\mathbf{b}. \end{aligned} \tag{A.24}$$

By using the above complete square formula, we can also derive the following formula when matrices $\mathbf{A}_1$ and $\mathbf{A}_2$ are symmetric:

$$\begin{aligned} &(\mathbf{x} - \mathbf{b}_1)^\mathsf{T}\mathbf{A}_1(\mathbf{x} - \mathbf{b}_1) + (\mathbf{x} - \mathbf{b}_2)^\mathsf{T}\mathbf{A}_2(\mathbf{x} - \mathbf{b}_2) \\ &= \mathbf{x}^\mathsf{T} \underbrace{(\mathbf{A}_1 + \mathbf{A}_2)}_{\triangleq \mathbf{A}} \mathbf{x} - 2\mathbf{x}^\mathsf{T} \underbrace{(\mathbf{A}_1\mathbf{b}_1 + \mathbf{A}_2\mathbf{b}_2)}_{\triangleq \mathbf{b}} + \underbrace{\mathbf{b}_1^\mathsf{T}\mathbf{A}_1\mathbf{b}_1 + \mathbf{b}_2^\mathsf{T}\mathbf{A}_2\mathbf{b}_2}_{\triangleq c} \\ &= (\mathbf{x} - \mathbf{u})^\mathsf{T} (\mathbf{A}_1 + \mathbf{A}_2) (\mathbf{x} - \mathbf{u}) + v \end{aligned} \tag{A.25}$$

where

$$\begin{aligned} \mathbf{u} &= (\mathbf{A}_1 + \mathbf{A}_2)^{-1}(\mathbf{A}_1\mathbf{b}_1 + \mathbf{A}_2\mathbf{b}_2) \\ v &= \mathbf{b}_1^\mathsf{T}\mathbf{A}_1\mathbf{b}_1 + \mathbf{b}_2^\mathsf{T}\mathbf{A}_2\mathbf{b}_2 - (\mathbf{A}_1\mathbf{b}_1 + \mathbf{A}_2\mathbf{b}_2)^\mathsf{T}(\mathbf{A}_1 + \mathbf{A}_2)^{-1}(\mathbf{A}_1\mathbf{b}_1 + \mathbf{A}_2\mathbf{b}_2). \end{aligned} \tag{A.26}$$

A.8 TENSOR ALGEBRA

- Norm of a given tensor $\mathcal{X} \in \mathbb{R}^{I_1 \times I_2 \times \cdots \times I_N}$ is the square root of the sum of the square of elements

$$\begin{aligned} \|\mathcal{X}\| &= \sqrt{\sum_{i_1} \sum_{i_2} \cdots \sum_{i_N} x_{i_1 i_2 \cdots i_N}^2} \\ &= \sqrt{\langle \mathcal{X}, \mathcal{X} \rangle}. \end{aligned} \tag{A.27}$$

- Inner product of two tensors $\mathcal{X}, \mathcal{Y} \in \mathbb{R}^{I_1 \times I_2 \times \cdots \times I_N}$ is the sum of element-wise product between their entries

$$\langle \mathcal{X}, \mathcal{Y} \rangle = \sum_{i_1} \sum_{i_2} \cdots \sum_{i_N} x_{i_1 i_2 \cdots i_N} \, y_{i_1 i_2 \cdots i_N}. \tag{A.28}$$

- If a tensor $\mathcal{X} \in \mathbb{R}^{I_1 \times I_2 \times \cdots \times I_N}$ can be written as the outer product of several vectors, then we called it a rank-one tensor

$$\mathcal{X} = \mathbf{u}^{(1)} \circ \mathbf{u}^{(2)} \circ \cdots \circ \mathbf{u}^{(N)} \tag{A.29}$$

where $\circ$ denotes the outer product.

- If a tensor $\mathcal{X} \in \mathbb{R}^{I_1 \times I_2 \times \cdots \times I_N}$ only has values whenever $i_1 = i_2 = \cdots = i_N$, other elements being all zeros, it's a diagonal tensor.
- The mode-n (matrix) product of a tensor $\mathcal{X} \in \mathbb{R}^{I_1 \times I_2 \times \cdots \times I_N}$ by a matrix $\mathbf{U} \in \mathbb{R}^{J \times I_n}$ is denoted by $\mathcal{X} \times_n \mathbf{U}$. Each mode-$n$ fiber is multiplied by the matrix $\mathbf{U}$. The equation can be expressed as

$$(\mathcal{X} \times_n \mathbf{U})_{i_1 \cdots i_{n-1} j i_{n+1} \cdots i_N} = \sum_{i_n=1}^{I_n} x_{i_1 i_2 \cdots i_N} u_{j i_n}. \tag{A.30}$$

PROBABILISTIC DISTRIBUTION FUNCTIONS

This appendix lists the probabilistic distribution functions (pdfs) used in the main discussions in this book. Each pdf section also provides values of mean, mode, variance, etc., which will be used in the book, although it does not include a complete set of distribution values to avoid too complicated appendix. The section partially provides some derivations of these values.

B.1 MULTIVARIATE GAUSSIAN DISTRIBUTION

- pdf (with covariance matrix $\boldsymbol{\Sigma}$):

$$
\begin{aligned}
\mathcal{N}(\mathbf{x}|\boldsymbol{\mu}, \boldsymbol{\Sigma}) &\triangleq C_{\mathcal{N}}(\boldsymbol{\Sigma}) \exp\left(-\frac{1}{2}(\mathbf{x}-\boldsymbol{\mu})^{\mathsf{T}}\boldsymbol{\Sigma}^{-1}(\mathbf{x}-\boldsymbol{\mu})\right) \\
&= C_{\mathcal{N}}(\boldsymbol{\Sigma}) \exp\left(-\frac{1}{2}\mathrm{tr}\left[\boldsymbol{\Sigma}^{-1}(\mathbf{x}-\boldsymbol{\mu})(\mathbf{x}-\boldsymbol{\mu})^{\mathsf{T}}\right]\right)
\end{aligned}
\tag{B.1}
$$

where

$$
\mathbf{x} \in \mathbb{R}^D
\tag{B.2}
$$

and

$$
\boldsymbol{\mu} \in \mathbb{R}^D, \quad \boldsymbol{\Sigma} \in \mathbb{R}^{D \times D},
\tag{B.3}
$$

$\boldsymbol{\Sigma}$ being positive definite.
- Normalization constant (with covariance matrix $\boldsymbol{\Sigma}$):

$$
C_{\mathcal{N}}(\boldsymbol{\Sigma}) \triangleq (2\pi)^{-\frac{D}{2}}|\boldsymbol{\Sigma}|^{-\frac{1}{2}}
\tag{B.4}
$$

- pdf (with precision matrix $\mathbf{R}$):

$$
\begin{aligned}
\mathcal{N}(\mathbf{x}|\boldsymbol{\mu}, (\mathbf{R})^{-1}) &\triangleq C_{\mathcal{N}}(\mathbf{R}^{-1}) \exp\left(-\frac{1}{2}(\mathbf{x}-\boldsymbol{\mu})^{\mathsf{T}}\mathbf{R}(\mathbf{x}-\boldsymbol{\mu})\right) \\
&= C_{\mathcal{N}}(\mathbf{R}^{-1}) \exp\left(-\frac{1}{2}\mathrm{tr}\left[\mathbf{R}(\mathbf{x}-\boldsymbol{\mu})(\mathbf{x}-\boldsymbol{\mu})^{\mathsf{T}}\right]\right)
\end{aligned}
\tag{B.5}
$$

- Normalization constant (with precision matrix $\mathbf{R}$):

$$C_{\mathcal{N}}(\mathbf{R}^{-1}) \triangleq (2\pi)^{-\frac{D}{2}} |\mathbf{R}|^{\frac{1}{2}} \tag{B.6}$$

- Mean:

$$\mathbb{E}_{(\mathbf{x})}[\mathbf{x}] = \boldsymbol{\mu} \tag{B.7}$$

- Variance:

$$\mathbb{E}_{(\mathbf{x})}[(\mathbf{x} - \boldsymbol{\mu})(\mathbf{x} - \boldsymbol{\mu})^{\mathsf{T}}] = \boldsymbol{\Sigma} = \mathbf{R}^{-1} \tag{B.8}$$

- Mode:

$$\boldsymbol{\mu} \tag{B.9}$$

B.2 LAPLACE DISTRIBUTION

- pdf:

$$\mathrm{Lap}(x|\mu, \beta) \triangleq C_{\mathrm{Lap}}(\beta) \exp\left(-\frac{|x - \mu|}{\beta}\right)$$

$$= C_{\mathrm{Lap}}(\beta) \begin{cases} \exp\left(-\frac{x-\mu}{\beta}\right) & \text{if } x \geq \mu, \\ \exp\left(-\frac{\mu-x}{\beta}\right) & \text{if } x < \mu \end{cases} \tag{B.10}$$

where

$$x \in \mathbb{R} \tag{B.11}$$

and

$$\mu \in \mathbb{R}, \quad \beta > 0. \tag{B.12}$$

- Normalization constant:

$$C_{\mathrm{Lap}}(\beta) \triangleq \frac{1}{2\beta} \tag{B.13}$$

- Mean:

$$\mathbb{E}_{(x)}[x] = \mu \tag{B.14}$$

- Mode:

$$\mu \tag{B.15}$$

Note that the probabilistic distribution function is not continuous and differentiable at μ.

B.3 STUDENT'S *t*-DISTRIBUTION

- pdf:

$$\text{St}(x|\mu, \lambda, \kappa) \triangleq C_{\text{St}} \left(1 + \frac{1}{\kappa\lambda}(x - \mu)^2\right)^{-\frac{\kappa+1}{2}} \tag{B.16}$$

where

$$x \in \mathbb{R} \tag{B.17}$$

and

$$\mu \in \mathbb{R}, \quad \kappa > 0, \quad \lambda > 0. \tag{B.18}$$

- Normalization constant:

$$C_{\text{St}}(\kappa, \lambda) \triangleq \frac{\Gamma\left(\frac{\kappa+1}{2}\right)}{\Gamma\left(\frac{\kappa}{2}\right)\Gamma\left(\frac{1}{2}\right)}\left(\frac{1}{\kappa\lambda}\right)^{\frac{1}{2}} \tag{B.19}$$

- Mean:

$$\mathbb{E}_{(x)}[x] = \mu \tag{B.20}$$

- Mode:

$$\mu \tag{B.21}$$

Parameter κ is called the number of degrees of freedom, and if κ is large, the distribution approaches a Gaussian distribution.

B.4 GAMMA DISTRIBUTION

A Gamma distribution is used as a prior/posterior distribution of precision parameter r of a Gaussian distribution.

- pdf:

$$\text{Gam}(x|\alpha, \beta) \triangleq C_{\text{Gam}}(\alpha, \beta)x^{\alpha-1}\exp(-\beta x) \tag{B.22}$$

where

$$x > 0 \tag{B.23}$$

and

$$\alpha, \beta > 0. \tag{B.24}$$

- Normalization constant:

$$C_{\mathrm{Gam}}(\alpha, \beta) \triangleq \frac{\beta^\alpha}{\Gamma(\alpha)} \tag{B.25}$$

- Mean:

$$\mathbb{E}_{(x)}[x] = \frac{\alpha}{\beta} \triangleq \mu \tag{B.26}$$

- Variance:

$$\mathbb{E}_{(x)}[(x - \mu)^2] = \frac{\alpha}{\beta^2} \tag{B.27}$$

- Mode:

$$\frac{\alpha - 1}{\beta}, \quad \alpha > 1 \tag{B.28}$$

which is derived from

$$\frac{d}{dx}\mathrm{Gam}(x|\alpha, \beta)$$
$$= C_{\mathrm{Gam}}(\alpha, \beta)(\alpha - 1 - \beta x)x^{\alpha - 2}\exp(-\beta x) = 0. \tag{B.29}$$

The shape of the Gamma distribution is not symmetric, and the mode and mean values are different.

To make the notation consistent with the Wishart distribution in Section B.5, we also use the following definition for the Gamma distribution by $\alpha \to \frac{\phi}{2}$ and $\beta \to \frac{r^0}{2}$ in the original Gamma distribution defined in Eq. (B.22):

- pdf (with $\frac{1}{2}$ factor):

$$\mathrm{Gam}_2(x|\phi, r^0) \triangleq \mathrm{Gam}\left(x \left| \frac{\phi}{2}, \frac{r^0}{2}\right.\right)$$
$$= C_{\mathrm{Gam}_2}\left(\phi, r^0\right)x^{\frac{\phi}{2}-1}\exp\left(-\frac{r^0 x}{2}\right) \tag{B.30}$$

- Normalization constant (with $\frac{1}{2}$ factor):

$$C_{\mathrm{Gam}_2}(\phi, r^0) \triangleq \frac{\left(\frac{r^0}{2}\right)^{\frac{\phi}{2}}}{\Gamma\left(\frac{\phi}{2}\right)} \tag{B.31}$$

- Mean (with $\frac{1}{2}$ factor):

$$\mathbb{E}_{(x)}[x] = \frac{\frac{\phi}{2}}{\frac{r^0}{2}} = \frac{\phi}{r^0} \triangleq \mu \tag{B.32}$$

- Variance (with $\frac{1}{2}$ factor):

$$\mathbb{E}_{(x)}[(x - \mu)^2] = \frac{\frac{\phi}{2}}{\left(\frac{r^0}{2}\right)^2} = \frac{2\phi}{\left(r^0\right)^2} \tag{B.33}$$

- Mode (with $\frac{1}{2}$ factor):

$$\frac{\frac{\phi}{2} - 1}{\frac{r^0}{2}} = \frac{\phi - 2}{r^0}, \quad \phi > 2 \tag{B.34}$$

It is shown in Section B.5 that this Gamma distribution ($\text{Gam}_2(\cdot)$) with $\frac{1}{2}$ factor is equivalent to the Wishart distribution when the number of dimensions is 1 ($D = 1$).

B.5 WISHART DISTRIBUTION

- pdf:

$$\mathcal{W}(\mathbf{X}|\phi, \mathbf{R}^0) \triangleq C_{\mathcal{W}}(\phi, \mathbf{R}^0)|\mathbf{X}|^{\frac{\phi - D - 1}{2}} \exp\left(-\frac{1}{2}\text{tr}\left[\mathbf{R}^0\mathbf{X}\right]\right) \tag{B.35}$$

where

$$\mathbf{X} \in \mathbb{R}^{D \times D} \tag{B.36}$$

and

$$\mathbf{R}^0 \in \mathbb{R}^{D \times D}, \quad \phi > D - 1, \tag{B.37}$$

$\mathbf{X}$ and $\mathbf{R}^0$ being positive definite.
- Normalization constant:

$$C_{\mathcal{W}}(\phi, \mathbf{R}^0) \triangleq \frac{|\mathbf{R}^0|^{\frac{\phi}{2}}}{(2)^{\frac{D\phi}{2}} \Gamma_D\left(\frac{\phi}{2}\right)} \tag{B.38}$$

where $\Gamma_D(\cdot)$ is a multivariate Gamma function.
- Mean:

$$\mathbb{E}_{(\mathbf{X})}[\mathbf{X}] = \phi(\mathbf{R}^0)^{-1} \tag{B.39}$$

- Mode:

$$(\phi - D - 1)(\mathbf{R}^0)^{-1}, \quad \phi \geq D + 1 \tag{B.40}$$

When $D \to 1$, the Wishart distribution becomes equivalent to the Gamma distribution defined in Eq. (B.30).

B.6 POISSON DISTRIBUTION

- pdf:

$$\text{Pois}(x|\theta) = \frac{\theta^x e^{-\theta}}{\Gamma(x+1)} \tag{B.41}$$
$$= \exp(x \log \theta - \theta - \log \Gamma(x+1))$$

where

$$x \geq 0 \tag{B.42}$$

and

$$\theta > 0. \tag{B.43}$$

- Mean:

$$\mathbb{E}_{(x)}[x] = \theta \triangleq \mu \tag{B.44}$$

- Variance:

$$\mathbb{E}_{(x)}[(x - \mu)^2] = \theta \tag{B.45}$$

- Mode:

$$\lfloor \theta \rfloor, \tag{B.46}$$

which means the floor value of θ.

B.7 EXPONENTIAL DISTRIBUTION

- pdf:

$$\text{Exp}(x; \theta) = \theta e^{-\theta x} \tag{B.47}$$
$$= \theta \exp(-\theta x)$$

where

$$x \geq 0, \tag{B.48}$$
$$\theta > 0. \tag{B.49}$$

- Mean:

$$\mathbb{E}_{(x)}[x] = \theta^{-1} \triangleq \mu \tag{B.50}$$

- Variance:

$$\mathbb{E}_{(x)}[(x - \mu)^2] = \theta^{-2} \qquad \text{(B.51)}$$

- Mode:

$$0 \qquad \text{(B.52)}$$

Bibliography

Ahmed, A., Andrieu, C., Doucet, A., Rayner, P.J.W., 2000. On-line non-stationary ICA using mixture models. In: Proc. of IEEE International Conference on Acoustics, Speech, and Signal Processing (ICASSP), vol. 5, pp. 3148–3151.

Amari, S., 1985. Differential–Geometrical Methods in Statistics. Springer.

Amari, S., 1998. Natural gradient works efficiently in learning. Neural Computation 10, 251–276.

Anderson, T.W., 1984. An Introduction to Multivariate Statistical Analysis. John Wiley & Sons, Inc.

Araki, S., Mukai, R., Makino, S., Nishikawa, T., Saruwatari, H., 2003. The fundamental limitation of frequency domain blind source separation for convolutive mixtures of speech. IEEE Transactions on Speech and Audio Processing 11 (2), 109–116.

Araki, S., Nakatani, T., Sawada, H., Makino, S., 2009a. Blind sparse source separation for unknown number of sources using Gaussian mixture model fitting with Dirichlet prior. In: Proc. of IEEE International Conference on Acoustics, Speech, and Signal Processing (ICASSP), pp. 33–36.

Araki, S., Nakatani, T., Sawada, H., Makino, S., 2009b. Stereo source separation and source counting with MAP estimation with Dirichlet prior considering spatial aliasing problem. In: Proc. of International Symposium on Independent Component Analysis and Blind Signal Separation (ICA), pp. 742–750.

Araki, S., Sawada, H., Mukai, R., Makino, S., 2007. Underdetermined blind sparse source separation for arbitrarily arranged multiple sensors. Signal Processing 87, 1833–1847.

Attias, H., 1999. Inferring parameters and structure of latent variable models by variational Bayes. In: Proc. of Conference on Uncertainty in Artificial Intelligence (UAI), pp. 21–30.

Bach, F.R., Jordan, M.I., 2003. Finding clusters in independent component analysis. In: Proc. of International Symposium on Independent Component Analysis and Blind Signal Processing (ICA), pp. 891–896.

Bahdanau, D., Cho, K., Bengio, Y., 2014. Neural machine translation by jointly learning to align and translate. arXiv preprint, arXiv:1409.0473.

Barker, T., Virtanen, T., 2013. Non-negative tensor factorisation of modulation spectrograms for monaural sound source separation. In: Proc. of Annual Conference of International Speech Communication Association (INTERSPEECH), pp. 827–831.

Bartlett, M.S., Movellan, J.R., Sejnowski, T.J., 2002. Face recognition by independent component analysis. IEEE Transactions on Neural Networks 13 (6), 1450–1464.

Basilevsky, A., 1994. Statistical Factor Analysis and Related Methods – Theory and Applications. John Wiley & Sons.

Belhumeur, P.N., Hespanha, J.P., Kriegman, D.J., 1997. Eigenfaces vs. fisherfaces: recognition using class specific linear projection. IEEE Transactions on Pattern Analysis and Machine Intelligence 19 (7), 711–720.

Bell, A.J., Sejnowski, T.J., 1995. An information-maximization approach to blind separation and blind deconvolution. Neural Computation 7 (6), 1129–1159.

Benesty, J., Chen, J., Huang, Y., 2008. Microphone Array Signal Processing. Springer.

Bengio, S., Pereira, F., Singer, Y., Strelow, D., 2009. Group sparse coding. In: Advances in Neural Information Processing Systems (NIPS), pp. 82–89.

Bengio, Y., Simard, P., Frasconi, P., 1994. Learning long-term dependencies with gradient descent is difficult. IEEE Transactions on Neural Networks 5 (2), 157–166.

Bishop, C.M., 2006. Pattern Recognition and Machine Learning. Springer.

Blanco, Y., Zazo, S., 2003. New Gaussianity measures based on order statistics: application to ICA. Neurocomputing 51, 303–320.

Blei, D.M., Ng, A.Y., Jordan, M.I., 2003. Latent Dirichlet allocation. Journal of Machine Learning Research 3, 993–1022.

Boll, S.F., 1979. Suppression of acoustic noise in speech using spectral subtraction. IEEE Transactions on Acoustics, Speech, and Signal Processing 27 (2), 113–120.

Boscolo, R., Pan, H., Roychowdhury, V.P., 2004. Independent component analysis based on nonparametric density estimation. IEEE Transactions on Neural Networks 15 (1), 55–65.

Boulanger-Lewandowski, N., Bengio, Y., Vincent, P., 2012. Discriminative non-negative matrix factorization for multiple pitch estimation. In: Proc. of Annual Conference of International Society for Music Information Retrieval Conference (ISMIR), pp. 205–210.

Boulanger-Lewandowski, N., Mysore, G., Hoffman, M., 2014. Exploiting long-term temporal dependencies in NMF using recurrent neural networks with application to source separation. In: Proc. of IEEE International Conference on Acoustics, Speech, and Signal Processing (ICASSP), pp. 7019–7023.

Boyd, S., Vandenberghe, L., 2004. Convex Optimization. Cambridge University Press.

Bregman, L.M., 1967. The relaxation method of finding the common point of convex sets and its application to the solution of problems in convex programming. U.S.S.R. Computational Mathematics and Mathematical Physics 7 (3), 200–217.

Cardoso, J.-F., 1999. High-order contrasts for independent component analysis. Neural Computation 11 (1), 157–192.

Carroll, J.D., Chang, J.-J., 1970. Analysis of individual differences in multidimensional scaling via an N-way generalization of Eckart–Young decomposition. Psychometrika 35 (3), 283–319.

Cemgil, A.T., 2009. Bayesian inference for nonnegative matrix factorisation models. Computational Intelligence and Neuroscience, 785152.

Cemgil, A.T., Févotte, C., Godsill, S.J., 2007. Variational and stochastic inference for Bayesian source separation. Digital Signal Processing 17 (5), 891–913.

Chan, K., Lee, T.-W., Sejnowski, T.J., 2002. Variational learning of clusters of undercomplete nonsymmetric independent components. Journal of Machine Learning Research 3, 99–114.

Chan, W., Jaitly, N., Le, Q., Vinyals, O., 2016. Listen, attend and spell: a neural network for large vocabulary conversational speech recognition. In: Proc. of IEEE International Conference on Acoustics, Speech and Signal Processing (ICASSP), pp. 4960–4964.

Chen, Y., 2005. Blind separation using convex functions. IEEE Transactions on Signal Processing 53 (6), 2027–2035.

Chen, J., Wang, D., 2017. Long short-term memory for speaker generalization in supervised speech separation. The Journal of the Acoustical Society of America 141 (6), 4705–4714.

Chien, J.-T., 2015a. Hierarchical Pitman–Yor–Dirichlet language model. IEEE/ACM Transactions on Audio, Speech and Language Processing 23 (8), 1259–1272.

Chien, J.-T., 2015b. Laplace group sensing for acoustic models. IEEE/ACM Transactions on Audio, Speech and Language Processing 23 (5), 909–922.

Chien, J.-T., 2016. Hierarchical theme and topic modeling. IEEE Transactions on Neural Networks and Learning Systems 27 (3), 565–578.

Chien, J.-T., 2018. Bayesian nonparametric learning for hierarchical and sparse topics. IEEE/ACM Transactions on Audio, Speech and Language Processing 26 (2), 422–435.

Chien, J.-T., Bao, Y.-T., 2018. Tensor-factorized neural networks. IEEE Transactions on Neural Networks and Learning Systems 29 (5), 1998–2011.

Chien, J.-T., Chang, Y.-C., 2016. Bayesian learning for speech dereverberation. In: Proc. of International Workshop on Machine Learning for Signal Processing (MLSP), pp. 1–6.

Chien, J.-T., Chen, B.-C., 2006. A new independent component analysis for speech recognition and separation. IEEE Transactions on Audio, Speech, and Language Processing 14 (4), 1245–1254.

Chien, J.-T., Hsieh, H.-L., 2012. Convex divergence ICA for blind source separation. IEEE Transactions on Audio, Speech, and Language Processing 20 (1), 290–301.

Chien, J.-T., Hsieh, H.-L., 2013a. Bayesian group sparse learning for music source separation. EURASIP Journal on Audio, Speech, and Music Processing, 18.

Chien, J.-T., Hsieh, H.-L., 2013b. Nonstationary source separation using sequential and variational Bayesian learning. IEEE Transactions on Neural Networks and Learning Systems 24 (5), 681–694.

Chien, J.-T., Ku, Y.-C., 2016. Bayesian recurrent neural network for language modeling. IEEE Transactions on Neural Networks and Learning Systems 27 (2), 361–374.

Chien, J.-T., Kuo, K.-T., 2017. Variational recurrent neural networks for speech separation. In: Proc. of Annual Conference of International Speech Communication Association (INTERSPEECH), pp. 1193–1197.

Chien, J.-T., Kuo, K.-T., 2018. Spectro-temporal neural factorization for speech dereverberation. In: Proc. of IEEE International Conference on Acoustics, Speech and Signal Processing (ICASSP), pp. 5449–5453.

Chien, J.-T., Lai, J.-R., Lai, P.-Y., 2001. Microphone array signal processing for far-talking speech recognition. In: Proc. of IEEE Signal Processing Workshop on Signal Processing Advances in Wireless Communications (SPAWC), pp. 322–325.

Chien, J.-T., Lee, C.-H., 2018. Deep unfolding for topic models. IEEE Transactions on Pattern Analysis and Machine Intelligence 40 (2), 318–331.

Chien, J.-T., Sawada, H., Makino, S., 2013. Adaptive processing and learning for audio source separation. In: Proc. of Asia-Pacific Signal and Information Processing Association Annual Summit and Conference (APSIPA-ASC), pp. 1–6.

Chien, J.-T., Ting, C.-W., 2008. Factor analyzed subspace modeling and selection. IEEE Transactions on Audio, Speech, and Language Processing 16 (1), 239–248.

Chien, J.-T., Tsou, K.-W., 2018. Recall neural network for source separation. In: Proc. of IEEE International Conference on Acoustics, Speech and Signal Processing (ICASSP), pp. 2956–2960.

Chien, J.-T., Wu, C.-C., 2002. Discriminant waveletfaces and nearest feature classifiers for face recognition. IEEE Transactions on Pattern Analysis and Machine Intelligence 24 (12), 1644–1649.

Chien, J.-T., Yang, P.-K., 2016. Bayesian factorization and learning for monaural source separation. IEEE/ACM Transactions on Audio, Speech and Language Processing 24 (1), 185–195.

Choudrey, R.A., Roberts, S.J., 2003. Bayesian ICA with hidden Markov sources. In: Proc. of International Workshop on Independent Component Analysis and Blind Signal Separation (ICA), pp. 809–814.

Choudrey, R., Penny, W.D., Roberts, S.J., 2000. An ensemble learning approach to independent component analysis. In: Proc. of IEEE Workshop on Neural Networks for Signal Processing (NNSP), vol. 1, pp. 435–444.

Chung, J., Kastner, K., Dinh, L., Goel, K., Courville, A.C., Bengio, Y., 2015. A recurrent latent variable model for sequential data. In: Advances in Neural Information Processing Systems (NIPS), pp. 2980–2988.

Cichocki, A., Amari, S., 2002. Adaptive Blind Signal and Image Processing: Learning Algorithms and Applications, Vol. 1. John Wiley & Sons.

Cichocki, A., Douglas, S.C., Amari, S., 1998. Robust techniques for independent component analysis (ICA) with noisy data. Neurocomputing 22 (1), 113–129.

Cichocki, A., Lee, H., Kim, Y.-D., Choi, S., 2008. Non-negative matrix factorization with α-divergence. Pattern Recognition Letters 29, 1433–1440.

Cichocki, A., Zdunek, R., 2006. Multilayer nonnegative matrix factorization. Electronics Letters 42, 947–948.

Cichocki, A., Zdunek, R., Amari, S., 2006a. Csiszar's divergences for non-negative matrix factorization: family of new algorithms. In: Proc. of International Conference on Independent Component Analysis and Blind Signal Separation (ICA), pp. 32–39.

Cichocki, A., Zdunek, R., Amari, S., 2006b. New algorithms for non-negative matrix factorization in applications to blind source separation. In: Proc. of IEEE International Conference on Acoustics, Speech and Signal Processing (ICASSP), vol. 5, pp. 621–624.

Cichocki, A., Zdunek, R., Phan, A.H., Amari, S., 2009. Nonnegative Matrix and Tensor Factorizations – Applications to Exploratory Multi-Way Data Analysis and Blind Source Separation. Wiley.

Comon, P., 1994. Independent component analysis, a new concept? Signal Processing 36 (3), 287–314.

Costagli, M., Kuruoğlu, E.E., 2007. Image separation using particle filters. Digital Signal Processing 17 (5), 935–946.

Csiszar, I., Shields, P.C., 2004. Information theory and statistics: a tutorial. Foundations and Trends in Communications and Information Theory 1 (4), 417–528.

De Lathauwer, L., 2008. Decompositions of a higher-order tensor in block terms – part II: definitions and uniqueness. SIAM Journal on Matrix Analysis and Applications 30 (3), 1033–1066.

De Lathauwer, L., De Moor, B., Vandewalle, J., 2000a. A multilinear singular value decomposition. SIAM Journal on Matrix Analysis and Applications 21 (4), 1253–1278.

De Lathauwer, L., De Moor, B., Vandewalle, J., 2000b. On the best rank-1 and rank-(r 1, r 2,..., rn) approximation of higher-order tensors. SIAM Journal on Matrix Analysis and Applications 21 (4), 1324–1342.

Dempster, A.P., Laird, N.M., Rubin, D.B., 1977. Maximum likelihood from incomplete data via the EM algorithm. Journal of the Royal Statistical Society (B) 39 (1), 1–38.

Dikmen, O., Fevotte, C., 2012. Maximum marginal likelihood estimation for nonnegative dictionary learning in the Gamma-Poisson model. IEEE Transactions on Signal Processing 10 (60), 5163–5175.

Ding, C., Li, T., Jordan, M.I., 2010. Convex and semi-nonnegative matrix factorizations. IEEE Transactions on Pattern Analysis and Machine Intelligence 32 (1), 45–55.

Doucet, A., Gordon, N.J., Krishnamurthy, V., 2001. Particle filters for state estimation of jump Markov linear systems. IEEE Transactions on Signal Processing 49 (3), 613–624.

Douglas, S.C., Gupta, M., 2007. Scaled natural gradient algorithms for instantaneous and convolutive blind source separation. In: Proc. of IEEE International Conference on Acoustics, Speech, and Signal Processing (ICASSP), vol. 2, pp. 637–640.

Douglas, S., Gupta, M., Sawada, H., Makino, S., 2007. Spatio-temporal FastICA algorithms for the blind separation of convolutive mixtures. IEEE Transactions on Audio, Speech, and Language Processing 15 (5), 1511–1520.

Du, J., Tu, Y., Dai, L.R., Lee, C.H., 2016. A regression approach to single-channel speech separation via high-resolution deep neural networks. IEEE/ACM Transactions on Audio, Speech and Language Processing 24 (8), 1424–1437.

Durrieu, J.-L., David, B., Richard, G., 2011. A musically motivated mid-level representation for pitch estimation and musical audio source separation. IEEE Journal of Selected Topics in Signal Processing 5 (6), 1180–1191.

Eggert, J., Körner, E., 2004. Sparse coding and NMF. In: Proc. of IEEE International Joint Conference on Neural Networks, vol. 4, pp. 2529–2533.

Elman, J.L., 1990. Finding structure in time. Cognitive Science 14 (2), 179–211.

Erdogan, H., Hershey, J.R., Watanabe, S., Le Roux, J., 2015. Phase-sensitive and recognition-boosted speech separation using deep recurrent neural networks. In: Proc. of IEEE International Conference on Acoustics, Speech, and Signal Processing (ICASSP).

Eriksson, J., Karvanen, J., Koivunen, V., 2000. Source distribution adaptive maximum likelihood estimation of ICA model. In: Proc. International Workshop on Independent Component Analysis and Blind Signal Separation (ICA), pp. 227–232.

Eriksson, J., Koivunen, V., 2003. Characteristic-function-based independent component analysis. Signal Processing 83 (10), 2195–2208.

Everson, R., Roberts, S., 2000. Blind source separation for non-stationary mixing. Journal of VLSI Signal Processing Systems for Signal, Image, and Video Technology 26 (1), 15–23.

Falk, T.H., Zheng, C., Chan, W.-Y., 2010. A non-intrusive quality and intelligibility measure of reverberant and dereverberated speech. IEEE Transactions on Audio, Speech, and Language Processing 18 (7), 1766–1774.

Févotte, C., 2007. Bayesian audio source separation. In: Makino, S., Lee, T.-W., Sawada, H. (Eds.), Blind Speech Separation. Springer, pp. 305–335.

Févotte, C., Bertin, N., Durrieu, J.-L., 2009. Nonnegative matrix factorization with the Itakura–Saito divergence: with application to music analysis. Neural Computation 21 (3), 793–830.

Fujihara, H., Goto, M., Ogata, J., Okuno, H.G., 2011. Lyricsynchronizer: automatic synchronization system between musical audio signals and lyrics. IEEE Journal of Selected Topics in Signal Processing 5 (6), 1252–1261.

Fukada, T., Yoshimura, T., Sagisaka, Y., 1999. Automatic generation of multiple pronunciations based on neural networks. Speech Communication 27 (1), 63–73.

Fukunaga, K., 1990. Introduction to Statistical Pattern Recognition. Academic Press.

Garrigues, P., Olshausen, B.A., 2010. Group sparse coding with a Laplacian scale mixture prior. In: Advances in Neural Information Processing Systems (NIPS), pp. 676–684.

Gaussier, E., Goutte, C., 2005. Relation between PLSA and NMF and implications. In: Proc. of Annual International ACM SIGIR Conference on Research and Development in Information Retrieval, pp. 601–602.

Geman, S., Geman, D., 1984. Stochastic relaxation, Gibbs distributions, and the Bayesian restoration of images. IEEE Transactions on Pattern Analysis and Machine Intelligence 6 (1), 721–741.

Gençağa, D., Kuruoğlu, E.E., Ertüzün, A., 2010. Modeling non-Gaussian time-varying vector autoregressive processes by particle filtering. Multidimensional Systems and Signal Processing 21 (1), 73–85.

Gilks, W.R., Richardson, S., Spiegelhalter, D.J., 1996. Markov Chain Monte Carlo in Practice. Chapman & Hall/CRC Interdisciplinary Statistics.

Goodfellow, I., Bengio, Y., Courville, A., 2016. Deep Learning. MIT Press.

Goodfellow, I., Pouget-Abadie, J., Mirza, M., Xu, B., Warde-Farley, D., Ozair, S., Courville, A., Bengio, Y., 2014. Generative adversarial nets. In: Advances in Neural Information Processing Systems (NIPS), pp. 2672–2680.

Grais, E.M., Erdogan, H., 2013. Discriminative nonnegative dictionary learning using cross-coherence penalties for single channel source separation. In: Proc. of Annual Conference of the International Speech Communication Association (INTERSPEECH), pp. 808–812.

Grais, E.M., Sen, M.U., Erdogan, H., 2014. Deep neural networks for single channel source separation. In: Proc. of IEEE International Conference on Acoustics, Speech, and Signal Processing (ICASSP), pp. 3762–3766.

Graves, A., Jaitly, N., Mohamed, A.-R., 2013. Hybrid speech recognition with deep bidirectional LSTM. In: Prof. of IEEE Workshop on Automatic Speech Recognition and Understanding (ASRU), pp. 273–278.

Graves, A., Wayne, G., Danihelka, I., 2014. Neural Turing machines. arXiv preprint, arXiv:1410.5401.

Hastings, W.K., 1970. Monte Carlo sampling methods using Markov chains and their applications. Biometrika 57, 97–109.

Hershey, J.R., Chen, Z., Roux, J.L., Watanabe, S., 2016. Deep clustering: discriminative embeddings for segmentation and separation. In: Proc. of IEEE International Conference on Acoustics, Speech and Signal Processing (ICASSP), pp. 31–35.

Hinton, G., Deng, L., Yu, D., Dahl, G., Mohamed, A., Jaitly, N., Senior, A., Vanhoucke, V., Nguyen, P., Sainath, T., Kingsbury, B., 2012. Deep neural networks for acoustic modeling in speech recognition – four research groups share their views. IEEE Signal Processing Magazine 29 (6), 82–97.

Hinton, G.E., Osindero, S., Teh, Y.-W., 2006. A fast learning algorithm for deep belief nets. Neural Computation 18 (7), 1527–1554.

Hinton, G.E., Salakhutdinov, R.R., 2006. Reducing the dimensionality of data with neural networks. Science 313, 504–507.

Hirayama, J., Maeda, S., Ishii, S., 2007. Markov and semi-Markov switching of source appearances for nonstationary independent component analysis. IEEE Transactions on Neural Networks 18 (5), 1326–1342.

Hochreiter, S., Schmidhuber, J., 1997. Long short-term memory. Neural Computation 9 (8), 1735–1780.

Hofmann, T., 1999. Probabilistic latent semantic indexing. In: Proc. of ACM SIGIR Conference on Research and Development in Information Retrieval, pp. 50–57.

Honkela, A., Valpola, H., 2003. On-line variational Bayesian learning. In: Proc. of International Workshop on Independent Component Analysis and Blind Signal Separation (ICA), pp. 803–808.

Hoyer, P.O., 2004. Non-negative matrix factorization with sparseness constraints. Journal of Machine Learning Research 5, 1457–1469.

Hsieh, H.-L., Chien, J.-T., 2010. Online Bayesian learning for dynamic source separation. In: Proc. of IEEE International Conference on Acoustics, Speech and Signal Processing (ICASSP), pp. 1950–1953.

Hsieh, H.-L., Chien, J.-T., 2011. Nonstationary and temporally correlated source separation using Gaussian process. In: Proc. of IEEE International Conference on Acoustics, Speech and Signal Processing (ICASSP), pp. 2120–2123.

Hsieh, H.-L., Chien, J.-T., Shinoda, K., Furui, S., 2009. Independent component analysis for noisy speech recognition. In: Proc. of International Conference on Acoustics, Speech, and Signal Processing (ICASSP), pp. 4369–4372.

Hsu, C.-C., Chi, T.-S., Chien, J.-T., 2016. Discriminative layered nonnegative matrix factorization for speech separation. In: Proc. of Annual Conference of the International Speech Communication Association (INTERSPEECH), pp. 560–564.

Hsu, C.-C., Chien, J.-T., Chi, T.-S., 2015. Layered nonnegative matrix factorization for speech separation. In: Proc. of Annual Conference of the International Speech Communication Association (INTERSPEECH), pp. 628–632.

Hsu, C.-L., Jang, J.-S.R., 2010. On the improvement of singing voice separation for monaural recordings using the MIR-1K dataset. IEEE Transactions on Audio, Speech, and Language Processing 18 (2), 310–319.

Hu, Y., Loizou, P.C., 2008. Evaluation of objective quality measures for speech enhancement. IEEE Transactions on Audio, Speech, and Language Processing 16 (1), 229–238.

Hu, K., Wang, D., 2013. An iterative model-based approach to cochannel speech separation. EURASIP Journal on Audio, Speech, and Music Processing 2013 (1), 14.

Huang, P.-S., Chen, S.D., Smaragdis, P., Hasegawa-Johnson, M., 2012. Singing-voice separation from monaural recordings using robust principal component analysis. In: Proc. of IEEE International Conference on Acoustics, Speech and Signal Processing (ICASSP), pp. 57–60.

Huang, P.-S., Kim, M., Hasegawa-Johnson, M., Smaragdis, P., 2014a. Deep learning for monaural speech separation. In: Proc. of IEEE International Conference on Acoustics, Speech, and Signal Processing (ICASSP), pp. 1562–1566.

Huang, P.-S., Kim, M., Hasegawa-Johnson, M., Smaragdis, P., 2014b. Singing-voice separation from monaural recordings using deep recurrent neural networks. In: Proc. of Annual Conference of International Society for Music Information Retrieval (ISMIR).

Huang, P.-S., Kim, M., Hasegawa-Johnson, M., Smaragdis, P., 2015. Joint optimization of masks and deep recurrent neural networks for monaural source separation. IEEE/ACM Transactions on Audio, Speech and Language Processing 23 (12), 2136–2147.

Huang, Q., Yang, J., Wei, S., 2007. Temporally correlated source separation using variational Bayesian learning approach. Digital Signal Processing 17 (5), 873–890.

Hyvärinen, A., 1999. Fast and robust fixed-point algorithms for independent component analysis. IEEE Transactions on Neural Networks 10 (3), 626–634.

Hyvärinen, A., Karhunen, J., Oja, E., 2001. Independent Component Analysis. John Wiley & Sons.

Hyvärinen, A., Oja, E., 2000. Independent component analysis: algorithms and applications. Neural Networks 13 (4), 411–430.

Ichir, M.M., Mohammad-Djafari, A., 2006. Hidden Markov models for wavelet-based blind source separation. IEEE Transactions on Image Processing 15 (7), 1887–1899.

Itakura, F., Saito, S., 1968. Analysis synthesis telephony based on the maximum likelihood method. In: Proc. of International Congress on Acoustics, pp. 17–20.

Jaiswal, R., FitzGerald, D., Barry, D., Coyle, E., Rickard, S., 2011. Clustering NMF basis functions using shifted NMF for monaural sound source separation. In: Proc. of IEEE International Conference on Acoustics, Speech and Signal Processing (ICASSP), pp. 245–248.

Jang, G.-J., Lee, T.-W., Oh, Y.-H., 2002. Learning statistically efficient features for speaker recognition. Neurocomputing 49 (1), 329–348.

Jiang, Y., Wang, D., Liu, R., Feng, Z., 2014. Binaural classification for reverberant speech segregation using deep neural networks. IEEE/ACM Transactions on Audio, Speech and Language Processing 22 (12), 2112–2121.

Jolliffe, I.T., 1986. Principal component analysis and factor analysis. In: Principal Component Analysis. Springer, pp. 115–128.

Jordan, M., Ghahramani, Z., Jaakkola, T., Saul, L., 1999. An introduction to variational methods for graphical models. Machine Learning 37 (2), 183–233.

Kabal, P., 2002. TSP speech database. Database Version 1. McGill University.

Kailath, T., Sayed, A.H., Hassibi, B., 2000. Linear Estimation, Vol. 1. Prentice Hall, Upper Saddle River, NJ.

Kameoka, H., Nakatani, T., Yoshioka, T., 2009. Robust speech dereverberation based on non-negativity and sparse nature of speech spectrograms. In: Proc. of IEEE International Conference on Acoustics, Speech, and Signal Processing (ICASSP), pp. 45–48.

Kim, M., Smaragdis, P., 2013. Collaborative audio enhancement using probabilistic latent component sharing. In: Proc. of IEEE International Conference on Acoustics, Speech, and Signal Processing (ICASSP), pp. 896–900.

Kim, M., Yoo, J., Kang, K., Choi, S., 2011. Nonnegative matrix partial co-factorization for spectral and temporal drum source separation. IEEE Journal of Selected Topics in Signal Processing 5 (6), 1192–1204.

Kim, Y.-D., Choi, S., 2009. Weighted nonnegative matrix factorization. In: Proc. of IEEE International Conference on Acoustics, Speech, and Signal Processing (ICASSP), pp. 1541–1544.

Kingma, D.P., Ba, J., 2014. Adam: a method for stochastic optimization. arXiv preprint, arXiv:1412.6980.

Kingma, D.P., Welling, M., 2013. Auto-encoding variational Bayes. arXiv preprint, arXiv:1312.6114.

Kinoshita, K., Delcroix, M., Nakatani, T., Miyoshi, M., 2009. Suppression of late reverberation effect on speech signal using long-term multiple-step linear prediction. IEEE Transactions on Audio, Speech, and Language Processing 17 (4), 534–545.

Kinoshita, K., Delcroix, M., Yoshioka, T., Nakatani, T., Sehr, A., Kellermann, W., Maas, R., 2013. The REVERB challenge: a common evaluation framework for dereverberation and recognition of reverberant speech. In: Proc. of IEEE Workshop on Applications of Signal Processing to Audio and Acoustics (WASPAA), pp. 1–4.

Kırbız, S., Günsel, B., 2014. A multiresolution non-negative tensor factorization approach for single channel sound source separation. Signal Processing 105, 56–69.

Kolbæk, M., Tan, Z.-H., Jensen, J., 2018. Monaural speech enhancement using deep neural networks by maximizing a short-time objective intelligibility measure. In: Proc. of IEEE International Conference on Acoustics, Speech and Signal Processing (ICASSP), pp. 5059–5063.

Kolbæk, M., Yu, D., Tan, Z.-H., Jensen, J., 2017. Multitalker speech separation with utterance-level permutation invariant training of deep recurrent neural networks. IEEE/ACM Transactions on Audio, Speech and Language Processing 25 (10), 1901–1913.

Koldovský, Z., Málek, J., Tichavský, P., Deville, Y., Hosseini, S., 2009. Blind separation of piecewise stationary non-Gaussian sources. Signal Processing 89 (12), 2570–2584.

Kuhn, R., Junqua, J.C., Nguyen, P., Niedzielski, N., 2000. Rapid speaker adaptation in eigenvoice space. IEEE Transactions on Speech and Audio Processing 8 (6), 695–707.

Kullback, S., Leibler, R.A., 1951. On information and sufficiency. The Annals of Mathematical Statistics 22 (1), 79–86.

Lawrence, N.D., Bishop, C.M., 2000. Variational Bayesian Independent Component Analysis. Technical report. University of Cambridge.

LeCun, Y., Bottou, L., Bengio, Y., Haffner, P., 1998. Gradient-based learning applied to document recognition. In: Proceedings of the IEEE, pp. 2278–2324.

Lee, D.D., Seung, H.S., 1999. Learning the parts of objects by non-negative matrix factorization. Nature 401, 788–791.

Lee, D.D., Seung, H.S., 2000. Algorithms for non-negative matrix factorization. In: Advances in Neural Information Processing Systems (NIPS), pp. 556–562.

Lee, H., Choi, S., 2009. Group nonnegative matrix factorization for EEG classification. In: Proc. of International Conference on Artificial Intelligence and Statistics (AISTATS), pp. 320–327.

Lee, T.-W., Jang, G.-J., 2001. The statistical structures of male and female speech signals. In: Proc. of IEEE International Conference on Acoustics, Speech, and Signal Processing (ICASSP), vol. 1, pp. 105–108.

Li, Y., Wang, D., 2007. Separation of singing voice from music accompaniment for monaural recordings. IEEE Transactions on Audio, Speech, and Language Processing 15 (4), 1475–1487.

Li, Y., Yu, Z.L., Bi, N., Xu, Y., Gu, Z., Amari, S., 2014. Sparse representation for brain signal processing – a tutorial on methods and applications. IEEE Signal Processing Magazine 31 (3), 96–106.

Liang, D., Hoffman, M.D., Mysore, G.J., 2015. Speech dereverberation using a learned speech model. In: Proc. of IEEE International Conference on Acoustics, Speech and Signal Processing (ICASSP), pp. 1871–1875.

Lin, J., 1991. Divergence measures based on the Shannon entropy. IEEE Transactions on Information Theory 37 (1), 145–151.

Lincoln, M., McCowan, I., Vepa, J., Maganti, H.K., 2005. The multi-channel Wall Street Journal audio visual corpus (MC-WSJ-AV): specification and initial experiments. In: Proc. of IEEE Workshop on Automatic Speech Recognition and Understanding (ASRU), pp. 357–362.

Liu, J.S., 2008. Monte Carlo Strategies in Scientific Computing. Springer.

Lyu, S., Wang, X., 2013. On algorithms for sparse multi-factor NMF. In: Advances in Neural Information Processing Systems (NIPS), pp. 602–610.

Ma, Z., Leijon, A., 2011. Bayesian estimation of beta mixture models with variational inference. IEEE Transactions on Pattern Analysis and Machine Intelligence 33 (11), 2160–2173.

Ma, Z., Teschendorff, A.E., Leijon, A., Qiao, Y., Zhang, H., Guo, J., 2015. Variational Bayesian matrix factorization for bounded support data. IEEE Transactions on Pattern Analysis and Machine Intelligence 37 (4), 876–889.

MacKay, D.J., 1992. Bayesian interpolation. Neural Computation 4 (3), 415–447.

MacKay, D.J.C., 1995. Probable networks and plausible predictions – a review of practical Bayesian methods for supervised neural networks. Network Computation in Neural Systems 6 (3), 469–505.

Mak, B., Kwok, J.T., Ho, S., 2005. Kernel eigenvoice speaker adaptation. IEEE Transactions on Speech and Audio Processing 13 (5), 984–992.

Mak, M.W., Pang, X., Chien, J.T., 2016. Mixture of PLDA for noise robust i-vector speaker verification. IEEE/ACM Transactions on Audio, Speech and Language Processing 24 (1), 130–142.

Makino, S., Sawada, H., Mukai, R., Araki, S., 2005. Blind source separation of convolutive mixtures of speech in frequency domain. IEICE Transactions on Fundamentals of Electronics, Communications and Computer Science 88 (7), 1640–1655.

Meinicke, P., Ritter, H., 2001. Independent component analysis with quantizing density estimators. In: Proc. of International Workshop on Independent Component Analysis Blind Signal Separation (ICA), pp. 224–229.

Mesaros, A., Virtanen, T., Klapuri, A., 2007. Singer identification in polyphonic music using vocal separation and pattern recognition methods. In: Proc. of Annual Conference of International Society for Music Information Retrieval (ISMIR), pp. 375–378.

Metropolis, N., Rosenbluth, A.W., Rosenbluth, M.N., Teller, A.H., Teller, E., 1953. Equation of state calculations by fast computing machines. Journal of Chemical Physics 21 (6), 1087–1092.

Miao, Y., Ox, C., Yu, L., Blunsom, P., 2016. Neural variational inference for text processing. In: Proc. of International Conference on Machine Learning (ICML), pp. 1727–1736.

Miskin, J.W., 2000. Ensemble Learning for Independent Component Analysis. PhD thesis. University of Cambridge.

Mohammad-Djafari, A., Knuth, K., 2010. Bayesian approaches. In: Handbook of Blind Source Separation: Independent Component Analysis and Applications, pp. 467–513.

Mohammadiha, N., Smaragdis, P., Doclo, S., 2015. Joint acoustic and spectral modeling for speech dereverberation using non-negative representations. In: Proc. of IEEE International Conference on Acoustics, Speech, and Signal Processing (ICASSP), pp. 4410–4414.

Mørup, M., 2011. Applications of tensor (multiway array) factorizations and decompositions in data mining. Wiley Interdisciplinary Reviews: Data Mining and Knowledge Discovery 1 (1), 24–40.

Mørup, M., Schmidt, M.N., 2006a. Sparse Non-negative Matrix Factor 2-D Deconvolution. Technical report. Technical University of Denmark.

Mørup, M., Schmidt, M.N., 2006b. Sparse Non-negative Tensor 2D Deconvolution (SNTF2D) for Multi Channel Time-Frequency Analysis. Technical report. Technical University of Denmark.

Moussaoui, S., Brie, D., Mohammad-Djafari, A., Carteret, C., 2006. Separation of non-negative mixture of non-negative sources using a Bayesian approach and MCMC sampling. IEEE Transactions on Signal Processing 54 (11), 4133–4145.

Muller, M., Konz, V., Bogler, W., Arifi-Muller, V., 2011. Saarland music data (SMD). In: Proc. of Annual Conference of International Society for Music Information Retrieval (ISMIR).

Murata, N., 2001. Properties of the empirical characteristic function and its application to testing for independence. In: Proc. of International Workshop on Independent Component Analysis and Signal Separation (ICA), pp. 295–300.

Murata, N., Ikeda, S., Ziehe, A., 2001. An approach to blind source separation based on temporal structure of speech signals. Neurocomputing 41 (1), 1–24.

Naqvi, S.M., Zhang, Y., Chambers, J.A., 2009. Multimodal blind source separation for moving sources. In: Proc. of IEEE International Conference on Acoustics, Speech and Signal Processing (ICASSP), pp. 125–128.

Narayanan, A., Wang, D., 2013. Ideal ratio mask estimation using deep neural networks for robust speech recognition. In: Proc. of IEEE International Conference on Acoustics, Speech and Signal Processing (ICASSP), pp. 7092–7096.

Neal, R.M. Probabilistic Inference Using Markov Chain Monte Carlo Methods, 1993.

Ngiam, J., Coates, A., Lahiri, A., Prochnow, B., Le, Q.V., Ng, A.Y., 2011. On optimization methods for deep learning. In: Proc. of International Conference on Machine Learning (ICML), pp. 265–272.

OGrady, P.D., Pearlmutter, B.A., 2006. Convolutive non-negative matrix factorisation with a sparseness constraint. In: Proc. of IEEE Workshop on Machine Learning for Signal Processing (MLSP), pp. 427–432.

Oh, J.-H., Seung, H.S., 1998. Learning generative models with the up propagation algorithm. In: Advances in Neural Information Processing Systems (NIPS), pp. 605–611.

Ozerov, A., Fevotte, C., 2010. Multichannel nonnegative matrix factorization in convolutive mixtures for audio source separation. IEEE Transactions on Audio, Speech, and Language Processing 18 (3), 550–563.

Ozerov, A., Philippe, P., Bimbot, F., Gribonval, R., 2007. Adaptation of Bayesian models for single-channel source separation and its application to voice/music separation in popular songs. IEEE Transactions on Audio, Speech, and Language Processing 15 (5), 1564–1578.

Park, S., Choi, S., 2008. Gaussian processes for source separation. In: Proc. of IEEE International Conference on Acoustics, Speech, and Signal Processing (ICASSP), pp. 1909–1912.

Plumbley, M.D., Blumensath, T., Daudet, L., Gribonval, R., Davies, M.E., 2010. Sparse representations in audio and music: from coding to source separation. Proceedings of the IEEE 98 (6), 995–1005.

Principe, J.C., Xu, D., Fisher, J.W., 2000. Information theoretic learning. In: Unsupervised Adaptive Filtering. Wiley, New York, pp. 265–319.

Rabiner, L.R., Juang, B.-H., 1986. An introduction to hidden Markov models. IEEE ASSP Magazine 3 (1), 4–16.

Rafii, Z., Pardo, B., 2011. A simple music/voice separation method based on the extraction of the repeating musical structure. In: Proc. of IEEE International Conference on Acoustics, Speech and Signal Processing (ICASSP), pp. 221–224.

Rafii, Z., Pardo, B., 2013. Repeating pattern extraction technique (REPET): a simple method for music/voice separation. IEEE Transactions on Audio, Speech, and Language Processing 21 (1), 73–84.

Raj, B., Smaragdis, P., Shashanka, M., Singh, R., 2007. Separating a foreground singer from background music. In: Proc. of International Symposium on Frontiers of Research on Speech and Music.

Rasmussen, C.E., Williams, C.K., 2006. Gaussian Processes for Machine Learning. MIT Press.

Rennie, S.J., Hershey, J.R., Olsen, P.A., 2010. Single-channel multitalker speech recognition – graphical modeling approaches. IEEE Signal Processing Magazine 27 (6), 66–80.

Rezende, D.J., Mohamed, S., Wierstra, D., 2014. Stochastic backpropagation and approximate inference in deep generative models. In: Proc. of International Conference on Machine Learning (ICML), pp. 1278–1286.

Rissanen, J., 1978. Modeling by shortest data description. Automatica 14 (5), 465–471.

Roman, N., Wang, D., Brown, G.J., 2003. Speech segregation based on sound localization. The Journal of the Acoustical Society of America 114 (4), 2236–2252.

Roux, J.L., Hershey, J.R., Weninger, F., 2015. Deep NMF for speech separation. In: Proc. of IEEE International Conference on Acoustics, Speech and Signal Processing (ICASSP), pp. 66–70.

Rowe, D.B., 2002. Multivariate Bayesian Statistics: Models for Source Separation and Signal Unmixing. CRC Press.

Rumelhart, D.E., Hinton, G.E., Williams, R.J., 1986. Learning internal representation by backpropagating errors. Nature 323, 533–536.

Salakhutdinov, R., Mnih, A., 2008. Bayesian probabilistic matrix factorization using Markov chain Monte Carlo. In: Proc. of International Conference on Machine Learning (ICML), pp. 880–887.

Saon, G., Chien, J.-T., 2012a. Bayesian sensing hidden Markov models. IEEE Transactions on Audio, Speech, and Language Processing 20 (1), 43–54.

Saon, G., Chien, J.-T., 2012b. Large-vocabulary continuous speech recognition systems: a look at some recent advances. IEEE Signal Processing Magazine 29 (6), 18–33.

Sawada, H., Araki, S., Makino, S., 2007. Frequency-domain blind source separation. Springer, pp. 47–78.

Sawada, H., Araki, S., Makino, S., 2011. Underdetermined convolutive blind source separation via frequency bin-wise clustering and permutation alignment. IEEE Transactions on Audio, Speech, and Language Processing 19 (3), 516–527.

Sawada, H., Kameoka, H., Araki, S., Ueda, N., 2013. Multichannel extensions of non-negative matrix factorization with complex-valued data. IEEE Transactions on Audio, Speech, and Language Processing 21 (5), 971–982.

Sawada, H., Mukai, R., Araki, S., Makino, S., 2003. Polar coordinate based nonlinear function for frequency-domain blind source separation. IEICE Transactions on Fundamentals of Electronics, Communications and Computer Science 86 (3), 590–596.

Schmidt, M., Mohamed, S., 2009. Probabilistic non-negative tensor factorisation using Markov chain Monte Carlo. In: Proc. of European Signal Processing Conference (EUSIPCO), pp. 24–28.

Schmidt, M.N., Morup, M., 2006. Non-negative matrix factor 2-D deconvolution for blind single channel source separation. In: Proc. of International Conference on Independent Component Analysis and Blind Signal Separation (ICA), pp. 700–707.

Schmidt, M.N., Olsson, R.K., 2007. Single-channel speech separation using sparse non-negative matrix factorization. In: Proc. of Annual Conference of International Speech Communication Association (INTERSPEECH), pp. 2614–2617.

Schmidt, M.N., Winther, O., Hansen, L.K., 2009. Bayesian non-negative matrix factorization. In: Proc. of International Conference on Independent Component Analysis and Signal Separation (ICA), pp. 540–547.

Schobben, D., Torkkola, K., Smaragdis, P., 1999. Evaluation of blind signal separation methods. In: Proc. International Workshop on Independent Component Analysis and Blind Signal Separation (ICA), pp. 261–266.

Schuster, M., Paliwal, K.K., 1997. Bidirectional recurrent neural networks. IEEE Transactions on Signal Processing 45, 2673–2681.

Schwarz, G., 1978. Estimating the dimension of a model. The Annals of Statistics 6 (2), 461–464.

Seki, S., Ohtani, K., Toda, T., Takeda, K., 2016. Stereo channel music signal separation based on non-negative tensor factorization with cepstrum regularization. The Journal of the Acoustical Society of America 140 (4), 2967.

Settle, S., Roux, J.L., Hori, T., Watanabe, S., Hershey, J.R., 2018. Generative adversarial source separation. In: Proc. of IEEE International Conference on Acoustics, Speech and Signal Processing (ICASSP), pp. 4819–4823.

Shannon, C.E., 1948. A mathematical theory of communication. The Bell System Technical Journal 27, 379–423.

Shao, Y., Wang, D., 2006. Model-based sequential organization in cochannel speech. IEEE Transactions on Audio, Speech, and Language Processing 14 (1), 289–298.

Shashanka, M., Raj, B., Smaragdis, P., 2008. Probabilistic latent variable models as nonnegative factorizations. Computational Intelligence and Neuroscience, 947438. pp. 1–8.

Singh, R., Raj, B., Stern, R.M., 2002. Automatic generation of subword units for speech recognition systems. IEEE Transactions on Speech and Audio Processing 10 (2), 89–99.

Smaragdis, P., 2007. Convolutive speech bases and their application to supervised speech separation. IEEE Transactions on Audio, Speech, and Language Processing 15 (1), 1–12.

Smaragdis, P., Fevotte, C., Mysore, G.J., Mohammadiha, N., Hoffman, M., 2014. Static and dynamic source separation using nonnegative factorization – a unified view. IEEE Signal Processing Magazine 31 (3), 66–75.

Smaragdis, P., Raj, B., 2008. Shift-invariant probabilistic latent component analysis. Journal of Machine Learning Research.

Smaragdis, P., Raj, B., Shashanka, M., 2006. A probabilistic latent variable model for acoustic modeling. In: Advances in Models for Acoustic Processing Workshop (NIPS).

Snoussi, H., Mohammad-Djafari, A., 2004a. Bayesian unsupervised learning for source separation with mixture of Gaussians prior. Journal of VLSI Signal Processing Systems for Signal, Image, and Video Technology 37 (2–3), 263–279.

Snoussi, H., Mohammad-Djafari, A., 2004b. Fast joint separation and segmentation of mixed images. Journal of Electronic Imaging 13 (2), 349–361.

Spiertz, M., Gnann, V., 2009. Source-filter based clustering for monaural blind source separation. In: Proc. of International Conference on Digital Audio Effects, pp. 1–7.

Spragins, J., 1965. A note on the iterative application of Bayes' rule. IEEE Transactions on Information Theory 11 (4), 544–549.

Stone, J.V., 2001. Blind source separation using temporal predictability. Neural Computation 13 (7), 1559–1574.

Sübakan, Y.C., Smaragdis, P., 2018. Generative adversarial source separation. In: Proc. of IEEE International Conference on Acoustics, Speech and Signal Processing (ICASSP), pp. 26–30.

Sukhbaatar, S., Szlam, A., Weston, J., Fergus, R., 2015. End-to-end memory networks. In: Advances in Neural Information Processing Systems (NIPS), pp. 2440–2448.

Sutskever, I., Vinyals, O., Le, Q.V., 2014. Sequence to sequence learning with neural networks. In: Advances in Neural Information Processing Systems (NIPS), pp. 3104–3112.

Tachioka, Y., Hanazawa, T., Iwasaki, T., 2013. Dereverberation method with reverberation time estimation using floored ratio of spectral subtraction. Acoustical Science and Technology 34 (3), 212–215.

Talmon, R., Cohen, I., Gannot, S., 2009. Relative transfer function identification using convolutive transfer function approximation. IEEE Transactions on Audio, Speech, and Language Processing 17 (4), 546–555.

Teh, Y.W., Jordan, M.I., Beal, M.J., Blei, D.M., 2006. Hierarchical Dirichlet processes. Journal of the American Statistical Association 101 (476), 1566–1581.

Tibshirani, R., 1996. Regression shrinkage and selection via the lasso. Journal of the Royal Statistical Society. Series B 58 (1), 267–288.

Tipping, M.E., 2001. Sparse Bayesian learning and the relevance vector machine. Journal of Machine Learning Research 1, 211–244.

Trigeorgis, G., Bousmalis, K., Zaferiou, S., Schuller, B.W., 2014. A deep semi-NMF model for learning hidden representations. In: Proc. of International Conference on Machine Learning, pp. 1692–1700.

Tsou, K.-W., Chien, J.-T., 2017. Memory augmented neural network for source separation. In: Proc. of IEEE International Workshop on Machine Learning for Signal Processing (MLSP), pp. 1–6.

Tucker, L.R., 1966. Some mathematical notes on three-mode factor analysis. Psychometrika 31 (3), 279–311.

Vincent, E., Bertin, N., Gribonval, R., Bimbot, F., 2014. From blind to guided audio source separation – how models and side information can improve the separation of sound. IEEE Signal Processing Magazine 31 (3), 107–115.

Vincent, E., Gribonval, R., Fevotte, C., 2006. Performance measurement in blind audio source separation. IEEE Transactions on Audio, Speech, and Language Processing 14 (4), 1462–1469.

Virtanen, T., 2007. Monaural sound source separation by nonnegative matrix factorization with temporal continuity and sparseness criteria. IEEE Transactions on Audio, Speech, and Language Processing 15 (3), 1066–1074.

Vlassis, N., Motomura, Y., 2001. Efficient source adaptivity in independent component analysis. IEEE Transactions on Neural Networks 12 (3), 559–566.

Wang, G.-X., Hsu, C.-C., Chien, J.-T., 2016. Discriminative deep recurrent neural networks for monaural speech separation. In: Proc. of IEEE International Conference on Acoustics, Speech and Signal Processing (ICASSP), pp. 2544–2548.

Wang, L., Ding, H., Yin, F., 2011. A region-growing permutation alignment approach in frequency-domain blind source separation of speech mixtures. IEEE Transactions on Audio, Speech, and Language Processing 19 (3), 549–557.

Wang, Y., Wang, D., 2012. Cocktail party processing via structured prediction. In: Advances in Neural Information Processing Systems (NIPS), vol. 25. MIT Press, pp. 224–232.

Wang, Y., Wang, D., 2013. Towards scaling up classification-based speech separation. IEEE Transactions on Audio, Speech, and Language Processing 21 (7), 1381–1390.

Wang, Z., Sha, F., 2014. Discriminative non-negative matrix factorization for single-channel speech separation. In: Proc. of IEEE International Conference on Acoustics, Speech, and Signal Processing (ICASSP), pp. 3749–3753.

Watanabe, S., Chien, J.-T., 2015. Bayesian Speech and Language Processing. Cambridge University Press.

Welling, M., Weber, M., 2001. Positive tensor factorization. Pattern Recognition Letters 22 (12), 1255–1261.

Weninger, F., Hershey, J.R., Roux, J.L., Schuller, B., 2014a. Discriminatively trained recurrent neural networks for single-channel speech separation. In: Proc. of IEEE Global Conference on Signal and Information Processing (GlobalSIP), pp. 577–581.

Weninger, F., Le Roux, J., Hershey, J.R., Watanabe, S., 2014b. Discriminative NMF and its application to single-channel source separation. In: Proc. of Annual Conference of International Speech Communication Association (INTERSPEECH), pp. 865–869.

Weston, J., Chopra, S., Bordes, A., 2014. Memory networks. arXiv preprint, arXiv:1410.3916.

Williams, R.J., Zipser, D., 1989. A learning algorithm for continually running fully recurrent neural networks. Neural Computation 1 (2), 270–280.

Williams, R.J., Zipser, D., 1995. Gradient-based learning algorithms for recurrent networks and their computational complexity. In: Back-Propagation: Theory, Architectures and Applications, pp. 433–486.

Winter, S., Kellermann, W., Sawada, H., Makino, S., 2007. MAP-based underdetermined blind source separation of convolutive mixtures by hierarchical clustering and ℓ_1-norm minimization. EURASIP Journal on Advances in Signal Processing 2007, 1–12.

Xiong, L., Chen, X., Huang, T.-K., Schneider, J., Carbonell, J.G., 2010. Temporal collaborative filtering with Bayesian probabilistic tensor factorization. In: Proc. of SIAM International Conference on Data Mining, pp. 211–222.

Xu, D., Principe, J.C., Fisher, J., Wu, H.-C., 1998. A novel measure for independent component analysis (ICA). In: Proc. of IEEE International Conference on Acoustics, Speech, and Signal Processing (ICASSP), vol. 2, pp. 1161–1164.

Xu, K., Ba, J., Kiros, R., Cho, K., Courville, A., Salakhudinov, R., Zemel, R., Bengio, Y., 2015. Show, attend and tell: neural image caption generation with visual attention. In: Proc. of International Conference on Machine Learning (ICML), pp. 2048–2057.

Xu, M., Golay, M.W., 2006. Data-guided model combination by decomposition and aggregation. Machine Learning 63 (1), 43–67.

Xu, Y., Du, J., Dai, L.R., Lee, C.H., 2014. An experimental study on speech enhancement based on deep neural networks. IEEE Signal Processing Letters 21 (1), 65–68.

Yang, D., Lee, W., 2004. Disambiguating music emotion using software agents. In: Proc. of Annual Conference of International Society for Music Information Retrieval (ISMIR), pp. 52–57.

Yang, H.H., Amari, S.-i., 1997. Adaptive online learning algorithms for blind separation: maximum entropy and minimum mutual information. Neural Computation 9 (7), 1457–1482.

Yang, P.-K., Hsu, C.-C., Chien, J.-T., 2014a. Bayesian factorization and selection for speech and music separation. In: Proc. of Annual Conference of International Speech Communication Association (INTERSPEECH), pp. 998–1002.

Yang, P.-K., Hsu, C.-C., Chien, J.-T., 2014b. Bayesian singing-voice separation. In: Proc. of Annual Conference of International Society for Music Information Retrieval (ISMIR), pp. 507–512.

Yang, Y.-H., 2012. On sparse and low-rank matrix decomposition for singing voice separation. In: Proc. of ACM International Conference on Multimedia, pp. 757–760.

Yoo, J., Kim, M., Kang, K., Choi, S., 2010. Nonnegative matrix partial co-factorization for drum source separation. In: Proc. of IEEE International Conference on Acoustics, Speech and Signal Processing (ICASSP), pp. 1942–1945.

Yoshii, K., Goto, M., 2012. A nonparametric Bayesian multipitch analyzer based on infinite latent harmonic allocation. IEEE Transactions on Audio, Speech, and Language Processing 20 (3), 717–730.

Yoshii, K., Tomioka, R., Mochihashi, D., Goto, M., 2013. Infinite positive semidefinite tensor factorization for source separation of mixture signals. In: Proc. of International Conference on Machine Learning (ICML), pp. 576–584.

Yoshioka, T., Nakatani, T., Miyoshi, M., Okuno, H.G., 2011. Blind separation and dereverberation of speech mixtures by joint optimization. IEEE Transactions on Audio, Speech, and Language Processing 19 (1), 69–84.

Yoshioka, T., Sehr, A., Delcroix, M., Kinoshita, K., Maas, R., Nakatani, T., Kellermann, W., 2012. Making machines understand us in reverberant rooms – robustness against reverberation for automatic speech recognition. IEEE Signal Processing Magazine 29 (6), 114–126.

Yu, D., Hinton, G., Morgan, N., Chien, J.-T., Sagayama, S., 2012. Introduction to the special section on deep learning for speech and language processing. IEEE Transactions on Audio, Speech, and Language Processing 20 (1), 4–6.

Zafeiriou, S., Tefas, A., Buciu, I., Pitas, I., 2006. Exploiting discriminant information in nonnegative matrix factorization with application to frontal face verification. IEEE Transactions on Neural Networks 17 (3), 683–695.

Zhang, J., 2004. Divergence function, duality, and convex analysis. Neural Computation 16 (1), 159–195.

Zhang, X.-L., Wang, D., 2016. A deep ensemble learning method for monaural speech separation. IEEE/ACM Transactions on Audio, Speech and Language Processing 24 (5), 967–977.

Zhu, B., Li, W., Li, R., Xue, X., 2013. Multi-stage non-negative matrix factorization for monaural singing voice separation. IEEE Transactions on Audio, Speech, and Language Processing 21 (10), 2096–2107.

Index

Symbols

α divergence, 56, 118, 123, 219
β divergence, 30, 162
f divergence, 56, 118
k-means algorithm, 114, 272

A

Absolute error of predictability, 157
Adam optimization, 282, 295
Adaptive signal processing, 17, 321
Addressing mechanism, 298, 310
Affinity matrix, 271, 284
Approximate inference, 75, 85, 95, 139, 153, 184
Audio source separation, 5, 14, 17, 25, 59, 202, 221, 231, 321
Auditory filterbank, 234, 268
Automatic relevance determination, 59, 133, 136, 151
Autoregressive process, 147

B

Back-end learning, 17, 321
Backpropagation through time, 45, 294, 313
Basis representation, 17, 25, 31, 60, 101, 166, 204
Bayesian group sparse learning, 60, 205
Bayesian information criterion, 101, 106
Bayesian kernel method, 147
Bayesian learning, 11, 17, 54, **57**, 132, 134, 142, 165, 173, 182, 201, 246, 250
Bayesian NMF, 60, 182, 183, 186, 190, 197, 208, 216, 247
Bayesian nonparametrics, 251, 255
Bayesian source separation, 59, 182
Bayesian speech dereverberation, 60, 162, 166
Bayesian tensor factorization, 249
Bidirectional long short-term memory, 273, 295, 314
Bidirectional recurrent neural network, 296
Binaural impulse response, 269
Biomedical source separation, 14
Blind source separation, **3**, 21, 55, 59, 99, 113, 117, 127, 135, 145, 150, 156
Bregman matrix divergence, 253

C

Cauchy–Schwartz divergence, 55, 118
Central limit theorem, 106
Cepstrum distance, 171
Cocktail party problem, 3, 21
Collaborative audio enhancement, 181
Component importance measure, 102
Computational auditory scene analysis, 14, 268
Concave function, 57, 88, 325
Conditional independence, 87, 92
Conjugate prior, 67, 138, 150, 167, 186, 190, 207, 246, 250
Constrained optimization, 170, 175
Contrast divergence, 43, 263
Convex divergence, 118, 127, 322
Convex function, 56, 78, 119, 253
Convex-exponential divergence, 120
Convex-logarithm divergence, 120
Convex-Shannon divergence, 120
Convolutional neural network, 277
Convolutive mixing system, 8, 131
Convolutive mixtures, 16, 132
Convolutive NMF, 26, 161
Convolutive NTF, 237, 242
CP decomposition, 36, 233
Cross-correlation function, 269

D

Deep belief network, 43, 263
Deep bidirectional long short-term memory, 297
Deep clustering, 259, 270, 275, 294
Deep ensemble learning, 265
Deep layered NMF, 218
Deep learning, 18, 37, 41, 68, 218, 221, 265, 271, 278, 284, 287, 323
Deep neural network, 6, **37**, 68, 259, 271, 276, 294, 322
Deep recurrent neural network, 7, 21, 48, 73, 278, 284, 294
Deep semi-NMF, 220
Deep spectral mapping, 260
Demixing matrix, 5, 22, 55, 75, 102, 106, 112, 122, 127, 132

Determined system, 5, 24, 135
Digamma function, 186
Direction of arrival, 6, 17
Discriminative deep recurrent neural network, 284
Discriminative DNN, 73
Discriminative embedding, 271
Discriminative layered NMF, 223
Discriminative learning, 54, 68, 160, 223, 287
Discriminative NMF, 68
Divergence measure, **55**, 69, 118, 121, 222, 238

E

EM algorithm, **77**, 83, 90, 177, 298
Encoder–decoder network, 316
End-to-end learning, 285, 324
End-to-end memory network, 307, 313
Ensemble learning, 259, 265
Error backpropagation, 40, 44, 52, 220, 262, 269, 280, 288, 299
Euclidean divergence, 55, 118
Evidence lower bound, 87, 255
Expectation–maximization algorithm, 28, 175
Exponential distribution, 167, 183, 191, 214, 247, 334
Exponential-quadratic kernel function, 148

F

Factor analysis, 22, 34, 221
Factorized error backpropagation, 281
Factorized neural network, 280
Factorized posterior distribution, 87
Factorized variational inference, 87, 140
Feedforward neural network, 38, 48, 291
Finite impulse response, 11, 16, 131
Fisher discriminant function, 69
Frequency-weighted segmental SNR, 172
Front-end processing, 17, 278, 321
Full Bayesian, 67, 76, 171, 190, 202, 209, 250, 284

G

Gamma distribution, 137, 150, 186, 192, 206, 213, 214, 254, **331**
Gamma function, 64, 82, **325**
Gamma process, 255
Gammatone filterbank, 232, 268, 295
Gammatone frequency cepstral coefficient, 269
Gated attention mechanism, 317
Gaussian distribution, 22, 62, 81, 106, 125, 137, 148, 183, 210, 212, 243, 250, 273, 303, 331

Gaussian mixture model, 6, 75, 111, 134, 154, 264
Gaussian process, 19, 59, 99, 147
Gaussian process prior, 59, 147, 149
Gaussian–Wishart distribution, 250
Gaussian-exponential BNMF, 183, 195
Generalized inverse-Gaussian distribution, 168, 255
Gibbs sampling, 75, 94, 185, 195, 251
Gradient descent algorithm, 23, 29, 55, 67, 112, 128, 164
Gram matrix, 147
Group basis representation, 202, 218
Group sparse NMF, 202, 247
Group-based NMF, 203, 216

H

Heterogeneous environments, 17, 57, 205
Hidden Markov model, 75, 100, 114, 264
Hierarchical Bayesian model, 246, 249
Higher-order orthogonal iteration, 35
Human-machine communication, 13
Hypothesis test, 99, 109, 117

I

Ideal amplitude mask, 274
Ideal binary mask, 268, 272
Ideal phase sensitive mask, 274
Ideal ratio mask, 261, 274, 305
Image enhancement, 15
Image separation, 15, 153
Importance sampling, 91, 145, 155
Importance weight, 92
Incomplete data, 76, 82, 123
Independent component analysis, 5, 16, **22**, 55, 99, 100, 106, 118, 132, 298, 322
Independent voices, 100, 113
Infinite tensor factorization, 251
Infomax principle, 108
Information redundancy, 15, 100
Information retrieval, 13, 174
Information-theoretic learning, 29, 54, 117, 132
Instantaneous mixing system, 4, 130
Instrumental distribution, 211
Interaural level difference, 269
Interaural time difference, 269
Inverse Gamma distribution, 183, 246, 250
Itakura–Saito divergence, 30, 162, 251

J

Jensen–Shannon divergence, 57, 118

Jensen's inequality, 76, 88, 119, 168, **325**

K

Kalman filter, 154
Kullback–Leibler divergence, 11, 29, 55, 86, 118, 162, 183, 224, 233, 239, 240, 286, 302
Kurtosis, 22, 101, 135

L

L-BFGS optimization, 263, 290
Laplace distribution, 62, 127, 191, 206, **330**
Laplacian scale mixture distribution, 60, 202, 206
Lasso regularization, 32, 61, 164, 191, 204, 240
Layered NMF, 221
Likelihood ratio, 110
Linear discriminant analysis, 69, 265, 286
Linear prediction coefficient, 172
Local gradient, 40, 46, 281, 288
Log likelihood ratio, 113, 172
Log-determinant divergence, 253
Logistic sigmoid function, 39, 279
Long short-term memory, **49**, 291, 309, 322
Low-rank embedding, 272

M

Machine learning, 17, 53, 57, 65, 75, 85, 161, 265, 278, 321
Markov chain Monte Carlo, 92, 153, 209, 231, 242, 250, 322
Maximum a posteriori, 53, 62, 83, 191, 243, 264
Maximum entropy, 108
Maximum likelihood, 31, 60, **76**, 81, 107, 175, 183
Maximum likelihood eigen-decomposition, 103
Maximum likelihood independent decomposition, 103
Mel-frequency cepstral coefficient, 27, 103, 116, 199, 234
Memory augmented neural network, 307
Metropolis–Hastings algorithm, 93, 211
Microphone array signal processing, 15
Minimum description length, 101
Minimum mutual information, 22, 108
Mixing matrix, 4, 65, 102, 117, 127, 136, 142, 148, 220
Model clustering, 114
Model complexity, 17, 58, 166, 191, 197, 231, 251, 323
Model regularization, 31, 54, **57**, 132, 174, 182, 191, 203, 243

Model selection, 17, 28, 58, 101, 191
Model uncertainty, 17, 58, 100, 231
Modern sequence learning, 298
Modified Bessel function, 168
Modulation spectrogram, 232
Monaural source separation, **6**, 13, 21, 68, 81, 183, 200, 218, 263, 277, 284, 306, 312, 318
Multi-channel source separation, 3, 16, 59, 106, 113, 117, 124
Multi-layer NMF, 219
Multicontext averaging, 266
Multicontext network, 265
Multicontext stacking, 266
Multilayer perceptron, 38
Multinomial distribution, 188
Multiplicative updating rule, 28, 29, 162, 219, 224, 234, 256
Multiresolution spectrogram, 234
Multivariate Gaussian distribution, 145, 152, **329**
Music information retrieval, 13, 16, 37
Music separation, 13, 196, 321

N

Natural gradient algorithm, 23, 128
Neural Turing machine, 309, 318
Noise adaptation, 99
Noiseless mixing system, 4
Noisy ICA model, 75, 134, 148, 153
Noisy ICA system, 134, 148
Non-Gaussianity, 22, 106, 118, 131, 156
Nonnegative convolutive transfer function, 11, 165, 174, 284
Nonnegative matrix factorization, 6, **25**, 32, 57, 81, 123, 161, 179, 202, 231, 251, 263
Nonnegative matrix factorization deconvolution, 162, 178, 234
Nonnegative matrix partial co-factorization, 202
Nonnegative tensor factor deconvolution, 237
Nonnegative tensor factorization, **32**, 181, 231
Nonparametric distribution function, 111, 117
Nonparametric likelihood ratio, 106, 110
Nonstationary Bayesian ICA, 132, 135
Nonstationary Bayesian learning, 141
Nonstationary environment, 54
Nonstationary source separation, 65, 99, 133, 153
Number of sources, 5, 16, 59, 107, 124, 132

O

Online Bayesian learning, 67, 135, 147

Online Gaussian process, 147
Online Gaussian process ICA, 145, 149
Online learning, 54, 65, 132, 145, 158
Overdetermined system, 5, 24

P
Parallel factor analysis, 233
Particle filter, 154
Parts-based representation, 25, 35, 161, 182, 202, 246
Parzen window density function, 111, 118, 125
Perceptual evaluation of speech quality, 264, 277, 282
Permutation ambiguity, 16, 276
Permutation-invariant training, 275
Poisson distribution, 82, 334
Poisson–exponential BNMF, 190
Poisson–Gamma BNMF, 186
Positive definite, 329
Positive semidefinite tensor factorization, 251
Principal component analysis, 22, 100
Probabilistic latent component analysis, 174
Probabilistic latent component sharing, 181
Probabilistic matrix factorization, 242
Probabilistic nonnegative factorization, 174
Probabilistic NTF, 242
Probabilistic tensor factorization, 244
Proposal distribution, 91

R
Rank-one tensor, 36, 327
Reading and writing mechanism, 311
Recall neural network, 314
Rectified Gaussian distribution, 247
Rectified linear unit, 39, 279
Rectified normal distribution, 184, 247
Recurrent neural network, **45**, 275, 292, 299, 307, 314
Recursive Bayesian learning, 134
Regularized least-squares estimation, 62
Restricted Boltzmann machine, 43, 263, 269
Reverberant source separation, 8, 16, 171, 268, 277
Reverberation time, 10, 171, 270
Room impulse response, 8, 165
Room reverberation, 16

S
Sampling method, 91, 161, 202, 211
Scaling ambiguity, 16
Semi-NMF, 220
Sequence-to-sequence learning, 314, 316

Sequential Monte Carlo ICA, 145, 153
Shannon entropy, 55
Shift invariance, 176
Shift invariant PLCA, 176, 181
Short-time Fourier transform, 6, 27, 131, 165, 197, 232, 251, 260, 307
Short-time objective intelligibility, 263, 267, 294, 313
Signal processing, 17, 161, 265, 278
Signal separation, 117
Signal-to-distortion ratio, 73, 197, 234, 263, 274, 295, 304
Signal-to-interference ratio, 24, 71, 117, 132, 216, 228, 236, 263, 285
Signal-to-noise ratio, 103, 264, 272, 295
Singing voice separation, 7, 13, 27, 48, 161, 183, 196, 284
Single-channel source separation, 6, 24, 39, 48, 72, 160, 196, 234, 260, 285, 295, 307, 322
Soft mask function, 26, 48, 261, 274, 285, 295, 315
Softmax function, 52
Source separation, 3
Sparse Bayesian learning, 62, 136, 185
Sparse coding, 60, 101, 204
Sparse learning, 31, **60**
Sparse nonnegative tensor factor deconvolution, 240
Sparse prior, 63, 206
Speaker adaptation, 15, 99
Speaker generalization, 293
Speaker-independent speech separation, 270
Spectral clustering, 271
Spectro-temporal neural factorization, 280
Speech dereverberation, 8, 165, 278
Speech enhancement, 13, 260, 313, 318
Speech recognition, 8, 99, 113, 264, 298
Speech segregation, 268
Speech separation, 13, 219, 260, 270, 284, 293, 298
Speech-to-reverberation modulation energy ratio, 172
Squared Euclidean distance, 28, 55, 222, 238, 240
Stochastic gradient descent, 39, 109, 262, 288
Student's t-distribution, 64, **331**
Sub-Gaussian distribution, 127
Sum-of-squares error function, 39, 46, 262, 266, 275, 280, 285, 307, 318
Super-Gaussian distribution, 127
Supervised learning, 6, 21, 26, 53, 57, 260, 269
Supervised source separation, 196

T
Temporal predictability, 157

Tensor, 33, 233, 325
Text mining, 15
Truncated Gaussian distribution, 211, 247
Tucker decomposition, 34, 278

U
Underdetermined system, 5, 24, 160
Unsupervised learning, 6, 12, 22, 26, 53, 99, 116, 216, 271
Unsupervised source separation, 199

V
Variational auto-encoder, 299
Variational Bayesian learning, 90, 99, 138, 152, 187, 189, 255, 299
Variational expectation step, 191

Variational inference, 75, 85, 139, 168, 186, 255, 298, 322
Variational lower bound, 87, 139, 151, 169, 187, 255, 302
Variational maximization step, 193
Variational recurrent neural network, 299
VB-EM algorithm, 85, 90, 138, 152, 166, 187, 299

W
Weight-decay regularization, 32, 204
Weighted nonnegative matrix factorization, 123
Whitening transformation, 23, 106, 110
Wiener filtering, 26, 48, 72, 171, 236, 261
Wishart distribution, 138, **333**
Word error rate, 106, 263